Microscale
Organic
Laboratory

Microscale
Organic
Laboratory

Second Edition

DANA W. MAYO
Professor of Chemistry
Bowdoin College

RONALD M. PIKE
Professor of Chemistry
Merrimack College

SAMUEL S. BUTCHER
Professor of Chemistry
Bowdoin College

WILEY

JOHN WILEY & SONS

New York   Chichester   Brisbane   Toronto   Singapore

All experiments contained herein have been performed several times by students in college laboratories under the supervision of the authors. If performed with the materials and equipment specified in this text, in accordance with the methods developed in this text, the authors believe the experiments to be a safe, valuable educational experience. However, all duplication or performance of these experiments is conducted at one's own risk. The authors do not warrant or guarantee the safety of individuals performing these experiments. The authors hereby disclaim any liability for any loss or damage claimed to have resulted from or related in any way to the experiments, regardless of the form of action.

Cover design by Kevin Murphy
Cover photo by Robert Wolfson

Library of Congress Cataloging-in-Publication Data

Mayo, Dana W.
  Microscale organic laboratory.

  Includes index.
  1. Chemistry, Organic—Laboratory manuals. I. Pike,
Ronald M. II. Butcher, Samuel S., 1936–
III. Title.
QD261.M38 1989   547'.007'8   88-33913
ISBN 0-471-63629-0

Printed in the United States of America

10 9 8 7 6 5 4 3 2

To Jeanne d'Arc, Marilyn, and Sally

Preface

In the three years since 1985, when bound xerox copies of *Microscale Organic Laboratory—1 Preliminary*, the manuscript for the later *Microscale Organic Chemistry*, first became available, nearly three hundred institutions have moved to convert to this style of program. The magnitude and rapidity of adoptions really does mean that we are now in the midst of an educational revolution. Most impressive to us, however, is that of all the institutions initiating trial microscale programs, none to our knowledge has later dropped the microscale approach. Those areas in need of further refinement are being attacked by an ever-widening and skillful group of enthusiastic evangelists. Indeed, it is our view that we have just skimmed the surface of fruitful micro techniques and experiments, and we look forward in the next several years to bringing into sharper focus the direction in which this exciting approach will be heading. It is particularly gratifying to see the program being explored by a cross section of institutions ranging from small community colleges to large research-oriented universities. This variety of interest has in turn led to a wide range of useful and ingenious comments and suggestions, many of which we are pleased to be able to incorporate in this latest edition.

In response to favorable reviews, we have retained the basic format of the first edition. Thus, we continue to integrate illustrative experimental examples in the techniques section; to incorporate keyed marginal illustrations of apparatus set-ups throughout the most of the experimental section; and to describe the experiments in formal journal style prose. We have extended our energies during the last few years to refining a number of key experimental techniques and to improving the reliability, flexibility, and variety of reactions to be studied. For a number of experiments in which clarity needed to be improved, we have rewritten procedures.

We are now convinced that by the end of the coming decade the great majority of chemical educational programs in the United States will have converted to the microscale approach. The question is not whether to miniaturize but how best to go about the job. We feel that *flexibility* in program design is vitally important in order to accommodate the wide variety of local educational environments. In our first edition we sent the message "Smaller is better," and all the experiments utilized approximately 150 mg or less of starting material. In this second edition we feel that it is now appropriate to build more flexibility of scale into the experimental design. We have, therefore spent a good deal of time modifying ten experiments and three of our four Sequential Sequences to include *Optional Scaleup* procedures. These range from twofold to two hundred-fold above the microscale level. We have two reasons for this change in content. First, although many students have the opportunity to experience multigram reaction scales in upper-level laboratories or independent study, a significant number depend on a single year of experimental organic chemistry to gain acquaintance with the field. We agree that this second category of students, and indeed perhaps many of the first, should experience the excitement of undertaking a set of sequential steps (starting with multigram quantities) that clearly emulates a research laboratory synthesis. These experiments form a fitting conclusion to the end of the microscale laboratory. Second, for programs that are interested in "going micro," but need time and equipment to effect the transition, these optional experiments offer a temporary bridge to begin moving in the micro direction. In this regard we should emphasize that for programs that have reached the microscale level, the recommended time for incorporation of the Optional Upscale procedures is late in the second semester.

The rapid, almost breathtaking speed of technique innovation taking place in the microscale laboratory program is best exemplified by the changes found in the completely rewritten section on fractional distillation. The development of bottom-driven, low-cost micro spinning band distillation columns in the authors' laboratories over the past three years has resulted in a major breakthrough in the technique for illustrating classic fractional distillation. These systems are unquestionably the most powerful stills ever to become available as instructional equipment at the introductory level. In particular, the Hinkle modification of the classic Hickman still will clearly have a major impact on the entire field of micro distillation, including the advanced research laboratory.

Developments in chemical instrumentation also continue to accelerate. The availability of low-cost Fourier transform–infrared systems has broken the jam of making infrared data available to students in real time and has provoked us into adding new detailed interpretations of the FT–IR spectra of reactants and products in fourteen experiments. We hope that these discussions can be used as a means of leading the student into a deeper appreciation of the value of these spectroscopic data. Fourier transform–nuclear magnetic resonance (^{1}H, ^{13}C) and ultraviolet–visible data have been introduced for the first time in modest fashion where they are appropriate for the characterization of reaction products. It is clear that the instrumental and cost barriers to the utilization of experimentally derived NMR data in the introductory organic laboratory are finally beginning to be bridged. Although earlier we had significant reservations about incorporating NMR data in the microscale experiments, we now expect their use in these programs to expand. The ability of the student to collect and interpret spectroscopic data is an essential aspect of "characterization science," which is one of the most vital elements of the modern-day academic and industrial laboratory

To further the flexibility of the approach, we have added keyed references in each experiment to the appropriate qualitative identification tests and derivative preparations in Chapter 7. These operations can be effectively utilized by programs looking to expand the amount of microscale chemistry involved in a particular experiment or as an alternative to instrumental characterization.

The very successful utilization of the chromatographic techniques given in the first edition has lead us to expand the coverage and application of these powerful experimental routines. Gas chromatography and thin-layer chromatography are effectively utilized in a number of new locations.

Several other areas of the text have been improved. For example, the section on acid-base extraction has been expanded and in addition includes a sequence that illustrates the separation of a three-component mixture. The variety of alternate methods to illustrate a particular reaction continues to increase. For example, an improved extraction technique for the isolation of caffeine, the Horner–Emmons modification of the Wittig reaction, and a new nitrating agent prepared from silica gel and nitric acid are now included in the reaction selections.

The authors wish to acknowledge the many helpful suggestions from Professors Charles E. Sundin, University of Wisconsin–Platteville, Chaim N. Sukenik, Case Western Reserve University, and Bruce Ronald of Idaho State University. We are also indebted to all those who have attended the Bowdoin College Microscale Summer Institutes and our workshops across the country. Your enthusiasm, interest, and insight have led to many of the improvements in procedures and techniques now incorporated into the program. The many contributors to the newsletter *Smaller Is Better* have significantly helped to sustain the momentum of the program. We especially thank Paulette Fickett and Lauren Bartlett, Laboratory Instructors at Bowdoin, for their dedication, hard work, and eternal optimism. The microscale program could not have evolved in such a successful fashion without the continued contributions of John Ryan of Ace Glass, Stephen Cantor of Pfaltz and Bauer, Robert Stevens of J. J. Stevens, and of Henry Horner and Thomas Tarrant, all of whom have given encouragement and helpful guidance at crucial stages along the way. Judy Foster again waved her magic wand and this time transformed a first-edition Apple Writer manuscript into Microsoft Word. We continue to marvel at the patience of Dennis Sawicki, our Chemistry Editor. We would like to acknowledge continued support from the Surdna Foundation in developing the microscale program. The Sloane Foundation and the National Science Foundation have supported two of the summer institutes.

We are particularly indebted to our colleague Peter Trumper for injecting his expertize in FT-NMR into the microscale program. His enthusiastic application of high-field NMR to the introductory organic program promises future exciting developments in this area.

As we stand on the threshold of the second decade of the microscale revolution, we wish to thank our students. Especially for them do we feel the program has meaning: a cleaner atmosphere in which to work, an awareness of the toxic factors associated with the chemical workplace, and a commitment to detail not hitherto fostered in the introductory laboratory. For their perseverance and eagerness to learn and adapt to new ideas, for the freshness that each new class brings, and for their willingness to grow, we are indeed most grateful.

DANA W. MAYO
RONALD M. PIKE
SAMUEL S. BUTCHER

Brunswick, Maine
October 30, 1988

Preface

This introductory organic laboratory textbook is a major departure from all other modern texts dealing with this subject. For a number of very cogent reasons, we have chosen to introduce experimental organic chemistry at the microscale level. Currently, beginning students perform the large majority of their experimental work at least two orders of magnitude above that described in this manual. Although contraction from the multigram to the milligram scale is the most obvious change, there are a number of other very unique aspects to our approach.

1. The laboratory environment has been made a distinct part of the experimental process. The student is given the means of easily determining his or her exposure to all volatile substances employed in the experiments. These calculations can be carried out for any laboratory by utilizing instructor-supplied ventilation rates.

2. Chemical instrumentation is given very high priority in the laboratory. The text avoids emphasizing data not directly determined by the student. Product characterization by infrared spectroscopy is routine, and detailed means for interpreting such data are provided.

3. Modern separation and purification techniques, including preparative gas chromatography, thin-layer chromatography, and column chromatography, are extensively utilized in product workups.

4. Over a third of the 82 reaction products are new to the undergraduate laboratory. Many of the reactions involve reagents or substrates that would present potential safety problems or entail exorbitant costs in a macroscale laboratory program; nevertheless, the use of these materials becomes safe and practical for experimentation at the microscale level. Reagents such as anhydrous nitric acid, diborane, chloroplatinic acid, "instant ylids," silver persulfate, chromium trioxide resin, tetrabutylammonium bromide, and triflic acid are representative of the materials that the student will encounter during the year.

Why are we committed to the goal of attempting to reduce significantly the scale of starting materials in the introductory organic laboratory? The academic community has become increasingly aware of the necessity of improving air quality in instructional laboratories. The standard solution to this problem, a costly upgrading of the ventilation system, has generally been considered the most reasonable answer.

Our study of the problem led us to the following conclusions. First, although current organic laboratory texts are filled with details of product characterization employing the latest spectroscopic methods, the descriptions of the techniques of preparing compounds have changed very little from those of a century ago. In particular, the scale of synthesis has changed very little over this period. (Indeed, the quantities of materials employed have only decreased modestly, and in some cases have actually increased!) Clearly, the strategy of introducing the student to organic chemical laboratory techniques, originally and still today, is centered on the multigram level (see Table 1).

Table 1 Starting Materials Employed in Classical Organic Laboratory Syntheses, 1902–1980

		Acetanilide	**4-Bromoacetanilide**	**Benzoin**
		Starting materials required (grams)		
Date	**Author**	**Aniline**	**Acetanilide**	**Benzaldehyde**
1902	Levy, 4th ed.	46.2	—	50.0
1915	Cohen, 3rd ed.	25.0	5.0	25.0
1933	Adkins	25.0	13.5	10.0
1941	Fieser, 2nd ed.	18.2	13.5	25.0
1963	Adams	20.0	13.5	16.0
1980	Durst	10.0	5.2	10.0

We seriously question the wisdom of maintaining the introduction of laboratory work at conventional levels. Is this approach relevant when one considers the quantities of materials commonly used in natural products, pharmaceutics, biochemistry, and other fields of modern research in which costly substances are employed?

Second, it is now fully recognized that there has been a very serious decrease in undergraduate laboratory contact time over the past two decades. It seems to us that an increasingly important question should be asked in evaluating the introductory organic program: At what scale of introductory laboratory work can the student gain the maximum ability to handle organic materials within the shortest period of time?

We now firmly believe that the microscale laboratory approach resolves both these concerns and, in addition, affords a number of significant bonuses. The immediate result of "going micro" is that there is a change in the laboratory air quality that can be described only as spectacular! Of greater importance, and a point that we have come to appreciate in retrospect, is that many reactions and operations carried out at the micro level require far less time to reach completion. Indeed, *many of the time-consuming aspects of current instructional experiments are dramatically shortened.* We believe that the concomitant advantage of the microscale approach, the substantial increase in the number of manipulations possible per laboratory period, will have a major pedagogic impact. Herein lies the significant advantage of this approach to teaching laboratory technique.

We also see the increase in use of chemical instrumentation as a further pedagogic advantage of operating at the microscale level. Routine use of gas chromatography may be avoided in macro experiments; however, if it is not employed at the microscale level, successful experimentation with liquid substances is limited. The capital investment that would allow routine use of this equipment in many undergraduate laboratories might appear to elevate the microscale program out of the reach of institutions with very limited budgets. For many institutions, however the very substantial savings in chemical costs (75–90% of current expenditures), if carefully managed, can offer a rapid payback period (2–4 years) for expanded gas chromatography capacity and for other pieces of equipment. The end result of conversion to microscale is a far more effective integration of modern instrumentation into the organic laboratory program.

Once the microscale approach is initiated, the advantages are endless. We will not dwell on them here, as many of the following points vary in importance with each institution: (1) major reduction in cost of chemicals, (2) elimination of fire or explosion danger, (3) elimination of chemical waste disposal costs, (4) expansion in variety and sophistication of experiments, (5) elimination of dependence on commercially available starting materials, and (6) more durable and less expensive glassware.

These advantages are highly compelling reasons for advocating the microscale approach. Initially, however, we had serious reservations. First, the microscale approach appeared to involve a sophisticated set of techniques and manipulations too advanced for sophomore undergraduates to master. Second, a student having experience only at the microscale level might be expected to encounter problems when larger-scale preparations were undertaken. Third, significantly less organic chemistry would be covered during the year because increased attention would have to be paid to the development of micro techniques. Fourth, certain classical procedures such as fractional distillation would have to be abandoned.

As of the completion of this text, we have conducted eight semesters of microorganic chemistry laboratory assessment. The experience with the test laboratory groups (made up of a cross section of volunteers) has been a revelation. It is clear from our observations that the entire range of the class achieves significantly better results on micro experiments. Better yields are realized and, in particular, the class appears to master experimental details and procedures more effectively. It is felt that these results arise from several causes, one of which is the increased attention to detail required in the laboratory. We also sense a synergistic influence with the analytical chemistry laboratory, which is often scheduled for the sophomore year concurrently with the organic course. Analytical chemistry is carried out at a scale not unlike that employed in the microscale organic laboratory.

At present there is no indication that the learning of micro techniques during the introduction to organic chemistry causes any adverse effects when scaleup work is introduced in advanced laboratory courses or in research areas. To the contrary, our students appear to be performing significantly better in upper-level work.

The development of the experimental section of the text, which contains eighty reactions, has been a major effort. The conditions for a large proportion of these reactions have been optimized to maximize yields at the micro level. We have chosen to describe the experimental work in language similar to formal journal style. At first, this impersonal construction would appear to be "user unfriendly" and to make the text a bit boring and difficult to read. On field testing with students, however, we have found that they quickly adapt to the style and soon come to appreciate the use of precise routine terminology. This introduction to formal style pays substantial dividends in upper-level courses where students are expected to consult the original literature.

The philosophy of this text is to focus the student, to a very large extent, on the *experimental* aspects of organic chemistry. We have purposely attempted to keep to a minimum the theoretical discussions, supplying only sufficient background material to cover potential discontinuities between lecture and laboratory. We want the student to become comfortable, and to develop a substantial degree of independence with the use of chemical instrumentation. To meet this objective, we have centered attention specifically on infrared spectroscopy and gas chromatography. We do this because we feel that these basic instrumental techniques can be utilized to a much larger extent by sophomores than the more expensive techniques.

We recognize the dominant role of nuclear magnetic resonance in modern organic chemistry. The current real-life situation is, however, that only in a very limited number of cases do second-year students ever have the opportunity to generate their own data with this instrumentation. Because we are concerned primarily with focusing student excitement and interest on gathering actual laboratory data, we feel that artificially incorporating material from outside the laboratory as a means of including the nuclear magnetic resonance or mass spectroscopy experiment may not enhance but divert attention away from the laboratory experience. We are seriously exploring ways to overcome this particular problem and hope to address this issue successfully in future editions.

The six-year road to completion of this text has been a long, but rewarding one. Many individuals have made vital contributions. Arnold Brossi, David Brooks, Miles Pickering, Lea Clapp, Eugene Cordes, and Henry Horner provided encouragement and sound advice at crucial points along the way.

We gratefully acknowledge that the large majority of the infrared and nuclear magnetic resonance spectra not recorded at Bowdoin College were obtained from the Aldrich Libraries of FT-IR and NMR Spectra through the courtesy of Dr. Charles Pouchert and the Aldrich Chemical Company.

John Ryan, Larry Reilly, Don Sellars, and Hugh Bowie of Ace Glass are primarily responsible for the development of the novel microglassware. They exhibited considerable patience throughout the ordeal. Ed Hollenbach and Lyle Phifer of Chem Services saw the advantage of making available small quantities of high-purity reagents and starting materials as a way of ensuring success at the microscale level.

The vast amount of experimental development was shared by Teaching Research Fellows Janet Hotham, David Butcher, Paulette Fickett, and Caroline Foote, plus a number of Bowdoin students—Mark Bowie, Sandy Hebert, Rob Hinkle, Marcia Meredith, and Gregory Merklin. The experimental ground covered by this group, much of which required ingenious solutions, is remarkable. The breadth of experiments available is a tribute to their dedication. Janet deserves a special thanks. A Merrimack College graduate adopted by Bowdoin College, she has been with the program almost from the beginning. Her thoughtful suggestions based on experience in the trenches and her willing contributions in any area of need at any time are most gratefully remembered. The enthusiasm of the sophomore organic students who volunteered for the initial pilot sections at Bowdoin and the field testing sections at Merrimack played a key role in encouraging us to continue these efforts.

Judy Foster's talent with a Macintosh enabled her to generate the majority of the illustrations and reaction schemes. Judy's efforts have greatly enhanced our descriptions of the techniques involved in microscale work. Her ingenuity also led to the pictorial keying of the equipment setups. The patience, understanding, and thoughtful advice of Dennis Sawicki, Chemistry Editor for Wiley, has been particularly valuable.

We thank Dean Alfred Fuchs for his constant encouragement of this program during its development. The initial exploratory work was supported by a grant from Bowdoin College and a department grant from the du Pont Company. A semester leave given to D.W.M. (spring 1984) was funded by a grant from the

ARCO Corporation. A semester sabbatic leave was granted to R.M.P. by Merrimack College (spring 1985), as was an appointment as Visiting Charles Weston Pickard Professor of Chemistry at Bowdoin College (1980–1981, spring 1984). The Surdna Foundation awarded two major grants which allowed the complete development and implementation of the program at Bowdoin College and the field testing of experiments at Merrimack College. The support and faith in this educational concept by these institutions is gratefully acknowledged.

DANA W. MAYO
RONALD M. PIKE
SAMUEL S. BUTCHER

March 1985
Brunswick, Maine

Contents

About the Authors

Dana W. Mayo holds the Charles Weston Pickard Professor of Chemistry Chair at Bowdoin College. A former Fellow of the School for Advanced Study at MIT and a Special Fellow of the National Institute of Health at the University of Maryland, he received his PhD in Chemistry from Indiana University. Professor Mayo is Director of the Bowdoin College Summer Course in Infrared Spectroscopy. His research interests include the application of vibrational spectroscopy to molecular structure determination, natural products chemistry, and environmental studies of oil pollution.

Samuel S. Butcher is Professor of Chemistry at Bowdoin College. His research interests lie in the areas of atmospheric chemistry and air pollution. He is the co-author of a textbook on air chemistry and has written several papers on air pollution and the effects of small wood-burning stoves. He has taught a wide range of undergraduate chemistry and science courses for nonspecialists. Professor Butcher received his PhD in Chemistry from Harvard University.

Ronald M. Pike is Professor of Chemistry at Merrimack College. He received his PhD from MIT. His main research interests involve the synthesis of organofunctional silanes and related silicone polymers. He is the author of numerous papers and patents in this area. Professor Pike was previously associated with Union Carbide Corporation and the Lowell Technological Institute. He has been a Visiting Charles Weston Pickard Professor of Chemistry at Bowdoin College.

*__Dana Mayo and Ronald Pike__ jointly received the James Flack Norris Award (1988) for outstanding achievement in the teaching of chemistry, and the John A. Timm Award (1987) for their work in developing the microscale instructional program. Together with **Samuel Butcher,** they were co-recipients of the first Charles A. Dana Foundation Award (1986) for pioneering Achievement in Health and Higher Education and of the American Chemical Society Division of Chemical Health and Safety Award (1987).*

Chapter 1

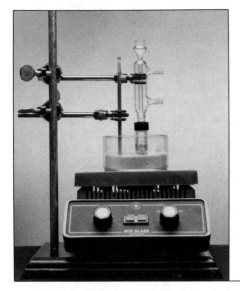

Introduction

You are breaking new ground in the organic chemistry laboratory!

Your course is going to be quite different from the conventional manner in which this laboratory has been taught. You will be learning the experimental side of organic chemistry from the microscale level. Surprising as it may seem, because you will be working with very small amounts of materials, you will be able to observe and learn more organic chemistry than many of your predecessors did in nearly two years of laboratory work. You will find this laboratory an exciting and interesting place to be. Although we cannot guarantee it for you individually, the majority of students who have been through the program during its development have found the microscale organic laboratory to be a pleasant adventure.

At the very beginning we want to acquaint you with the organization and contents of the text. We will then give you a few words of advice, which, if they are heeded, will allow you to avoid many of the sand traps you will find as you work your way through the course. Finally, we will turn philosophical and attempt to describe what we think you should derive from this experience.

After this brief introduction, the second chapter is concerned with safety in the laboratory. This chapter is unique. For the first time in a laboratory text, information is supplied that will allow you to easily calculate your maximum possible exposure to any volatile chemical employed in any of the experiments. Chapter 2 also discusses general safety protocol for the laboratory. It is vitally important that you become familiar with the details of the material contained in this chapter; your health and safety depend on this knowledge.

The next section of the text consists of three chapters that are concerned primarily with the development of experimental technique. Chapter 3 describes in detail the glassware employed in microorganic chemistry: the logic behind its construction, tips on its usage, the common arrangements of equipment, and various other laboratory manipulations, including techniques for transferring microquantities of materials. Suggestions for the organization of your laboratory notebook are considered at the end of this chapter.

Chapter 4 deals with equipment and techniques for determining a number of physical properties of microscale samples. Chapter 5 is divided into 11 experimental sections. These exercises, and the detailed discussions accompanying them, develop the major areas of experimental technique to be used in the microscale organic laboratory. A number of specific reactions incorporated into this section illustrate the use of the techniques being introduced. At the discretion of your instructor, these particular examples may be easily replaced to correspond more closely to local development of the subject, so do not be surprised if you find an experiment from Chapter 6 mated to a discussion in Chapter 5.

Chapters 6 and 7 encompass the major experimental section of the text. Chapter 6 is focused on preparative organic chemistry and consists of 41 experiments. Many of these exercises involve a number of optional parts. Experiments 47 to 52 are made up of four sets of sequentially arranged experiments in which the product from one experiment is utilized in the next (a situation very similar to the real-life research laboratory). For convenience in organization, the reactions contained in Chapter 6 are grouped according to reaction mechanism, and the sequential experiments by mechanism and group. (The list of Experiments Classified by Mechanism follows the Contents.) Chapter 7 develops the characterization of organic materials at the microscale level by the use of classical organic reactions and vibrational spectroscopy. Tables of derivative data for use in compound identification by these techniques are given in Appendix D. A detailed discussion of the interpretation of infrared group frequencies and brief introductions to the interpretation of ultraviolet-visible and nuclear magnetic resonance spectral data make up a significant portion of Chapter 7. The theoretical basis for these spectroscopic sections is considered in Appendixes A, B and C. Appendix B also contains a convenient collection of infrared group frequency data derived from discussions in Chapter 7.

The experimental procedures in Chapter 5 and 6 are organized in the following fashion. A short opening statement describing the reaction to be studied is followed by the reaction scheme. Generally, a brief discussion of the reaction follows, including a mechanistic interpretation. In a few cases of particularly important reactions, or where the experiment is likely to precede consideration in lecture, a more detailed description is given. The estimated time to complete the work and a table of reactant data come next. For ease in organizing your laboratory time, the experimental section is divided into four subsections: reagents and equipment, reaction conditions, isolation of product, and purification and characterization. A table of relevant environmental data follows the experimental procedure.

We then introduce a series of questions and problems designed to enhance and focus your understanding of the chemistry and the experimental procedures involved in a particular laboratory exercise. Finally, a list of literature references is given. Although this list comes at the end of the experimental section, we view it as a very important part of the text. The discussion of the chemistry involved in each experiment is necessarily brief. We hope that you will take time to read and expand your knowledge about the particular experiment that you are conducting. (You may, in fact, find that some of these references become assigned reading.) The formal language used in the experimental sections has been intentionally introduced to ease your transition into the chemical literature.

The experimental apparatus and materials involved at important stages of an experiment are indicated by a prompt sign (■) in the text and are shown in the margin. Important comments are italicized in the text, and **Warnings** and **Cautions** are given in boxes.

GENERAL RULES FOR THE MICROSCALE LABORATORY

1. *Study the experiment before you come to lab.* This rule is a historical plea from all laboratory instructors. In the microscale laboratory it takes on a more important meaning. You will not survive if you do not prepare ahead of time. In microscale experiments, operations happen much more quickly than in the macroscale laboratory. Your laboratory time will be overflowing with many more events. If you are not familiar with the sequences you are to follow, you will be in deep trouble. Although the techniques employed at the microscale level are not particularly difficult to acquire, they do demand a significant amount of attention. For you to reach a successful and happy conclusion, you cannot afford to have the focus of your concentration broken by having to constantly refer to the text during the experiment. Disaster is ever present for the unprepared.

2. *Always work with clean equipment.* You must take the time to scrupulously clean your equipment before you start any experiment. Contaminated glassware ultimately will cost you additional time, and you will face the frustration of experiencing inconsistent results and lower yields. Dirty equipment is the primary cause of reaction failure at the microscale level.

3. *Carefully measure the quantities of materials to be used in the experiments.* A little extra time at the beginning of the laboratory can speed you on your way at the end of the day. A great deal of time has been spent optimizing the conditions employed in these experiments to maximize yields. Many organic reactions are very sensitive to the relative quantities of substrate (the material on which the reaction is taking place) and reagent (the reactive substance or substances that bring about the change in the substrate). The second largest cause of failed reactions, after equipment contamination, is attempting to run a reaction with incorrect quantities of the reactants present. Do not be hurried or careless at the balance.

4. *Clean means DRY.* Water or cleaning solution can be as detrimental to the success of a reaction as dirt or sludge in the system. You often will be working with very small quantities of moisture-sensitive reagents. The glass surface areas with which these reagents come in contact, however, are relatively large. A slightly damp piece of glassware can rapidly deactivate a critical reagent and result in reaction failure. *This rule must be strictly followed.*

5. *ALWAYS work on a clean laboratory bench surface,* preferably glass!

6. *ALWAYS protect the reaction product* that you are working with from a disastrous spill by carrying out all solution or solvent transfers over a crystallizing dish.

7. *ALWAYS place reaction vials or flasks in a clean beaker* when standing them on the laboratory bench.

8. *NEVER use cork rings to support round-bottom flasks,* particularly if they contain liquids. You are inviting disaster to be a guest at your laboratory bench.

9. *ALWAYS think through the next step* you are going to perform *before* starting it. Once you have added the wrong reagent, it's back to square one.

10. *ALWAYS save everything* you have generated in an experiment until it is successfully completed. You can retrieve a mislabeled chromatographic fraction from your locker, but not from the waste container!

THE ORGANIC CHEMISTRY LABORATORY

This laboratory experience will give you a brief glimpse of how organic chemistry operates as an experimental science. Historically, the organic lab has had a reputation of being smelly, long, tedious, and pockmarked with fires and explosions. Modern organic chemistry still has trouble shaking this image, but present-day organic chemistry is undergoing a revolution at the laboratory bench. New techniques are sweeping away many of the old complaints, as an increasing fraction of industrial and academic research is being carried out at the microscale level.

This text allows the interested beginning student to rapidly develop the necessary skills to slice more deeply into organic chemistry, as a sophomore, than ever before. The attendant benefits are greater confidence and independence in acquired laboratory techniques. The happy result is that in the microscale organic chemistry laboratory, you are more likely to have a satisfying first encounter with the experimental side of this fascinating field of knowledge.

Chapter 2

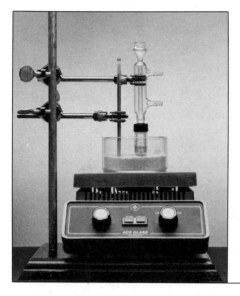

Safety and the Environment of the Laboratory

The responsibilities that one assumes while working in an organic chemistry laboratory are not unlike those taken on when driving a car. We are clearly responsible for injuries to ourselves or others that may arise from improper use of the car. In another sense, as we drive we also add pollution to the environment in which everyone must live and work. Although we cannot reduce this pollution to zero, we are responsible for keeping our car in proper operating condition to minimize adverse effects.

Similar risks are associated with working in an organic chemistry laboratory (or any chemistry laboratory). The focus in this chapter and the text as a whole is on the nature of these risks and the ways in which one can minimize the risks while still learning something about chemistry. In the laboratory, as in a car, we cannot reduce to zero the risk of an acute injury that might be associated with a fire or an explosion, nor can we reduce to zero the risk of chronic effects resulting from exposure to certain toxic chemicals. What we can do is to minimize these risks by *thinking safety* and working to maintain a reasonably safe laboratory for ourselves and for others in the lab.

NATURE OF HAZARDS The chemistry laboratory presents a wide range of risks. They are outlined briefly here so that you can begin to realize the steps necessary to reduce the risks.

1. *Physical hazards.* Injuries resulting from flames, explosions, and equipment (cuts from glass, electrical shocks from faulty instrumentation, or improper use of instruments).

2. *External exposure to chemicals.* Injuries to skin and eyes resulting from contact with chemicals that have splashed or have been left on the bench top or on equipment.

3. *Internal exposure.* Longer-term (usually) health effects resulting from breathing hazardous vapors or ingesting chemicals.

5

Reduction of Risks

Your instructor may have a more specific list of rules to be observed in your laboratory section. The following rules can serve as a starting point.

1. *Stick to the procedures described by your instructor and to the materials available in the laboratory.*

2. *Wear safety glasses.* We often recover quickly from injuries affecting only a few square millimeters on our bodies, *unless* that area happens to be our eyes. Goggles provide added protection.

3. *Do not put anything in your mouth while in the laboratory.* This includes food, drinks, and pipets. There are countless ways in which surfaces can become contaminated in the lab. Because there are substances that must *never* be pipetted by mouth, one should get into the habit of *not* mouth pipetting *anything*.

4. *Be cautious with flames and flammable solvents.* Use them only under conditions specified by your instructor. Remember that the flame at one end of the bench may ignite the flammable liquid at the other end in the event of a spill or improper disposal. Flames must never be used when certain liquids are present in the laboratory, and they must always be used with care.

5. *Minimize the loss of chemicals to air or water and dispose of waste properly.* Some water-soluble materials may be safely disposed of in the water drains. Special receptacles will be provided for other wastes, sometimes in the hoods. Pay attention to the labels on these receptacles.

6. *Minimize skin contact with any chemicals.* Use impermeable gloves, when directed, and wash any chemical off your body promptly. If you have to wash something off with water, use lots of it. Be sure that you know where the nearest water spray device is located.

7. *Tie back or confine long hair and loose items of clothing.* You don't want things to be falling into reagents or getting near flames.

8. *Don't work alone.* Too many things can happen to a person working alone that might leave that person unable to obtain assistance. In the rare case that you have permission to work alone, be sure that someone checks on you at regular intervals.

9. *Exercise care in assembling glass apparatus.* Separating standard taper glassware, pushing thermometers through rubber stoppers, and other operations with glassware all involve the risk that the glass may break and that lacerations or punctures may result. Make sure that thermometers and glassware are lubricated so that a great deal of force need not be used. Seek advice when unsure or for difficult cases.

10. *Report any injury or accident to your instructor.* This is important so that medical assistance can be requested, if necessary. It is also essential for the instructional staff to be made aware of any safety problems. Many of these are correctable.

11. *Keep things clean.* Put unused apparatus away. Wipe or care for spills on the bench top or on the floor immediately.

Precautionary Measures

Locate the nearest

- Fire extinguisher (understand uses of different types)
- Eye wash
- Spray shower
- Fire blanket
- Exit

THINKING ABOUT THE RISKS OF USING CHEMICALS

The reduction in quantities used in the microscale laboratory carries with it a reduction in hazards caused by fires and explosions. Hazards associated with skin contact are also reduced; however, care must be exercised when working in close proximity to the small quantities involved.

There is a great potential for reducing the exposure to chemical vapors, but these reductions will be realized only if everyone in the laboratory is careful. One characteristic of vapors is that they mix rather rapidly throughout the lab and quickly reach the nose of the student on the other side of the room. For this reason some operations may have to be performed in hoods. In some institutions the majority of the experiments may be carried out in hoods. In the open laboratory each experimenter becomes a "polluter" whose emissions affect the people nearby the most, but become added to the laboratory air and to the burden each of us must bear.

The concentration of the vapor in the laboratory will depend on the vapor pressure of the liquid, the area of liquid or solid exposed, the nature of air currents near the liquid, and the ventilation characteristics of the laboratory. One factor over which each individual has control is the evaporation process, which can be affected by the following procedures:

- Certain liquids must remain in hoods.
- Reagent bottles must be recapped when not in use.
- Spills must be quickly cleaned up and the waste properly discarded.

Material Safety Data Sheets

Although risks are associated with the use of most chemicals, the magnitudes of these risks vary greatly. A short description of these risks is provided by a Material Safety Data Sheet, commonly referred to as an *MSDS*. These sheets are normally provided by the manufacturer or vendor of the chemical and many departments keep files of MSDSs.

As an example, the MSDS for chloroform is shown here. This sheet is provided by the J. T. Baker Chemical Company. Sheets from other sources will be very similar. Much of the information on these sheets is self-explanatory, but let's review the major sections of the chloroform example. (See pages 10 and 11.)

Section I provides identification numbers and codes for the compound and includes a summary of the risks associated with the use of chloroform. Because these sheets are available for many thousands of compounds and mixtures, there must be a means of unambiguously identifying the substance. One of the more standard reference numbers for chemists is the CAS No., or Chemical Abstract Service Number.

A quick review of the degree of the risks is given by the numerical scale under Precautionary Labeling. This particular scale is a proprietary scale that ranges from 0 (very little or nonexistent risk) to 4 (extremely large risk). The National Fire Protection Association (NFPA) uses a similar scale but the risks considered are different. Other systems may use slightly different scales and there are some scales that represent low risks with the highest number! Be sure that you understand the scale being used. Perhaps some day one scale will become standard.

Section II covers risks from mixtures. Because a mixture is not considered here, the section is empty. Selected physical data are included in Section III. Section IV contains fire and explosion data, including a description of the toxic gases produced when chloroform is exposed to a fire. MSDSs are routinely made available to fire departments who may be faced with fighting a fire in a building where large amounts of chemicals are stored.

Health hazards are described in Section V. The entries here of most significance for evaluating the risks from vapors are the threshold limit value (or TLV) and the short-term exposure limit (STEL). The TLV is a term used by the

American Conference of Governmental Industrial Hygienists (ACGIH). This organization examines the toxicity literature for a compound and establishes the TLV. This standard is designed to protect the health of workers exposed to the vapor eight hours per day, five days per week. The Occupational Safety and Health Administration (OSHA) adopts a value to determine the safety of workplaces in the United States. Their value is termed the time weighted average (TWA) and in many cases is numerically equal to the TLV. The STEL should not be exceeded for even a 15-minute averaging time. TLV/TWA and STEL values for many chemicals are summarized in a little handbook available from the ACGIH, 6500 Glenway Avenue, Building D-5, Cincinnati, OH 45211. They are also collected in the *CRC Handbook of Chemistry and Physics.*

The toxicity of chloroform is also described in terms of the toxic oral dose. In this case, the LD50 is the dose that will cause the death of 50% of the mice or rats given that dose. The dose is expressed as milligrams of chloroform per kilogram body weight of the subject animal. The figures for small animals are often used to estimate the effects on humans. If, for example, we used the mouse figure of 80 mg/kg and applied it to a 60-kg student, a dose of 4800 mg, or about 3 mL, would cause fairly immediate death to 50% of the subjects receiving that dose. The effects of exposure of skin to the liquid or vapor are also described.

Section VI describes the reactivity of chloroform and the classes of compounds with which it should not come into contact. For example, sodium metal reacts violently with a number of substances (including water) and should not come into contact with them. Strong oxidizing agents (such as nitric acid) should not be mixed with organic compounds (among other things). The final sections are self-explanatory.

ESTIMATING RISKS FROM VAPORS

Other things (availability, suitability) being equal, one would, of course, choose the least toxic chemical for a given experiment. Some chemicals that play very important roles in synthetic organic chemistry are quite toxic, and the toxicity of chemicals in regular use in the laboratory varies greatly. Bromine and benzene have TLV values of 0.7 and 30 mg/m^3, respectively, and are at the more toxic end of the spectrum of chemicals used. Acetone has a TLV of 1780 mg/m^3. These representative figures do not mean that acetone is "harmless" or that bromine cannot be used. In general, one should exercise care at all times (make a habit of good laboratory practice) and should take special precautions when working with highly toxic materials.

This text provides some quantitative information on the risks of exposure to the vapors of each of the compounds used in experiments where a TLV has been established for the compound. These numbers are provided to give some guidance to instructors as they plan experiments. Instructors must evaluate whether or not use of a chemical should be confined to a hood, or the number of students who may reasonably perform the experiment in a laboratory section, or the feasibility of performing the experiment at all. The numbers are also made available to *you* to provide some insight into the *relative* risks resulting from exposure to vapors of the chemicals you will be working with.

Three important variables are needed to estimate the risk from exposure to vapors in the open laboratory. By open laboratory we mean one in which the experiments are done on open bench tops and in which any vapors emitted are removed by the general room ventilation (which may be provided by hoods). The important variables are

- TLV or other health standard
- Amount of chemical vaporized
- Effective ventilation rate of the room

The main problem we have in making a semiquantitative estimate of the risk is how to compare the risk from evaporation of a small amount (50 mg, say) of a chemical with a very low TLV (a very toxic material) with the risk from evaporation of a large amount (20 g, for instance) of a less toxic substance.

We have chosen to tabulate the ratio of the amount of material emitted (in a reasonable worst-case situation) to the TLV. If the amount emitted (m) is expressed in milligrams and the TLV (L) in milligrams per cubic meter, the ratio has the dimensions of cubic meters. In fact, m/L describes the volume of air required to dilute the emission to the TLV. If m/L is small, say 1 m^3, much less air is required to achieve this dilution and the hazard will be less than if m/L is large, say 100 m^3. The m/L value is given for each chemical used for which a TLV has been established. The relative risk from exposure to vapors for the entire experiment is the sum of m/L figures for the compounds used.

The m/L figures may also be used to assess the relative risk of doing the experiment outside a hood. As m/L represents the volume of air required for *each* student, this may be compared with the volume of air actually available for each student. If the ventilation rate for the entire laboratory is Q (in cubic meters per minute) for a section of n students meeting for t minutes, the volume for each student is kQt/n cubic meters. Here k is a factor that allows for the fact that the ventilation air will not be perfectly mixed in the laboratory before it is exhausted. In a reasonable worst-case mixing situation a k value of 0.3 seems reasonable. Laboratories with modest ventilation rates supplied by 15–20 lineal feet of hoods can be expected to provide 30–100 m^3 per student over a three-hour laboratory period if the hoods are working properly. Let us take the figure of 50 m^3 per student as an illustration. If the value of m/L for a compound (or a group of compounds in a reaction) is substantially less than 50 m^3 it may be safe to do that series of operations in the open laboratory. If m/L is comparable to or greater than 50 m^3, a number of options are available. (1) Steps using that compound may be restricted to a hood. (2) The instructional staff may satisfy themselves that much less than the assumed value is actually evaporated under conditions present in their laboratory. (3) The number of individual repetitions of this experiment may be reduced. The size of the section can be reduced or the experiment may be done in pairs or in trios.

EMISSIONS AND ENVIRONMENTAL DATA

Emissions in the laboratory result from a wide range of processes involving many chemicals. Emissions from some of these processes have been estimated from measurements of surrogate procedures. In other cases estimates have been made by interpolation, or maximum losses have been assumed. The environmental data presented with each experiment have been developed in the following way.

Amount Used

The figure for the amount used is obtained directly from the description of the experiment. Often, when the same material is used more than once in an experiment (i.e., methylene chloride as a reaction solvent and as the mobile phase for column chromatography), the uses are listed separately. This is done to permit an evaluation of the effects of an experiment when experiments are modified by elimination of some steps.

Emissions

The estimated emissions are given in milligrams. The assumptions used in making these estimates and other data for the compounds are given in the line under the numbers. Some notes on the abbreviations follow.

J. T. BAKER CHEMICAL CO. 222 RED SCHOOL LANE, PHILLIPSBURG, NJ 08865
M A T E R I A L   S A F E T Y   D A T A   S H E E T
24-HOUR EMERGENCY TELEPHONE — (201) 859-2151
CHEMTREC # (800) 424-9300 — NATIONAL RESPONSE CENTER # (800) 424-8802

A0446 –01	ACETONE	PAGE: 1
EFFECTIVE: 10/11/85		ISSUED: 01/23/86

SECTION I – PRODUCT IDENTIFICATION

PRODUCT NAME: ACETONE
FORMULA: (CH3)2CO
FORMULA WT: 58.08
CAS NO.:   00067-64-1
NIOSH/RTECS NO.:   AL3150000
COMMON SYNONYMS: DIMETHYL KETONE; METHYL KETONE; 2-PROPANONE
PRODUCT CODES: 9010,9006,9002,9254,9009,9001,9004,5356,A134,9007,9005,9008

PRECAUTIONARY LABELLING

BAKER SAF-T-DATA(TM) SYSTEM

HEALTH   – 1
FLAMMABILITY   – 3 (FLAMMABLE)
REACTIVITY   – 2
CONTACT   – 1

LABORATORY PROTECTIVE EQUIPMENT

SAFETY GLASSES; LAB COAT; VENT HOOD; PROPER GLOVES; CLASS B EXTINGUISHER

PRECAUTIONARY LABEL STATEMENTS

DANGER
EXTREMELY FLAMMABLE
HARMFUL IF SWALLOWED OR INHALED
CAUSES IRRITATION
KEEP AWAY FROM HEAT, SPARKS, FLAME. AVOID CONTACT WITH EYES, SKIN, CLOTHING.
AVOID BREATHING VAPOR. KEEP IN TIGHTLY CLOSED CONTAINER. USE WITH ADEQUATE
VENTILATION. WASH THOROUGHLY AFTER HANDLING. IN CASE OF FIRE, USE WATER SPRAY,
ALCOHOL FOAM, DRY CHEMICAL, OR CARBON DIOXIDE. FLUSH SPILL AREA WITH WATER
SPRAY.

SECTION II – HAZARDOUS COMPONENTS

COMPONENT	%	CAS NO.
ACETONE	90-100	67-64-1

SECTION III – PHYSICAL DATA

BOILING POINT:	56 C (133 F)	VAPOR PRESSURE(MM HG):	181
MELTING POINT:	−95 C (−139 F)	VAPOR DENSITY(AIR=1):	2
SPECIFIC GRAVITY:	0.79	EVAPORATION RATE:	5.6
(H2O=1)		(BUTYL ACETATE=1)	

SOLUBILITY(H2O):   COMPLETE (IN ALL PROPORTIONS) % VOLATILES BY VOLUME: 100

APPEARANCE & ODOR: CLEAR, COLORLESS LIQUID WITH FRAGRANT SWEET ODOR.

SECTION IV – FIRE AND EXPLOSION HAZARD DATA

FLASH POINT: −18 C (0 F)   NFPA 704M RATING: 1-3-0

FLAMMABLE LIMITS: UPPER –   13 %   LOWER –   2 %

FIRE EXTINGUISHING MEDIA
    USE ALCOHOL FOAM, DRY CHEMICAL OR CARBON DIOXIDE.
    (WATER MAY BE INEFFECTIVE.)

SPECIAL FIRE-FIGHTING PROCEDURES
    FIREFIGHTERS SHOULD WEAR PROPER PROTECTIVE EQUIPMENT AND SELF-CONTAINED
    (POSITIVE PRESSURE IF AVAILABLE) BREATHING APPARATUS WITH FULL FACEPIECE.
    MOVE EXPOSED CONTAINERS FROM FIRE AREA IF IT CAN BE DONE WITHOUT RISK.
    USE WATER TO KEEP FIRE-EXPOSED CONTAINERS COOL.

UNUSUAL FIRE & EXPLOSION HAZARDS
    VAPORS MAY FLOW ALONG SURFACES TO DISTANT IGNITION SOURCES AND FLASH BACK.
    CLOSED CONTAINERS EXPOSED TO HEAT MAY EXPLODE. CONTACT WITH STRONG
    OXIDIZERS MAY CAUSE FIRE.

SECTION V – HEALTH HAZARD DATA

THRESHOLD LIMIT VALUE (TLV/TWA):   1780 MG/M3 (750 PPM)

SHORT-TERM EXPOSURE LIMIT (STEL): 2375 MG/M3 (1000 PPM)

TOXICITY: LD50 (ORAL-RAT)(MG/KG)   – 9750
              LD50 (IPR-MOUSE)(G/KG)   – 1297

EFFECTS OF OVEREXPOSURE
    CONTACT WITH SKIN HAS A DEFATTING EFFECT, CAUSING DRYING AND IRRITATION.
    OVEREXPOSURE TO VAPORS MAY CAUSE IRRITATION OF MUCOUS MEMBRANES, DRYNESS
    OF MOUTH AND THROAT, HEADACHE, NAUSEA AND DIZZINESS.

EMERGENCY AND FIRST AID PROCEDURES
    CALL A PHYSICIAN.
    IF SWALLOWED, IF CONSCIOUS, IMMEDIATELY INDUCE VOMITING.
    IF INHALED, REMOVE TO FRESH AIR. IF NOT BREATHING, GIVE ARTIFICIAL
    RESPIRATION. IF BREATHING IS DIFFICULT, GIVE OXYGEN.
    IN CASE OF CONTACT, IMMEDIATELY FLUSH EYES WITH PLENTY OF WATER FOR AT
    LEAST 15 MINUTES. FLUSH SKIN WITH WATER.

SECTION VI – REACTIVITY DATA

STABILITY: STABLE   HAZARDOUS POLYMERIZATION: WILL NOT OCCUR

CONDITIONS TO AVOID: HEAT, FLAME, SOURCES OF IGNITION

INCOMPATIBLES: SULFURIC ACID, NITRIC ACID, STRONG OXIDIZING AGENTS

SECTION VII – SPILL AND DISPOSAL PROCEDURES

STEPS TO BE TAKEN IN THE EVENT OF A SPILL OR DISCHARGE
WEAR SUITABLE PROTECTIVE CLOTHING.   SHUT OFF IGNITION SOURCES; NO FLARES, SMOKING, OR FLAMES IN AREA.   STOP LEAK IF YOU CAN DO SO WITHOUT RISK.   USE WATER SPRAY TO REDUCE VAPORS.   TAKE UP WITH SAND OR OTHER NON-COMBUSTIBLE ABSORBENT MATERIAL AND PLACE INTO CONTAINER FOR LATER DISPOSAL.   FLUSH AREA WITH WATER.

J. T. BAKER SOLUSORB(R) SOLVENT ADSORBENT IS RECOMMENDED FOR SPILLS OF THIS PRODUCT.

DISPOSAL PROCEDURE
DISPOSE IN ACCORDANCE WITH ALL APPLICABLE FEDERAL, STATE, AND LOCAL ENVIRONMENTAL REGULATIONS.

EPA HAZARDOUS WASTE NUMBER: U002 (TOXIC WASTE)

SECTION VIII – PROTECTIVE EQUIPMENT

VENTILATION: USE GENERAL OR LOCAL EXHAUST VENTILATION TO MEET TLV REQUIREMENTS.

RESPIRATORY PROTECTION: RESPIRATORY PROTECTION REQUIRED IF AIRBORNE CONCENTRATION EXCEEDS TLV.   AT CONCENTRATIONS UP TO 5000 PPM, A GAS MASK WITH ORGANIC VAPOR CANNISTER IS RECOMMENDED.   ABOVE THIS LEVEL, A SELF-CONTAINED BREATHING APPARATUS WITH FULL FACE SHIELD IS ADVISED.

EYE/SKIN PROTECTION: SAFETY GLASSES WITH SIDESHIELDS, POLYVINYL ACETATE GLOVES ARE RECOMMENDED.

SECTION IX – STORAGE AND HANDLING PRECAUTIONS

SAF-T-DATA(TM) STORAGE COLOR CODE: RED

SPECIAL PRECAUTIONS
BOND AND GROUND CONTAINERS WHEN TRANSFERRING LIQUID.   KEEP CONTAINER TIGHTLY CLOSED.   STORE IN A COOL, DRY, WELL-VENTILATED, FLAMMABLE LIQUID STORAGE AREA.

SECTION X – TRANSPORTATION DATA AND ADDITIONAL INFORMATION

DOMESTIC (D.O.T.)

PROPER SHIPPING NAME	ACETONE
HAZARD CLASS	FLAMMABLE LIQUID
UN/NA	UN1090
LABELS	FLAMMABLE LIQUID

INTERNATIONAL (I.M.O.)

PROPER SHIPPING NAME	ACETONE
HAZARD CLASS	3.1
UN/NA	UN1090
LABELS	FLAMMABLE LIQUID

(TM) AND (R) DESIGNATE TRADEMARKS.
N/A = NOT APPLICABLE OR NOT AVAILABLE

THE INFORMATION PUBLISHED IN THIS MATERIAL SAFETY DATA SHEET HAS BEEN COMPILED FROM OUR EXPERIENCE AND DATA PRESENTED IN VARIOUS TECHNICAL PUBLICATIONS. IT IS THE USER'S RESPONSIBILITY TO DETERMINE THE SUITABILITY OF THIS INFORMATION FOR THE ADOPTION OF NECESSARY SAFETY PRECAUTIONS. WE RESERVE THE RIGHT TO REVISE MATERIAL SAFETY DATA SHEETS PERIODICALLY AS NEW INFORMATION BECOMES AVAILABLE.

max.

Max. indicates that all the material used is assumed to be emitted. This assumption is generally used for experiments in which the solvent is evaporated to recover a solute. Emissions of many processes have not been evaluated, and the maximum assumption has been used where there is reason to believe that a substantial fraction of the material is emitted. The emissions for many of these processes will be much less if adequate hood space is available. The assumption that all of a compound may be emitted to the room may seem extreme; however, measurements of inputs and outputs of solvents in the microscale laboratory have indicated that only about 50% of the material used in the experiment ends up in a waste container. Even though some of the solvent ends up in waste water, the assumption that all solvent is vaporized may only err by a factor of 2 and the use of the maximum figure provides a degree of conservatism.

open container

The following losses from 50-mL beakers (ambient temp.) have been measured.

Compound	Loss (mg/hr)
Acetone	920
2-Butanone	390
3-Pentanone	140

reflux

The loss of acetone is about 75 mg/hour for a water-cooled condenser. In cases involving other compounds the 75 mg/hour figure is multiplied by the ratio of open container losses for appropriate compounds (those having volatilities similar to those of the ketones listed).

transfer

Losses from liquid transfer are 30 and 75 mg for 3-pentanone (bp = 103°C) and acetone (bp = 56°C), respectively, when the transfer is performed at the macro level with a graduated cylinder. Losses from liquid transfer are 1 and 9.4 mg for the same two compounds when the transfers are done with a micropipet.

treated as hexane

Petroleum ether and ligroin are hydrocarbon mixtures for which TLVs have not been assigned. As a conservative step, these solvents have been treated as hexane, the substance with the lowest TLV likely to be present.

volume

This is the figure for emissions divided by TLV.

Abbreviations

col. chromatog.	purification by column chromatography
Craig	recrystallization with a Craig tube
evap.	evaporation of nearly all of the solvent
extract	liquid–liquid extraction
react.	substance used as a reactant
recryst.	recrystallization
solv.	substance used as a solvent
vac. filtr.	vacuum filtration

MTDS	mutation dose
IRDS	irritation dose
TXDS	toxic dose
LDLo	lethal dose low
LD50	lethal dose for 50%
TFX : Car	carcinogenic
ihl	inhalation
orl	oral
unk	unknown
mam	mammal
mus	mouse
rat	rat
ham	hamster
ipr	intraperitoneal
hmn	human
inv	intravenous
SKIN	skin
rbt	rabbit
ims	intramuscular
scu	subcutaneous

QUESTIONS

2-1. Think about what you would do in the following cases. (Note that you may need more information for some of these problems.)

a. A hot solution "bumps," splashing your face.

b. A breaker of solvent catches fire.

c. A reagent bottle falls, spilling concentrated sulfuric acid.

d. You hear a sizzle as you pick up a hot test tube.

2-2. A laboratory has four hoods each of which is 39 in. wide. When the hood door is open to a height of 8 in. and the hoods are operating, the average air velocity through the hood face is 170 ft/minute.

a. Evaluate the total ventilation rate for this room assuming that there are no other exhausts.

b. The laboratory is designed for use by 30 students. Evaluate the air available per student if the mixing factor is 0.3 and experiments last three hours.

c. An experiment is considered in which each student would be required to evaporate 7 mL of methylene chloride. Estimate the average concentration of methylene chloride. Look up the TLV (or TWA) for methylene chloride and consider how the evaporation might be performed.

2-3. A laboratory has a ventilation system that provides 20 m^3 for each student during the laboratory period. (This figure includes the mixing factor.) An experiment is considered for this program that uses the following quantities of materials A, B, and C. The TLV is also listed for each compound.

Substance	Quantity (mg)	TLV (mg/m^3)
A	400	1200
B	500	200
C	200	5

Assess the relative risks of these three compounds. Is there a likely need for operations to be conducted in a hood if the compounds are assumed to be entirely evaporated?

2-4. An experiment is considered in which 1 mL of diethylamine would be used by each student. The ventilation rate for the laboratory is 5 m^3/minute. Look up the TLV (or TWA) for diethylamine. What restrictions might be placed on the laboratory to keep the average concentration over a three-hour period less than one-third of the TWA? Assume a mixing factor of 0.3.

Chapter 3

Introduction to Microscale Organic Laboratory Equipment and Technique

We will first describe special pieces of glassware that you will use to perform the experiments described in this text. Most of this equipment will be found in your laboratory locker. The following descriptions should be helpful when checking your equipment list at the beginning of the semester. Next, we will consider a series of standard experimental apparatus setups that utilize this equipment. At the end of each of these short discussions, you will find a listing of arrangements of the particular setup and the experiment in which their use is described.

Most of the equipment listed below is glassware that has been specially developed and tested by students at Bowdoin and Merrimack for use in the microscale organic laboratory program. We feel that this particular collection offers significant advantages over earlier versions available for student use at this scale of laboratory operation. This is our opinion, but it is only an opinion. As you explore more deeply into the experimental side of chemistry you will find in laboratory technique a component that is very much *artistic* in nature. Two chemists may do exactly the same experiment and describe the experimental procedures in very nearly the same terms, but in fact in the laboratory they may go about the work in rather different ways. This is the *art* of doing research. The esthetics of laboratory experimentation touches chemists in very personal ways. Thus, the equipment you find in your locker may be identical to, similar but not identical to, or only vaguely reminiscent of the components described in this chapter. Consult your laboratory instructor for further details on those items that are not obvious extensions of apparatus we have discussed. Remember your instructional laboratory is in large part an extension of your instructor's own personal research laboratory. Thus, you will be using equipment that he or she feels comfortable having you work with, as you begin learning the art of experimental organic chemistry on their home ground (see Figs. 3.1–3.7).

STUDENT MICROGLASSWARE EQUIPMENT

Listed below are the items and quantities of glassware that we have installed in our laboratories over the last few years. We list these mainly as a guide to instructors who are considering conversion to this type of program. This collection obviously is not meant to imply that it is the only type and number of apparatus that must be present in a student locker. It represents, in our current view, the optimum content when balancing cost against experimental breadth.

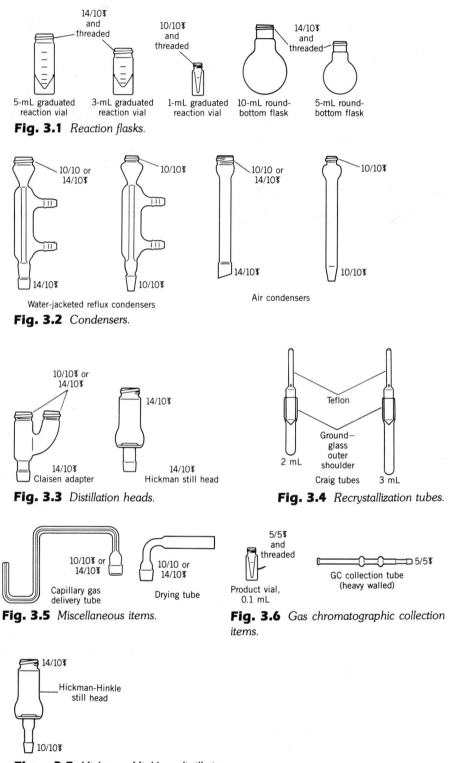

Fig. 3.1 *Reaction flasks.*

Fig. 3.2 *Condensers.*

Fig. 3.3 *Distillation heads.*

Fig. 3.4 *Recrystallization tubes.*

Fig. 3.5 *Miscellaneous items.*

Fig. 3.6 *Gas chromatographic collection items.*

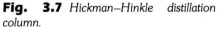

Fig. 3.7 *Hickman–Hinkle distillation column.*

No.	Glassware Item
2	conical vial, 5-mL, 14/10ℑ and threaded (one thin-walled)
2	conical vial, 3-mL, 14/10ℑ and threaded
(1)	conical vial, 1-mL, 10/10ℑ and threaded (optional)
1	round-bottom flask, 5- or 10-mL, 14/20ℑ
1	water-jacketed reflux condenser, male 14/10ℑ, female 14/10ℑ or water-jacketed reflux condenser, male 10/10ℑ, female 10/10ℑ
1	air condenser, male 14/10ℑ or 10/10ℑ, female 14/10ℑ or 10/10ℑ
1	Claisen head, male 14/10ℑ threaded, female 14/10ℑ or 10/10ℑ
1	Craig tube, 3-mL, with Teflon head
1	Craig tube, 2-mL, accepts above head
1	drying tube, male 10/10ℑ or 14/10ℑ
1	capillary gas delivery tube, male 10/10ℑ or 14/10ℑ
1	Hickman still head, male 14/10ℑ, female 14/10ℑ, or Hickman–Hinkle Spinning Band Head, male 14-10/10ℑ, female 14/10ℑ
(1)	double O-ring Teflon reducing bushing 14/10ℑ to 10/10ℑ (optional)

No.	Gas Chromatographic Collection Item
2	GC collection tube, male 5/5ℑ
2	conical vial, 0.1-mL, 5/5ℑ and threaded

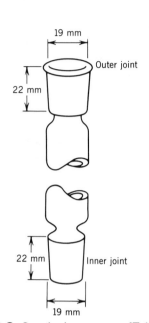

Fig. 3.8 *Standard taper joints [ℑ 19/22]. From Zubric, James W. The Organic Chem Lab Survival Manual, 2nd ed.; Wiley: New York, 1988. (Reprinted by permission of John Wiley & Sons, New York.)*

Fig. 3.9 *Threaded female joint.*

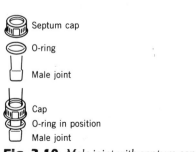
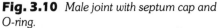

Fig. 3.10 *Male joint with septum cap and O-ring.*

Pieces of glassware are most efficiently assembled into a functional apparatus using standard taper ground-glass joints (see Fig. 3.8). The symbol ℑ is commonly used to indicate the presence of this type of connector. Normally, ℑ is either followed or preceded by #/#. The first # refers to the maximum inside diameter of a female (outer) joint or the maximum outside diameter of a male (inner) joint, measured in millimeters. The second number corresponds to the total length of the ground surface of the joint. The advantage of this type of connection is that if the joint surfaces are lightly greased, a vacuum seal is achieved. One of the drawbacks in employing these joints, however, is that contamination of the reacting system readily occurs through extraction of the grease by the solvents present in the reaction vessel. In carrying out microscale reactions, this is particularly troublesome.

An alternative to using greased joints as connectors is to employ either outside or inside screw-threaded glass systems. This type of joint utilizes plastic Teflon-lined screw-cap connectors. These threaded glass joints are commercially available; they originally suffered from the same problem as the ungreased standard taper joint, that is, they were not gastight (at least not in the hands of most undergraduates). The more recent adaptations of these seals have minimized this problem to a reasonable extent. The glassware joints developed in the microscale experimental organic laboratory programs at Bowdoin and Merrimack, however, have the ease and physical integrity of research-grade standard taper ground-glass joints along with a number of extremely important added features. The dimensions usually are either ℑ 14/10 or 10/10, male or female (ℑ 7/10 systems are optional for some setups). The conical vials in which the large majority of reactions are carried out also employ this alternative type of sealing system. You will note that in addition to being ground to a standard taper on the inside surface these vials also possess a screw thread on the outside surface (Fig. 3.9). This arrangement allows a standard taper male joint to be sealed to the reaction flask by a septum-type (open) plastic screw cap. The screw cap applies compression to a retaining silicone rubber O-ring positioned just above the male joint (Fig. 3.10). The compression of the O-ring thereby achieves a *greaseless* gastight seal while at the same time clamping the two pieces of equipment together. The ground joint provides protection from intimate solvent contact with the O-ring. This type of joint connection (now available on all joints on this glassware) provides a quick, easy, and reliable mechanism for assembling the glassware required for carrying out the microscale reactions. The use of this type of connector leads to a further bonus during construction of an experimental setup. Because the individual sections are small, light, and firmly sealed together the entire arrangement most often can be mounted on the support rack by a single clamp. In conventional systems it is necessary in the majority of cases to employ at least two clamps. This latter arrangement can easily lead to points of high strain in the glass components unless considerable care is taken in the assembly process. Clamp strain is one of the major sources of experimental glassware breakage. The ability to single clamp essentially all microscale setups effectively eliminates this problem.

It should be emphasized that the ground joint surfaces are grease free; therefore, it is important to disconnect joints soon after use or they may become locked or "frozen" together.

Joints of the size employed in these experiments, however, seldom are a problem to separate, if given proper care (keep them clean!).

A complete set of glassware would involve 14 different components and a total of 17 items grouped according to function as follows (see Figs. 3.1–3.7):

5 Reaction Flasks

2   conical vial, 5-mL, 14/10$ and threaded (one thin-walled)
2   conical vials, 3-mL, 14/10$ and threaded
(1)   conical vial, 1-mL, 10/10$ and threaded (optional)
1   round-bottom flask, 10-mL, 14/20$
(1)   round-bottom flask, 5-mL, 14/20$ (optional)

Both the conical vials and the round-bottom flasks are designed to be connected via an O-ring compression cap installed on the male joint of the adjacent part of the system (Fig. 3.1).

2 Condensers

1   water-jacketed reflux condenser, male 14/10$, female 14/10$ or
(1)   water-jacketed reflux condenser, male 10/10$, female 10/10$
1   air condenser, male 14/10$, female 14/10$ or
(1)   air condenser, male 10/10$, female 10/10$

These items form two sets of condensers for use with 14/10$ or 10/10$ jointed reaction flasks. The female joints allow connection of the condenser to the 10/10 or 14/10$ drying tube and the 14/10 or 10/10$ capillary gas delivery tube (Fig. 3.2).

2 Distillation Heads

1   Hickman still head, male 14/10$, female 14/10$ or
(1)   Hickman–Hinkle Spinning Band Head, male 10/10$, female 14/10$ (optional)

The simple Hickman still is used with an O-ring compression cap to carry out semimicro simple or crude fractional distillations. The newly developed Hickman–Hinkle is a powerful modification of the simple system. The 3-cm fractionating column of this still has been shown to develop more than six theoretical plates. The Hickman–Hinkle is currently available with either a male 10/10$ or 10-14/10$ joint. The former can be conveniently operated with the 14/10$, 3- and 5-mL conical vials employing the 14/10$ to 10/10$ Teflon reducing bushing. See Figures 3.3 and 3.7.

2 Recrystallization Tubes

1   Craig tube, 2-mL
1   Craig tube, 3-mL

Craig tubes are a particularly effective method for recrystallizing small quantities of reaction products. These tubes possess a nonuniform ground joint in the outer section. The Teflon head modification (earlier models used glass male sections) has made these systems much more rugged and much less susceptible to breakage during centrifugation (Fig. 3.4).

3 Miscellaneous Items

1   Claisen head, male 14/10$ or 10/10$, female 14/10$ or 10/10$
1   drying tube, male 10/10$ or 14/10$
1   capillary gas delivery tube, male 10/10$ or 14/10$

The Claisen head (Fig. 3.3) is often employed to facilitate the syringe addition of reagents to closed moisture-sensitive systems (such as Grignard reactions) via a septum seal in the vertical upper joint. This joint can also function to position the thermometer in the Hickman–Hinkle well. The Claisen adapter is also used to mount the drying tube in a protected and remote position from the reaction chamber. The drying tube in turn is used to protect moisture-sensitive reaction components from atmospheric water vapor while allowing a reacting system to remain unsealed. The capillary gas delivery tube is employed in transferring gases formed during reactions to storage containers (Figs. 3.5 and 3.18).

4 Gas Chromatographic Collection Items

2   GC collection tubes, male 5/5$
2   conical vials, 0.1-mL, 5/5$ and threaded

The collection tube is connected directly to the gas chromatographic column for fraction collection followed by transfer of the sample to a 0.1-mL conical vial for storage. The system is conveniently employed in the resolution and isolation of two-component mixtures (Fig. 3.6).

STANDARD EXPERIMENTAL APPARATUS

Heating and Stirring Reactions

It is important to be able to carry out microscale experiments at accurately determined temperatures. Very often, successful transformations require rather precise temperature control. In addition, many reactions require that reactants be intimately mixed to obtain a substantial yield of product. Therefore, the majority of the reactions you perform in this laboratory will be conducted with rapid stirring of the reaction mixture.

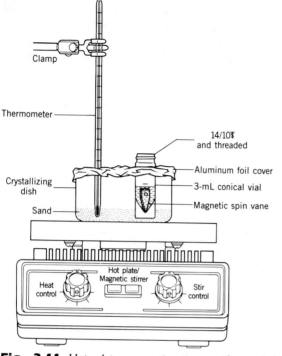

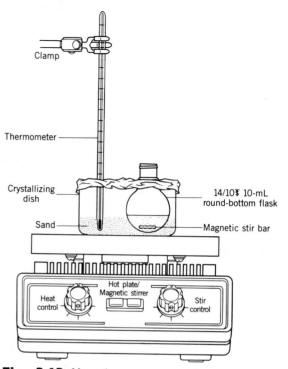

Fig. 3.11 *Hot plate—magnetic stirrer with sand bath and reaction vial.*

Fig. 3.12 *Hot plate—magnetic stirrer with sand bath and reaction flask.*

The most convenient piece of equipment for heating or stirring or for performing both operations simultaneously on a microscale level is the hot plate stirrer. Heat transfer from the hot surface to the reaction flask is generally accomplished by employing a crystallizing dish containing a **shallow** layer of sand which can conform to the size and shape of the particular vessel employed. The temperature of the system is most conveniently monitored by embedding a thermometer in the sand near the reaction vessel. (A successful procedure for determining the actual temperature inside the vial relative to the bath temperature is to mount a second thermometer in the vial after first adding 2 mL of high-boiling silicone oil. Then measure vial temperature at various sand bath temperatures, and enter on the graph of temperature versus hot plate setting. The weight of the amount of sand used will help to make the values more reproducible.) The high sides of the crystallizing dish act to protect the apparatus from air drafts, and so the dish also operates somewhat as a hot air bath. Heating can be made even more uniform by covering the crystallizing dish with aluminum foil (Fig. 3.11 and marginal figure, page 20). This procedure works well, but is a bit awkward and is required only in a few instances.

Stirring the reaction mixture in a conical vial is carried out with Teflon-coated magnetic spin vanes, and in round-bottom flasks with Teflon-coated magnetic stirring bars (see Figs. 3.11 and 3.12).

It is important that you become acquainted with the settings on the controls that adjust the current to the heating element and the motor that spins the magnet. You can make a rough graph of heat control setting versus temperature (Fig. 3.13) for your particular hot plate system (see the suggestions just given). These data will save considerable time when you bring a reaction apparatus to operating temperature. As your first step in the laboratory, it is advisable to adjust the temperature setting on the hot plate stirrer with the sand bath in place. The setting is determined from your control—temperature calibration curve. This procedure will allow the heated bath to reach a relatively constant temperature by the time it is required. You will then readily be able to make small final adjustments.

It should be emphasized that heavy layers of sand act as an insulator on the hot plate surface, which can result in damage to the heating element at high temperature settings. Therefore, when temperatures over 150°C are required, remember to use the minimum amount of sand.

It is important to position the reaction flask close to the bottom surface of the crystallizing dish in those experiments that depend on magnetic stirring. This arrangement is good practice, in general, as it leads to use of the minimum amount of sand.

If the reaction does not require elevated temperatures, but needs only to be stirred, the system is assembled without the sand bath. Some stirred reactions, on the other hand, require cooling. In this case the sand bath may be replaced with a crystallizing dish filled with ice water or ice water–salt, if lower temperatures are required.

Reflux Apparatus

We often find that to bring about a successful reaction between two substances, it is necessary to intimately mix the materials together and to maintain a specific temperature. The mixing operation is conveniently achieved by dissolution of the materials in a solvent in which they are mutually soluble. If the reaction is carried out in solution under reflux conditions, the choice of solvent can be used to control the temperature of the reaction. The very large majority of all organic reactions described in this text involve the use of a reflux apparatus in one or another arrangement. What do we mean by *reflux?* The term means to "return" or "run back." This is exactly how the reflux apparatus functions. When the temperature of the reaction system is raised to the boiling point (constant temperature), *all* solvent vapors are condensed and returned to the reaction flask or vial (this operation is not a distillation). In the microscale reactions described, we employ two basic types of reflux condensers: the *air condenser* and the *water-jacketed condenser.* The air condenser operates, as its name implies, by condensing solvent vapors on the cool vertical wall of an extended glass tube that dissipates the heat by contact with laboratory room air. This simple arrangement functions quite effectively with liquids boiling above 150°C. Indeed, a simple test tube can act as a reaction chamber and air condenser all in one unit, and many simple reactions can be most easily carried out in test tubes. (See, for example, Experiment 51 for the use of a test-tube-type system in a rather high-temperature reaction.) Air condensers can occasionally be used with lower-boiling systems; however, the water-jacketed condenser, which employs cold water to remove heat from the vertical column and thus facilitates condensation, is more often employed in these situations. This latter apparatus is highly efficient at condensing vapor from low-boiling liquids. To accommodate various-size reaction flasks, both styles of condensers are available in two different male standard taper joint sizes: 10/10⊤ and 14/10⊤. the top of both condenser columns possesses the female 10/10⊤ and 14/10⊤ joints. One fixable arrangement is to use the 10/10⊤ condensers with both 10/10⊤ and 14/10⊤ reaction vessels via the 14/10 double O-ring Teflon reducing bushing.

In refluxing systems that do not require agitation, the stirrer (magnetic spin vane or bar) usually is replaced by a "boiling stone." These sharp-edged stones possess highly fractured surfaces that are very efficient at initiating bubble formation as the reacting medium approaches the boiling point. The boiling stone acts to protect the system from disastrous boilovers and also reduces "bumping." (They should be used only once and *never* should a boiling stone be added to a hot solution. Why?)

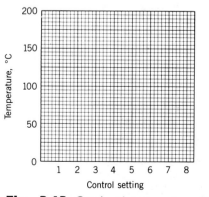

Fig. 3.13 *Graph of temperature (°C) versus hot plate control setting.*

Various Arrangements of Reflux Apparatus (see Figs. 3.14–3.16)

Air condenser with 3- or 5-mL conical vial, arranged for heating and magnetic stirring (Experiments 19; 34B; 45; 48C; 52).

Air condenser with 3- or 5-mL conical vial, arranged for magnetic stirring (Experiments 21A, C; 22; 26A, B; 27; 34A, C, D; 42A, B; 50A).

Air condenser with 3- or 5-mL conical vial, arranged for cooling and magnetic stirring (Experiment 31).

Air condenser with 1-mL conical vial, arranged for heating and magnetic stirring (Experiment 49).

Air condenser with 1-mL conical vial, arranged for heating (Experiment 50B).

Water-jacketed condenser with 3- or 5-mL conical vial, arranged for heating and magnetic stirring (Experiments 7A, B; 13; 20; 46; 48D).

Water-jacketed condenser with 1-mL conical vial, arranged for heating and magnetic stirring (Experiments 25A, B; 37).

Water-jacketed condenser with 10-mL round-bottom blask, arranged for heating and stirring (Experiments 14; 15; 19; 20; 21B, D; 22; 25A, B; 26A; 40A; 41; 42; 43B; 47A, C; 48A, B, C; 49; 50A, B; 52).

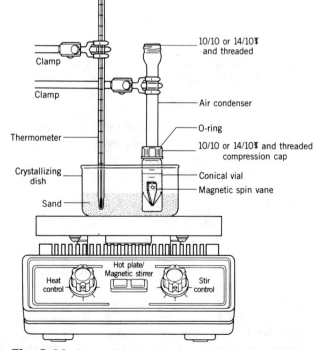

Fig. 3.14 *Air condenser with conical vial, arranged for heating and magnetic stirring.*

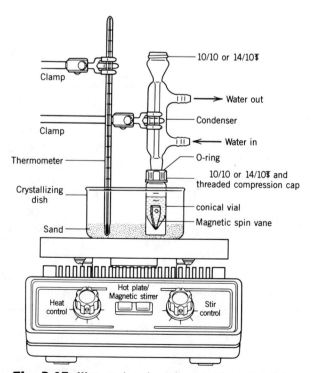

Fig. 3.15 *Water-jacketed condenser with conical vial, arranged for heating and magnetic stirring.*

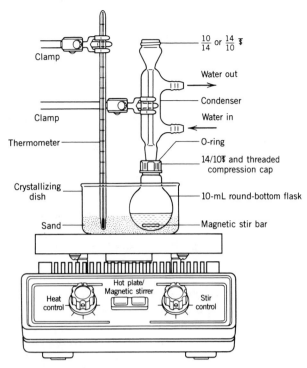

Fig. 3.16 *Water-jacketed condenser with 10-mL round-bottom flask, arranged for heating and magnetic stirring.*

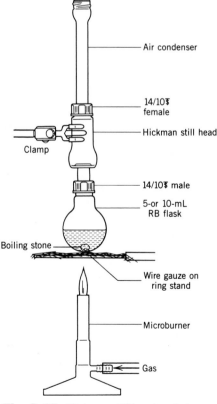

Fig. 3.17 *Hickman still head and air condenser with 5-mL round-bottom flask, arranged for microburner heating.*

Distillation Apparatus

Distillation is a laboratory operation used to separate substances that have different boiling points. The mixture is heated, vaporized, and then condensed, with the more volatile component being enriched in the early fractions of condensate. Unlike the reflux operation, in distillations, none or only a portion of the condensate is returned to the flask where vaporization is taking place. A large number of distillation apparatus have been designed to carry out this basic operation. They differ mainly in small features used to solve particular types of separation problems. In the present microscale experiments, a number of simple semimicroscale distillations are required. For these distillations the Hickman still head (Fig. 3.3) is ideally suited. This system has a 14/10Ŧ male joint for connection to either the 3- or 5-mL conical vials or the 5- or 10-mL round-bottom flasks. The still head functions as both an air condenser and a condensate trap. For a detailed discussion of this piece of equipment see Experiments 2 and 3A. During 1987 a new modification of the basic Hickman still was developed, the Hickman–Hinkle Spinning Band Still (Fig. 3.18), named for a Bowdoin undergraduate who first proposed this arrangement. The modified still functions in much the same way as the simple Hickman, except that a Teflon spinning band element is mounted in the slightly extended section between the male joint and the collection collar. In addition, this system has a built-in thermometer well that allows modestly accurate measurement of vapor temperature. When the band is spun at 1500 rpm by a magnetic stirring hot plate, this still is transformed into a very effective short-path fractional distillation column.

The most powerful system currently available for the instructional laboratory, however, is the 2.5-in. vacuum-jacketed microscale spinning band distillation column. (See Experiment 3B and Figure 5.19 for description and details.) This still utilizes the same type of band drive as the Hickman–Hinkle columns, but in addition the system is designed for conventional downward distillate collection, nonstopcock reflux control, and reasonably accurate temperature sensing. The column, which can develop nearly twelve theoretical plates, possesses the highest resolution of any standard distillation apparatus presently in use in the sophomore introductory organic laboratory.

Various Distillation Apparatus (see Figs. 3.17 and 3.18)

Hickman still head and air condenser with 5- and 10-mL round-bottom flasks, or 3- and 5-mL conical vials arranged for microburner or hot plate heating (Experiments: 2A; 30; 34A, B, C; 39; 48A).

Hickman–Hinkle still head with 5-mL conical vial (thin-walled), arranged for heating and magnetic spinning (Experiments 2C; 30; 39; 48A).

Moisture-Protected Reaction Apparatus

Many organic reagents react rapidly and preferentially with water. *The success or failure of many of the following experiments depends to a large degree on how well you can exclude atmospheric moisture from your reaction system.* The "drying tube," which is packed with a desiccant such as anhydrous calcium chloride, is a handy way to carry out a reaction in apparatus that is not closed to the atmosphere but is reasonably well protected from water vapor. The microscale apparatus described here are designed to be used with mainly the 10/10 or 14/10ℑ drying tube. The reflux condensers discussed earlier are constructed with female 10/10 or 14/10ℑ joints at the top of the column, which allows convenient connection of the drying tube if the refluxing system is moisture sensitive.

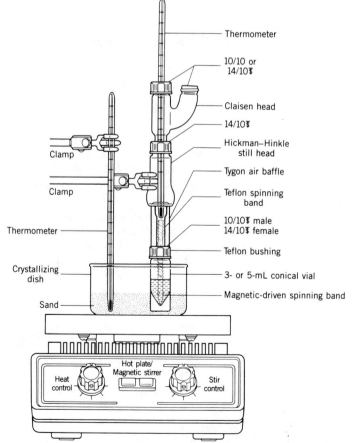

Fig. 3.18 *Hickman–Hinkle still head with 3- or 5-mL conical vial and Teflon spinning band, arranged for heating and magnetic stirring.*

A number of reactions to be studied are highly sensitive to moisture; thus successful operation at the microscale level can be rather challenging. If reagents are to be added after an apparatus has been assembled and dried, it is important to be able to add these reagents without exposing the system to the atmosphere, particularly when operating under humid conditions. In reactions not requiring reflux conditions and conducted at room temperature, this addition procedure is best accomplished by use of the microscale Claisen head adapter. The adapter has a vertical screw-threaded standard taper joint that will accept a septum cap. The septum seal allows syringe addition of reagents and avoids the necessity of opening the apparatus to the laboratory atmosphere.

In a few instances reactions are unusually moisture sensitive. In this situation, reactions are best carried out in completely sealed systems that are scrupulously dry. The use of the Claisen head adapter with a balloon substituted for the drying tube provides a satisfactory solution to the problem. Occasionally, it becomes important to maintain dry conditions during a distillation. The Hickman stills are constructed with a 14/10$\overline{S}$ joint at the top of the head which readily accepts the drying tube.

Various Moisture-Protected Reaction Apparatus (see Figs. 3.19–3.22)

Moisture-protected air condenser with 1-mL conical vial, arranged for heating and magnetic stirring (Experiments 8B, C, E; 27).

Moisture-protected water-jacketed condenser with 3- or 5-mL conical vial, arranged for heating and stirring (Experiments 8A, D; 24; 47B).

Moisture-protected water-jacketed condenser with 1-mL conical vial, arranged for heating and stirring (Experiments 28A, B; 29; 38).

Moisture-protected Claisen head with 3- or 5-mL conical vials, arranged for syringe addition and magnetic stirring (Experiments 13; 17; 18; 23; 32; 35; 48B).

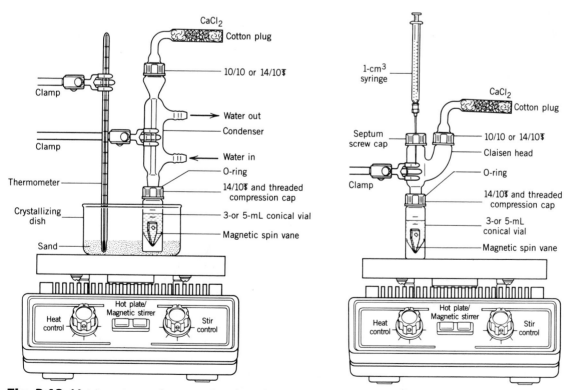

Fig. 3.19 *Moisture-protected water-jacketed condenser with 3- or 5-mL conical vials, arranged for heating and magnetic stirring.*

Fig. 3.20 *Moisture-protected Claisen head with 3- or 5-mL conical vial, arranged for syringe addition and magnetic stirring.*

Sealed Claisen head with 3- or 5-mL conical vials, arranged for N₂ flushing, heating, and magnetic stirring (Experiment 16A, B).

Moisture-protected Hickman still head with 10-mL round-bottom flask, arranged for heating and stirring (Experiment 30).

Specialized Pieces of Equipment

In a few experiments the reaction leads to gaseous products. The collection or trapping of these products is conveniently carried out using the capillary gas delivery tube. This item is designed to be attached directly to a 1- or 3-mL conical vial or to the female 10/10 or 14/10 ꭓ joint of a condenser connected to a reaction flask or vial. The tube leads to the collection system, which is a simple inverted graduated cylinder, blank threaded septum joint or air condenser. The 0.1-mm capillary bore considerably reduces dead volume and increases efficiency of product transfer.

The trapping and collection of gas chromatographic liquid fractions becomes particularly important at the microscale level of experimentation. A number of the experiments depend on this type of substance purification and isolation. The ease and efficiency of carrying out this operation are greatly facilitated by employing the 5/5 ꭓ collection tube and 0.1-mL 5/5 ꭓ conical collection vial.

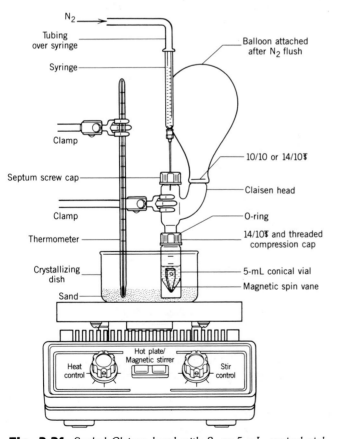

Fig. 3.21 *Sealed Claisen head with 3- or 5-mL conical vial, arranged for N₂ flushing, heating, and magnetic stirring.*

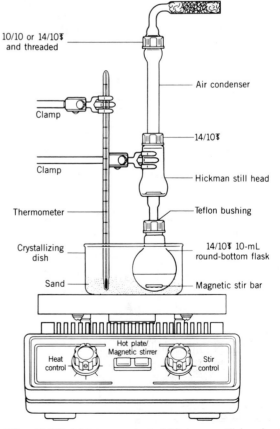

Fig. 3.22 *Moisture-protected Hickman still head with 10-mL round-bottom flask, arranged for heating and magnetic stirring.*

Various Sample Collection Apparatus (see Figs. 3.23–3.25)

Conical vial (1 mL) and capillary gas delivery tube, arranged for heating and stirring (Experiment 9).

Water-jacketed condenser with 3-mL conical vial and capillary gas delivery tube, arranged for heating and stirring (Experiment 41).

Gas chromatographic collection tube and 0.1-mL conical vial (Experiments 1; 5A, B; 13; 18; 39).

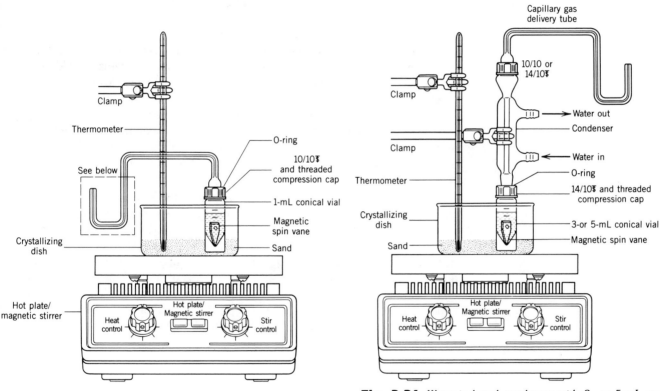

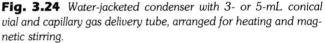

Fig. 3.24 *Water-jacketed condenser with 3- or 5-mL conical vial and capillary gas delivery tube, arranged for heating and magnetic stirring.*

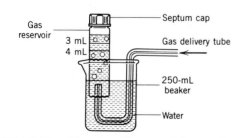

Fig. 3.23 *Vial (1-mL) and capillary gas delivery tube, arranged for heating and magnetic stirring.*

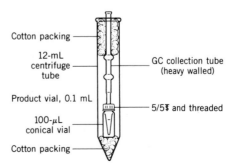

Fig. 3.25 *Gas chromatographic collection tube and 0.1-mL conical vial.*

MICROSCALE LAWS   **Rules of the Trade for Handling Organic Materials at the Microscale Level**

Now that we have briefly looked at the equipment we will be using to carry out the microscale organic reactions, let us examine specific techniques used to deal with the small quantities of material involved. These reactions by our definition start with 15 to 150 mg of the limiting reagent. These quantities sound small, and they are. Although 150 mg of a light powdery material will fill half of a 1-mL conical vial, you will have a hard time observing 15 mg of a clear liquid in the same container even with magnification. On the other hand, this volume of liquid is reasonably easy to observe when placed in a 0.1-mL conical vial. A vital part of the game of working with small amounts of materials is to become familiar with microscale techniques and to practice them as much as possible in the laboratory.

Rules for Working with Liquids at the Microscale Level

1. *Liquids are never poured at the microscale level.* Liquid substances are transferred by pipet or syringe. As we are working with small easy-to-hold glassware at the microscale level, the best technique for transfer is to hold both containers with the fingers of one hand, with the mouths as close together as possible. The free hand is then used to operate the pipet (syringe) to withdraw the liquid and make the transfer. This approach reduces to a minimum the time that the open tip is not in or over the reservoir or the reaction flask. We employ three different pipets and two standard syringes for most experiments in which liquids are involved. This equipment is a prime source of contamination. *Be very careful to thoroughly clean the equipment after each use.*

a. *Pasteur pipet.* Often called a capillary pipet, the Pasteur pipet is a simple glass tube with the end drawn to a fine capillary. These pipets can hold several milliliters of liquid (Fig. 3.26a) and are filled using a small rubber bulb or one of the very handy commercially available pipet pumps. You will do many transfers using the Pasteur pipet. It is suggested that you calibrate several of them for approximate delivery of 0.5, 1.0, 1.5, and 2.0 mL of liquid. This is easily done by drawing the measured amount of a liquid from a 10-mL graduated cylinder and marking the level of the liquid in the pipet. This can be done with transparent tape or by scratching with a file. Indicate the level with a marking pen before trying to tape or file the pipet.

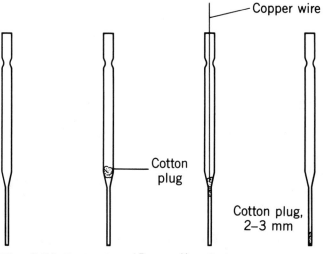

Fig. 3.26 *Preparation of Pasteur filter pipet.*

b. *Pasteur filter pipet.* A very handy adaptation of the Pasteur pipet is a filter pipet. This pipet is constructed by taking a small cotton ball and placing it in the large open end of the standard Pasteur pipet. Hold the pipet vertically and tap gently to position the cotton ball in the drawn section of the tube (Fig. 3.26b). Now form a plug in the capillary section by pushing the cotton ball down the pipet with a piece of copper wire (Fig. 3.26c). Finish by seating the plug flush with the end of the capillary (Fig. 3.26d). The optimum-size plug will allow easy movement along the capillary while it is being positioned by the copper wire. Compression of the cotton will build enough pressure against the walls of the capillary (once the plug is in position) to prevent plug slippage while the pipet is filled with liquid. If the ball is too big, it will wedge in the capillary before the end is reached, and wall pressure will be so great that liquid flow will be shut off. Even some plugs that are loose enough to be positioned at the end of the capillary still will have developed sufficient lateral pressure to make the filling rate unacceptably slow. With a little practice, however, these plugs can be quickly and easily inserted. Once in place the plug is rinsed with 1 mL of methanol and 1 mL of hexane and dried before use.

The purpose of placing the cotton plug in the pipet is twofold: First, a particular problem with the transfer of volatile liquids via the standard Pasteur pipet is the rapid buildup of back-pressure from solvent vapors in the rubber bulb. This pressure quickly tends to force the liquid back out of the pipet, and can cause valuable product to drip on the bench top. The cotton plug tends to resist this back-pressure and allows much easier control of the solution once it is in the pipet. The time delay factor becomes particularly important when the Pasteur filter pipet is employed as a microseparatory funnel (see the discussion on extraction techniques, Experiment 4).

Second, each time a transfer of material is made, the material is automatically filtered. This process effectively removes dust and lint, which are a constant problem when working at the microscale level with unfiltered room air.

c. *Automatic pipet (considered the Cadillac of pipets).* Automatic pipets quickly, safely, and reproducibly measure and dispense specific volumes of liquids. They are particularly valuable at the microscale level, as they generate the precise, accurate liquid measurements that are absolutely necessary when handling microliter volumes of reagent. The automatic pipet adds considerable insurance for the success of an experiment, as any liquid can be efficiently measured, transferred, and delivered to the reaction flask. They become almost an essential instrument in laboratory sections with large numbers of students.

The automatic pipet system consists of a calibrated piston pipet with a specially designed disposable plastic tip. You may encounter any one of three pipet styles: single volume, multirange, continuously adjustable (see Fig. 3.27). The first type is calibrated to deliver only a single volume. The second type is adjustable to two or three predetermined delivery volumes. The third type is the most versatile and can be user set to deliver any volume within the range of the pipet. Obviously, the price of these valuable laboratory tools goes up with increasing attributes. They are expensive and their use must be shared in the laboratory. Treat them with respect!

The automatic pipet is designed so that the liquid comes in contact with the special tip only. Never load the pipet without the tip in place. Never immerse the tip completely in the liquid that is being pipetted. Always keep the pipet vertical when the tip is attached. Follow these three rules and most automatic pipets will give many years of reliable service. The following suggestions are a few general rules for improving reproducibility with an automatic pipet.

Try to effect the same uptake and delivery motion for all samples. Smooth depression and release of the piston will give the most consistent results. Never allow the piston to snap back.

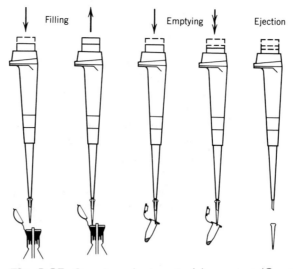

Fig. 3.27 *Operation of automatic delivery pipet.* (Courtesy of Brinkmann Instruments Co., Westbury, N.Y.)

Always depress the piston to the first stop before inserting the tip into the liquid. If the piston is depressed after submersion, formation of an air bubble in the tip becomes likely. Bubble formation will result in a filling error.

Never insert the tip more than 5 mm into the liquid. It is good practice not to allow the body of the pipet to contact any surface or bottleneck that might be wet with a corrosive chemical.

If an air bubble forms in the tip during uptake, return the fluid, discard the tip, and repeat the sampling process.

d. *Syringes.* Syringes are particularly helpful pieces of equipment when transferring liquid reagents or solutions to sealed reaction systems from sealed reagent or solvent reservoirs. They can be inserted through a septum, which avoids opening the apparatus to the atmosphere (see Experiment 13). They are also routinely employed in the determination of ultramicro boiling points (10-μL GC syringe). It is critically important to clean the syringe needle after each use. Effective cleaning of a syringe requires as many as a dozen flushes. The microscale laboratory utilizes a low-cost glass 1-mL insulin syringe in which the rubber plunger seal is replaced with a Teflon seal. For gas chromatographic separation work, the standard 50- or 100-μL syringes are preferred (see Experiment 1).

2. *Liquid volumes may be converted easily to weight measure by the following relationship:*

$$\text{Volume (mL)} = \frac{\text{weight (g)}}{\text{density (g/mL)}}$$

3. *Work with liquids in conical vials* and work in vials that are approximately double the volume of the material. The trick here is to reduce the surface area of the flask in contact with the sample to an absolute minimum. Conical systems are far superior to the spherical surface of the conventional round-bottom flask.

Rules for Working with Solids at the Microscale Level

1. *General considerations.* Working with a crystalline solid is much easier than working with the equivalent quantity of a liquid. Unless the solid is in solution, a spill on a clean glass working surface usually can be recovered quickly and efficiently. *Be careful, however, when working in solution. ALWAYS use the same precautions that you would use if you were handling a pure liquid.*

2. *The transfer of solids.* Solids are normally transferred with microspatulas, a technique which is not difficult to develop.

3. *Weighing solids at the milligram level.* The current generation of single-pan electronic balances has removed much of the drudgery from weighing solids. These systems can automatically tare an empty vial. Once the vial is tared, the reagent is added in small portions. The weight of each addition is instantly registered; material is added until the desired quantity has been transferred.

THE LABORATORY NOTEBOOK   Written communication is the most important method by which a chemist transmits his or her work to the scientific community. It begins with the record kept in a laboratory notebook. The reduction to practice of an experiment recorded in the laboratory notebook is the source of information used to prepare scientific papers published in journals or presented at meetings. For the industrial chemist it is especially critical in obtaining patent coverage.

It is important that potential scientists, whatever the field, learn to keep a detailed account of their work. A laboratory notebook has several key components. Note how each component is incorporated into the example that follows.

Key Components of a Laboratory Experimental Writeup

1. Date experiment was conducted.
2. Title of experiment.
3. Purpose for running the reaction.
4. Reaction scheme.
5. Table of reagents and products.
6. Details of procedure used.
7. Characteristics of the product(s).
8. References to product or procedure (if any).
9. Analytical and spectral data.
10. Signature of person performing the experiment and that of a witness, if required.

In reference to point 6, it is the obligation of the person doing the work to list the equipment, the amounts of reagents, the experimental conditions, and the method used to isolate the product. Any color or temperature changes should be carefully noted and recorded.

Several additional points can be made with respect to the proper maintenancy of a laboratory record.

1. A hardbound, permanent notebook is essential.
2. Each page of the notebook should be numbered in consecutive order. For convenience, an index at the beginning or end of the book is recommended and blank pages should be retained for this purpose.
3. If a page is not completely filled, an × should be used to show that no further entry was made.
4. Always record your data in ink. If a mistake is made, draw a neat line through the word or words so that they remain legible. Data are always recorded directly into the notebook, *never* on scrap paper!

5. Make the record clear and unambiguous. Pay attention to grammar and spelling.
6. In industrial research laboratories, your signature as well as that of a witness is required, because the notebook may be used as a legal document.
7. Always write and organize your work so that someone else could come into the laboratory and repeat your directions without confusion or uncertainty. *Completeness* and *legibility* are key factors.

Since the reactions you will be carrying out in this laboratory have been worked out and checked in detail, your instructor may not require you to keep your notebook in such a meticulous fashion. For example, when you describe the procedure (item 6), it may be acceptable to make a clear reference to the material in the laboratory manual and to note any modifications or deviations from the prescribed procedure. In some cases, it may be more practical to use an outline method. In any event, the following example should be studied carefully. It may be used as a reference when detailed records are important in your work.

Note. *Because of its length, the example is presented typed. Normally, the entry in the notebook is handwritten. However, now that computers are gaining wide acceptance in laboratory work, many chemists are using this means to record their data.*

EXAMPLE OF A LABORATORY NOTEBOOK ENTRY[1]

① 16 AUG. 1985

② PREPARATION OF DIPHENYL SUCCINATE

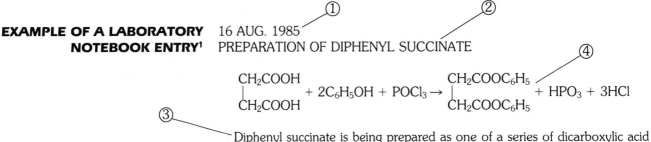

$$\begin{matrix} CH_2COOH \\ | \\ CH_2COOH \end{matrix} + 2C_6H_5OH + POCl_3 \rightarrow \begin{matrix} CH_2COOC_6H_5 \\ | \\ CH_2COOC_6H_5 \end{matrix} + HPO_3 + 3HCl$$

④

③ Diphenyl succinate is being prepared as one of a series of dicarboxylic acid esters that are to be investigated as growth stimulants for selected fungi species.

⑧ This procedure was adapted from that reported by Daub, G. H.; Johnson, W. S. *Organic Syntheses*; Wiley: New York, 1963; Collect. Vol. IV, p 390.

Physical Properties of Reactants and Products

⑤

Compound	MW	Wt/Vol	mmol	mp(°C)	bp(°C)
Succinic acid	118.09	118 mg	1	182	
Phenol	94.4	188 mg	2		182
Phosphorus oxychloride	153.33	84 μL	0.9		105.3
Diphenyl succinate	270.29			121	

In a 3.0-mL conical vial containing a magnetic spin vane and equipped with a reflux condenser protected by a calcium chloride drying tube were placed succinic acid (118 mg, 1 mmol), phenol (188 mg, 2 mmol), and phosphorous oxychloride (84 μL, 0.9 mmol). The reaction mixture was heated with stirring at

HOOD 115°C in a sand bath in the **hood** for 1.25 hr. It was necessary to conduct the reaction in the **hood** as HCl gas evolved during the course of the reaction. The
⑥ drying tube was removed, toluene (0.5 mL) added through the top of the

---

[1] Circled numbers refer to list on p. 31.

condenser using a Pasteur pipet, and the drying tube replaced. The mixture was then heated for an additional hour at 115°C.

The hot toluene solution was separated from the red syrupy residue of phosphoric acid using a Pasteur pipet. The toluene extract was filtered by gravity using a fast-grade filter paper, and the filtrate collected in a 10-mL Erlenmeyer flask. The phosphoric acid residue was then extracted with two additional 1.0-mL portions of hot toluene. These extracts were also separated using the Pasteur pipet and filtered, and the filtrate was collected in the same Erlenmeyer flask. The combined toluene solutions were concentrated to a volume of approximately 0.6 mL by warming them in a sand bath under a gentle stream of nitrogen gas in the **hood**. The pale yellow liquid residue was then allowed to cool to room temperature. Diphenyl succinate separated as colorless crystals. The solid was collected by vacuum filtration using a Hirsch funnel and the filter cake washed with three 0.5-mL portions of cold diethyl ether. The product was dried in a vacuum oven at 30°C (3 mm) for $\frac{1}{2}$ hour.

There was obtained 181 mg (67%) of diphenyl succinate having mp 120–121°C (Lit. value 121°C; *CRC Handbook of Chemistry and Physics*, 60th ed.; CRC Press: Boca Raton, FL, 1979; #S197, p C-501).

The IR spectrum exhibits the expected peaks for the compound. [*At this point the data may be listed or the spectrum pasted on a separate page of the notebook.*]

⑦
HOOD

⑦
⑧

⑨

Marilyn C. Waris
⑩

witnessed by
D. Jeanne d'Arc Mailhiot   16/Aug./1985

CALCULATION OF YIELDS   Almost without exception, in each of the experiments presented in this text, you are requested to calculate the percentage yield. For any reaction, it is always important for the chemist to know how much of a product is *actually* produced (experimental) in relation to the *theoretical* amount (maximum) that could have been formed. The percentage yield is calculated on the basis of the relationship

$$\% \text{ yield} = \frac{\text{actual yield (experimental)}}{\text{theoretical yield (calculated maximum)}} \times 100$$

The percentage yield is generally calculated on a weight (gram or milligram) or on a mole basis. In the present text, the calculations are made using milligrams. Several steps are involved in calculation of the percentage yield.

1. Write a *balanced* equation for the reaction. For example, consider Experiment 25A, the Williamson synthesis of propyl *p*-tolyl ether.

$$CH_3\text{—}\bigcirc\text{—}OH + CH_3CH_2CH_2\text{—}I \xrightarrow[\text{(C}_4\text{H}_9)_4\text{N}^+,\ \text{Br}^-]{\text{NaOH}} CH_3\text{—}\bigcirc\text{—}O\text{—}(CH_2)_2CH_3 + Na^+, I^-$$

p-Cresol   Propyl iodide   Propyl *p*-tolyl ether

2. Identify the *limiting* reactant. The ratio of reactants is calculated on a millimole (or mole) basis. In the example, 0.78 mmol of *p*-cresol is used compared with 0.77 mmol of propyl iodide. Note that the sodium hydroxide is used as a reagent but *does not* appear in the product, propyl *p*-tolyl ether. Therefore, it is not considered in the calculations. Nor is the quaternary salt considered; it is used as a phase transfer catalyst. The reaction is run essentially on a 1 : 1 molar ratio and, thus, calculation of the theoretical yield can be based on the *p*-cresol or the propyl iodide.

3. Calculate the *theoretical* (maximum) amount of the product that could be obtained for the conversion, based on the limiting reactant. In the present case, referring to the balanced equation, one mole of propyl iodide affords one mole of the ether product. Therefore, if we start with 0.77 mmol of the propyl iodide, the maximum amount of propyl *p*-tolyl ether that can be produced is 0.77 mmol, or 115.7 mg.

4. Determine the *actual* (experimental) yield (milligrams) of product isolated in the reaction. This amount is invariably less than the theoretical quantity, unless the material is impure (one common contaminant is water). For example, student yields for the preparation of propyl *p*-tolyl ether average 70 mg.

5. Calculate the *percentage yield* using the weights determined in steps 3 and 4. The percentage yield is then

$$\% \text{ yield} = \frac{70 \text{ mg (actual)}}{115.7 \text{ mg (theoretical)}} \times 100 = 60.5\%$$

As you carry out each reaction in the laboratory, strive to obtain as high a percentage yield of product as possible. The reaction conditions have been carefully developed and, therefore, it is essential that you master the microscale techniques concerned with transfer of reagents and the isolation of products as soon as possible.

Chapter 4

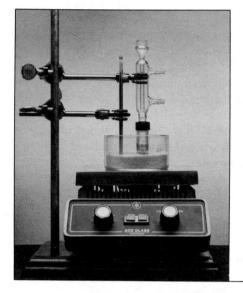

Determination of Physical Properties

The physical properties of a compound are those measurable characteristics of a material that are independent of the influence of any other substance. Determination of these properties is important for (1) substance identification and (2) an indication of material purity. Historically, the physical constants of prime interest for determination in liquids include boiling point, density, and refractive index, and in solids, melting point. In special cases, optical rotation and molecular weight determinations may be required. Today, with the widespread availability of spectroscopic instrumentation, particularly infrared spectrometers, the absorption characteristics of a substance can be quickly recorded and interpreted (see Chapter 5, Experiment 5A,B, for a discussion of IR spectra, now routinely generated in 45 seconds). The infrared spectrum obtained with one data point per wavenumber can add more than 4000 transmission measurements to the few classically determined properties. Indeed, even with development of high-resolution mass spectrometry, and proton and ^{13}C nuclear magnetic resonance (NMR) spectroscopy, the infrared spectrum of a material still remains the most powerful set of physical properties (transmission elements) available to the organic chemist for the *identification* of an unknown compound.

At the present time, simple physical constants are determined mainly to assist in establishing the purity of *known* compounds. The boiling point or the melting point of a material is very sensitive to small quantities of impurities. These data can be particularly helpful in determining whether or not a starting material needs further purification, or whether a product has been isolated in acceptable quality. When a new compound has been synthesized, an elemental (combustion) analysis is normally included also, if sufficient material is available. In the case of a new substance, one is interested in establishing not only the identity but also the molecular structure of the substance. In this situation, other modern techniques such as proton and ^{13}C NMR, high-resolution mass spectrometry, and single-crystal X-ray diffraction can provide powerful structural information.

When comparisons are made between experimental data and values obtained from the literature, it is essential that the latter information be obtained from the most reliable sources available. Certainly, judgment, which improves with experience, must be exercised in accepting any value as a standard. The known classical properties of the large majority of the compounds prepared in this text are taken from the *CRC Handbook of Chemistry and Physics*. This reference work is a valuable source that lists the physical constants of a very large number of inorganic, organic, and organometallic compounds. The handbook is kept current and a new edition is published each year.

ULTRAMICRO BOILING POINT

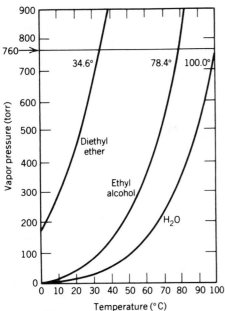

Fig. 4.1 *Vapor–pressure curves. From Brady, J. E.; Humiston, G. E. General Chemistry, 3rd ed.; Wiley: New York, 1982. (Reprinted by permission of John Wiley & Sons, New York.)*

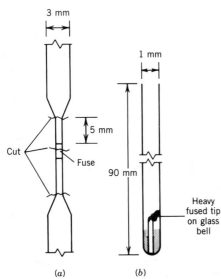

Fig. 4.2 *(a) Preparation of small glass bell for ultramicro boiling point determination. From Mayo, D. W.; Pike, R. M.; Butcher, S. S.; Meredith, M. L. J. Chem. Educ.* **1985,** *62, 1114. (b) Ultramicro boiling point assembly. From Mayo, D. W.; Pike, R. M.; Butcher, S. S.; Meredith, M. L. J. Chem. Educ.* **1985,** *62, 1114.*

The vapor pressure of a liquid increases in a nonlinear fashion on heating. When the pressure reaches the point where it matches the local atmospheric or applied pressure, the liquid boils. That is, internally it spontaneously begins to form large vapor bubbles which rapidly rise to the surface. If heating is continued, both the vapor pressure and the temperature of the liquid will remain constant until the substance has been completely vaporized (Fig. 4.1).

Since microscale preparations generally yield quantities of liquid products in the range 30–70 μL, the allocation of 5 μL or less to boiling point measurements becomes highly desirable. The modification of the Wiegand ultramicro boiling point procedure described here has established that reproducible and reasonably accurate (±1°C) boiling points can be observed on 3–4 μL of many liquids.[1]

Procedure

Ultramicro boiling points can be conveniently determined in standard (90-mm length) Pyrex glass capillary melting point tubes. The melting point tube replaces the conventional 3- to 4-mm (o.d.) tubing used in the Siwoloboff[2] procedure. The sample (3–4 μL) is loaded into the melting point capillary via a 10-μL syringe and centrifuged to the bottom.

A small glass bell (which replaces the conventional melting point tube as the bubble generator in micro boiling point determinations) is formed by heating 3-mm (o.d.) Pyrex tubing with a microburner and drawing it out to a diameter small enough to be readily accepted by the melting point capillary. A section of the drawn capillary is fused and then cut to yield two small glass bells approximately 5 mm long (Fig. 4.2a). It is important that the fused section be reasonably large. This section is more than just a seal. The fused glass must add sufficient weight to the bell so that it will firmly seat itself in the bottom of the melting point tube.

One of the glass bells is inserted into the loaded melting point capillary, open end first (down), and allowed to fall to the bottom. The assembled system (Fig. 4.2b) is then inserted into the stage of a Thomas–Hoover Uni-Melt Capillary Melting Point Apparatus[3] or similar system (Fig. 4.3).

The temperature is rapidly raised to 15–20°C below the expected boiling point (the temperature should be monitored carefully in the case of unknown substances) and then adjusted to a 2°/minute rise rate until a fine stream of bubbles is emitted from the glass bell. The heat control is then adjusted to drop the temperature. The boiling point is taken at the point where the last escaping bubble collapses (i.e., when the vapor pressure of the substance equals the atmospheric pressure). The heater is then rapidly adjusted to again raise the temperature at 2°/minute and induce a second stream of bubbles. This procedure may then be repeated several times. *It should be emphasized that the precise and sensitive temperature control provided by the Thomas–Hoover system is essential to the successful application of this cycling technique. Although this control is a desirable feature it is not, however, essential for obtaining satisfactory boiling point data.*

[1] Wiegand, C. *Angew. Chem.* **1955,** *67,* 77. Mayo, D. W.; Pike, R. M.; Butcher, S. S.; Meredith, M. L. *J. Chem. Educ.* **1985,** *62,* 1114.
[2] Siwoloboff, A. *Ber.* **1886,** *19,* 795.
[3] Thomas Scientific, 99 High Hill Road, P.O. Box 99, Swedesboro, NJ 08085.

Fig. 4.3 *Thomas–Hoover melting point determination device. (Courtesy of Arthur Thomas Scientific, Swedesboro, NJ.)*

Utilization of the conventional melting point capillary as the "boiler" tube has the particular advantage that the system is ideally suited for observation in a conventional melting point apparatus. The illumination and magnification available make the observation of rate changes in the bubble stream readily apparent. Economical gas chromatographic syringes (10 μL) appear to be the most successful instrument for dealing with the small quantities of liquids involved in these transfers. The 3-in. needles normally supplied with the 10-μL barrels will not reach the bottom of the capillary; however, liquid samples deposited on the walls of the tube are easily and efficiently moved to the bottom by centrifugation. After the sample is packed in the bottom of the capillary tube, the glass bell is introduced. Use of the glass bell is necessary because, if a conventional Siwoloboff fused capillary insert is employed (it would extend beyond the top of the melting point tube), capillary action between the "boiler" tube wall and the capillary insert would draw the majority of the sample from the bottom of the tube up onto the walls. This effect often precludes the formation of the requisite bubble stream.

Little loss of low-boiling liquids occurs (see Table 4.1). Furthermore, if the boiling point is overrun, and the sample is flashed from the bottom section of the "boiler" capillary, it rapidly will condense on the upper cooler sections of the tube which extend above the heat transfer liquid. The sample can easily be recentrifuged to the bottom of the tube and a new determination of the boiling point commenced. Indeed, if the bell cavity fills completely during the cooling point of a cycle, it is often difficult to reinitiate the bubble stream without first emptying the entire cavity by overrunning the boiling point.

Observed boiling points for a series of compounds which boil over a wide range of temperatures are summarized in Table 4.1.

Materials that are thermally stable at their boiling point will give identical values on repeat determinations. Substances that begin to decompose will give values that slowly drift after the first few measurements. The observation of color and/or viscosity changes plus a variable boiling point all signal the need for caution in making extended repeat measurements.

Table 4.1 Observed Boiling Points (°C)[a]

Compound	Observed	Literature Value	Reference
Methyl iodide	42.5	42.4	b
Isopropyl alcohol	82.3	82.4	c
2,2-Dimethoxypropane	80.0	83.0	d
2-Heptanone	149–150	151.4	e
Cumene	151–153	152.4	f
Mesitylene	163	164.7	g
p-Cymene	175–178	177.1	h
Benzyl alcohol	203	205.3	i
Diphenylmethane	263–265	264.3	j

[a] Observed values are uncorrected for changes in atmospheric pressure (corrections all estimated to be less than ±0.5°C).

[b] *CRC Handbook of Chemistry and Physics,* 62nd ed.; CRC Press: Boca Raton, FL, 1981; #9082, p C-373.

[c] Ibid. #11971, p C-470.

[d] *Dictionary of Organic Compounds,* 4th ed.; Oxford University Press: London, 1965; Vol. I, p 11.

[e] *CRC Handbook of Chemistry and Physics,* 62nd ed.; CRC Press: Boca Raton, FL, 1981, #7627, p C-321.

[f] Ibid. #5394, p C-244.

[g] Ibid. #8987, p C-370.

[h] Ibid. #2192, p C-138.

[i] Ibid. #3160, p C-169.

[j] Ibid. #6282, p C-274.

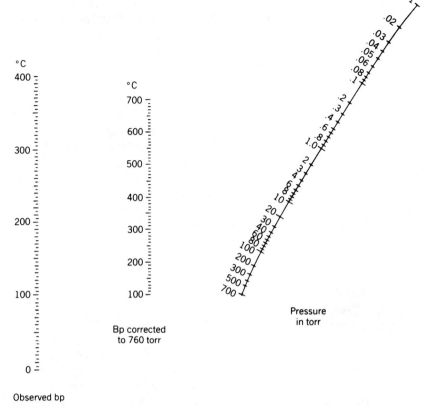

°C
400

300

200

100

0

Observed bp

°C
700

600

500

400

300

200

100

Bp corrected
to 760 torr

.02
.03
.04
.05
.06
.08
.1

.2
.3
.4
.6
.8
1.0

2
3
4
6
8
10

20
30
40
60
80
100

200
300
500
700

Pressure
in torr

Fig. 4.4 *Pressure–temperature nomograph,*

Comparison of the boiling points obtained experimentally at various atmospheric pressures with reference boiling points at 760 torr is greatly facilitated by the use of pressure–temperature nomographs such as shown in Fig. 4.4. A straight line from the observed boiling point to the observed pressure will pass through the corrected boiling point value.

DENSITY   Density, defined as mass per unit volume, is generally expressed as grams per milliliter (g/mL) or grams per cubic centimeter (g/cm³) for liquids. Accurate procedures have been developed for the measurement of this physical constant at the micro level. A micropycnometer (density meter), developed by Clemo and McQuillen,[4] requires approximately 2 μL (Fig. 4.5). This very accurate device gives the density to three significant figures. The system is self-filling, and the fine capillary ends do not need to be capped while coming to temperature equilibrium or while weighing (the measured values tend to degrade for substances boiling under 100°C and when room temperatures rise much above 20°C). In addition, the apparatus must first be tared, filled, and then reweighed on an *analytical* balance. A technique that results in less precise densities (good

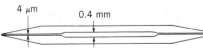

4 μm   0.4 mm

Fig. 4.5 *Pycnometer of Clemo and McQuillen. From Schneider, F. L. Monographien aus dem Gebiete der qualitativen Mikroanalyse, Qualitative Organic Microanalysis, Vol. II; Benedetti-Pichler, A. A., Ed.; Springer-Verlag: Vienna, Austria, 1964.*

---

[4] Clemo, G. R.; McQuillen, A. *J. Chem. Soc.* **1935,** 1220.

to about two significant figures) but is far easier to use is simply to substitute a 50- or 100-μL syringe for the pycnometer. The method simply requires weighing the syringe before and after filling it to a measured volume as in the conventional technique. With the volume of the pycnometer and the weight of the liquid known, the density can be calculated. (A further advantage of the syringe technique is that in this case the pycnometer is not limited to a fixed volume. Although much larger samples are required, it is not inconvenient to utilize the entire sample obtained in the reaction for this measurement, as the material can be efficiently recovered from the syringe for additional characterization studies.) Since density changes with temperature, these measurements are made at a constant temperature. If, in the table of data given for each experiment, no temperature is indicated, the value was measured at 20°C.

REFRACTIVE INDEX

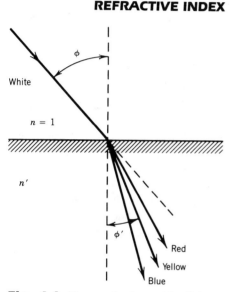

Fig. 4.6 *Upon refraction white light is spread out into a spectrum. This is called dispersion.*

It is commonly observed that a beam of light "bends" as it passes from one medium to another. For example, an oar looks bent as one views the portion under the water. This effect is a consequence of the refraction of light. It results from the change in velocity of the radiation at the interface of the media, and the angle of refraction is related to the velocity change as follows (see Fig. 4.6).

$$\frac{\sin \phi}{\sin \phi'} = \frac{\text{velocity in vacuum}}{\text{velocity in sample}} = n \text{ (refractive index)}$$

Since the velocity of light in a medium must be less than that in a vacuum, the index of refraction n will always be greater than one. In practice, n is taken as the ratio of the velocity of light in air relative to the medium being measured. The refractive index also is wavelength dependent.

The wavelength dependence gives rise to the effect of dispersion or the spreading of white light into its component colors. When we measure n, therefore, we must specify the wavelength at which the measurement is made. The standard wavelength for refractive index determinations has become the bright yellow sodium 589-nm emission, the sodium D line. Sodium, unfortunately, is a poor choice of wavelength for these measurements with organic substances, but as the sodium lamp represents one of the easiest monochromatic sources of radiation to obtain experimentally, it is widely used. Because the density of the medium is sensitive to temperature, the velocity of radiation also changes with temperature, and therefore, refractive index measurements must be made at constant temperatures. Many values in the literature are reported at 20°C. In the table of data given for each experiment, no record of temperature indicates a value measured at 20°C, as in the case of density values. The index can be measured optically quite accurately to four decimal places. Since this measurement is particularly sensitive to the presence of impurities, the refractive index can be a valuable physical constant for tracking the purification of liquid samples.

The measurement (for example) is reported as

$$n_D^{20} = 1.4628$$

Procedure

In the Abbe-3L refractometer (Fig. 4.7), white light is used as the source, but compensating prisms give indices for the D line. This refractometer is commonly used in many undergraduate organic laboratories.

Fig. 4.7 *Abbe-3L refractometer.* (Courtesy of Milton Roy Co., Rochester, NY.)

Samples (approximately 10 μL) are applied between the horizontal surfaces of a pair of hinged prisms (Fig. 4.8). A sampling procedure recently developed by B. P. Ronald significantly reduces the amount of sample required and allows accurate measurements of highly volatile materials. The technique involves placing a small precut 6-mm disk of good-quality lens paper at the center of the bottom prism. The sample is loaded onto the disk with a micro-Pasteur pipet or microliter syringe.[5]

> **CAUTION:** *Do not touch the prisms with the Pasteur pipet or syringe tip as they may be easily marred and scratched and will then give erroneous results.*

Refractive Index Measurements Utilizing the Lens Paper Disk Technique

Substance	T (°C)	n^t (normal) 100 μL	n^t (microdisk) 2–4 μL
Water	24.5	1.3224	1.3226
Diethyl ether	24	1.3508	1.3505
Chlorobenzene	24.5	1.5225	1.5219
Iodobenzene	24.5	1.6151	1.6151

The refractometer is adjusted so that the field of view has a well-defined light and dark split image (see your instructor for the correct routine for making instrument adjustments on your particular refractometer).

When using the refractometer, always clean the prisms with alcohol and lens paper before and after use. Record the temperature at which the reading is taken. A reasonably good extrapolation of temperature effects can be obtained by assuming that the index of refraction changes 0.0004 unit per degree Centigrade and varies inversely with the temperature.

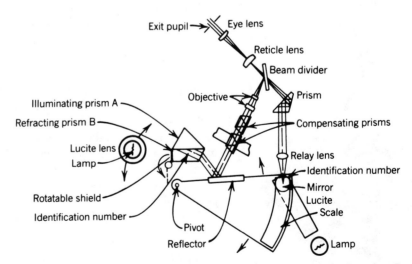

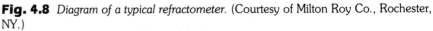

Fig. 4.8 *Diagram of a typical refractometer.* (Courtesy of Milton Roy Co., Rochester, NY.)

---

[5] B. P. Ronald, Department of Chemistry, Idaho State University, Pocatello, Idaho, Personal Communication.

MELTING POINTS   In general, the crystalline lattice forces holding organic solids together are distributed over a relatively narrow energy range. The melting points of organic compounds are usually relatively sharp, that is, less than 2°C. The range and maximum temperature of the melt, however, are very sensitive to impurities. Small amounts of sample contamination by soluble impurities nearly always will result in melting point depressions. A useful application of this effect involving inorganic compounds is the addition of calcium chloride to the sand used to clear ice-covered roads.

The drop in melting point is also accompanied by an expansion of the melt range. Thus, although the melting point is a useful guide in identification, it also is a particularly effective indication of sample purity.

Procedure

In the microscale laboratory program, two different types of melting point determinations are carried out: (1) simple capillary melting points and (2) evacuated melting points.

Simple Capillary Melting Point

Because the microscale laboratory utilizes the Thomas–Hoover Uni-Melt Apparatus or a similar system for determining boiling points, melting points are conveniently obtained on the same apparatus. The Uni-Melt system utilizes an electrically heated and stirred silicone oil bath. The temperature readings require no correction in this case as the depth of immersion is held constant. (This assumes, of course, that the thermometer is calibrated to the operational immersion depth.) Melting points are determined in the same capillaries as boiling points. The capillary is loaded by introducing ~1 mg of material into the open end. The sample is then tightly packed (~2 mm) into the closed end by dropping the capillary down a length of glass tubing held vertically to the bench top. The melting point tube is then ready for mounting on the metal stage, which is immersed in the silicone oil bath of the apparatus. If the melting point of the substance is expected to occur in a certain range, the temperature can be rapidly raised to ~20° below the expected value. At that point the temperature rise should be adjusted to 2°/minute, which is the standard rate of change at which the reference determinations are obtained. The melting point range is recorded from the temperature at which the first drop of liquid forms to that at which the last crystal melts.

Evacuated Melting Points

Many organic compounds begin to decompose at their melting points. This decomposition often begins as the melting point is approached and may adversely affect the values measured. The decomposition can be invariably traced to reaction with oxygen at elevated temperatures. If the melting point is obtained in an evacuated tube, much more accurate melting points can be obtained. These more reliable values arise not only from increased sample stability, but because several repeat determinations can often be made on the same sample. The multiple measurements then may be averaged to provide more accurate data.

Evacuated melting points are quickly and easily obtained with a little practice. The procedure is as follows: Shorten the capillary portion of a Pasteur pipet to approximately the same length as a normal melting point tube (see Fig. 4.9a). Seal the capillary end by rotating in a microburner flame. Touch the pipet only to the very edge of the flame, and keep the large end at an angle below the end being sealed (see Fig. 4.9b). This will prevent water from the flame being carried into the tube where it will condense in the cooler sections. Then load 1–2 mg of

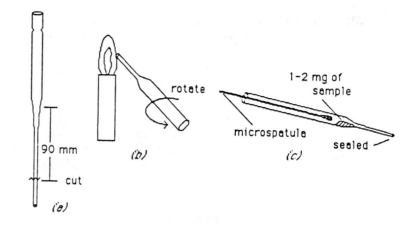

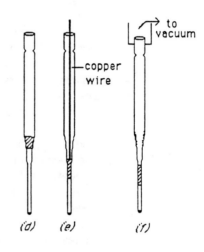

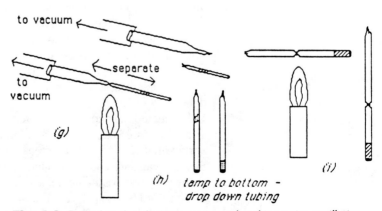

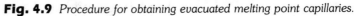

Fig. 4.9 *Procedure for obtaining evacuated melting point capillaries.*

sample into the drawn section of the pipet with a microspatula (see Fig. 4.9c). Tap the pipet gently to seat the solid powder as far down the capillary as it can be worked (see Fig. 4.9d). Then push the majority of the sample part way down the capillary with the same-diameter copper wire that you used to seat the cotton plug in constructing the Pasteur filter pipet (see Fig. 4.9e). Next, connect the pipet to a mechanical high-vacuum pump with a piece of vacuum tubing. Turn on the pump and evacuate the pipet for 30 seconds (see Fig. 4.9f). Then with a microburner gently warm the surface of the capillary tubing just below the drawn section. On warming, the remaining fragments of the sample (the majority of which has been forced further down in the tube) will sublime in either direction away from the hot section. Once the traces of sample have been "chased" away, the heating is increased, and the capillary tube collapsed, fused, and separated from the shank, which remains connected to the vacuum system (see Fig. 4.9g). The pump is then turned off and the shank discarded. The sample is tightly packed into the initially sealed end of the evacuated capillary by dropping down a section of glass tubing, as in the case of packing open melting point samples. After the sample is packed (~2 mm in length), a section of the evacuated capillary is once more quickly heated and collapsed by the microburner flame about 10–15 mm above the sample (see Fig. 4.9h).

This procedure is required to trap the sample and avoid sublimation up the tube during measurement of the melting point. The operation is a little tricky and should be practiced a few times. It is very important that complete fusion take place. Now the sample is ready to be placed in the melting point apparatus. The procedure beyond this point is the same as in the open capillary case, except that after it melts, the sample can be cooled, crystallized, and remelted several times and the average value of the range reported. If these latter values begin to drift downward, the sample can be considered to be decomposing even under evacuated, deoxygenated conditions. In this case the first value observed should be recorded as the melting point and decomposition noted (mp xx dec.).

MIXTURE MELTING POINTS

Advantage can be taken of the sensitivity of the melting point to impurities. In the case where two different substances possess identical melting points, it would be impossible to identify an unknown sample as either material based on the melting point alone. If reference standards of the two compounds are available,

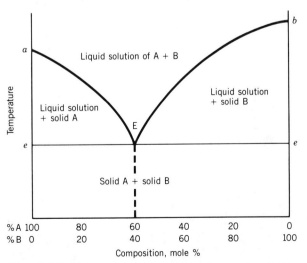

Fig. 4.10 *Melting point composition diagram for the binary mixture,* **A** + **B.** *In this diagram,* **a** *is the melting point of the solid* **A,** **b** *of solid* **B,** *and* **e** *of eutectic mixture* **E.**

however, then mixtures of the unknown and the two standards can be prepared. It is important to prepare several mixtures of varying concentrations for melting point comparisons, as the point of maximum depression need not occur on the phase diagram at the 50:50 ratio (see Fig. 4.10). The melting points of the unknown and the mixed samples are conveniently obtained simultaneously (the Uni-Melt stage will accept up to seven capillaries at one time). The unknown sample and the mixture of the unknown with the correct reference will have identical values, but the mixture with the reference of a different substance will give a depressed melting point. This procedure is the classical route to positive identification of a crystalline solid.

Chapter 5

Development of Experimental Technique

As pointed out in Chapter 1, Chapter 5 begins the laboratory work, which necessarily is centered on the development of experimental technique. This chapter is divided into 11 experimental sections. These exercises, and the detailed discussions of them, develop the major areas of experimental technique to be used during this first year in the microscale organic laboratory. The study of a number of specific organic reactions is simultaneously incorporated into this section to illustrate the use of the techniques being considered. *These particular examples which the authors have found attractive in their own laboratory programs may be easily replaced to correspond more closely to local development of the subject in lecture, so do not be surprised if you find a substitute experiment from Chapter 6 mated to a particular technique discussion given in Chapter 5.*

TECHNIQUE 1: MICROSCALE SEPARATION OF LIQUID MIXTURES BY PREPARATIVE GAS CHROMATOGRAPHY

Theory

One of the principal hurdles in dealing with experimental chemistry is the isolation of materials in their pure state. To characterize a substance fully, we require a pure sample of the material. In organic chemistry this is a particularly difficult demand since most organic reactions generate several products. We are generally satisfied if the desired product is the major component of the mixture obtained. As you proceed down the trail of developing experimental technique, you will note the heavy emphasis placed on separation techniques. Experiments 2 and 3, for example, deal directly with distillation routines that focus on the separation of mixtures involving one to several milliliters of material. In Experiment 1 we will start to develop separation techniques with an introduction to microscale preparative gas chromatography.

The methods of chromatography have revolutionized experimental organic chemistry over the past 30-odd years. It is by far the most powerful technique for separating mixtures and isolating pure substances. Chromatography is defined as the resolution of a multicomponent mixture (several hundred in some cases) by distribution between two phases, one stationary and one moving. The various methods of chromatography are categorized by the phases involved: column, thin-layer, and paper (solid–liquid); partition (liquid–liquid); and vapor

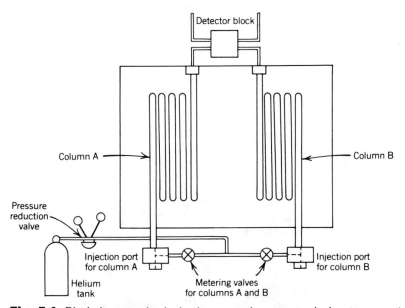

Fig. 5.1 *Block diagram of a dual-column gas chromatograph showing essential parts.* (Courtesy of Gow-Mac Instrument Co., Bound Brook, NJ.)

phase (gas–liquid). The principal mechanism on which these separations depend is differential solubility or adsorptivity of the mixture components with respect to the two phases involved. That is, the components exhibit different partition coefficients.

Gas chromatography is an extraordinarily powerful technique for the separation of mixtures. In this case the stationary phase is a liquid and the moving phase is a gas (the carrier gas). Fairly sophisticated instrumentation is required to carry out this type of chromatography. A diagram of a common laboratory gas chromatograph is given in Figure 5.1.

The system starts with a heated injection port. The sample mixture is introduced by syringe through a high-temperature septum in the port, into the heated chamber through which the carrier gas (the gas phase) is flowing. The solubility of the sample in the partitioning gas depends to a large extent on the vapor pressure of the substances in the mixture. Helium and nitrogen are commonly used as carrier gases. Heating the injection system ensures vaporization of the sample. Thus, two major constraints of gas chromatography are (1) that the sample must be stable at the temperature required to cause vaporization and (2) that the sample possess sufficient vapor pressure to be completely soluble in the gas phase at the operating temperature.

The vaporized mixture is swept from the injection block onto the column. The next step constitutes the key component of the separation process. The stationary liquid phase, in which the sample will dissolve and partition with the moving gas phase, is physically and/or chemically bound to inert packing material contained in the column. Gas chromatographic columns come in a variety of sizes and shapes. In the diagram of the Gow-Mac instrument (Fig. 5.1), two parallel coiled columns are mounted in a well-insulated oven. Considerable oven space may be saved and better temperature regulation achieved if the columns are coiled. This is important because the column must be kept under very precise temperature control. Most liquid mixtures will require elevated oven temperatures to maintain reasonable vapor pressures in the gas phase during the course of the separation. Separation of the mixture occurs as the carrier gas sweeps the sample through the column. Most columns are constructed of stainless steel, glass, or fused silica. The diameter and length of the column are critical factors in determining how the internal part of the column is designated to achieve separation of the sample mixture. The surface area of the liquid phase, in contact with the sample contained in the moving gas phase, is maximized by coating the finely divided inert support with a nonvolatile liquid material (stationary phase). The coated support is carefully loaded into the column to avoid channeling. Columns prepared in this fashion are termed "packed columns." Packed columns are usually $\frac{1}{4}$ or $\frac{1}{8}$ in. in diameter and are 4 to 12 ft long. These columns are particularly attractive for use in the microscale laboratory. Numerous examples are known of simple mixtures in the range 20–80 μL that can be resolved into their pure components and collected at the exit port of the column.

Research columns are also available that have no packing; the liquid phase is simply applied directly to the walls of the column. These are referred to as wall-coat or open-tubular columns. The reduction in surface area is compensated for by making the diameter very small (0.1 mm) and the length very long (several hundred meters would not be uncommon). Termed *capillary columns*, they are the most efficient columns employed for analytical separations. Mixtures of several hundred compounds can be completely resolved in a single pass through one of these systems. Capillary columns, however, are complicated to operate and require very small samples (0.1 μL or less). Thus, they cannot be used in a preparative routine.

Once introduced on the column, the sample will undergo partition with the liquid phase. The choice of the liquid phase is particularly important since it directly affects the relative distribution coefficients. The temperature of the column will also affect the separation in that the retention time of the lower-boiling component will decrease as the temperature is increased. That is, the residence time of the component in the column will decrease. If the oven temperature is too high, equilibration with the stationary phase will not be established and the component mixture may elute together or at best undergo incomplete resolution. The flow rate of the carrier gas constitutes another important parameter. The rate must be slow enough to allow equilibration but sufficiently rapid to ensure that diffusion will not overcome resolution of the components. As noted, the length of the column also is an important factor in separation performance. We will see in the distillation experiment that distillation column efficiency is proportional to column height and determines the number of evaporation–condensation cycles. In a similar manner, increasing the length of the gas chromatographic column allows more partition cycles to occur. Difficult-to-separate mixtures, such as the xylenes (very similar boiling points), will have a better chance of being resolved on longer columns.

A successfully resolved mixture will elute as individual components, sequentially with time, at the exit port of the instrument. To monitor the exit vapors, a detector is placed in the gas stream (Fig. 5.1). After passing through the detector, the carrier gas and the separated sample components are vented through a heated exit port. Sequential collection of the separated materials can be made by attaching suitable sample-condensing tubes to the exit port (see Fig. 3.25).

A widely used detector is the nondestructive thermal conductivity sensor often referred to as a hot-wire detector. A heated element in the gas stream changes electrical resistance when a substance dilutes the carrier gas and changes its thermal conductivity. Helium possesses a higher thermal conductivity than most organic substances. When samples other than helium are present, the conductivity of the gas stream decreases and the resistance of the heated wire changes. The change in resistance is measured by differences (Wheatstone bridge), with a reference detector mounted on the second (parallel) column. This signal is plotted by a recorder. The horizontal axis is time and the vertical axis is the magnitude of the resistance difference. This plot is the chromatogram. Retention time t_R is defined as the time from sample injection to the time of

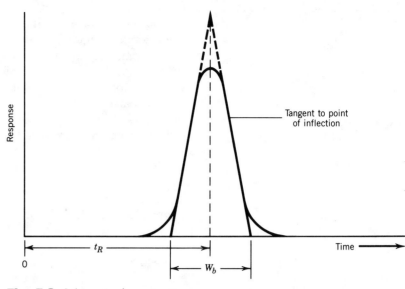

Fig. 5.2 *Schematic chromatogram.*

maximum recorder pen deflection. The baseline width W_b of a peak is defined as the distance between two points where tangents to the points of inflection cross the baseline (Fig. 5.2).

It is possible to estimate the number of theoretical plates (directly related to the number of distribution cycles) present in a column for a particular substance. The parameters are given in the relationship[1]

$$n = 16[t_R/W_b]^2$$

The units of t_R and W_b are the same (minutes, seconds, centimeters). As in distillation columns, the larger the number of theoretical plates, n, the higher the resolution of the column.

The efficiency of a system may also be expressed as the *height equivalent to a theoretical plate*, HETP. This parameter is related to the number of theoretical plates n by

$$\text{HETP} = \frac{L}{n}$$

where L is the length of the column, usually reported in centimeters. The smaller the HETP, the more efficient the column.

The number of theoretical plates available in fractional distillation columns is limited by column holdup (see Experiments 2 and 3). Gas chromatographic columns, on the other hand, operate most *efficiently* at the *microscale level*. The power of this technique is demonstrated in the following experiment.

Experiment 1

The Separation of a 25-μL Mixture of Heptanal (bp = 153°C) and Cyclohexanol (bp = 160°C) by Gas Chromatography

In this experiment you will attempt to separate 25 μL of a mixture consisting of heptanal and cyclohexanol into the pure components. The volume of the mixture is approximately that of a single drop. The materials boil within 7°C of each other. This mixture would be difficult to separate even by the best microdistilla-

---

[1] Berg, E. W. *Physical and Chemical Methods of Separation*; McGraw–Hill: New York, 1963; p 111.

tion techniques available. The purity of the fractions collected will be assessed by boiling points and refractive indices.

Before starting the experiment, let us observe the sensitivity of gas chromatographic separations to experimental conditions. For example, two sets of experimental data on the heptanal–cyclohexanol mixture are included to demonstrate the results of variations in oven temperature on retention time.

In Data Set A, the oven temperature was allowed to rise slowly from 160°C to ~170°C during a series of sample collections. The retention time of the first component, heptanal, dropped from ~3 minutes to close to 2 minutes, whereas the retention time of the second component, cyclohexanol, was reduced from about 5.5 minutes to nearly 4 minutes. The significant decrease in resolution over this series of collections is reflected in the number of theoretical plates calculated, which was over 300 for heptanal and about 500 for cyclohexanol in the first trial, but declined to below 200 for both compounds toward the last run (see Fig. 5.3 and Data Set A).

DATA SET A

	Heptanal				Cyclohexanol			
Trial No.	**Retention Time (min)**	**Baseline Width (min)**	**Number of Theoretical Plates**	**Recovery (mg)**	**Retention Time (min)**	**Baseline Width (min)**	**Number of Theoretical Plates**	**Recovery (mg)**
1	3.1	0.7	314	8.0	5.6	1.0	502	8.0
2	2.9	0.7	275	8.0	5.3	1.0	449	8.0
3	3.0	0.7	294	7.0	5.7	1.0	520	8.0
4	2.8	0.7	256	8.0	5.1	1.1	344	8.0
5	2.5	0.6	278	8.0	4.3	1.1	244	9.0
6	2.7	0.5	467	7.0	4.6	1.0	339	10.0
7	2.5	0.6	278	10.0	4.2	1.0	282	8.0
8	2.2	0.5	310	9.0	3.5	1.0	196	8.0
9	1.8	0.5	207	8.0	3.0	1.0	144	8.0
10	2.3	0.7	173	8.0	3.9	1.0	243	8.0
Ave.	2.6 ± 0.4	0.6 ± 0.09	285 ± 78	8.1 ± 0.8	4.5 ± 0.9	1.0 ± 0.05	326 ± 129	8.3 ± 0.7

Fig. 5.3 *Data Set A: separation of heptanal–cyclohexanol.*

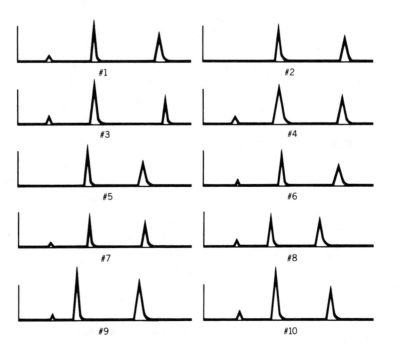

#1   #2

#3   #4

#5   #6

#7   #8

#9   #10

COLLECTION YIELD   **Cyclohexanol**

Density of cyclohexanol = 0.963 mg/μL.

In 25 μL of 1:1 cyclohexanol–heptanal, there are 12.5 μL of cyclohexanol.

Therefore, 12.5 μL × 0.963 mg/μL = 12 mg of cyclohexanol injected.

Percentage recovered = (8.3 mg/12.0 mg) × 100 = 69% cyclohexanol collected.

Heptanal

Density of heptanal = 0.850 mg/μL.

Therefore, 12.5 μL × 0.85 mg/μL = 10.6 mg of heptanal injected.

Percentage recovered = (8.1 mg/10.6 mg) × 100 = 76% heptanal.

In the B series of collections, stable oven temperatures and flow rates were maintained, and the data exhibit excellent reproducibility. Oven temperature was held at 155°C isothermal throughout the sampling process. The retention time of heptanal was observed to be slightly longer than 3 minutes with a variance of 6 seconds, whereas the cyclohexanol retention time was found to be slightly longer than 6 minutes with a variance of 12 seconds. The resolution remained essentially constant throughout the series, and the number of theoretical plates calculated was ~350 for heptanal and ~500 for cyclohexanol (see Fig. 5.4 and Data Set B).

DATA SET B

Trial No.	Heptanal				Cyclohexanol			
	Retention Time (min)	Baseline Width (min)	Number of Theoretical Plates	Recovery (mg)	Retention Time (min)	Baseline Width (min)	Number of Theoretical Plates	Recovery (mg)
1	3.5	0.7	400	8.0	6.6	1.1	576	8.0
2	3.2	0.7	334	9.0	6.0	1.1	476	7.0
3	3.5	0.7	400	7.0	6.6	1.2	484	10.0
4	3.2	0.7	334	9.0	6.1	1.0	595	9.0
5	3.1	0.6	427	8.0	6.0	1.1	476	8.0
6	3.2	0.7	334	9.0	6.0	1.1	476	9.0
7	3.3	0.8	272	9.0	6.1	1.1	492	8.0
8	3.1	0.7	313	8.0	6.0	1.1	476	10.0
9	3.2	0.7	334	8.0	6.1	1.1	492	8.0
10	3.2	0.7	334	8.0	6.2	1.1	508	8.0
Ave.	3.2 ± 0.1	0.7 ± 0.05	348 ± 47	8.3 ± 0.7	6.2 ± 0.2	1.1 ± 0.05	505 ± 44	8.5 ± 1.0

COLLECTION YIELD   **Cyclohexanol**

Density of cyclohexanol = 0.963 mg/μL.

In 25 μL of 1:1 cyclohexanol–heptanal, there are 12.5 μL of cyclohexanol.

Therefore, 12.5 μL × 0.963 mg/μL = 12 mg of cyclohexanol injected.

Percentage recovered = (8.5 mg/12.0 mg) × 100 = 71% cyclohexanol collected.

Heptanal

Density of heptanal = 0.850 mg/μL.

Therefore, 12.5 μL × 0.85 mg/μL = 10.6 mg of heptanal injected.

Percentage recovered = (8.3 mg/10.6 mg) × 100 = 78% heptanal.

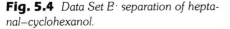

Fig. 5.4 *Data Set B· separation of hepta-nal–cyclohexanol.*

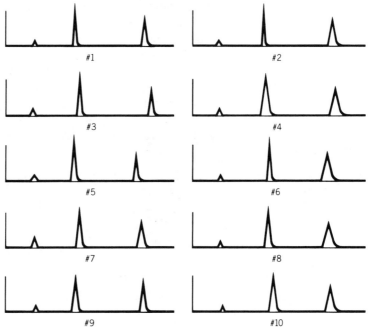

The results just described demonstrate that the resolution of gas chromatographic peaks may be very sensitive to changes in retention time resulting from instability in oven temperatures. As the calculated number of theoretical plates is highly dependent on resolution values, significant degradation in column plate values can occur with variations in oven temperatures. When you compare the time and effort required to obtain a two-plate fractional distillation on a 2-mL mixture in Experiment 3A with the speed and ease used to obtain a 500-plate separation on 12.5 μL of cyclohexanol in this experiment, it is hard not to be impressed with the enormous power of this technique.

EXPERIMENTAL

Estimated time to complete the experiment: 2.5 hours.

Physical Properties of Reactants and Products

Compound	MW	Wt/Vol	bp(°C)	Density	n_D
Heptanal	114.19	12.5 μL	153	0.85	1.4113
Cyclohexanol	100.16	12.5 μL	160	0.96	1.4641

PROCEDURE

This procedure involves injection of 25 μL heptanal–cyclohexanol 1:1 (v/v) onto a $\frac{1}{4}$ in. × 8 ft stainless-steel column packed with 10% Carbowax 80/100 20M PAW-DMS and installed in a gas chromatograph (GC). Experimental conditions are: He flow rate, 50 mL/minute; current, 150 mA; attenuator setting, 256; chart speed, 1 cm/minute; temperature, 155°C.

The liquid effluents are collected in an uncooled 4-mm-diameter collection tube (double reservoirs; overall tube length 40–50 mm). See Figures 3.25 and 5.5.

The collection tube (oven-dried until 5 minutes before use) is attached to the heated exit port by the 5/5$ joint. Sample collection is initiated 0.5 minutes prior to detection on the recorder of the expected peak (time based on previously determined retention values; see your lab instructor) and continued until 0.5 minutes following return to baseline. After the collection tube is detached, the sample is transferred to the 0.1-mL conical GC collection vial. The transfer is

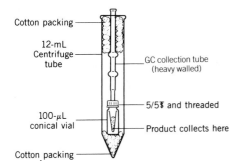

Cotton packing

12-mL
Centrifuge
tube

GC collection tube
(heavy walled)

5/5$ and threaded

100-μL
conical vial

Product collects here

Cotton packing

Fig. 5.5 *Gas chromatographic collection tube and 0.1-mL conical vial.*

facilitated by the 5/5$ joint on the conical vial. After the collection tube is joined to the vial (preweighed with stopper), the system is centrifuged. The collection tube is then removed, and the vial stoppered and reweighed.

Calculate the percentage recovery. Amounts should range between 7 and 10 mg for each component. Determine the boiling point (see Chapter 4) of each fraction and obtain the refractive index (see Chapter 4) if possible. This latter measurement will require most, if not all, of the remaining sample.

Assess the purity and efficiency of the separation from your tabulated data and the GC chromatogram.

Alternative Mixture Pairs for Preparative Collection

a. Separation of a 40-μL mixture of
(1S)-(−)-α-pinene (bp = 156°C, n_D = 1.4650, d = 0.855) and
(1S)-(−)-β-pinene (bp = 165°C, n_D = 1.4782, d = 0.859)

Chromatographic Parameters
40-μL injection

Flow rate: 50 mL/minute

Column temperature: 120°C

Column: 20% Carbowax

Elution time
α-Pinene: ~8 minutes
β-Pinene: ~12 minutes

Average recovery
α-Pinene: 8.3 μL (40%, n_D = 1.4662)
β-Pinene: 10.6 μL (53%, n_D = 1.4773)

b. Separation of a 40-μL mixture of
2-heptanone (bp = 149–150°C, n_D = 1.4085, d = 0.820) and
cyclohexanol (bp = 160–161°C, n_D = 1.4641, d = 0.963)

Chromatographic Parameters
40-μL injection

Flow rate: 50 mL/minute

Column temperature: 145°C

Column: 20% Carbowax

Elution time
2-Heptanone: ~5.5 minutes
Cyclohexanol: ~10.0 minutes

Average recovery
2-Heptanone: 8.1 μL (41%)
Cyclohexanol: 11.4 μL (57%)

c. Separation of a 40-μL mixture of
d-limonene (bp = 175–176°C, n_D = 1.4743, d = 0.8402) and
cyclohexyl acetate (bp = 173°C, n_D = 1.4401, d = 0.9698)

Chromatographic Parameters
40-μL injection

Flow rate: 50 mL/minute

Column temperature: 170°C

Column: 20% Carbowax

Elution time
  d-Limonene: ~5.5 minutes
  Cyclohexyl acetate: ~7.5 minutes

Average recovery
  d-Limonene: 8.7 μL (44%)
  Cyclohexyl acetate: 10.0 μL (50%)

ENVIRONMENTAL DATA

Substance	Amount (μL)	TLV (mg/m³)	Emissions (mg)	Volume (m³)
Heptanal (GC, refractive index)	12.5	—	12.5	
Cyclohexanol (GC, refractive index)	12.5	200	11.3	0.1
Cyclohexyl acetate	20	TXDS:		
2-Heptanone	20	ihl-hmn TCLo: 3000 mg/m²/45M		
d-Limonene	20	orl-rat LD50: 1670 mg/kg		
α-Pinene	20	orl-hmn LDLo: 50 mg/kg		
β-Pinene	20			

QUESTIONS   **5-1.** On the basis of the data presented in the series A chromatographic separation, can you explain why there is such a steep decline in column efficiency with temperature change?

5-2. Consider the following gas chromatogram for a mixture of analytes X and Y.

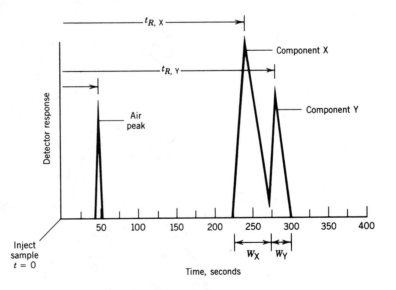

a. Calculate the number of theoretical plates for the column in reference to the peaks of each component (X and Y).

b. If the column is 12 ft long, calculate the HEPT for this column.

5-3. The number of theoretical plates a column has is important, but the crucial factor is the ability to separate two or more substances. That is, how well resolved are the peaks? The resolution of two peaks depends not only on how far apart they are (t_R) but also on the peak width (W).

Baseline resolution (R) is defined by the equation

$$R = \frac{2\Delta t_R}{(W_X + W_Y)}$$

Because of the tailing of most species on the column, a value of 1.5 is required to give baseline resolution.

  a. Calculate the resolution for the peaks in Question 5-2.
  b. Do you think a quantitative separation of the mixture is possible based on your answer?
  c. Has baseline resolution been achieved?

5-4. Discuss at least two techniques you might employ to increase the resolution of the column in Question 5-3 without changing the column.

5-5. Retention times for several organic compounds separated on a gas chromatographic column are listed here:

Compound	t_R (sec)
Air	75
Pentane	190
Heptane	350
2-Pentene	275

  a. Calculate the relative retention of 2-pentene with respect to pentane.
  b. Calculate the relative retention of heptane with respect to pentane.

TECHNIQUES 2 AND 3: DISTILLATION

Distillation is the process of heating a liquid to the boiling point, condensing the heated vapor by cooling it, and returning only a portion or none of the condensed vapors to the distillation flask. Distillation varies from the process of reflux (see p. 21) only in that a fraction of the condensate is diverted from the boiling system. Distillations in which a fraction of the condensed vapors are returned to the boiler are often referred to as being under "partial reflux."

Theory

Distillation techniques often can be used for separating two or more components on the basis of their differences in vapor pressure. Separation can be accomplished by taking advantage of the fact that the vapor phase is generally richer in the more volatile (lower-boiling) component of the liquid mixture. Molecules in a liquid are in constant motion and possess a spectrum of kinetic energies. Those with higher energies (a larger fraction for the lower-boiling component) moving near the surface have a greater tendency to escape into the vapor (gas) phase. If a pure liquid (hexane, for example) is in a closed container, eventually hexane molecules in the vapor phase will reach equilibrium with hexane molecules in the liquid phase. The pressure exerted by the hexane vapor molecules at a given temperature is called the *vapor pressure* and is represented by the symbol P_H^o, where the superscript o indicates a pure component. For any pure component A, the vapor pressure would be P_A^o.

Suppose a second component (toluene, for example) is added to the hexane. The total vapor pressure (P_{total}) is then the sum of the individual component *partial vapor pressures* (P_H, P_T) as given by *Dalton's law*

$$P_{total} = P_H + P_T$$

or in general

$$P_{total} = P_A + P_B + P_C + \ldots + P_n$$

Assuming that the vapors are ideal, the mole fraction of hexane in the **vapor phase** is given by

$$Y_H = P_H/P_{total} \tag{1}$$

It is important to realize that the vapor pressure (P_A°) and the partial vapor pressure (P_A) are not equivalent, since the presence of a second component in the liquid system has an effect on the vapor pressure of the first component. If the solution is ideal, the partial vapor pressure of hexane is given by *Raoult's law,*

$$P_H = P_H^\circ X_H \tag{2}$$

where X_H is the mole fraction of hexane in the **liquid system.**

For ideal solutions, Equations 1 and 2 may be combined to obtain the *phase diagram* shown in Figure 5.6. In this figure and elsewhere we will drop the subscripts from X_H and Y_H. X and Y will represent the mole fractions of the *most volatile component* (MVC) (hexane in this case) in the liquid and vapor phases, respectively.

Figure 5.6 describes hexane and toluene mixtures at a fixed temperature. For the region above the X curve, there will be only liquid present. For the region below the Y curve, there will be only vapor. In the area between (the sloping lens-shaped region), liquid and vapor will be present in equilibrium. This is the only region of interest to us in examining the distillation process.

To see what the phase diagram tells us about the composition of the liquid and vapor phases, imagine that the total pressure of the system is 500 torr, shown by the horizontal line in Figure 5.7. At this pressure a liquid of composition X_1 will be in equilibrium with a vapor of composition Y_1. These two points are defined by the intersection of the constant-pressure line with the X and Y

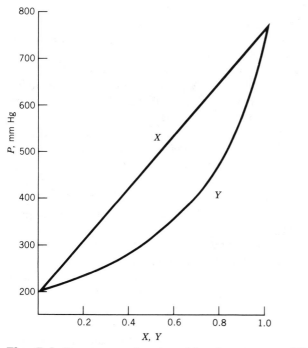

Fig. 5.6 *Pressure as a function of liquid composition (X) and vapor composition (Y), for n-hexane and toluene (temperature held constant at 69°C).*

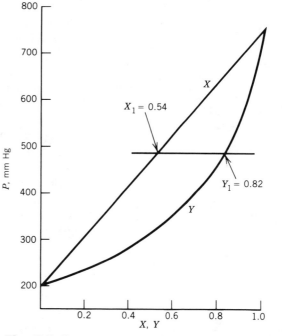

Fig. 5.7 *Pressure as a function of liquid composition (X) and vapor composition (Y), for n-hexane and toluene (temperature held constant at 69°C).*

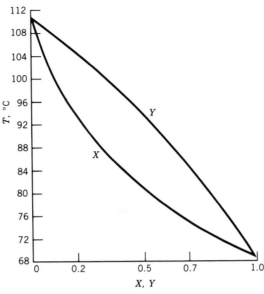

Fig. 5.8 *Temperature as a function of liquid composition (X) and vapor composition (Y).*

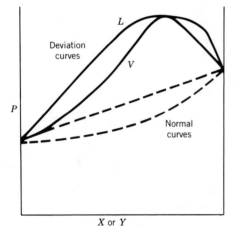

Fig. 5.9 *Positive deviation from Raoult's law;* L = *liquid,* V = *vapor.*

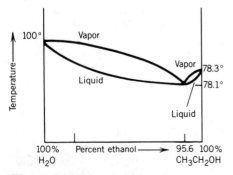

Fig. 5.10 *Ethanol–water minimum-boiling-point phase diagram.*

curves. It is important to note here that Y will be greater than X for the equilibrium system. That is, the vapor in equilibrium with a given liquid will be richer in the more volatile component than in the liquid.

Diagrams such as Figure 5.7 are not very useful in describing the distillation process. We need a phase diagram for the mixture at constant pressure instead of constant temperature. Figure 5.6 may be transformed to the desired diagram if we know how P_H^o and P_T^o depend on T. This information may be supplied by the Clausius–Clapeyron equation or by appropriate experimental data.

We will obtain a qualitative diagram for temperature as a function of composition by the following reasoning. (1) The substance having the *higher* vapor pressure at a given temperature will have the *lower* boiling point at a given pressure. (2) At *low* temperatures only the liquid phase will be present, and at *high* temperatures only the vapor phase will be present. Thus, the temperature–composition diagram is shown by Figure 5.8. Note that this figure may also be obtained (qualitatively) by turning Figure 5.6 upside down.

Many pairs of liquids do not obey Raoult's law. Often, pairs of liquids encountered in organic chemistry exhibit a *positive deviation* from Raoult's law. The positive deviation means that the pressure above the solution is *greater* than would be predicted by Raoult's law. If this deviation is large, the pressure–composition curve may exhibit a maximum, as shown in Figure 5.9. Here the curves for normal Raoult's law behavior are shown as dashed lines for reference. Mixtures in which one of the components is polar and the other component is at least partly nonpolar often exhibit positive deviations from Raoult's law.

The temperature–composition diagram for systems showing a positive deviation from Raoult's law is again obtained by the simple inversion process. Such a diagram is shown in Figure 5.10. In this diagram at a temperature of T_1 a liquid of composition X_1 will be in equilibrium with a vapor of Y_1. At a temperature of T_{az}, however, the composition of the liquid and vapor will be the same. This mixture is an azeotropic or constant-boiling mixture. Water and ethanol form one of the more familiar azeotropic systems. They exhibit a positive deviation from Raoult's law and have a minimum boiling azeotrope at 78.1°C and 95.6% ethanol by volume.

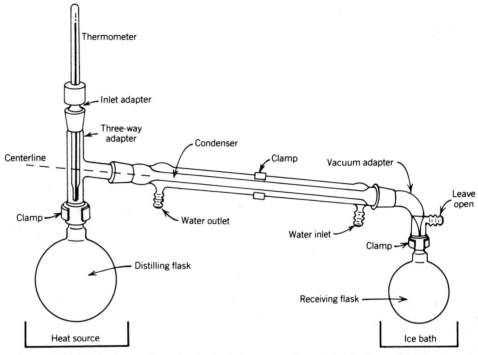

Fig. 5.11 *A complete, simple distillation setup. From* Zubrick, James W. *The Organic Chem Lab Survival Manual,* 2nd ed.; Wiley: New York, 1988. (Reprinted by permission of John Wiley & Sons, New York.)

TECHNIQUE 2: SIMPLE DISTILLATION AT THE SEMIMICROSCALE LEVEL

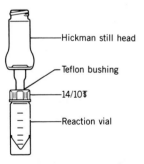

Fig. 5.12 *Hickman still.*

Simple distillation involves the use of the distillation process to separate a liquid from minor components that are nonvolatile or that have boiling points over 30–40°C above that of the major component. A typical setup for a macroscale distillation of this type is shown in Figure 5.11. At the microscale level, when working with volumes smaller than 100 μL, gas chromatographic techniques (see Experiment 1) have replaced conventional microdistillation processes.[2] Semimicroscale simple distillation in the volume range 0.1–2 mL remains an effective separation technique. Apparatus have been developed that achieve good separation of mixture samples smaller than 2.0 mL in volume. One of the most useful of these designs is the Hickman still, shown in Figure 5.12.

This still is employed in several of the microscale experiments to purify solvents, carry out reactions, and concentrate solutions for recrystallization. An introduction to the use of the Hickman still is given in Experiments 2 and 3A.

In the first distillation experiment described here a liquid is separated from an involatile solute. In this case the pressure of the liquid is lowered by the presence of the solute, but the vapor phase consists only of one component. Thus, except for the transfer of involatile material by splashing, the material condensed should consist only of the volatile component.

We can understand what is going on in a simple distillation of two volatile components by referring to the phase diagrams shown in Figures 5.13 and 5.14. Figure 5.13 is the phase diagram for hexane and toluene. The boiling points of these liquids are separated by 42°C. Figure 5.14 is the phase diagram for

[2] Schneider, F. L. *Monographien aus dem Gebiete der qualitativen Mikroanalyse,* Vol. II: *Qualitative Organic Microanalysis*; Benedetti-Pichler, A. A., Ed.; Springer-Verlag: Vienna, Austria, 1964; p 31.

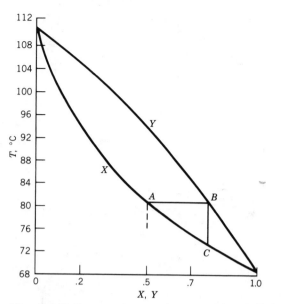

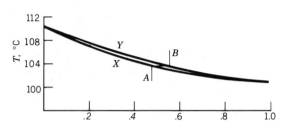

Fig. 5.13 *Temperature as a function of liquid composition (X) and vapor composition (Y): hexane and toluene.*

Fig. 5.14 *Temperature as a function of liquid composition (X) and vapor composition (Y): methylcyclohexane and toluene.*

methylcyclohexane and toluene. Here the boiling points are separated by only 9.7°C.

Imagine a simple distillation of the hexane–toluene pair in which the liquid in the pot is 50% hexane. In Figure 5.13, when the liquid reaches 80.8°C it will be in equilibrium with a vapor having a composition of 77% hexane. This is indicated by the line A–B. If this vapor is condensed to a liquid of the same composition, as shown by line B–C, we will have achieved a significant enrichment of the condensate with respect to hexane. This is referred to as a simple distillation. The process of evaporation and condensation is achieved by the theoretical construction known as a *theoretical plate*. When this distillation is actually done with a Hickman still, some of the mixture will go through one evaporation and condensation cycle, some will go through two of these cycles, and some may be splashed more directly into the collar. A resolution of between one and two theoretical plates is generally obtained.

Referring to Figure 5.14, if we consider the same process for a 50% mixture of methylcyclohexane and toluene, the methylcyclohexane composition will increase to 58% for one theoretical plate. Thus, simple distillation may provide adequate enrichment of the MVC *if* the boiling points of the two liquids are reasonably well separated as they are for hexane and toluene. If the boiling points are close together, as they are for methylcyclohexane and toluene, the simple distillation will not provide much improvement.

As we continue the distillation process and remove some of the MVC by condensing it, the residue in the heated flask becomes less rich in the MVC. This means that the next few drops of condensate will be less rich in the MVC. As the distillation is continued the condensate becomes less and less rich in the MVC.

We can improve upon simple distillation by repeating the process. For example, we could collect the condensate until about one third is obtained. Then we could collect a second one-third aliquot in a separate container. Our original mixture would then be separated into three fractions. The first third would be richest in the MVC and the final third in the pot would be richest in the least volatile component. If the MVC was the compound of interest we could redistill the first fraction collected (in a clean flask!) and collect the first third of the material condensing in that process. This procedure is used in Experiment 3A.

Experiment 2

Simple Distillation at the Semimicroscale Level: Separation of Ethyl Acetate from *trans*-1,2-Dibenzoylethylene

Physical Properties of Reactants and Products

Compound	MW	Wt/Vol	mp(°C)	bp(°C)	Density	n_D
Ethyl acetate	88.12	1.0 mL		77	0.90	1.3723
trans-1,2-Dibenzoylethylene	236.27	50 mg	111			

Transfer 1 mL of the yellow stock solution (*trans*-1,2-dibenzoylethylene/ethyl acetate, 50 mg/mL) to a 3-mL conical vial by automatic delivery pipet (remember to place the vial in a small beaker to prevent tipping during the transfer). Place a boiling stone in the vial and assemble the Hickman still head. The still assembly is mounted in a sand bath and placed on a hot plate (see Fig. 5.15).

The temperature of the bath is raised to 90–100°C at a rate of 5°C/minute.

> **CAUTION: *Do not let the temperature of the still rise too rapidly.***

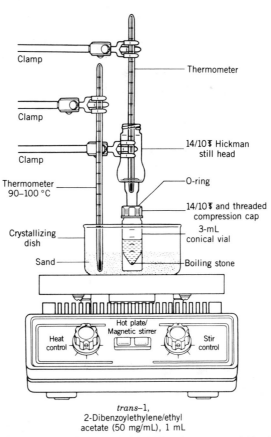

Clamp

Clamp

Clamp

Thermometer 90–100 °C

Crystallizing dish

Sand

Thermometer

14/10 Hickman still head

O-ring

14/10 and threaded compression cap

3-mL conical vial

Boiling stone

Hot plate/ Magnetic stirrer

Heat control

Stir control

trans–1, 2-Dibenzoylethylene/ethyl acetate (50 mg/mL), 1 mL

Fig. 5.15 *Hickman still 4/10 with conical vial (3 mL).*

Once boiling commences, the rate of heating should be lowered to 2–3°C/min. A slow distillation rate is very important in establishing equilibrium between the vapor and liquid components in the mixture. Follow the course of the distillation by the rise of condensate on the sides of the Hickman column. When the condensate reaches the trap, adjust the bath temperature so that liquid is removed from the column slowly (approximately 100 μL/minute). A smooth slow distillation will provide a cleaner separation of the components and will also avoid mechanical transfer of nonvolatile components via splattering to the condensate trap (if the condensate appears yellow, mechanical contamination has occurred).

Collect approximately 50–150 μL of the ester in the collar of the still (the first fraction collected is often referred to as the forerun; give the temperature range). As the distillation continues, remove the forerun with a Pasteur pipet having a slightly bent tip (microburner). Place the fraction in a clean, dry 1-dram screw-capped vial (use an aluminum foil liner to avoid cap contamination). Number the fraction with a marking pen. Collect a second fraction of ester (400–500 μL, which may require combining two or even three collections from the collar; give the temperature range), which should be clear and colorless. Remove and store as before. Discontinue the distillation. Allow the distilling flask to cool slowly by leaving it in the warm sand bath while measuring the physical properties of the distillate fractions. Three physical properties of the ester will be measured to establish the identity and purity of the compound by comparison with known literature values.

Determine the refractive index (see Chapter 4) of the two fractions collected. Compare the experimental values to those found in the literature for ethyl acetate. If the values are within 0.0010 unit of each other, the fractions are considered to have the same constitution. Are the values for the two fractions the same? If not, which one deviates the most from the reference data? Attempt to explain the result?

Determine the density (see Chapter 4) of the ester using material contained in the second fraction. This measurement is nondestructive. The material employed may be recovered for use in further tests. Compare your results with those values found in the literature.

Determine the boiling point of the second fraction by the ultramicro boiling point procedure (see Chapter 4). Compare your result with the literature value. Does this fraction appear to be pure ethyl acetate?

In the next step, disconnect and cool the 3-mL conical vial in an ice water bath for 10 min. *trans*-1,2-Dibenzoylethylene will crystallize from the concentrated solution. Remove the remaining solvent from the distillation vial with a Pasteur filter pipet and place the crystals on a porous clay plate to air dry. The melting point of the crystalline material is obtained by the evacuated capillary method and compared with the literature value.

Reference values of the physical constants are available in the *CRC Handbook of Chemistry and Physics*. Submit a copy of the table prepared in your laboratory notebook to the instructor after tabulating the experimentally measured values of the physical properties in addition to those reported in the literature for ethyl acetate (see "acetic acid, ethyl ester").

TECHNIQUE 3: FRACTIONAL SEMIMICROSCALE DISTILLATION

Process

Fractional distillation is the application of a distillation system containing more than one theoretical plate. This process must be used if fairly complete separation is desired when the boiling points of the components differ by less than 30–40°C. In this situation a fractionating column is required to increase the efficiency

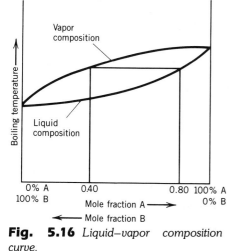

Fig. 5.16 *Liquid–vapor composition curve.*

of the separation. As discussed earlier, it may be seen from a liquid–vapor composition curve (Fig. 5.16) that the lower-boiling component of a binary mixture makes a larger contribution to the vapor composition than does the higher-boiling component. On condensation the liquid formed will be richer in the lower-boiling component. This condensate will not be pure, however, and in the case of closely boiling components it may show only slight enrichment. If the condensate is volatilized a second time, the vapor in equilibrium with this liquid will show a further enrichment in the lower-boiling component. Thus, the trick to separating liquids that possess similar boiling points is repeating the vaporization–condensation cycle many times. Each cycle is termed a *theoretical plate*. A number of different column designs are available for use at the macro level that achieve varying numbers of theoretical plates (see Fig. 5.17).

In most columns the design is such that increased fractionation efficiency is dependent on a very large increase in the surface area in contact with the vapor phase. This is normally accomplished by packing the fractionating column with wire gauze or glass beads. Unfortunately, a large volume of liquid must be distributed over the column surface in equilibrium with the vapor. Furthermore, the longer the column the more efficient it becomes, but longer columns also require additional liquid phase. The column requirement of the liquid phase is termed *column holdup*. Column holdup is defined as the amount of liquid distributed over the column packing required to maintain the system in equilibrium. This material is essentially lost from the liquid phase held in the distillation pot. The amount of column holdup can be large compared with the total volume of material available for the distillation. With mixtures of less than 2 mL, column holdup precludes the use of the most common fractionation columns. Microfractionating columns constructed of rapidly spinning bands of metal or Teflon gauze have very low column holdup and have a large number of plates relative to their height (Fig. 5.18). They are, however, rather expensive and normally are available only for research purposes.

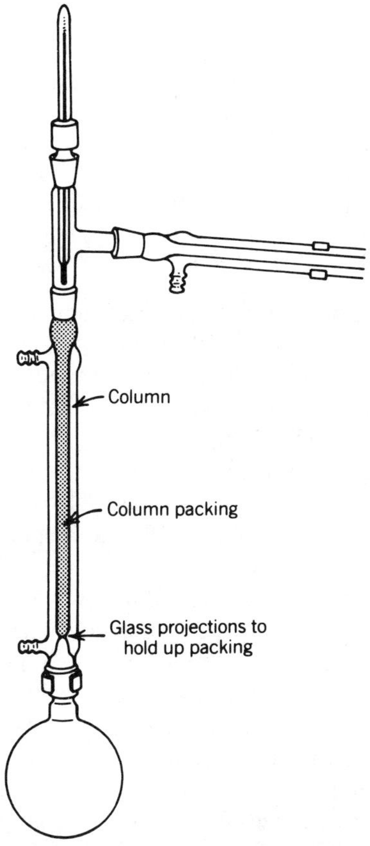

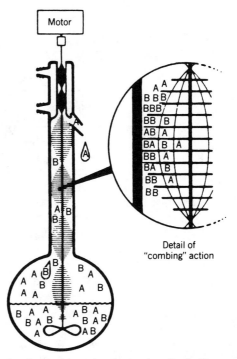

Fig. 5.18 *Schematic of a metal mesh spinning band still.* (Courtesy of Perkin–Elmer Corp., Norwalk, CT.)

Fig. 5.17 *The fractional distillation setup. From* Zubrick, James W. *The Organic Chem Lab Survival Manual,* 2nd ed.; Wiley: New York, 1988. (Reprinted by permission of John Wiley & Sons, New York.)

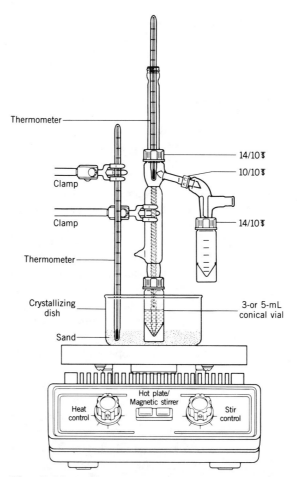

Thermometer

Clamp

Clamp

Thermometer

Crystallizing
dish

Sand

14/10 ⅋

10/10 ⅋

14/10 ⅋

3-or 5-mL
conical vial

Hot plate/
Magnetic stirrer

Heat
control

Stir
control

Fig. 5.19 *Micro spinning band distillation column (2.5 inch).*

In the development of the microscale laboratory, several new distillation systems have been designed. The microscale spinning band distillation apparatus (Fig. 5.19) achieves up to twelve theoretical plates in careful hands and may be used in the instructional laboratory. This system contains a Teflon band that fits rather closely inside an insulated glass tube. The Teflon band has spiral grooves to return the liquid to the distillation pot and is spun at about 1000 rpm. A modification of this apparatus uses a spinning band inside a Hickman still head (see Fig. 3.18). Experiments 3B and 3C involve fractional distillation with spinning band columns.

An alternative to the microspinning band distillation column is the concentric tube column. In these columns the fractionating section is constructed of two concentric tubes in which the vapor–liquid equilibrium is established within the annular space between the two columns. The resolution of the concentric tube system is inversely proportional to the thickness of the annular ring. Columns of this type can achieve very good separations with the number of theoretical plates approaching 3 per centimeter. In addition, column holdup can be close to 10 μL per theoretical plate. The major constraint in the use of these columns is the very low throughput, which can be as little as 100 μL/hour. Concentric tube columns also have the nasty habit of flooding in inexperienced hands. These factors cause long residence times at elevated temperatures for the liquid components. When time and thermal stability are not a problem, the concentric tube column can be a powerful system for mixture separations.

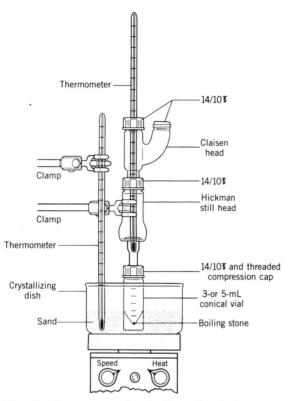

Fig. 5.20 *Hickman still with Claisen head adapter.*

Experiment 3A utilizes the Hickman still in a simple example of fractional distillation. The system is arranged so that the thermometer is positioned directly down the center of the still column with the bulb extending just to the bottom of the well. As assembled, the system functions as a rough concentric tube fractionating column. It is very important to position both the still and the thermometer as close to the vertical as possible, and in no circumstances should the two elements come into direct contact (see Fig. 5.20). A clean, two-theoretical-plate distillation is obtained with a two-component mixture by carrying out sequential fractional distillations with this system.

To understand how the spinning band improves the performance of the column and to see some of the important characteristics of the column, we will analyze it a bit further. Figure 5.21 shows a very simple distillation column possessing just one plate. Overall, however, this system has two theoretical plates.

The compositions of the liquid in the pot and of the infinitesimal amount of liquid at the first plate are X_0 and X_1, respectively. Y_0 and Y_1 are compositions of the vapor in equilibrium with each of the liquids, X_0 and X_1, respectively. Since we are talking about the MVC, $Y_0 > X_0$, $Y_1 > X_1$, and we may assume that $Y_1 > Y_0$ (that is, we have already established that the vapor will be enriched in the MVC). V is the *rate* at which vapor is transported upward, L is the rate of downward transport of liquid back to the pot, and D is the rate at which material is distilled. L, D, and V are related by $V = L + D$. (What goes up must come down.)

We will first look at the process qualitatively. If D and V are comparable (L is small), as vapor at composition Y_1 is removed, X_1 becomes smaller—the liquid in the first plate becomes less rich in the MVC as the MVC-rich vapor is removed. If, on the other hand, L and V are both large compared with D, the composition at plate 1 will be maintained at Y_0 by the large supply of incoming

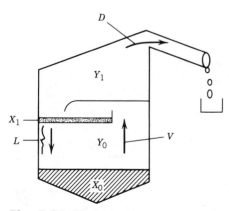

Fig. 5.21 *Model for fractional distillation with partial reflux.*

vapor with composition Y_0. This may also be seen in the following quantitative discussion.

The first relationship we have already seen:

$$V = L + D \tag{1}$$

In addition, if an insignificant amount of material is held up at plate 1, the moles of MVC going into plate 1 must equal the moles of MVC leaving plate 1;

$$V * Y_0 = L * X_1 + D * Y_1 \tag{2}$$

We may eliminate V $(=L + D)$ from this expression and rearrange to get

$$X_1 = Y_0 - (D/L) * (Y_1 - Y_0) \tag{3}$$

Now, in Equation 3, $Y_1 - Y_0$ will always be greater than zero under ideal conditions (since they both represent the MVC). D and L are both positive. Therefore, the *upper limit* for X_1 is Y_0. This would be the case when $D/L = 0$ (and nothing is being distilled). In the real world something is being distilled ($D > 0$) and X_1 will be less than Y_0 (and down the line the condensate will be less rich in the MVC; there will be less separation). You should be able to generalize this result for one plate to the situation in which we have n plates. When D/L is small the composition of the first plate is at a maximum; therefore, Y_1 is also a maximum and, further, X_2 will be a maximum and so on.

The purpose of the spinning band is to make D/L as small as possible so that X_1 approaches its upper limit of Y_0. The spinning band ensures that the optimum separation of a pair of liquids will be achieved. We will have a lot of vapor going up, a lot of liquid being returned by the spinning action of the spiral band, and a relatively small amount of material actually being distilled. This also implies that if we want to achieve a high degree of separation, the distillation rate should be low. There is, of course a compromise between the low rate of distillation required to obtain maximum separation and a rate that will allow someone else to use the apparatus and will allow you to get on to other things.

Experiment 3A   Fractional Semimicroscale Distillation: Separation of Hexane and Toluene

Hexane and toluene are liquid hydrocarbons that have boiling points approximately 40°C apart. The liquid–vapor composition curve in Figure 5.8 represents this system; therefore, it is apparent that a two-plate distillation should yield nearly pure components. The procedure to be outlined consists of two parts. The first deals with the initial distillation (first plate), which separates the liquid mixture into three separate fractions. The second deals with *redistillation* of the first and third fractions (second plate). Exercising careful technique during the first distillation should provide a fraction rich in the lower-boiling component, a middle fraction, and a fraction rich in the higher-boiling component. Then careful redistillation of these fractions can be expected to complete the separation of the two components and to produce fractions of relatively pure hexane and toluene. The Hickman still employed in the microscale laboratory is a simple, short-path column, and therefore, one would not expect complete separation of the hexane and toluene in one cycle.

Physical Properties of Reactants and Products

Compound	MW	Wt/Vol	bp(°C)	Density	n_D
Hexane	86.18	1.0 mL	69	0.66	1.3751
Toluene	92.15	1.0 mL	111	0.87	1.4961

In a clean, dry, stoppered 5-mL conical vial are placed 1.0 mL of hexane and 1.0 mL of toluene using an automatic delivery pipet.

Place the vial in a small beaker to prevent tipping.

A boiling stone is added, the Hickman still is assembled with the thermometer positioned directly down the center of the column (see previous discussion), and the system is mounted in a sand bath (see Fig. 5.20). The temperature of the sand bath is raised to 80–90°C at a rate of 5°C/min using a hot plate. *Do not let the temperature of the still rise too rapidly.*

Once gentle boiling begins, the heating rate should be lowered to 2°C/minute. It is absolutely crucial that the distillation rate be kept below 100 μL/3 min to achieve the necessary fraction enrichment that will permit good separation during the second stage of the experiment. The distillate is collected in *three fractions* over the temperature ranges (1) 65–85°C (bath temperature 95–110°C); (2) 85–105°C (bath temperature ~140°C); and (3) 105–110°C (bath temperature ~170°C) in amounts of approximately 800, 400, and 800 μL, respectively. Remove each fraction from the still with a bent-tip Pasteur pipet (as in Experiment 2). Store the liquid condensate (fractions) in clean dry 1-dram screw-cap vials. *Remember to number the vials in order and use an aluminum foil cap liner.*

For each of the three fractions, record the refractive index. Fraction 1 has been enriched in one of the two components. Which one? Does the refractive index agree with that found in the literature? Fraction 3 has been enriched in the other component. Does the refractive index of that fraction support your first conclusion? If partial enrichment has been achieved, proceed to the second part of the experiment.

REDISTILLATION OF FRACTION 1

Redistill fraction 1 in a clean Hickman still with a thermometer arranged as before (Fig. 5.20), using a 3-mL conical vial and the procedure just outlined. Collect an initial fraction over the boiling range, 68–71°C (~100–200 μL). Remove it from the collar using the Pasteur pipet, and place it in a 1-dram screw-cap vial.

Determine the ultramicro boiling point and the refractive index of this lower-boiling fraction. Compare the experimental values obtained with those of pure hexane reported in the literature. *If time permits, the boiling point and refractive index of pure hexane may be determined for comparison purposes.*

REDISTILLATION OF FRACTION 3

Fraction 3 is placed in a clean Hickman still, using a thermometer and a 3-mL conical vial (Fig. 5.20), and redistilled using the procedure outlined. Collect an initial fraction over the boiling range 95–108°C (~500 μL), and transfer this fraction by Pasteur pipet to a screw-cap vial. Collect a final fraction at 108–110°C (~250 μL), and transfer the material to a second vial. *This second fraction is the highest-boiling fraction to be collected in the three distillations and should be the richest in the high-boiling component.*

Determine the refractive index and boiling point of the second fraction, and compare your results with those found in the literature for toluene. *If time permits, determine the refractive index and boiling point of pure toluene for comparison purposes.*

Report the results obtained on the separation of hexane and toluene. Show a comparison of the experimental refractive index and boiling point data versus those values found in the literature for hexane and toluene. Give a brief discussion indicating the efficiency of the separation based on the data recorded in your laboratory notebook.

ENVIRONMENTAL DATA

Substance	Amount (mL)	TLV (mg/m^3)	Emissions (mg)	Volume (m^3)
Ethyl acetate (dist., transfer, ref. index)	1.0	1400	170	0.1
Hexane (fract. dist., transfer, ref. index, bp)	1.0	180	250	1.4
Toluene (fract. dist., transfer, ref. index, bp)	1.0	750	200	0.3

TXDS: 1,2-Dibenzoylethylene—ipr-mus LD50: 25 mg/kg.

Experiment 3B

Fractional Distillation of 2-Methylpentane and Cyclohexane with a Spinning Band Column

In this experiment you will separate 2 mL of a mixture of 2-methylpentane and cyclohexane using a 2.5-in. spinning band distillation column. The purity of the fractions will be determined by gas chromatography and by measurement of the refractive index. Finally, the number of theoretical plates will be estimated.

Physical Properties of Components in the Mixture

Compound	MW	bp (°C)	n_D
2-Methylpentane	86.18	60.3	1.3715
Cyclohexane	84.16	80.7	1.4266

Assemble the system as shown in Figure 5.19. In the process, make sure that the Teflon band is aligned as straight as possible in the column. In particular, the pointed section extending into the pot must be straightened to minimize vibration during spinning of the band.

Insert the Teflon stopper in the side arm. This stopper plays a very important function in the operation of the column. Insertion of the stopper creates a closed system. Suspension of the thermometer with a septum on the top of the condenser can act to release any buildup of pressure. THE SYSTEM MUST BE ABLE TO VENT AT THE THERMOMETER DURING OPERATION!

Once the spinning band has been tested and rotates freely, place 1.0 mL of 2-methylpentane and 1.0 mL of cyclohexane in the pot (to be delivered with a Pasteur pipet or an automatic delivery pipet). Reassemble the system and lower the column into the sand bath. It is important to note that the beveled edge on the air condenser should be rotated 180° from the collection arm.

Note. *It is important to make an aluminum foil sand bath cover which reflects heat and hot air away from the collection vial.*

Gently heat the pot until boiling occurs. The magnetic stirrer is turned to a low spin rate when heating commenses. When reflux is observed at the base of the column, the magnetic stirrer is adjusted to intermediate spin rate. Once liquid begins to enter the column the spin rate is increased to the maximum (1000 to 1500 rpm).

It is absolutely critical that the temperature of the pot be adjusted so that vapors in the column rise very slowly. It is possible for overheated vapors to be forced through the air condenser.

When the vapors slowly arrive in the unjacketed section of the head of the column the condenser joint acts as a vapor shroud to effectively remove vapors from the receiver cup area. During this total reflux period, maximum separation of the components is achieved. Once total reflux is obtained the system is left for 20–30 minutes to reach thermal equilibrium. During total reflux the head thermometer should read about 57–60°C (at least for the equilibration time).

After the end of equilibration you are ready for collection. Rotate the air condenser 180° so that the beveled edge is over the collection duct. By pulling the Teflon stopper in the side arm out, some of the condensate will be collected. (Putting it back in stops collection.) Collect six drops (approximately 0.30 mL), and put the stopper back into the arm. After removing the collection vial, transfer the contents into a covered shell vial using a Pasteur pipet. *Label all fractions.* Collect two 0.60-mL fractions; then turn off the heat and stirring motor and remove the pot from the sand bath. Transfer the pot solution using a Pasteur pipet to a fourth covered shell vial. You have now collected four fractions.

Characterization of the Fractions

The composition of each of the fractions may be determined by gas chromatography and the refractive index.

The Gow-Mac gas chromatograph should be set up as follows:

Column	DC 710
Attenuation	64
Injection	10 μL
Temperature	80°C
Flow rate (He)	55 mL/minute
Chart speed	1 cm/minute

If we assume that the refractive index is a linear function of the volume fraction, the following relationship gives us the volume fraction of 2-methylpentane in a mixture. X is the volume fraction and n_D is the measured refractive index.

$$X = (1.4266 - n_D)/(1.4266 - 1.3715)$$

The curve shown below in Figure 5.22 may be used to estimate the number of theoretical plates from the composition of the *first* 0.30-mL fraction. For example, if the composition of the first 0.30 mL is 0.89, we would infer that the system had a resolution equivalent to about four theoretical plates. Note that the number of plates cannot be determined with confidence if the composition is greater than about 0.97. If we really wanted to determine the number of theoretical plates for a system with more than five plates, we could start with a mixture only 10 or 20% in the MVC, rather than the 50% used in this experiment.

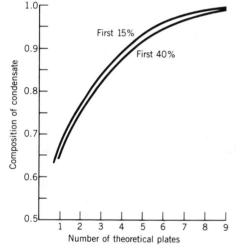

Fig. 5.22 *Composition of the first 15% and the first 40% of the volume collected in the distillation of a 50% (v/v) mixture of 2-methylpentane and cyclohexane.*

ENVIRONMENTAL DATA

Substance	Amount (mL)	TLV (mg/m³)	Emissions (mg)	Volume (m³)
2-Methylpentane[a]	1.0	1800	650	0.4
Cyclohexane	1.0	1050	780	0.7

[a] treated as hexane isomers other than *n*-hexane.

Experiment 3C

Fractional Distillation of 2-Methylpentane and Cyclohexane with a Spinning Band in a Hickman–Hinkle Still

The distillation will separate the same two compounds used in Experiment 3B. The distillate can be analyzed to determine the number of theoretical plates. If careful attention is given to the procedure, the spinning Hickman–Hinkle is capable of more than seven theoretical plates.

In this experiment separation of a 2-mL mixture of 2-methylpentane and cyclohexane is achieved. The purity of the fractions will be determined by gas chromatography and by measurement of the refractive index.

Physical Properties of Components in the Mixture

Compound	MW	bp (°C)	n_D
2-Methylpentane	86.18	60.3	1.3715
Cyclohexane	84.16	80.7	1.4266

Assemble the system as shown in Figure 3.18. In the process, make sure that the Teflon band is aligned as straight as possible in the column. In particular, the pointed section extending into the pot must be straightened to minimize vibration during spinning of the band.

Once the spinning band has been tested and rotates freely, place 1.0 mL of 2-methylpentane and 1.0 mL of cyclohexane in the pot (to be delivered with a Pasteur pipet or an automatic delivery pipet). Reassemble the system and lower the column into the sand bath.

Cover the sand bath with aluminum foil during the distillation to prevent the collar of the still from overheating.

Gently heat the pot until boiling occurs. When heating commences turn on the magnetic stirrer at a low setting. When reflux commences at the base of the column the magnetic stirrer is raised to intermediate settings. Once liquid begins to enter the column the spin rate is increased to the maximum (1000–1500 rpm). *It is extremely important that careful temperature control be exercised at this stage so that the condensing vapors ascend the column very slowly.* Vapor phase enrichment by the most volatile component is limited mainly to this period, as fraction collection commences immediately on arrival of the vapor column at the annular ring. Once condensation occurs fractions are collected by the same technique used in Experiment 3A. Characterization of the fractions, however, follows the procedure given for Experiment 3B.

Characterization of the Fractions

The composition of each of the fractions may be determined by gas chromatography, the refractive index, or both. See Experiment 3B for details.

An alternate approach to the procedures discussed in Experiment 3B is to establish the fraction volume by weight. The curves shown in Figure 5.22 again may be used to estimate the number of theoretical plates. The volume of the first fraction can be estimated or determined more accurately by weighing the fraction in a tared screw-cap vial. The composition of this fraction then may be determined and the fraction of the total represented by this portion calculated. If, for example, the first fraction has a volume of 0.4 mL (20% of the total) and has a composition 0.898 by volume of 2-methylpentane, we would infer that the system had a resolution equivalent to about four theoretical plates.

STEAM DISTILLATION   There are times when ordinary distillation may not be feasible for the separation of a liquid from dissolved impurities. The compound of interest may have a high boiling point difficult to control with a simple apparatus or it may tend to decompose or oxidize on being raised to a high temperature. If the compound is only sparingly soluble in water and any small amount of water can be removed with a drying agent, *steam distillation* may be the technique of choice. Compounds that are immiscible in water have very large positive deviations from Raoult's law. Therefore, the boiling temperature is generally lower than that of water and the compound. The phase diagrams of such systems are more involved than we want to consider right now, so we will use a different analysis.

If water and the compound are quite insoluble in one another, the boiling point is essentially the temperature at which the vapor pressures of the two *pure* materials add up to one atmosphere, or whatever the pressure of the system is. That is, $P_A^\circ + P_B^\circ = 1$ atmosphere.

Cyclohexanone is isolated from by-products by steam distillation in Experiment 39. The normal boiling point of pure cyclohexanone is 156°C and that for water is, of course, 100°C. At about 94.5°C the vapor pressure of cyclohexanone is 112 torr and the vapor pressure of water is 648 torr. The vapor pressures of cyclohexanone and water add up to 760 torr and thus the two compounds steam distill at this temperature. From Dalton's law, the condensate in a steam distillation will consist of water and the compound in the same *molar* ratio as the ratio of their vapor pressures at the steam distillation temperature. The ratio cyclohexanone/water in this example is 112/648, or 0.17. The cyclohexanone may be separated from the water in the condensate by any of the methods used to separate immiscible liquids. Steam distillation with toluene is also used in a later experiment (48B) to remove water from a higher-boiling product.

Experiment 3D

Isolation of a Natural Product by Steam Distillation: Cinnamaldehyde from Cinnamon

Cinnamaldehyde may be extracted from cinnamon by hot water. The cinnamaldehyde decomposes at its normal boiling point but it may then be separated from other cinnamon by-products by steam distillation.

$$\langle\!\!\!\bigcirc\!\!\!\rangle\text{—CH}=\text{CH—CHO}$$

Cinnamaldehyde

Physical Properties

Compound	MW	Wt/Vol	bp(°C)	Density	n_D
Cinnamaldehyde	132.16	—	246	1.048–1.052	1.618–1.623
Water	18.02	8.0 mL			

In a 10-mL round-bottom flask containing a boiling stone and equipped with a Hickman still head, place 1 g of chopped stick cinnamon and 4 mL of water. Place the apparatus on a sand bath maintained at 150–160°C. An aluminum foil shield is used to cover the sand bath to prevent the collar of the Hickman still from overheating.

The cinnamon tends to foam during the distillation, so care must be taken to prevent contamination of the Hickman still by cinnamon particles.

Position a thermometer in the throat of the Hickman still to record the distillation temperature, which should be very close to 100°C. Since this thermometer makes pipetting the condensate difficult, it is suggested that it be removed after the distillation temperature is recorded.

The milky cinnamaldehyde—water distillate collects in the collar of the still and may be removed using a 9-in. Pasteur pipet and transferred to a 12-mL centrifuge tube. The distillation is continued for approximately one hour. Approximately 5–6 mL of distillate is collected in the centrifuge tube.

An additional 4 mL of water is added during the course of the distillation to maintain the original volume of water in the flask. This water is added using a 9-in. Pasteur pipet inserted down the neck of the Hickman still.

The combined distillate in the centrifuge tube is now extracted with three successive 2-mL portions of methylene chloride. The first portion of methylene chloride is used to rinse the Hickman still. After each extraction, the lower methylene chloride layer is transferred (Pasteur filter pipet) to a 25-mL Erlenmeyer flask and the combined extracts are dried over sodium sulfate.

The dried methylene chloride solution is now transferred **in two portions,** using a Pasteur filter pipet, to a **tared** 5-mL conical vial and the solvent is evaporated in a warm sand bath using a slow stream of nitrogen gas. After all the solution has been transferred and the solvent evaporated, the sodium sulfate is rinsed with an additional two 0.5-mL portions of methylene chloride. The rinsings are transferred to the same vial and concentrated as before.

The refractive index is measured and the result compared with the literature values. The sample of aldehyde may be stored and used in Technique 5 as an optional material for characterization by infrared spectroscopy.

ENVIRONMENTAL DATA

Substance	Amount (mL)	TLV (mg/m³)	Emissions (mg)	Volume (m³)
Methylene chloride	7	350	9300	27

QUESTIONS

5-6. The boiling point of a liquid is affected by several factors. What effect does each of the following conditions have on the boiling point of a given liquid?
  a. The pressure of the atmosphere.
  b. Use of an uncalibrated thermometer.
  c. Rate of heating of the liquid in a distillation flask.

5-7. Calculate the vapor pressure of a solution containing 30 mol% hexane and 70 mol% octane at 90°C assuming Raoult's law is obeyed.

Given: vapor pressure of the pure compounds at 90°C: vapor pressure of hexane = 1390 torr; vapor pressure of octane = 253 torr.

5-8. In any distillation, for maximum efficiency of the column, the distilling flask should be approximately half full of liquid. Comment on this fact in terms of (a) a flask that is too full and (b) a flask that is nearly empty.

5-9. Occasionally during a distillation a solution will foam rather than boil. One way of avoiding this problem is to add a surfactant to the solution.
  a. What is the chemical constitution of a surfactant?
  b. How does a surfactant reduce the foaming problem?

5-10. Explain why packed and spinning band fractional distillation columns are more efficient at separating two liquids with close boiling points than are unpacked columns.

5-11. Why are boiling stones added to a liquid that is to be heated to the boiling point?

TECHNIQUE 4: SOLVENT EXTRACTION

Solvent extraction is a technique frequently used in the organic laboratory to separate or isolate a desired species from a mixture of compounds or from impurities. It is used extensively in the experiments given in this text. Solvent extraction methods are readily adapted to microscale work, since small quantities are easily manipulated in solution. This method is based on the solubility characteristics of the organic substances involved in relation to the solvents used in a particular separation procedure.

Solubility

Substances vary greatly in their solubility in various solvents. One of the most useful principles a chemist uses to predict the solubility of a substance is that *a substance tends to dissolve in a solvent that is chemically similar.* In other words, *like dissolves like.*

For example, solubility in water requires that a species have some characteristics of water. An important class of compounds, the organic alcohols, have the hydroxyl group —O—H bonded to a hydrocarbon chain or framework. The hydroxyl group is polar, because of the difference in electronegativity of the hydrogen and oxygen atoms. In other words, the bond has *partial ionic character.*

$$\overset{\delta^-}{\underset{\cdot\cdot}{\text{O}}}\!\!-\!\!\overset{\delta^+}{\text{H}}$$

Partial ionic character of the hydroxyl group

This *polar* or partial ionic character leads to relatively strong hydrogen bond formation between molecules having this entity. Strong hydrogen bonding is evident in molecules that contain groups having a hydrogen atom attached to an oxygen, nitrogen, or fluorine atom, as diagrammed here for the water system. This polar nature of a functional group is present in those compounds that have sufficient electronegativity difference between the atoms making up the group.

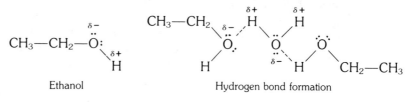

Ethanol   Hydrogen bond formation

In ethanol it is apparent that the hydroxyl end of the molecule is very similar to that of water. Therefore, when ethanol is added to water, they are miscible in all proportions. This is so because the attractive forces set up between the two molecules are nearly as strong as in water itself; however, the attraction is somewhat weakened by the presence of the nonpolar hydrocarbon unit, CH_3CH_2—. Hydrocarbon substrates attract each other only weakly as evidenced by their low melting and boiling points. Three examples of the contrast in boiling points between compounds of different structure, but similar molecular weight, are summarized in Table 5.1. Clearly, the nonassociative species have much lower boiling points.

Table 5.1 Comparison of Boiling Point Data

Name	Formula	MW	bp(°C)
Ethanol	CH_3CH_2OH	46	78.3
Propane	$CH_3CH_2CH_3$	44	−42.2
Methyl acetate	CH_3COOCH_3	74	54
Diethyl ether	$(CH_3CH_2)_2O$	74	34.6
Ethylene	$CH_2{=}CH_2$	28	−102
Methylamine	CH_3NH_2	31	−6

If the solubilities in water of ethanol and a higher member of the alcohol series, octanol, are compared, octanol is found to have less than 2% solubility whereas ethanol dissolves completely. Why the difference? It lies in the fact that the *dominant* structural feature in octanol is the nonpolar hydrocarbon unit:

$$CH_3—CH_2—CH_2—CH_2—CH_2—CH_2—CH_2—CH_2—\overset{\delta-}{\underset{\delta+}{\ddot{O}}}: \qquad\qquad CH_3—CH_2—\ddot{O}—CH_2—CH_3$$
$$\qquad\qquad\qquad\qquad\qquad\qquad\qquad\qquad H$$

Octanol   Diethyl ether

The attraction of the polar hydroxyl group in the alcohol for the water is not sufficiently strong to overcome the *hydrophobic character* of the nonpolar hydrocarbon unit. On the other hand, since octanol has such a large nonpolar group as its dominant structural feature, we would expect enhanced solubility in nonpolar solvents. Such is the case, as is demonstrated in Experiment 13, where octanol is prepared and isolated by an extraction procedure using diethyl ether, a common, relatively nonpolar solvent. The octanol is found to be completely miscible with diethyl ether since the nonpolar characteristics predominate in both molecules. Experimentation has demonstrated that, in general, if a compound has both polar and nonpolar groups present in its structure, those having five or more carbon atoms in the hydrocarbon portion of the molecule will be more soluble in nonpolar solvents such as diethyl ether, pentane, or methylene chloride. In Figure 5.23 the solubilities of a number of straight-chain alcohols, carboxylic acids, and hydrocarbons in water are summarized. Those compounds with more than five carbon atoms have solubilities similar to those of the hydrocarbons.

Several general observations should be made in relation to the preceding discussion.

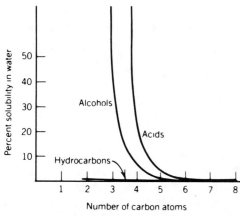

Fig. 5.23 *Solubility curve of acids, alcohols, and hydrocarbons. From Kamm, Oliver Qualitative Organic Analysis,* 2nd ed.; Wiley: New York, 1932. (Reprinted by permission of John Wiley & Sons, New York.)

1. Branched-chain compounds have greater water solubility than their straight-chain counterparts, as illustrated in Table 5.2 for a series of alcohols.
2. The presence of more than one polar unit in a compound increases solubility in water. For example, high-molecular-weight sugars like maltose, which contain multiple hydroxyl and/or acetal units, are water soluble and ether insoluble. On the other hand, cholesterol, having only one hydroxyl unit, is water insoluble and ether soluble.

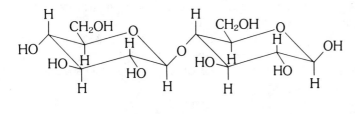

Maltose

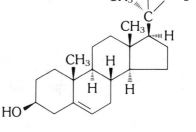

Cholesterol

Table 5.2 Water Solubility of Alcohols[a]

Name	Formula	Solubility (g/100 g H$_2$O)
Pentanol	$CH_3(CH_2)_3CH_2OH$	4.0
2-Pentanol	$CH_3(CH_2)_2CH(OH)CH_3$	4.9
2-Methyl-2-butanol	$(CH_3)_2C(OH)CH_2CH_3$	12.5

[a] Data at 20°C.

Table 5.3 Water Solubility of Amines[a]

Name	Formula	Solubility (g/100 g H₂O)
Ethylamine	$CH_3CH_2NH_2$	∞
Diethylamine	$(CH_3CH_2)_2NH$	∞
Trimethylamine	$(CH_3)_3N$	91
Triethylamine	$(CH_3CH_2)_3N$	14
Aniline	$C_6H_5—NH_2$	3.7
p-Phenylenediamine	$H_2N—C_6H_4—NH_2$	3.8

[a] Data at 25°C.

3. The presence of a chlorine atom, even though it lends some partial ionic character to the C—Cl bond, does not impart water solubility to a compound. In fact, such compounds as methylene chloride, chloroform, and carbon tetrachloride have long been used as extraction solvents. It should be noted that use of the latter two solvents has been curtailed, unless strict safety precautions are exercised, because of their carcinogenic nature.

4. Any functional group capable of forming a hydrogen bond with water, if it constitutes the dominant structural feature of the system, lends water-soluble characteristics to the substance (do not forget the five-carbon rule). For example, amine derivatives of the hydrocarbons (organic relatives of ammonia) would be expected to have water solubility. This is found to be the case. In Table 5.3 the water solubility data for a series of amines are summarized.

Partition Coefficient

A given substance, if placed in a mixture of two immiscible solvents, will distribute (partition) itself in a manner that is a function of its relative solubility in the two solvents. For example, a solute X will be distributed between two immiscible solvents according to the following equilibrium distribution.

$$X_{solvent\ 1} \rightleftharpoons X_{solvent\ 2}$$

Then

$$K_{equilibrium} = \frac{[X_{solvent\ 2}]}{[X_{solvent\ 1}]}$$

The *ratio* of the concentrations of the species in each solvent is a constant K_{eq} for a given system at a given temperature. This equilibrium constant expression is designated the *partition coefficient* (also referred to as the *distribution coefficient*). This coefficient is similar to the partitioning of a species that occurs in chromatographic separations. The procedure for determining the partition coefficient is demonstrated in the accompanying experiment (Experiment 4A) in which benzoic acid is partitioned between methylene chloride and water.

The basic equation used to express the coefficient K is

$$K = \frac{(g/100\ mL)_{organic\ layer}}{(g/100\ mL)_{water\ layer}}$$

This expression uses grams per 100 mL or grams per deciliter (g/dL), but grams per liter (g/L), parts per million (ppm), and molarity (M) are also valid. The partition coefficient is dimensionless so that any concentration units may be used, provided the units are the same for both phases. If equal volumes of both solvents are used, the equation reduces to the ratio of the weights of the given species in the two solvents.

$$K = \frac{g_{organic\ layer}}{g_{water\ layer}}$$

Determination of the partition coefficient for a particular compound in various immiscible solvent combinations can often give information valuable to the isolation and purification of the species using extraction techniques.

Let us now look at a typical calculation for the extraction of an organic compound "P" from an aqueous solution using diethyl ether. We will assume that the $K_{ether/water}$ value (*partition coefficient* of "P" between ether and water) is 3.5 at 20°C.

If an aqueous solution containing 100 mg of "P" in 300 μL of water is extracted at 20°C with 300 μL of ether, the following expression holds:

$$K_{ether/water} = \frac{C_e}{C_w} = \frac{W_e/300 \ \mu L}{W_w/300 \ \mu L}$$

where W_e = weight of "P" in the ether layer
W_w = weight of "P" in the water layer
C_e = concentration of "P" in the ether layer
C_w = concentration of "P" in the water layer

Since $W_w = 100 - W_e$, the preceding relationship can be written as

$$K_{ether/water} = \frac{W_e/300 \ \mu L}{(100 - W_e)/300 \ \mu L} = 3.5$$

Solving for the value of W_e one obtains 77.8 mg; the value for $W_w = 22.2$ mg. Thus, we see that after one extraction with 300 μL of ether, 77.8 mg of "P" (77.8% of the total) is removed by the ether and 22.2 mg (22.2% of the total) remains in the water layer.

The question often comes up whether it is preferable to make a single extraction with the total quantity of solvent available or to make multiple extractions with portions of the solvent. The second method is usually preferable in terms of efficiency of extraction. To illustrate, let us consider the following:

In relation to the foregoing example, let us now extract the 100 mg of "P" in 300 μL of water with **two** 150-μL portions of ether **instead of one** 300-μL portion as previously done.

For the first 150-μL extraction,

$$\frac{W_e/150 \ \mu L}{W_w/300 \ \mu L} = \frac{W_e/150 \ \mu L}{(100 - W_e)/300 \ \mu L}$$

Solving for the value of W_e, we obtain 63.6 mg. The amount of "P" remaining in the water layer (W_w) is then 36.4 mg. The aqueous solution is now extracted with the second portion of ether (150 μL). We then have

$$\frac{W_e/150 \ \mu L}{(36.4 - W_e)/300 \ \mu L} = 3.5$$

As before, by solving for W_e, we obtain 23.2 mg for the amount of "P" in the ether layer; $W_w = 13.2$ mg.

The two extractions, each of 150 μL of ether, removed a total of 63.6 mg + 23.2 mg = 86.8 mg of "P" (86.8% of the total). The "P" left in the water layer is then 100 - 86.8 or 13.2 mg (13.2% of the total).

Based on the preceding calculations, it can be seen that the multiple extraction technique is the more efficient. Whereas the single extraction removed 77.8% of "P", the double extraction increased this to 86.8%. To extend this relationship, multiple extractions with one-third of the total quantity of the ether solvent in **three** portions would be even more efficient. You might wish to calculate this extension to prove the point (See Question 5-17). Of course, there is a practical limit to the number of extractions that can be performed based on time and the degree of efficiency realized.

Extraction

The two major types of extraction utilized in the organic laboratory are the (1) solid–liquid and (2) liquid–liquid methods.

Solid–Liquid Extraction

A solid–liquid extraction is often used in studies of natural products. The extraction of usnic acid from its native lichen using acetone is described in Experiment 10. This approach is useful and simple, since only one main chemical species is soluble in the solvent. The extraction of caffeine from tea (Experiment 11) is accomplished by heating the tea in an aqueous solution of sodium carbonate. These methods work well because of the solubility relationships of the compounds involved.

It is also possible to carry out the extraction of a solid on a continuous basis at the microscale level. This approach depends on refluxing the solvent, followed by slow filtration of the condensate through the solid sample during its return to the reflux chamber. In this manner the desired species is concentrated in the refluxing solvent. This process, therefore, continuously reuses the same solvent to repeat the extraction over and over. The extractant solution is then concentrated to isolate the desired compound. A device for accomplishing such continuous extraction is shown in Figure 5.24. This apparatus, developed by Garner, consists of a small cold finger condenser inserted into a test tube. The test tube has indentations near the bottom to support a small funnel. The sample is carefully wrapped in filter paper and placed in the funnel.

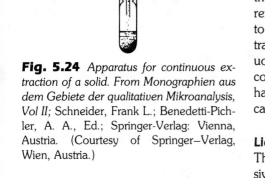

Fig. 5.24 *Apparatus for continuous extraction of a solid. From Monographien aus dem Gebiete der qualitativen Mikroanalysis, Vol II; Schneider, Frank L.; Benedetti-Pichler, A. A., Ed.; Springer-Verlag: Vienna, Austria. (Courtesy of Springer–Verlag, Wien, Austria.)*

Liquid–Liquid Extraction

The more common type of procedure, liquid–liquid extraction, is used extensively in the experiments presented in this text. It is a very powerful method for the separation and isolation of materials encountered at the microscale level.

In the majority of extractions, a capped centrifuge tube, a 10×75-mm test tube, or a conical vial is used as the container. In any extraction technique employed, it is essential that complete mixing of the two immiscible solvents be realized.

A typical extraction procedure is described in Experiment 46. Benzanilide, formed in an acidic aqueous solution, is separated from the product mixture by extraction with three 1.0-mL portions of methylene chloride solvent.

Note. *This wording is the accepted manner of indicating that three extractions are performed, each using 1.0 mL of methylene chloride.*

At the microscale level the extraction process consists of two parts: (1) mixing of the two immiscible solutions and (2) separation of the two layers after the mixing process.

1. *Mixing.* As outlined in Experiment 46, the methylene chloride solvent (1.0 mL) is added to the 5.0-mL conical vial containing the aqueous phase and product (~1.5 mL). The procedure is outlined in the following steps.
  a. The vial is capped.
  b. The vial is shaken gently to mix thoroughly the two phases.

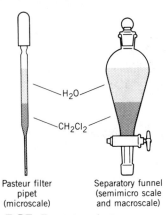

Fig. 5.25 *Extraction devices.*

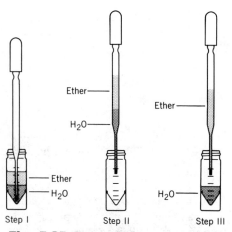

Fig. 5.26 *Pasteur filter pipet separation of two immiscible liquid phases, with the denser layer containing the product.*

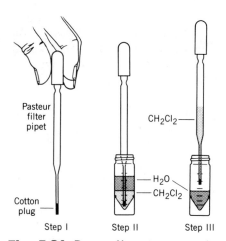

Fig. 5.27 *Pasteur filter pipet separation of two immiscible liquid phases, with the less dense layer containing the product.*

c. The vial is carefully vented by loosening the cap to release any pressure that may develop.

d. The vial is allowed to stand on a level surface to allow the two phases to separate. A sharp boundary should be evident.

Note. *Step 1, the mixing stage, may be carried out using a Vortex mixer. This alternative is mentioned frequently in the text.*

2. *Separation.* At the microscale level the two phases are separated with a Pasteur filter pipet (in some situations a Pasteur pipet can be used), which acts as a miniature separatory funnel. The separation of the phases is shown in Figure 5.25.

A major difference between macro and micro techniques is that at the microscale level the mixing and separation are done in two steps, whereas at the macroscale level with the separatory funnel, the mixing and separation are both done in the funnel in one step. It is important to note that the separatory funnel is an effective device for extractions at the semimicroscale or macroscale level, but at the microscale level it is not practical to use because the volumes are so small. The recommended procedures are diagrammed in Figures 5.26 and 5.27. Continuing the example, it is known that benzanilide is more soluble in methylene chloride than in water. Multiple extractions are performed to ensure complete removal of the benzanilide from the aqueous phase. The methylene chloride solution is the lower layer since it is heavier than water. The following steps outline the general method (refer to Fig. 5.26).

1. Squeeze the pipet bulb to force air from the pipet.
2. Insert the pipet into the vial until close to the bottom. Be sure to hold the pipet in a vertical position.
3. *Carefully* allow the bulb to expand, drawing only the lower methylene chloride layer into the pipet. This is done in a smooth, steady manner so as not to disturb the boundary between the layers. With practice, one can judge the amount that the bulb must be squeezed to just separate the layers.
4. Holding the pipet in a vertical position, place it over an empty vial and gently squeeze the bulb to transfer the methylene chloride solution into the vial. A second extraction can now be performed after addition of another portion of methylene chloride to the original vial. The identical procedure is repeated. In this manner multiple extractions can be performed, with each methylene chloride extract being transferred to the same vial; that is, the extracts are combined. The reaction product has now been transferred from the aqueous to the methylene chloride layer and the phases have been separated.

In a diethyl ether–water extraction, the ether layer is less dense and thus is the top phase. An example of this type of extraction is presented in Experiment 48D. Two extractions of the aqueous phase (~2.0 mL) using 0.5-mL portions of diethyl ether are carried out. The reaction product dissolves in the ether layer and is thus separated from by-products and other impurities. The procedure followed to separate the water–ether phases is identical to that outlined earlier, except that it is the top layer that is transferred to the new container. The steps are shown in Figure 5.27.

1. Draw *both* phases into the pipet as outlined before in steps 1 and 2. Try not to allow air to be sucked into the pipet, since this will tend to mix the phases in the pipet. If mixing does occur, allow time for the boundary to reform.
2. *Return* the *bottom aqueous layer* to the *original container* by gently squeezing the pipet bulb.
3. Transfer the separated ether layer to a new vial.

Separatory Funnel Extraction

As previously mentioned, the separatory funnel (Fig. 5.25) is an effective device for extractions carried out at the semimicroscale and macroscale levels. With the use of this funnel, the mixing and separation are done in the funnel itself in one step. Several of the Optional Scaleup procedures employ this extraction technique (see Experiments 25A, 25B, 43B, 48A, 48B, and 52).

The solution to be extracted is added to the funnel, making sure that the stopcock is in the closed position. The funnel is generally supported in an iron ring attached to a ring stand.

The proper amount of extracting solvent is now added (about one-third of the volume of the solution to be extracted is a good rule) and the stopper placed on the funnel.

Note. *The size of the funnel should be such that the total volume of solution is less than three-fourths the total volume of the funnel. The stopcock and stopper must be lightly greased to prevent sticking, leaking, or freezing. If Teflon stoppers and stopcocks are used, greasing is not necessary since they are self-lubricating.*

The funnel is removed from the ring stand, the stopper rested against the index finger of one hand, and the funnel held in the other with the fingers positioned so as to operate the stopcock (Fig. 5.28a). The funnel is carefully inverted, the liquid is allowed to drain away from the stopcock, and then the stopcock is opened slowly to release any built-up pressure (Fig. 5.28b).

Note. *Make sure the stem of the funnel is up and that it does not point at anyone.*

The stopcock is closed, the funnel again shaken for several seconds, the funnel positioned for venting, and the stopcock opened to release built-up pressure. This process is repeated several more times. After the final sequence, the stopcock is closed and the funnel returned upright to the iron ring.

The layers are allowed to separate, the stopper is removed, the stopcock is opened gradually, and the bottom layer is drained into a suitable container; the upper layer is removed by pouring from the top of the funnel.

When aqueous solutions are extracted with *a less dense solvent*, such as ether, the bottom aqueous layer is drained *into the original container* from which

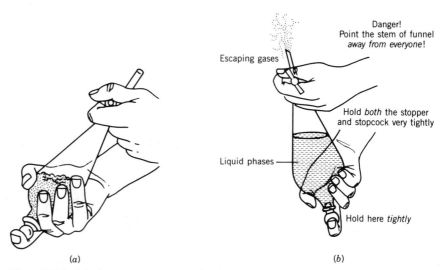

Fig. 5.28 *(a) Correct position for holding a separatory funnel while shaking. (b) Correct method for venting a separatory funnel.*

it was poured into the funnel. Once the top ether layer is removed from the funnel, the aqueous layer can then be returned for further extraction. Losses can be minimized by rinsing this original container with a small portion of the extracting solvent, which is then added to the funnel.

When the extracting solvent is *denser* than the aqueous phase, for example, methylene chloride, the aqueous phase is the top layer and is therefore retained in the funnel for subsequent extractions.

Drying of the Wet Organic Layer

It is important to realize that the organic extracts just separated (methylene chloride or diethyl ether) are wet. That is, water has some slight solubility in these solvents, and they are now saturated or "wet." Before evaporation of the solvent or before further purification, the extracts are dried to remove any residual water. This is achieved conveniently with an inorganic anhydrous salt such as magnesium, sodium, or calcium sulfate. These materials readily form insoluble hydrates, thus removing the water from the wet organic phase.

$$\text{Drying agent} + n\text{H}_2\text{O} \rightleftarrows \text{drying agent} \cdot (\text{H}_2\text{O})_n$$

Drying Agent	n
Na_2SO_4	10
$MgSO_4$	7
$CaCl_2$	6
$CaSO_4$	$\frac{1}{2}$

This table summarizes several of the more common neutral drying agents used in the organic laboratory. In addition silica gel beads and molecular sieves are now beginning to move out of the research laboratories and into the instructional laboratories. There are two basic requirements for an effective solid drying agent: (1) it should not react with the organic material in the system, and (2) it should be easily and completely separable from the dried liquid phase. The amount of drying agent used depends on the amount of water present and on the capacity of the solid desiccant to absorb water. If the solution is wet, the first amount of drying agent will clump. In this case, additional drying agent is added until the agent appears mobile on swirling the container. This amount should be sufficient. Swirling the contents of the container (by hand or by using a magnetic stirring apparatus) increases the rate of drying, because it aids in establishment of the equilibrium for hydration.

The drying agent may be added directly to the vial containing the organic extract (Experiment 4C, 48D) or the extract may be passed through a Pasteur filter pipet packed with the drying agent (Experiment 48B). A funnel fitted with a cotton, glass wool, or polyester plug to hold the drying agent may also be used (Experiment 47A).

Separation of Acids and Bases

The separation of organic acids and bases constitutes another important and extensive use of the extraction method. An organic acid *reacts* with dilute aqueous sodium hydroxide solution to form a salt. The salt, having an ionic charge, dissolves in the more polar aqueous phase.

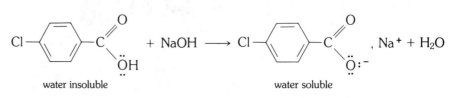

The reaction reverses the solubility characteristics of a water-insoluble acid. The water phase may then be extracted with an immiscible organic solvent to remove any impurities, leaving the acid salt in the water phase. Neutralization of the soluble salt with hydrochloric acid causes precipitation of the insoluble organic acid in a relatively pure state. An example of this sequence is demonstrated in Experiments 4C, 7A, 7B, 30, 43A, and 43B.

In a similar fashion, organic bases such as amines can be rendered completely water soluble by treatment with dilute hydrochloric acid to form hydrochloride salts.

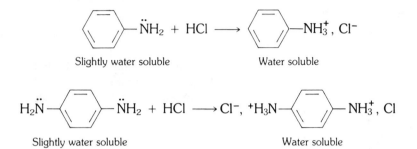

This technique is used to advantage in Experiments 4C, 26A, and 26B.

Experiment 4C demonstrates the extraction procedure used in the separation of a mixture of solids. In this example, the mixture is made up of an aromatic organic acid (Ar—COOH), base (Ar—NH₂), and neutral compound (Ar—H). A flow chart is given below that diagrams the sequence.

Separation of a Mixture by Extraction

(Ar—COOH, Ar—NH₂, Ar—H)

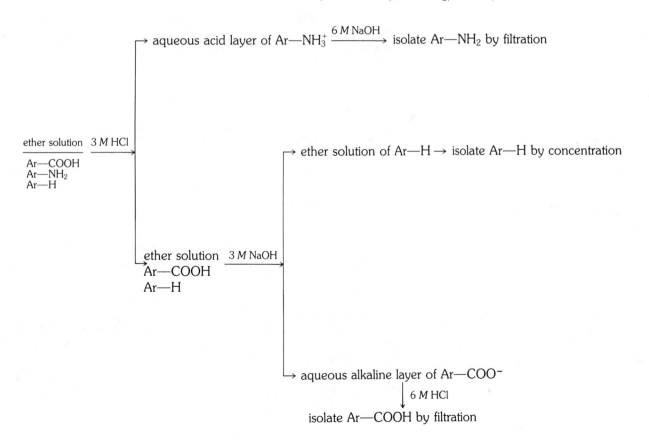

In the preceding example, the organic acid and base are solid compounds. If either is a liquid, an additional extraction of the final acidic aqueous or alkaline solution with ether, followed by drying and concentration, would be required to isolate the acid or base component.

Salting Out

It was emphasized under **Solubility** that organic compounds with fewer than five carbon atoms and also those with one or more polar groups tend to be water soluble. In these cases, the partition coefficient for the species would generally be near or lower than one. Transfer of this type of compound into an organic phase by extraction from water is then quite difficult. A technique often used to overcome this problem is the addition of an inorganic salt, such as sodium chloride, to the water phase. As a result of this addition, the attraction of the polar water molecules for the inorganic ions is much stronger than that for the organic species. The organic compound is then free to be extracted into an organic solvent. This approach is often referred to as *salting out*.

Experiment 4A, 4B   Solvent Extraction

PART A: Determination of a Partition Coefficient: Benzoic Acid—Methylene Chloride and Water

This experimental exercise illustrates the procedure that is used to determine a partition coefficient. Weighing small quantities on an electronic balance, the use of automatic delivery pipets, the transfer of solutions with the Pasteur pipet, and the use of a Vortex mixer (if available) are techniques encountered in this experiment.

Physical Properties of Reactants and Products: Parts A and B

Compound	MW	Wt/Vol		mmol	mp(°C)	bp(°C)	Density
Benzoic acid	122.13	50	mg	0.41	122		
Methylene chloride		2.4 mL				40	1.33
Water		600	μL			100	1.00
Sodium bicarbonate (10% solution)		600	μL				

To a 3.0-mL conical vial fitted with a screw cap, add 50 mg (0.41 mmol) of benzoic acid. There is then added 600 μL of methylene chloride followed by 600 μL of water.

These solvents are delivered by aid of automatic delivery pipets. The methylene chloride is dispensed in the hood.

The mixture is shaken (or a Vortex mixer is used) until the benzoic acid dissolves. The two layers are then allowed to separate after being thoroughly mixed (remember to vent the vial).

Draw the lower layer into a Pasteur filter pipet, carefully transfer the methylene chloride solution to a vial containing 100 mg of anhydrous, granular sodium sulfate, and then cap the vial.

If the methylene chloride layer is not transferred totally in the first operation, perform a second transfer.

After drying the methylene chloride solution for a few minutes, transfer it to a previously **tared** vial using a Pasteur filter pipet. The sodium sulfate is rinsed with an additional 600 μL of methylene chloride and the rinse combined with the solution in the tared vial. The solvent is now evaporated under a gentle

HOOD stream of nitrogen gas or in a sand bath in the **hood**.

If a sand bath is used, a boiling stone is placed in the vial before it is tared.

Weigh the vial and determine the weight of benzoic acid in the methylene chloride layer and in the water layer.

For the methylene chloride layer, the weight of the benzoic acid and vial minus the weight of the vial equals the weight of the benzoic acid.

For the water layer, the original weight of benzoic acid minus the amount of benzoic acid in the methylene chloride layer equals the weight of the benzoic acid.

Since equal volumes of both solvents were used, the partition coefficient may be determined from the ratio of the weight of benzoic acid in the methylene chloride solvent to the weight of benzoic acid in the water layer.

Calculate the partition coefficient for benzoic acid in the solvent pair (equal volumes) used in this exercise.

PART B: Benzoic Acid—Methylene Chloride and 10% Sodium Bicarbonate Solution: An Example of Acid–Base Extraction Techniques

Benzoic acid reacts readily with sodium bicarbonate to form sodium benzoate, carbon dioxide, and water. The sodium derivative has saltlike characteristics. It is very soluble in water and nearly insoluble in methylene chloride. **Repeat the identical procedure** carried out in **Part A** but replace the 600 μL of water used with 600 μL of 10% sodium bicarbonate solution. The efficiency of the basic aqueous extraction procedure can be measured by recovering any unreacted benzoic acid from the organic layer. Obtain the melting point of any recovered residue. Sodium benzoate has a melting point above 300°C, whereas benzoic acid melts near 122°C.

When a carboxylic acid is placed in a solution containing bicarbonate ion, bubbles of carbon dioxide are observed. This reaction may be used as a qualitative test for the presence of a carboxylic acid.

$$CH_3\overset{O}{\underset{||}{C}}\!\!-\!OH + HCO_3^- \rightleftharpoons CH_3\overset{O}{\underset{||}{C}}\!\!-\!O^- + H_2CO_3$$

$$H_2CO_3 \rightleftharpoons CO_2\uparrow + H_2O$$

Test for a Carboxylic Acid

Place 1 mL of 5% sodium or potassium bicarbonate on a small watch glass. Add the pure acid sample, one drop from a Pasteur pipet if the sample is a liquid (~5 mg if a solid), to the bicarbonate solution. Evolution of bubbles of carbon dioxide indicates the presence of an acid.

Perform the preceding test for carboxylic acids on several organic acids such as acetic, benzoic, propanoic, and chloroacetic.

PART C: A Three-Component Mixture: An Example of Separation of an Acid, a Base, and a Neutral Substance by Solvent Extraction

This experiment illustrates a further example of the solvent extraction technique as it is used in the organic laboratory to separate organic acids and bases. As outlined in the discussion section, the solubility characteristics of these important organic compounds in water are dependent on the pH of the solution. The extraction procedure employed for the separation of a mixture of an acid, a base, and a neutral substance takes advantage of this fact.

The components of the mixture to be separated in this experiment are benzoic acid, ethyl 4-aminobenzoate (a base), and 9-fluorenone (a neutral compound prepared in Experiment 40A).

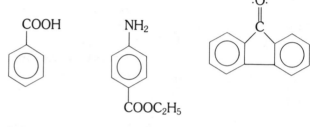

| Benzoic acid | Ethyl 4-aminobenzoate | 9-Fluorenone |

In carrying out the separation, you should keep a record or flow chart of your procedure (see discussion: separation of acids and bases) in the laboratory notebook and carefully label all flasks.

The estimated time of the reaction is 1.5 hours.

Physical Properties of Reactants and Products

Compound	MW	Wt/Vol	mmol	mp(°C)	bp(°C)	Density
Benzoic acid	122.13	50 mg	0.41	122		
Ethyl 4-aminobenzoate	165.19	50 mg	0.31	89		
9-Fluorenone	180.22	50 mg	0.27	154		
Diethyl ether		5 mL			40	1.33
3 M HCl		4 mL				
3 M NaOH		4 mL				
6 M HCl		—				
6 M NaOH		—				

In a stoppered or capped 15-mL centrifuge tube containing 4 mL of diethyl ether are added 50 mg (0.41 mmol) of benzoic acid, 50 mg (0.31 mmol) of ethyl 4-aminobenzoate, and 50 mg (0.27 mmol) of 9-fluorenone. Dissolution of the solids is accomplished by stirring with a glass rod or mixing on a Vortex mixer.

HOOD   *The diethyl ether is measured using a 10-mL graduated cylinder. It is dispensed in the* **hood**.

Separation of the Basic Component

Using a calibrated Pasteur pipet, 2 mL of 3 M hydrochloric acid are added to the centrifuge tube while cooled in an ice bath, and the resulting two-phase system mixed throughly for several minutes (a Vortex mixer is excellent for this purpose). After the layers have separated, the bottom aqueous layer is removed, using a Pasteur filter pipet, and transferred to a **labeled**, 10-mL Erlenmeyer flask.

Note. *A small amount of crystalline material may form at the interface between the layers. The second extraction dissolves this material.*

This step is now repeated with an additional 2 mL of the 3 *M* acid solution and the aqueous layer again transferred to the same Erlenmeyer flask. This flask is now stoppered (capped) and set aside.

Isolation of the Ethyl 4-Aminobenzoate

To the acidic aqueous solution, separated and set aside, add 6 *M* NaOH dropwise until the solution is distinctly alkaline to litmus paper. Cool the flask in an ice bath for about 10–15 minutes. Collect the solid precipitate by reduced-pressure filtration using a Hirsch funnel. Wash the precipitate with two 1-mL portions of distilled water. Dry the material on a clay plate, on filter paper, or in a vacuum drying oven. Weigh your product and calculate the percentage recovery. Obtain a melting point of the dried material and compare your result with the literature value. It is of interest to note that this material is used as a topical anesthetic.

Separation of the Acidic Component

To the remaining ether solution now add 2 mL of 3 *M* NaOH. The system is mixed as before and the aqueous layer is separated and transferred to a **labeled**, 10-mL Erlenmeyer flask.

This step is repeated and the aqueous layer again removed and transferred to the same Erlenmeyer flask. This flask is stoppered and set aside.

Separation of the Neutral Component

After washing with two 1-mL portions of distilled water, to the remaining wet, ether solution in the centrifuge tube add about 250 mg of anhydrous sodium sulfate. Set this mixture aside while working up the other extraction solution. This will allow sufficient time for the solution to dry. If the drying agent clumps, add additional sodium sulfate.

Isolation of the Benzoic Acid

To the aqueous alkaline solution, separated and set aside, add 6 *M* HCl dropwise until the solution is distinctly acidic to litmus paper. Cool the flask in an ice bath for about 10 minutes. Collect the precipitated benzoic acid by reduced-pressure filtration using a Hirsch funnel. Wash the precipitate with two 1-mL portions of distilled water. Dry the product using one of the techniques described earlier for the ethyl 4-aminobenzoate. Weigh the benzoic acid and calculate your percentage recovery. Obtain the melting point of the dried material and compare your result with the literature value. The qualitative test for organic carboxylic acids, given in the discussion section, may also be performed.

Isolation of the 9-Fluorenone

Transfer the dried ether solution by use of a Pasteur filter pipet to a tared 10-mL Erlenmeyer flask containing a boiling stone. Rinse the drying agent with an additional 1 mL of ether and also transfer this rinse to the same Erlenmeyer flask.

HOOD   Concentrate the solution on a warm sand bath in the **hood** using a **slow** stream of nitrogen gas. Obtain the weight of the isolated 9-fluorenone and calculate the percentage recovery. Obtain a melting point of the material and compare your result with the literature value.

ENVIRONMENTAL DATA

Substance	Amount	TLV (mg/m^3)	Emissions (mg)	Volume (m^3)
Methylene chloride (Max., extract, evap.)	2.4 mL	350	3192	9.1
Diethyl ether	5.0 mL	1200	3500	3

TLV: Hydrochloric acid—5 ppm
      Sodium hydroxide—2 $\mu g/m^3$
TXDS: Benzoic acid—orl-rat LD50: 2530 mg/kg
      Ethyl 4-aminobenzoate—unk-mus LD50: 261 mg/kg
      9-fluorenone—scu-rat TDLo: 300 mg/kg/26W-1
      Sodium sulfate—orl-mus LD50: 5989 mg/kg
      Sodium bicarbonate—orl-rat LD50: 4220 mg/kg

QUESTIONS

5-12. Explain why diethyl ether would be expected to be a satisfactory solvent for the straight-chain hydrocarbons hexane and heptane.

5-13. The solubility of p-dibromobenzene, BrC_6H_4Br, in benzene is 80 $\mu g/100$ μL at 25°C. Would you predict the solubility of the dibromo compound to increase, decrease, or remain the same in 100 μL of acetone solvent at this temperature? Explain.

5-14. Each of the following solvents is used in the experiments in this text to extract organic compounds from aqueous solutions.
   a. Methylene chloride
   b. Pentane
   c. Toluene
   d. Diethyl ether
Will the organic phase be the upper or lower layer when each of these organic solvents is mixed with water? Explain your answer for each situation.

5-15. A 36-mg sample of an organic compound (MW 84) is dissolved in 10 mL of water. This aqueous solution is extracted with 5.0 mL of hexane. Separation and analysis of the aqueous phase show that it now contains 12 mg of the organic compound. Calculate the partition coefficient for the compound.

5-16. A qualitative test often used to determine whether an organic compound contains oxygen is to test its solubility in concentrated sulfuric acid. Almost all oxygen-containing compounds are soluble in this acid. Explain.

5-17. In the discussion section on multiple extractions, it was suggested that you might extend the relationship in the example given to the next step by using one-third of the total quantity of the ether solvent in **three** portions. The reason for doing this is to determine whether this would increase the efficiency of the process even further. To determine whether this next step is worth the effort, perform the calculations for the extraction of 100 mg of "P" in 300 μL of water with **three** 100-μL portions of ether. Assume that the partition coefficient is 3.5 as before.

Compare the amounts of "P" extracted from the water layer using one, two, or three extractions.

Do you think that the additional amount of "P" extracted from the water layer using three extractions justifies the added effort?

TECHNIQUE 5: INTRODUCTION TO INFRARED AND NUCLEAR MAGNETIC RESONANCE SPECTROSCOPY

INFRARED SPECTROSCOPY

The infrared spectrometer is the most complex and expensive instrument that you will encounter on a regular basis in the microscale laboratory. TREAT THE IR SPECTROMETER WITH RESPECT. You are being introduced to this sophisticated piece of equipment early in the first semester because it is particularly adapted to the characterization of microscale products. Data obtained from the

IR spectrometer will be used many times throughout the year. You will become more proficient at obtaining this type of spectral information as you gain experience with the instrument. Practice in preparing the sample for instrumental analysis can significantly improve the quality of the spectra.

The late Robert B. Woodward, Nobel Laureate, one of the most outstanding synthetic organic chemists of this century, once stated: "But no single tool has had more dramatic impact upon organic chemistry than infrared measurements. The development [of easily] operated machines for the determination of infrared spectra has permitted a degree of immediate and continuous analytical and structural control in synthetic organic work which was literally unimaginable. . . . The power of the method grows with each day, and further progress may be expected for a long time to come. Nonetheless, its potentialities are even now greater than many realize."[3]

That statement was made some time ago, and although the dramatic impact has shifted to high-field NMR and high-resolution mass spectrometry for structural elucidation, infrared data and particularly gas chromatographic coupled infrared interferometers continue to play a dominant role in compound identification and characterization. The infrared spectrum is a reflection of the vibrational energy levels present in a molecule. As no two substances have the same set of vibrational frequencies, no two substances have identical infrared spectra. There are at present a number of excellent collections of infrared spectra that may be used for reference comparisons.[4]

The use of infrared spectra to identify the presence of particular functional groups in a molecule is one of the principal applications of this technique. Many of the absorption bands present in an infrared spectrum can be related to specific arrays of atoms in that material. These "group frequencies" can be extremely helpful in the interpretation of experimental results (see Experiment 5A).

An introduction to the theory of this effect is given in Appendix B and the interpretation of infrared spectroscopic data is developed in some detail in Chapter 7. Fifteen of the experiments (5A,B; 6; 7A; 8B,C,D; 11B; 14; 16B; 21C; 22; 36; 46; and 52) contain interpretations of the associated infrared spectra as a further guide to the application of this information.

Instrumentation

The workhorse infrared instrument used for routine characterization of materials in the undergraduate organic laboratory is the optical null double-beam grating spectrometer (Fig. 5.29). Although the large majority of undergraduate laboratories utilize this type of instrumentation, the winds of change are blowing. Over the last decade the development of interferometric nondispersive spectrometers, which depend on high-speed computer manipulation of interferograms to generate the data in spectral form, have invaded the research laboratory. It is now clear that "low-cost" infrared interferometers will become the next generation's infrared instructional instrumentation. The large majority of the spectra utilized in the interpretive discussions associated with a number of the experiments (see the list just given) were generated on a prototype of this kind of infrared instrumentation, the Perkin–Elmer model 1600. This instrument will acquire 16 scans and carry out the required calculations in 42 seconds. While the spectrum is being printed out (~40 seconds) the data on a second sample are being acquired. The 42-second acquisition data are significantly superior to those currently recorded

---

[3] Woodward, R. B. In *Perspectives in Organic Chemistry*; Todd, A., Ed.; Interscience; New York, 1956; p 157.

[4] For example, see References, p. 91.

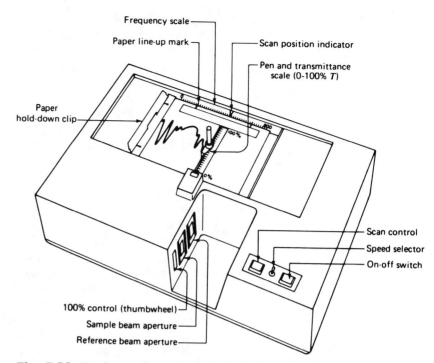

Fig. 5.29 *The Perkin–Elmer Model 710B IR. From* Zubrick, James W. *The Organic Chem Lab Survival Manual,* 2nd ed.; Wiley: New York, 1988. (Reprinted by permission of John Wiley & Sons, New York.)

by dispersive instruments which take from 5 to 8 minutes to scan a sample from 4000 to 600 cm^{-1}.

As the dispersive spectrometer, however, is still pervasive in the instructional laboratory we will describe how that system (Fig. 5.30) operates in some detail (but this is likely to be the last edition in which it will appear). The double-beam optical null grating infrared spectrophotometer operates as follows. A source of infrared radiation is generated by a hot wire. The source output is divided into the two parallel and converging beams by the two spherical mirrors, m_1 and m_2. (Mirrors are used throughout infrared instruments instead of lenses, as glass and quartz absorb in this region of the spectrum.)

The beam reflected from m_1 is the sample beam, and the beam from m_2 is the reference beam. The two beams then pass through the sample compartment of the instrument, with the beam from m_1 also passing through the sample to be analyzed. Most substances to be examined are located near the convergence point of the sample beam. At this position the instrument requires the minimum quantity of sample. The transmitted beam undergoes a certain amount of instrument distortion, but generally follows the Beer–Lambert absorption law:

$$I/I_0 = e^{-acl}$$

where I = intensity of the transmitted beam
I_0 = intensity of incident beam
a = apparent absorption coefficient
c = sample concentration, mol/L
l = path length (sample thickness)

The two beams are then combined at the chopper C. The chopper is a rotating sector mirror (either half or quarter), which alternately passes the sample

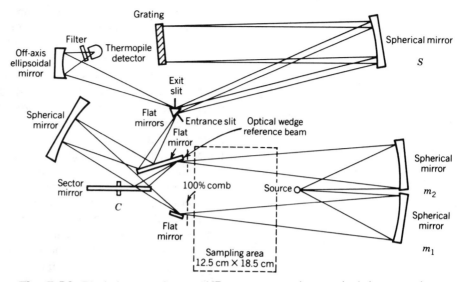

Fig. 5.30 *Block diagram of a typical IR spectrometer, showing the light source, beam chopper, diffraction grating, and attenuator wedge.* (Courtesy of Perkin–Elmer, Norwalk, CT.)

beam or reflects the reference beam to the entrance slit. The combined beams then follow the same path, but alternate with time. The frequency of alternation equals the frequency of chopping for the half-sector chopper. The chopping frequency generally is in the range 10–15 Hz. The beam then passes through the entrance slit (the slit forms a long and very narrow vertical rectangular image) and into that part of the instrument called the monochromator (the term literally means "making single colors"). In this section of the instrument the beam of radiation is dispersed. Thus, this type of spectrometer is often referred to as a dispersive spectrometer. That is, the different frequencies of electromagnetic radiation are separated from each other (or at least partially separated) in space. The dispersing agent in all modern instruments is a reflection grating. The beam enters the monochromator with the narrow image of the slit and is collected on a parabolic mirror or large spherical mirror S, which collimates the beam (makes all the beam rays parallel, just as a searchlight mirror converts all the light directed outward into parallel rays).

The collimated beam then strikes the reflection grating and is dispersed by diffraction. For one angle setting of the grating, a particular wavelength of radiation will be diffracted back onto the collimating mirror and out the exit slit. The grating is mounted so that it can be rotated. As the grating angle changes with respect to the collimating mirror, the dispersed radiation is sequentially swept by the exit slit, one wavelength (or frequency) after another. After the dispersed radiation exits the monochromator, it is tightly focused by an ellipsoidal mirror onto a small thermocouple detector. The thermocouple is evacuated and has a blackened junction that heats up as the infrared radiation strikes the black detector target.

When no sample is in either beam, the intensity of radiation reaching the detector and the temperature of the detector remain constant as the two beams alternately strike the detector. When a substance that absorbs radiation at a particular frequency is placed in the sample beam, that beam will be less intense than the reference beam at the frequency where absorption is taking place. When the radiant energy decreases, the detector will cool down and change the voltage potential across the junction. (The chopping frequency is dictated by the ability of the detector to heat up and cool down, or to "track" the chopping frequency.) When the reference beam is switched back on (by chopper rotation), the detector temperature goes up again. The detector is connected to a "tuned" amplifier that will respond only to an alternating current identical in phase and frequency to the chopper rotation.

The amplifier drives a servomotor that is mechanically connected to the pen drive on the spectrum chart. The pen drive is also connected to an "optical wedge" that is simultaneously driven into the reference beam. The wedge reduces the intensity of the reference beam. The wedge is shaped so that the linear travel across the beam corresponds to the percentage transmission recorded by the pen. That is, if the optical wedge blocks 10% of the reference beam, the pen reads 90% transmission on the chart. The servomotor drives the pen as long as it detects a signal. When the optical wedge is driven into the reference beam to the point where the two beams are of equal intensity, the signal at the detector drops to zero and the pen stops. As the grating continues to turn and the absorption of the sample diminishes, a new signal develops at the detector. This signal will be of the opposite phase since the reference beam is now less intense than the sample beam. The signal is amplified and a phase-sensitive recorder motor drives the pen in the opposite direction. At the same time, the wedge is pulled out of the reference beam. Thus, the instrument has recorded an absorption band in the sample over that particular set of frequencies. The servo system functions to drive the optical wedge to zero out or "null" the signal at the detector, hence the name *optical null spectrometer*.

The double-beam arrangement has some distinct advantages. For example, absorption signals from any substances present in both beams in equal quantities are automatically canceled. Since atmospheric water and carbon dioxide have strong absorption bands in the infrared, the double-beam instrument automatically subtracts the absorption of these substances from the recorded spectrum.

Sample Handling in the Infrared

For a spectrum to be obtained in the infrared region, the sample must be mounted in a cell that is transparent to the radiation. Glass and quartz absorb in this region and cells constructed of these materials cannot be employed. Alkali metal halides have large regions of transmission in the infrared as do silver halides. Of these materials, sodium chloride, potassium bromide, and silver chloride are most often used as cell windows in infrared sampling.

Liquid Samples

For materials boiling above 100°C, the procedure is very simple. Using a syringe or Pasteur pipet, place 3–5 μL of sample on a polished plate of NaCl or AgCl. Then cover it with a second plate of the same material and clamp it in a holder that can be mounted vertically in the instrument. *Be sure that the plates are clean when you start and when you are through!* Obviously the sodium chloride cannot be cleaned with water. Silver chloride is very soft and scratches easily; it also must be kept in the dark when not in use because it darkens quickly in direct light. Spectra obtained in this fashion are referred to as capillary film spectra (see Fig. 5.31).

Solution Spectra and the Spectra of Materials
Boiling Below 100°C

These samples generally require a sealed cell constructed of either NaCl or KBr. Such cells are expensive and need careful handling and maintenance. They are assembled as shown in Fig. 5.32. Only in rare cases do MOL experiments require the use of sealed infrared cells.

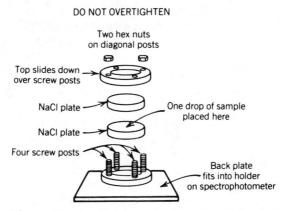

DO NOT OVERTIGHTEN

Two hex nuts on diagonal posts

Top slides down over screw posts

NaCl plate — One drop of sample placed here

NaCl plate

Four screw posts

Back plate fits into holder on spectrophotometer

Fig. 5.31 *IR salt plates and holder. From* Zubrick, James W. *The Organic Chem Lab Survival Manual,* 2nd ed.; Wiley: New York, 1988. (Reprinted by permission of John Wiley & Sons, New York.)

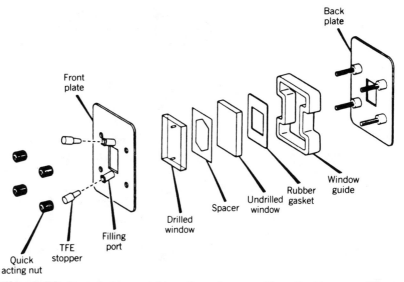

Back plate

Front plate

Quick acting nut   TFE stopper   Filling port   Drilled window   Spacer   Undrilled window   Rubber gasket   Window guide

Fig. 5.32 *Sealed demountable cell or demountable cell with ports.* (Courtesy of Perkin–Elmer, Norwalk, CT.)

Solid Samples

Solid powders can be mounted on horizontal NaCl plates, and the beam diverted through the sample by mirrors. This would make sample preparation very easy for solids. Unfortunately, powders tend to scatter the entering radiation very efficiently by reflection, refraction, and molecular scattering. Some of these effects become rapidly magnified at higher frequencies since they vary as the fourth power of the frequency. The result of solid-sample scattering is that large amounts of energy are removed from the sample beam. This results in very poor absorption spectra, as the instrument is forced to operate at very low energies. Remember that if there is a large reduction in energy in the sample beam, the optical wedge is driven a large distance into the reference beam. This results in very small amounts of energy reaching the detector from either beam. The detector cannot differentiate between a drop in energy from absorption and one from scattering.

GENTLY For materials melting below 80°C the simplest technique, however, is to mount the sample between two salt plates and *gently* apply a heat lamp until

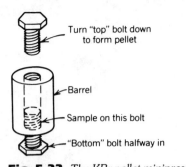

Turn "top" bolt down to form pellet

Barrel

Sample on this bolt

"Bottom" bolt halfway in

Fig. 5.33 *The KBr pellet minipress. From Zubrick, James W. The Organic Chem Lab Survival Manual, 2nd ed.; Wiley: New York, 1988. (Reprinted by permission of John Wiley & Sons, New York.)*

melting occurs. With the fast acquisition times of interferometers, the melting point range is now close to 100°C. For substances melting above 100°C the sampling routine most often employed to avoid these scattering problems is the potassium bromide disk. The sample (2–3 mg) is finely ground in a mortar, the finer the better (lower reflection or refraction losses). Then 150 mg of previously ground and dried KBr is added to the mortar and quickly mixed by stirring, *not* grinding, it with the sample. (KBr is very hygroscopic and will rapidly pick up water while being ground in an open mortar.) When mixing is complete the mixture is transferred to a die and pressed into a solid disk. Potassium bromide will flow under high pressure and seal the solid sample in the alkali metal halide matrix. Potassium bromide is transparent to infrared radiation in the region of interest. Most important, however, the KBr makes a much better match of the refractive indices between the sample and its matrix than does air. Thus, reflection and refraction effects at the crystal faces of the sample are greatly suppressed. Several styles of dies are commercially available. For routine student use a die consisting of two stainless-steel bolts and a barrel is the simplest to operate. The ends of the bolts are polished flat to form the die faces. The first bolt is seated to within a turn or two of the head. Then the sample mixture is added (avoid breathing over the die while adding the sample). The second bolt is firmly seated in the barrel, and then the clamped assembly is tightened by a torque wrench to 240 ft-lb. After standing for 1.5 minutes, the two bolts are removed, leaving the KBr disk mounted in the center of the barrel, which can then be mounted in the instrument. After the spectrum of the sample is run, the disk can be retrieved and the sample recovered if necessary (Fig. 5.33). *Always clean the die immediately after use. KBr is highly corrosive to steel.*

For a discussion of an alternate matrix material for use with solid samples, *see* Chapter 7, Question 7-27.

The standard techniques of sample preparation employed to obtain infrared spectra of microscale laboratory products are the use of capillary films with liquids on NaCl or AgCl plates and the use of KBr disks and melts in solids.

When infrared spectra are obtained, it is important to establish that the wavenumber values have been accurately recorded. Successful interpretation of the data often depends on very small shifts in these values. Calibration of the frequency scale is usually accomplished by obtaining the spectrum of a reference compound such as polystyrene film. To save time, record absorption peaks only in the region of particular interest. As these instruments are quite stable, determination of the calibration curve once a day is usually a satisfactory procedure.

NUCLEAR MAGNETIC RESONANCE

The NMR spectrometer is one of the most powerful experimental tools available to the modern organic chemist and is particularly useful in the discrimination and identification of diastereomers. Experiment 5B presents the option for students to obtain and interpret NMR data if the local opportunity exists. The two diastereomeric alcohols exhibit different splitting patterns for the proton on the carbon bearing the —OH group. Integration of these signals allows one to determine the ratio of the diastereomeric alcohols in a sample.

The theory of the effect and the basic instrument design are briefly introduced in Appendix C. Interpretation of the data is presented in Chapter 7.

NMR Sampling

NMR sample preparation is fairly routine. Most samples are measured in solution in thin-walled tubes approximately 5 mm in diameter and 18–20 cm long. The sample size compatible with CW spectrometers is in the range 30–50 mg dissolved in 0.3–0.5 mL of solvent. Fourier transform spectrometers require only 2–3 mg of sample in the same volume of solvent. The most practical solvent is

HOOD deuteriochloroform (DCCl$_3$). Handle this solvent with care, in the **hood**, as it is
TOXIC **toxic!** A number of other deuterated solvents are commercially available. The
universally accepted internal reference compound employed in making these
measurements is tetramethylsilane (TMS). The most convenient source of TMS
is deuteriochloroform, which contains about 1% TMS (0.03% solutions are most
appropriate for FT spectrometers). Since the TMS is particularly volatile
(bp 26.5°C), the tube should be capped immediately after addition or it will
vaporize. The sample tube is spun in the instrument to average out small
changes in field strength over the sample volume. The tube is inserted into the
magnetic field and the spectrum recorded. In some cases, your laboratory instructor or teaching assistant will actually record the data for you. If you do have
the opportunity to work directly with your instructor, detailed information will be
provided for operating the instrument. Controls vary depending on make or
model of the instrument; therefore, we will not present this information in the
text.

REFERENCES
1. Sadtler Library. About 80,000 spectra of single compounds; about 12,000 spectra of
commercial products. Sadtler Research Labs., 3316 Spring Garden Street, Philadelphia, PA 19014.
2. D.M.S. System (Documentation of Molecular Spectra). About 15,000 spectra. IFI/
Plenum Data Corp., 227 West 17th Street, New York, NY 10011.
3. Japanese collection. About 17,000 spectra. Good quality, but labeled in Japanese.
Infrared Data Committee of Japan. Nankodo Co., Haruki-Cho, Tokyo.
4. Coblentz Society. 10,000 spectra. Marketed through Sadtler Research Labs., 3316
Spring Garden Street, Philadelphia, PA 19104.
5. A.P.I. collection (American Petroleum Institute). About 4500 spectra. M.C.A. collection (Manufacturing Chemists' Association). About 3000 spectra. Chemical Thermodynamics Property Center, Texas A&M College, Department of Chemistry, College
Station, TX 77843.
6. Aldrich Library of Infrared Spectra, Aldrich Chemical Co., Inc., 940 West Saint Paul
Avenue, Milwaukee, WI 53233. 3rd ed., 1981, 10,000 spectra arranged by chemical
type.
7. Grasselli, J. G. *Atlas of Spectral Data and Physical Constants for Organic Compounds*; CRC Press: Cleveland, OH, 1973.
8. Pouchert, C. J., Ed. *The Aldrich Library of NMR Spectra*; Aldrich Chemical Co.;
Milwaukee, WI, 1983; Vols. 1, 2.

Experiments 5A, 5B

Reduction of Ketones Using a Metal Hydride Reagent: Cyclohexanol; *cis-* and *trans*-4-*tert*-Butylcyclohexanol

(cyclohexanol; cyclohexanol, 4-*tert*-butyl-)

These reactions illustrate the reduction of a ketone carbonyl to the corresponding alcohol by use of sodium borohydride. The cis and trans isomers formed in
the reduction of the substituted cyclohexanone are separated by gas chromatography.

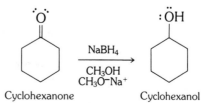

Cyclohexanone   Cyclohexanol

DISCUSSION   An important route for the synthesis of primary and secondary alcohols is the reduction of aldehydes and ketones, respectively. Reduction involves the addition of the elements of H—H across the carbonyl functional group.

There are numerous ways to accomplish the process, but the most common laboratory method involves the use of complex metal hydride reagents. The two commercially available reagents most often used are lithium aluminum hydride, $LiAlH_4$, and sodium borohydride, $NaBH_4$.

Lithium aluminum hydride is a very active reducing agent that reduces not only aldehydes and ketones but many other functional groups as well, such as esters, lactones, carboxylic acids and anhydrides, amides, alkyl halides, alkyl azides, alkyl isocyanates, and nitriles. Of great importance is the fact that it *must* be used in aprotic solvents such as diethyl ether or tetrahydrofuran. Lithium aluminum hydride reacts *violently* with water. It is not uncommon for the hydrogen gas generated in the reaction to ignite. *This reagent should not be employed unless specific instructions are available for its proper use.*

Sodium borohydride is a much more selective reducing reagent, and for this reason it is usually employed for the reduction of aldehydes and ketones. It does not react with the vast majority of common organic functional groups including $C=C$, $C\equiv C$, nitro, cyano, amide, and carboxylic acids and esters. Sodium borohydride reacts at an appreciable rate with water but slowly with aqueous alkaline solution or with methanol. For small-scale reactions an excess of the sodium borohydride reagent is generally used to compensate for the amount that reacts with the protic solvent (methanol). This technique is preferred to using a solvent in which the sodium borohydride is less soluble. On the other hand, sodium borohydride reacts rapidly with strong acid to generate hydrogen gas. This reaction can be used to advantage for the reduction of $C=C$ bonds (see Experiment 12). The relatively high cost of the metal hydride reducing agents is offset by their low molecular weight and the fact that one mole of reducing agent reduces four moles of aldehyde or ketone.

The key step in the reduction of a carbonyl group by sodium borohydride is the transfer of a hydride ion, $:H^-$, from boron to the carbon atom of the carbonyl unit. In the reaction the hydride ion is acting as a *nucleophile*.

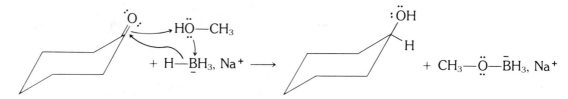

(The hydride may make an equatorial or axial attack depending on steric factors.)

The overall reduction process requires two hydrogen atoms, but only one comes from the borohydride reagent (attached to the carbon atom). The other hydrogen atom is derived from the protic solvent (methanol).

On reduction of an acyclic aldehyde or ketone, the attack of the hydride species occurs with equal probability from either side of the planar carbonyl unit.

In the 4-*tert*-butylcyclohexanone example, the steric environment is similar on each side of the carbonyl unit and thus the more stable equatorial alcohol (trans) is formed in larger amounts. This reduction occurs by *axial* attack of the hydride species on the carbonyl group, which is possible with the relatively small sodium borohydride (or lithium aluminum hydride) reagent. Reduction of this compound with sodium borohydride yields a mixture of cis and trans isomers of the corresponding alcohol. This type of reaction wherein more than one stereoisomeric product is formed, one of which predominates, is termed a *stereoselective* reaction.

Table 5.4 Reduction of 4-*t*-Butylcyclohexanone

Reagent	trans (%)	cis (%)
Sodium borohydride	80	20
Lithium aluminum hydride	92	8
Lithium tris(*sec*-butyl)borohydride	7	93

Highly substituted hydride reducing reagents, such as lithium tris(*sec*-butyl)borohydride, must make an equatorial attack, because of steric factors, and thus produce the less stable isomer (cis) in greater amounts.

Table 5.4 summarizes data relating to the stereochemistry of the reduction with metal hydride reagents.

PART A: Cyclohexanol

The reaction is shown in the preceding section

EXPERIMENTAL   Estimated time for the experiment: 1.5 hours.

Physical Properties of Reactants and Products

Compound	MW	Wt/Vol	mmol	bp(°C)	Density	n_D
Cyclohexanone	98.15	100μL	0.97	156	0.95	1.4507
Methanol		250 μL		65		
Sodium borohydride reducing solution		300 μL				
Cyclohexanol	100.16			161	0.96	1.4641

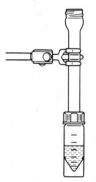

Cyclohexanone, 100 μL
+ CH₃OH, 250 μL +
NaBH₄ solution, 300 μL

HOOD

Reagents and Equipment

In a tared 5.0-mL conical vial equipped with an air condenser, place 100 μL (95 mg, 0.97 mmol) of cyclohexanone followed by 250 μL of methanol. (■) The vial is gently shaken to obtain a homogeneous solution.

Add 300 μL of sodium borohydride reducing solution dropwise, with swirling in the **hood**.

Note. *The reactants are dispensed using automatic delivery pipets. It is suggested that the cyclohexanone be weighed after delivery so as to get an accurate weight for the percentage yield calculations.*

The stock reducing solution should be prepared just prior to conducting the experiment.

Instructor Preparation

In a 10-mL Erlenmeyer flask is placed 50 mg of anhydrous sodium methoxide and 2.5 mL of methanol. To this solution is added 100 mg of sodium borohydride and the flask tightly stoppered and swirled gently to dissolve the solid phase (100 μL of this solution provides approximately 2.0 mg $NaOCH_3$ and 4.0 mg $NaBH_4$).

Note. *Test for the reducing solution: add 1–2 drops of the freshly prepared reducing solution to ~200 μL of concentrated hydrochloric acid. Generation of hydrogen gas bubbles is a positive test.*

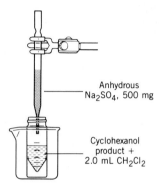

Anhydrous
Na₂SO₄, 500 mg

Cyclohexanol
product +
2.0 mL CH₂Cl₂

Reaction Conditions

The resulting solution is allowed to stand at room temperature for a period of 15 minutes.

Isolation of Product

Using a calibrated Pasteur pipet, 1.0 mL of cold dilute hydrochloric acid is added dropwise and the aqueous mixture extracted with three 0.5-mL portions of methylene chloride. On each addition of methylene chloride, the vial is capped, shaken gently, and then carefully vented by loosening the cap (a Vortex mixer may be used if available).

After separation of the layers, the bottom methylene chloride layer is removed using a Pasteur filter pipet. Transfer the organic phase to a Pasteur filter pipet containing 500 mg of anhydrous sodium sulfate.

The dried eluate is collected in a tared 5.0-mL conical vial containing a boiling stone. An additional 0.5 mL of methylene chloride is used to rinse the sodium sulfate.(■) *Additional rinsings of the sodium sulfate may be made if desired.* The methylene chloride is removed by careful evaporation in the hood by gentle warming in a sand bath. *Do not use a stream of nitrogen gas to hasten the evaporation. This results in loss of product.*

Purification and Characterization

The cyclohexanol product remaining after evaporation of the methylene chloride solvent is sufficiently pure for characterization.

Determine the weight of the liquid material and calculate the percentage yield. Determine the refractive index (optional), and boiling point of the cyclohexanol product and compare your results with the literature values.

Obtain the infrared spectrum of the crude (dry) reaction product by the capillary film sampling technique. Compare the spectrum of this material with that of the starting ketone (Fig. 5.34).

[It should be noted that the majority of the infrared spectra referred to in the experimental analysis sections are Fourier-transform-derived and have been plotted on a slightly different scale than the other spectra presented in the text. These spectra utilize a 12.5-cm^{-1}/mm format below 2000 cm^{-1} and undergo a 2:1 compression above 2000 cm^{-1} (25 cm^{-1}/mm).]

Infrared Analysis: A Comparison of Reactant and Product

The key absorption bands to examine in the spectrum of cyclohexanone occur at 3420, 3000–2850, 1715, and 1425 cm^{-1}. The lack of significant absorption between 3100–3000 and 1400–1350 cm^{-1} also should be noted. The sharp weak band at 3420 cm^{-1} is not a fundamental vibration (not O—H or N—H stretching), but arises from the first overtone of the very intense carbonyl stretching mode found at 1715 cm^{-1}. Note that the overtone does not fall exactly at double the frequency of the fundamental, but usually occurs slightly below that value because of anharmonic effects (see Appendix B). The lack of absorption in the region near 3100–3000 cm^{-1} and the presence of a series of very strong absorption bands at 3000–2850 cm^{-1} indicate that the only C—H stretching modes present are part of sp^3-type systems. Thus, the spectrum is typical of an aliphatic ketone. The occurrence of a band at 1425 cm^{-1} suggests the presence of at least one methylene group adjacent to the carbonyl group, whereas the 1450 cm^{-1} band requires other methylene groups more remote from the C=O group. The lack of absorption in the 1400–1375-cm^{-1} region indicates the absence of any methyl groups (a good indication of a simple aliphatic ring system) and that the absorption at 1450 cm^{-1} must arise entirely from methylene scissoring modes. The value of 1715 cm^{-1} for the C=O stretch supports the presence of a six-membered ring.

Now examine the spectrum of your reaction product (a typical example is given in Fig. 5.35). The spectrum is rather different from that of the starting material. The major differences are a new very strong broad band occurring between 3500–3100 cm^{-1} and a great drop in intensity of the band found at 1715 cm^{-1}. These changes indicate the

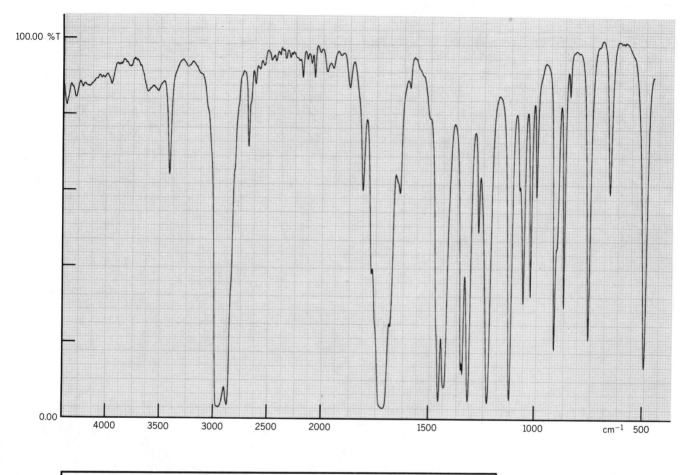

100.00 %T

0.00

4000   3500   3000   2500   2000   1500   1000   cm⁻¹ 500

Sample	Cyclohexanone

%T _X_ ABS — Background Scans _4_   Scans _____ 16 _____
Acquisition & Calculation Time _42 sec_   Resolution _4.0 cm⁻¹_
Sample Condition _liquid, neat_   Cell Window ___KBr___
Cell Path Length _capillary film_   Matrix Material _____

Fig. 5.34 *IR spectrum: cyclohexanone.*

reductive formation of an alcohol group from the carbonyl system. The band centered near 3300 cm⁻¹ results from the single highly polarized O—H stretching mode. The drop in intensity of the 1715 cm⁻¹ band indicates the loss of the carbonyl function. The exact amount of cyclohexanone remaining could very easily be determined by carrying out a Beer's law-type analysis, but here we will use gas chromatographic techniques to determine the value. Other bands of interest in the spectrum of cyclohexanol occur at 1069 and 1031 cm⁻¹. These can be assigned respectively to the equatorial and axial C—O stretching of the rotational conformers of this alicyclic secondary alcohol. A broad band (width ~300 cm⁻¹) can be found near 670 cm⁻¹ and arises from an O—H bending, out-of-plane mode of the associated alcohol. This band is generally identified only in neat samples where extensive H bonding occurs. Also note that the band at 1425 cm⁻¹ has vanished as there are no methylene groups α to carbonyl systems in the product.

Now proceed with purification of the reaction product by preparative gas chromatography. Use the following conditions and refer to Experiment 1 for the collection technique. Then if time permits, or later, determine the infrared spectrum of the purified product. Describe and explain the changes observed in the new spectrum compared with that of the crude product.

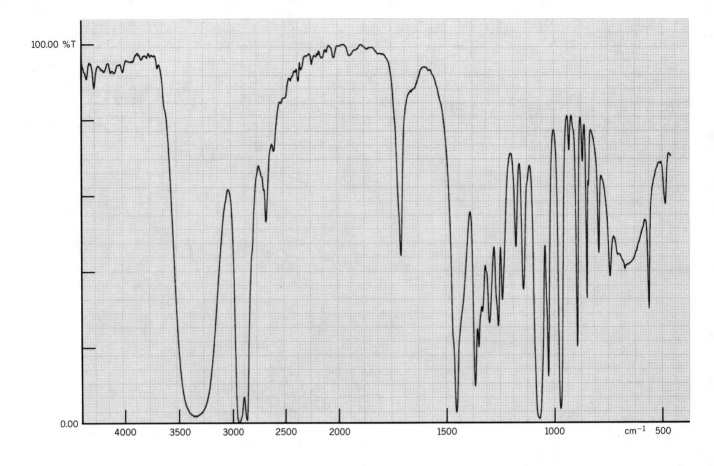

Sample	Cyclohexanol (reaction product)		
%T X ABS — Background Scans 4		Scans 16	
Acquisition & Calculation Time 42 sec		Resolution 4.0 cm⁻¹	
Sample Condition liquid, neat		Cell Window KBr	
Cell Path Length capillary film		Matrix Material	

Fig. 5.35 *IR spectrum: cyclohexanol (reaction product).*

Separation of Small Quantities of Cyclohexanone from Cyclohexanol

Example

9 : 1 (v/v) cyclohexanol/cyclohexanone

10% Carbowax 20M (stationary phase)

Injection volume: 15 μL

Temperature: 130°C

He flow rate: 50 mL/minute

Column: 1/4 in. × 8 ft stainless steel

Attenuator: 512

Chart speed: 1 cm/minute

| | Cyclohexanol | | Cyclohexanone |
Run	Retention Time (min)	Yield (mg)	Retention Time (min)
1	15.3	6.5	11.6
2	17.3	7.3	12.5
3	17.2	8.5	12.6
4	16.0	9.6	12.0
5	14.6	7.3	11.2
6	14.5	8.7	11.2
7	15.5	8.9	11.7
8	15.5	8.9	11.8
9	16.4	8.8	12.3
10	15.4	8.4	12.7
Ave.	15.8 ± 1	8.3 ± 0.9	12.0 ± 0.6

Cyclohexanol injected $= 0.9 \times 15~\mu L \times 0.963~mg/\mu L = 13.0~mg$

Percentage yield $= 8.3/13.0 \times 100 = 63.8\%$

Chemical Tests

Several chemical tests (see Chapter 7) may also be used to establish that an alcohol has been formed by the reduction of a ketone. Perform the ceric nitrate and 2,4-dinitrophenylhydrazine test on **both** the starting ketone and the alcohol product. Do your results demonstrate that the alcohol was obtained? You might also wish to prepare a phenyl or α-naphthyl urethane derivative of the cyclohexanol. Before the advent of IR spectroscopy, the formation of solid derivatives was used extensively to identify unknown compounds.

PART B: *cis*- and *trans*-4-*tert*-Butylcyclohexanol

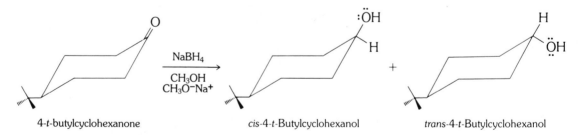

4-*t*-butylcyclohexanone   *cis*-4-*t*-Butylcyclohexanol   *trans*-4-*t*-Butylcyclohexanol

EXPERIMENTAL Estimated time of the experiment: 2.0 hours. For the GC analysis: 15 minutes per student.

Physical Properties of Reactants and Products

Compound	MW	Wt/Vol	mmol	mp(°C)	bp(°C)
4-*t*-Butylcyclohexanone	154.25	50 mg	0.33	47–50	
Methanol		50 μL			65
Sodium borohydride reducing solution		100 μL			
4-*t*-Butylcyclohexanol (mixed isomers)	156.27			62–70	

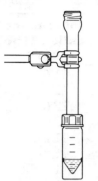

4-*t*-Butylcyclohexanone,
50 mg + CH$_3$OH, 50 μL +
NaBH$_4$ solution, 100 μL

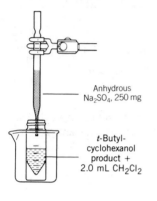

Anhydrous
Na$_2$SO$_4$, 250 mg

t-Butyl-
cyclohexanol
product +
2.0 mL CH$_2$Cl$_2$

HOOD

Reagents and Equipment

A tared 3.0-mL conical vial equipped with an air condenser is charged with 50 mg (0.33 mmol) of 4-*t*-butylcyclohexanone followed by 50 μL of methanol. (■) The vial is gently shaken to obtain a homogeneous solution. Then 100 μL of sodium borohydride reducing solution is added dropwise, with swirling.

Note. *The liquid reagents are dispensed by use of automatic delivery pipets.*

The preparation of the reducing solution is given in Part A, Reagents and Equipment.

Reaction Conditions

The solution is allowed to stand at room temperature for a period of 10 minutes.

Isolation of Product

The resulting solution is worked up using the procedure described in Part A, Isolation of Product, except that 250 mg of sodium sulfate is placed in the Pasteur filter pipet.(■)

The methylene chloride may be removed from the final solution by allowing a gentle stream of nitrogen gas to impinge on the surface while the vial is warmed in a sand bath in the **hood**.

Purification and Characterization

The product remaining after removal of the methylene chloride is sufficiently pure for characterization.

Weigh the solid product and calculate the percentage yield. Determine the melting point of your material and compare it with the value reported in the literature.

Obtain the IR and NMR spectra of the crude mixture of isomers. IR sampling in this instance is best accomplished by the capillary film melt (use the heat lamp) technique.

Infrared Analysis

Refer to the discussion in Part A for an interpretation of the absorption bands found at 3435, 3000–2850, 1715, and 1425 cm^{-1} in the starting material (Fig. 5.36), and at 3250, 3000–2850, 1715 (variable relative intensity—may be quite weak—why?), 1069, and 1031 cm^{-1} in the crude alcohol (Fig. 5.37). In addition, the ketone has bands at 1396 (weak) and 1369 (strong) cm^{-1}, and the alcohol has bands at 1399 (weak) and 1375 (strong) cm^{-1}. These two pairs of bands establish the presence of the tertiary butyl group in these compounds.

Note that (a) a weak band (3495 cm^{-1}) is present on the high wavenumber side of the 3250 cm^{-1} O—H stretching mode and (b) even in neat samples of the tertiary butyl derivative, the 670 cm^{-1} band, clearly evident in cyclohexanol, is difficult to observe.

The mixture of two diastereomeric alcohols you have synthesized provides an ideal opportunity for the introduction of nuclear magnetic resonance (NMR) spectroscopy, since this technique is an extremely powerful tool for the discrimination and characterization of diastereomeric compounds. As you will see, the two diastereomers have quite different NMR spectra which allow the determination of the ratio of the two isomers as well as the unambiguous assignment of relative stereochemistry.

This experiment presents the option for students to obtain and interpret NMR data if the local opportunity exists. The two diastereomeric alcohols exhibit different splitting patterns for the proton on the carbon bearing the —OH group.

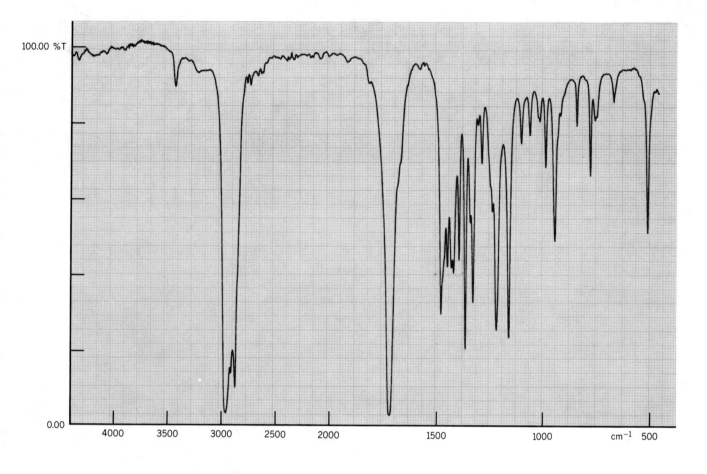

Sample ___4-tert-Butylcyclohexanone_____

%T _X_ ABS _ Background Scans _4_____   Scans _____16_____
Acquisition & Calculation Time _42 sec__   Resolution _4.0 cm⁻¹____
Sample Condition ___solid - melt_____   Cell Window ___KBr_____
Cell Path Length ___capillary film____   Matrix Material _____

Fig. 5.36 *IR spectrum: 4-t-butylcyclohexanone.*

Integration of these signals allows one to determine the ratio of the diastereomeric alcohols in the sample.

Nuclear Magnetic Resonance Analysis

Notice the expanded NMR spectrum in Figure 5.38. The signals at about 4.04 ppm and about 3.52 ppm correspond to the proton on the carbon bearing the OH group in the two diastereomers of 4-t-butylcyclohexanol shown. On closer inspection, the downfield signal (4.04 ppm) is a pentet and the upfield signal (3.52 ppm) is a triplet of triplets. The pentet implies that the proton in question is coupled with equal coupling constants (*J*) to four adjacent protons. The triplet of triplets implies that the proton in question is coupled to two adjacent protons with a large coupling constant and to two other adjacent protons with a smaller coupling constant. Specifically, the proton in the first case must be equatorial and the proton in the second case must be axial, because the dihedral angle between an equatorial proton and each of the four adjacent protons is the same, about 60°. When a proton is axial, the dihedral angle to the two adjacent equatorial protons is about 60°

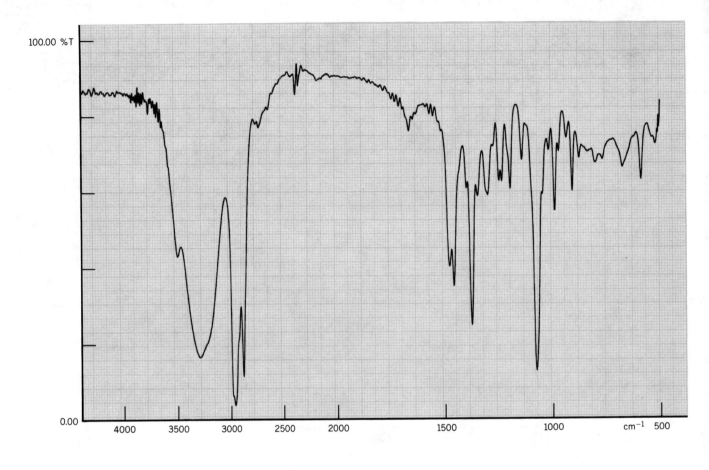

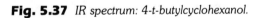

Sample	4-tert-Butylcyclohexanol

%T _X_ ABS _ Background Scans _4_   Scans _16_
Acquisition & Calculation Time _42 sec_   Resolution _4.0 cm⁻¹_
Sample Condition _solid_   Cell Window _____
Cell Path Length _____   Matrix Material _KBr_

Fig. 5.37 *IR spectrum: 4-t-butylcyclohexanol.*

and the dihedral angle to the two adjacent axial protons is about 180°, thus producing a triplet of triplets.

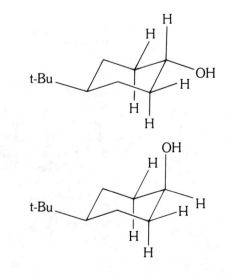

300 MHz ¹H NMR OF CIS + TRANS 4-*t*-BUTY*L*CYCLOHEXANOL IN CDCl₃

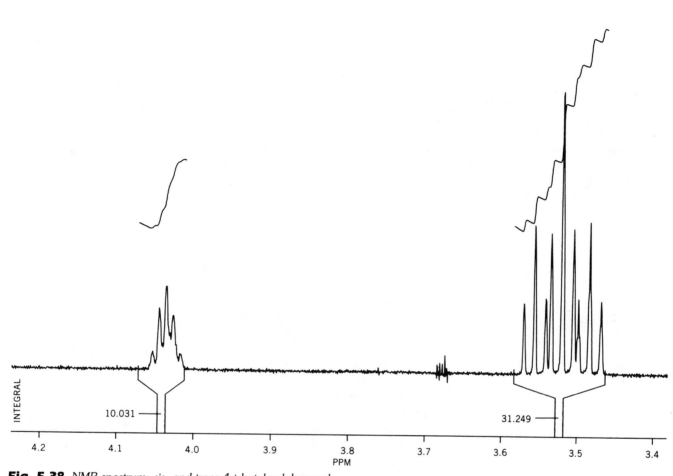

Fig. 5.38 *NMR spectrum: cis- and trans-4-t-butylcyclohexanol.*

GC Analysis

The cis and trans isomers of 4-*t*-butylcyclohexanol may be separated by gas chromatography using a ¼ in. × 8 ft 20% Carbowax column set at 170°C. A methylene chloride solution of the alcohol mixture of concentration 0.5 mg/µL is prepared and a 5.0-µL sample is injected into the gas chromatograph. At a flow rate of 50 mL/minute (He) the cis isomer has a retention time of 13 minutes; the trans isomer, 16 minutes. Determine the percentage of each isomer present in the sample by determining the area under the curve. The starting ketone has a retention time similar to that of the cis isomer alcohol. Therefore, if the reaction does not go to completion, the cis–trans ratio will not be accurate.

Area Under A Curve. *Several techniques may be used. The following method gives reproducible results of ±3–4%.*

Multiplication of the peak height (mm) by the width at half-height (mm), measured from the baseline of the curve, yields the area under the curve.

ENVIRONMENTAL DATA

Substance	Amount		TLV (mg/m³)	Emissions (mg)	Volume (m³)
Experiment 5A					
Methylene chloride (Max., extract, evap.)	2.0	mL	350	2670	7.6
Methanol (SKIN) (solv.)	0.55	mL	260	435	1.7
Cyclohexanone (Max. from yield, transfer)	100	μL	100	15	0.1
Experiment 5B					
Methanol (SKIN) (Max., solv.)	0.15	mL	260	120	0.5
Methylene chloride (Max., extract, evap.)	2.0	mL	350	2670	7.6
4-*t*-Butylcyclohexanone (Max., from yield, transfer)	50	mg	—	3.5	

TLV: Cyclohexanol—50 ppm
Hydrochloric acid—5 ppm

TXDS: Sodium borohydride—orl-rat LD50: 160 mg/kg
4-*t*-Butylcyclohexanone—orl-rat LD50: 5000 mg/kg
4-*t*-Butylcyclohexanol—orl-rat LD50: 4200 mg/kg

QUESTIONS

5-18. Suggest a chemical test that would allow you to distinguish between t-butyl alcohol and 1-butanol, both of which give a positive ceric nitrate test.

5-19. Which of the isomeric butyl alcohols, $C_4H_{10}O$, can be prepared by reduction of a ketone using the sodium borohydride reagent?

5-20. Why can the axial and equatorial hydroxyl isomers for 4-*t*-butylcyclohexanol be resolved, but not those for cyclohexanol itself?

5-21. What aldehyde or ketone would you reduce to prepare the following alcohols?
   a. Benzyl alcohol
   b. 3,3-Dimethyl-2-butanol
   c. 2,2-Dimethyl-1-pentanol

5-22. The *cis* and *trans*-4-*t*-butylcyclohexanols prepared in Part B have a plane of symmetry. Draw this symmetry element for each compound.

5-23. In the IR spectrum of the crude product obtained from the reduction of *t*-butyl-cyclohexanone, the fingerprint region appears to possess bandwidths that are slightly broader than those found in cyclohexanol. Explain.

5-24. The reduction of 4-*t*-butylcyclohexanone is a stereoselective reaction. Which isomer predominates? Assuming that NMR data are not available, it is still possible to arrive at a rough estimate of the product ratio. How would you go about making this measurement? Suggest a value.

5-25. In the IR spectrum of the crude *t*-butylcyclohexanols one observes (a) that a weak band (3495 cm^{-1}) on the high side of the 3250 cm^{-1} O—H stretching mode appears, and (b) that even in neat samples of this tertiary alcohol derivative, the 670 cm^{-1} band, clearly evident in cyclohexanol, is difficult to observe. Explain these observations. Is the same effect operating in both cases?

5-26. Sketch the NMR spectrum you would expect to observe for the following compounds.
   a. Acetone
   b. 1,1,2-Tribromoethane
   c. Ethyl chloride
   d. 2-Iodopropane
   e. 1-Bromo-4-methoxybenzene

REFERENCES

1. General references on metal hydride reduction:
   a. Walker, E. R. H. *Chem. Soc. Rev.* **1976**, *5*, 23.
   b. House, H. O. *Modern Synthetic Reactions*; Benjamin: Reading, MA, 1972.

2. Sodium borohydride as a reducing agent:
   a. Fieser, L. F.; Fieser, M. *Reagents for Organic Synthesis*; Wiley: New York, 1967; Vol. 1, p 1050, and subsequent volumes.
   b. Cragg, G. M. W. *Organoboranes in Organic Synthesis*; Marcel Dekker: New York, 1973.
   c. Brown, H. C. *Boranes in Organic Chemistry*; Cornell University Press: Ithaca, NY, 1972.
3. Lithium aluminum hydride as a reducing agent:
   a. Brown, W. G. *Org. React.* **1951,** *6,* 469.
   b. See reference 2a, p 581, and subsequent volumes.
4. *trans-t*-Butylcyclohexanol has been prepared from the ketone using LiAlH$_4$ as the reducing agent: Eliel, E. L.; Martin, R. J. L.; Nasipuri, D. *Organic Syntheses*; Wiley: New York, 1973; Collect. Vol. V, p 175.
5. *IR Spectral Collection: The Aldrich Library of IR Spectra,* 3rd ed.; Pouchert, C. J. Ed.; Aldrich Chemical Co.: Milwaukee, WI, 1981.
6. *IR Spectral Collection*; Sadtler Research Laboratory, Inc., Spectra; Philadelphia, PA.

TECHNIQUE 6: CRYSTALLIZATION; ULTRAVIOLET–VISIBLE SPECTROSCOPY

CRYSTALLIZATION

This experiment introduces the basic strategy involved in achieving the purification of solid organic substances by crystallization. The technique of crystallizing an organic compound is one of fundamental importance and must be mastered in order to deal successfully with the purification of these materials. *It is not an easy art to acquire.* Organic solids tend not to crystallize with the ease of inorganic substances.

Indeed, in earlier times an organic chemist occasionally would resist an invitation to leave a well-worn laboratory for new quarters. This concern arose from the suspicion that the older facility (in which many crystallizations had been carried out) harbored seed crystals for a large variety of substances in which the resident investigator had an interest. Carried by dust from the earlier work this trace of material presumably aided the successful initiation of crystallization of reluctant materials. Further support for this argument was gained by the often quoted (but not substantiated) observation that after a material was first crystallized in a particular laboratory subsequent crystallizations of the material irrespective of its purity or origin were always easier to carry out.

A reaction very often will be viewed as a failure unless an amorphous sludge can be enticed to become a collection of beautiful white crystals. The melting point of an amorphous substance is ill-defined, and if this material is mixed with a crystalline reference compound, large melting point depressions usually result.

In a number of areas of organic chemistry, particularly those dealing with natural products, the success or failure of an investigation can depend to a large extent on the ability of the research chemist to isolate tiny quantities of crystalline substances. Often the compounds of interest must be extracted from enormous amounts of extraneous material. In one of the more spectacular examples, Reed[5] in 1953 isolated 30 mg of the crystalline coenzyme lipoic acid from 10 tons of beef liver residue.

Lipoic acid

[5] Reed, L. J.; Gunsalus, I. C.; Schnakenberg, G. H. F.; Soper, Q. F.; Boaz, H. E.; Kern, S. F.; Parke, T. V. *J. Am. Chem. Soc.* **1953,** *75,* 1267.

Table 5.5 Common Solvents

Solvent	bp(°C)
Water	100
Methanol	65
Ethanol, 95%	78
Ligroin	60–90
Dioxane[a]	101
Acetone	56
Diethyl ether	35
Methylene chloride	41
Petroleum ether	30–60

[a] This solvent is no longer recommended for instruction purposes because of new toxicological data.

Procedure

The essentials of this purification technique are outlined as follows. First, dissolve the material (primarily made up of the compound of interest along with smaller quantities of contaminating substances) in a warm solvent. Second, once the solid mixture is fully dissolved, the heated solution is filtered and then brought to the point of saturation by evaporating a portion of the solvent. Third, cool the warm saturated solution to cause a drop in solubility of the dissolved substance. This results in precipitation of the solid material. Fourth, isolate the precipitate by filtration and remove the last traces of solvent.

The technique is considered successful if the solid is recovered in good yield and is obtained in a state of higher purity than the material initially dissolved. The cycle from solid state to solution and back to solid state is termed *recrystallization* if the initial and final solid materials are crystalline.

Although the technique sounds fairly simple, in reality it is demanding. The successful purification of microscale quantities of solids will require your utmost attention.

The first major problem to be faced is the choice of solvent system. In order to achieve *high recoveries*, the compound to be crystallized would ideally be very soluble in the solvent of choice at elevated temperatures, but nearly insoluble when cold. If the crystallization is to *increase the purity* of the compound, however, the impurities should be either very soluble in the solvent at all temperatures or not soluble at any temperature. In addition, the solvent should possess as low a boiling point as possible so that traces can be easily removed from the crystals after filtration.

Thus, the choice of solvent is critical to a good crystallization. Table 5.5 is a list of common solvents used in the purification of most organic solids. (The list has contracted significantly in the past few years as health concerns about these very volatile compounds have arisen.)

Seldom are the solubility relationships ideal for crystallization. Most often a compromise is made. If there is no suitable single solvent available, it is possible to employ a mixture of two solvents, termed a *solvent pair*. In this situation, a solvent is chosen that will readily dissolve the solid. After dissolution, the system is filtered. A second solvent miscible with the first, but in which the solute has lower solubility, is then added dropwise to the hot solution to achieve saturation. In general, polar organic molecules have higher solubilities in polar solvents, and nonpolar materials are more soluble in nonpolar solvents ("like dissolves like"). Considerable time can be spent in the laboratory working out an appropriate solvent system for a particular reaction product. In most instances with known compounds, the optimum solvent system has been established. Thus, in the large majority of cases in the microlab text, the best recrystallization solvent system will be suggested.

Because many impurities have solubilities similar to those of the compounds of interest, most crystallizations are not very efficient. Recoveries of 50–70% are not uncommon. It is important that the purest possible material be isolated prior to undertaking recrystallization.

A number of microscale crystallization routines are available.

Simple Crystallization

Simple crystallization works well with large quantities of material (100 mg and up), and it is essentially identical to that of the macrolaboratory program.

1. Place the solid in a small Erlenmeyer flask or test tube.
2. Add a minimum of solvent and bring the mixture to the boiling point in a sand bath.
3. Stir and add solvent dropwise with continued heating until all of the materials has dissolved.

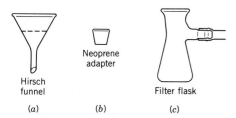

Fig. 5.39 *Component parts for vacuum filtration.*

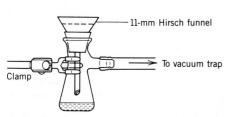

Fig. 5.40 *Vacuum filtration apparatus.*

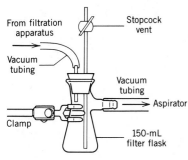

Fig. 5.41 *Vacuum trap.*

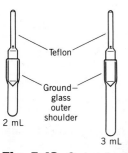

Fig. 5.42 *Craig tubes.*

4. Add a decolorizing agent, if necessary (powdered charcoal, or better charcoal pellets), to remove colored minor impurities and other resinous byproducts.

5. Filter hot into a second Erlenmeyer flask (preheat the funnel with hot solvent). This operation removes the decolorizing agent and any insoluble material initially present in the sample.

6. Evaporate enough solvent to reach saturation.

7. Cool to allow crystallization (crystal formation will be better if this step takes place slowly).

8. Collect the crystals by filtration.

9. Wash (rinse) the crystals.

10. Dry the crystals.

Filtration Techniques

Use of the Hirsch Funnel

The standard filtration system for collecting products purified by recrystallization in the microscale laboratory is vacuum filtration with an 11-mm Hirsch funnel. In addition, many reaction products that do not require the crystallization step are collected directly by this technique. The funnel is shown in Figure 5.39a.

The Hirsch filter funnel is composed of a ceramic cone with a circular flat bed perforated with small holes. The diameter of the bed is 11 mm, and in operation it is covered by a flat piece of filter paper of the same diameter. The funnel is sealed into a filter flask with a Neoprene adapter (see Fig. 5.39b).

The filter flask, which is heavy-walled and especially designed to operate under vacuum, is constructed with a side arm (they are often called "side-arm pressure flasks"; see Fig. 5.39c).

The side arm is connected with heavy-walled rubber vacuum tubing to a water aspirator or water pump. When water is running through the aspirator, a partial vacuum is formed, which creates a flow of air down the vacuum tubing from the filter flask. With the rubber adapter in place, the entering air is forced through the filter paper, which is held flat by suction. The mother liquors of the crystallization are rapidly forced into the filter flask, while the crystals retained by the filter are quickly dried by the stream of air passing through them (Fig. 5.40).

When you are using a water pump, it is very important to have a safety trap mounted in the vacuum line leading from the filter flask. Any drop in water pressure (easily created by one or two other students on the same water line turning on their aspirators at the same time) can result in the backup of water into the system as the flow through the aspirator decreases (see Fig. 5.41).

Craig Tube Crystallizations

The Craig tube is commonly used for microscale crystallizations in the range of 10–100 mg of material (see Fig. 5.42). The process consists of the following steps.

1. The sample is placed in a small test tube (10 × 75 mm).

2. The solvent (0.5–2 mL) of choice is added, and the sample dissolved by heating in the sand bath. Rapid stirring with a microspatula (roll the spatula rod between your fingers) greatly aids the dissolution and protects against boilover. A modest excess of solvent is added after the sample is completely dissolved. It will be easy to remove this excess at a later stage, as the volumes involved are very small. The additional solvent ensures that the solute will stay in solution during the hot transfer.

3. The heated solution is transferred to the Craig tube by Pasteur filter pipet (the pipet is preheated with hot solvent). This transfer automatically filters the solution (if decolorizing charcoal powder has been added, two filtrations by

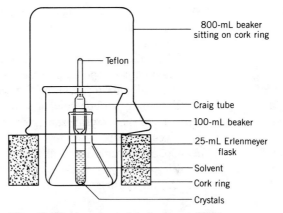

Fig. 5.43 *Apparatus for slow crystallization.*

Labels for Fig. 5.43:
- 800-mL beaker sitting on cork ring
- Teflon
- Craig tube
- 100-mL beaker
- 25-mL Erlenmeyer flask
- Solvent
- Cork ring
- Crystals

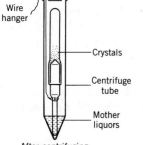

Labels for Fig. 5.44:
- Wire hanger
- Crystals
- Centrifuge tube
- Mother liquors
- After centrifuging

Fig. 5.44 *Crystal collection with a Craig tube.*

the pipet may be required. The second filtration is almost always avoided by the use of charcoal pellets.)

4. The hot filtered solution is then concentrated to saturation by gentle boiling in the sand bath. Constant agitation of the solution with a microspatula during this short period will avoid the use of a boiling stone and guarantee that a boilover will not occur. The ready crystallization of product on the microspatula just above the solvent surface serves as a good indication that saturation is close at hand.

5. The upper section of the Craig tube is set in place, and the system allowed to cool in a safe place. As cooling commences, seed crystals, if necessary, may be added by crushing them against the side of the Craig tube with a microspatula just above the solvent line. A good routine, if the time is available, is to place the assembly in a small Erlenmeyer, then place the Erlenmeyer in a beaker, and finally cover the first beaker with a second inverted beaker. This will ensure slow cooling, which will enhance good crystal growth (Fig. 5.43). A Dewar flask may be used when very slow cooling and large crystal growth are required (as in X-ray crystallography).

6. After the system reaches room temperature, cooling in an ice bath will further improve the yield.

7. Solvent is now removed by inverting the Craig tube assembly into a centrifuge tube and spinning the mother liquors away from the crystals (Fig. 5.44). This operation takes the place of the usual filtration step in simple crystallizations. It avoids another transfer of material and also avoids product contact with filter paper.

8. After removal from the centrifuge, the Craig tube is disassembled and any crystalline product clinging to the upper section is scraped into the lower section. If the lower section is tared it can be left to air dry to constant weight or placed in a warm vacuum oven (wrap a piece of filter paper over the open end secured by a rubber band to prevent dust from collecting on the product while drying). The yield can then be directly calculated.

The cardinal rule in carrying out the purification of small quantities of solids is *keep the transfers to an absolute minimum!*. The Craig tube is very helpful in this regard.

The preceding routine will maximize the crystallization yield. If time is important, the process can be shortened considerably. Shortcuts, however, invariably lead to a corresponding drop in yield.

In the following experiment the techniques of simple crystallization, vacuum filtration, and Craig tube recrystallization are introduced.

ULTRAVIOLET–VISIBLE SPECTROSCOPY

Organic compounds possessing *conjugated* systems have the potential to yield structural information through analysis of the electronic absorption spectrum. The technique is limited, however, to those materials that exhibit absorption in the *near* ultraviolet and visible regions of the spectrum, that is, the range 210–800 nm (where nm = nanometer or 10^{-9} m). At wavelengths below 210 nm oxygen in the atmosphere begins to absorb radiation. This region (below 210 nm) is known as the vacuum UV as the instrumentation required to examine the spectrum at these wavelengths must be evacuated. Thus, common sampling is limited to only those materials absorbing above 210 nm. As these substances are the materials possessing *conjugation* the scope of useful information that can be acquired is somewhat limited compared with the infrared and radio (NMR) frequency regions. Valuable information concerning the environment of the π-electron systems in complex molecules, however, often can be obtained utilizing this spectral region.

In ultraviolet–visible spectroscopy, the absorption of radiation in the near UV (wavelength range 210–350 nm) and the visible (wavelength range 350–800 nm) is recorded as a function of the wavelength, lambda (λ), expressed in nanometers. Appendix A presents an introduction to the theory of ultraviolet–visible spectroscopy.

Ultraviolet–visible spectra are generally recorded in solution with samples dissolved in a nonabsorbing solvent such as ethanol, methanol, hexane, or chloroform. Cells are commonly constructed of quartz, as glass absorbs ultraviolet radiation at a number of wavelengths in this region.

It is common practice to refer to the wavelength at which an absorption maximum of a particular substance occurs as a λ_{max} of the sample. The quantitative relationship of absorbance to concentration is expressed by the **Beer–Lambert** equation.

$$A = \varepsilon c l$$

where

A = absorbance
ε = molar absorptivity (a constant characteristic of the specific molecule being observed)
c = concentration (mol/liter)
l = length of sample path (cm)

Electronic transitions measured at room temperature give rise to spectra composed of absorption bands that are typically few in number and very broad in appearance. The peak position in the ultraviolet–visible spectrum is identified by assigning the wavelength to the point of maximum absorption, (λ_{max}). The calculated molar absorptivity and solvent are usually reported. For example, for methyl vinyl ketone,

λ_{max} 219 nm (ε_{max} = 3600, ethanol); λ_{max} 324 nm (ε_{max} = 24, ethanol)

The technique of serial dilution is generally used to prepare solutions for analysis. A sample of the material to be analyzed is accurately weighed, dissolved in the chosen solvent and diluted to volume in a volumetric flask. At the microscale level, 4- to 5-mg samples and 10-mL volumetric flasks are convenient to use. An aliquot is then taken from this original solution, transferred to a second volumetric flask, and diluted to volume as before. This sequence is repeated until the desired concentration is obtained.

An example of this procedure is outlined here for the sequence used in

Experiment 49 to prepare the solution from which the absorption spectrum of tetraphenylcyclopentadienone is obtained.[6]

1. Weigh a 4.0- to 4.7-mg sample of the product into a 10-mL volumetric flask.
2. Dissolve the sample in $CHCl_3$ and dilute to the 10-mL mark with this solvent.
3. Withdraw 2.0 mL of this solution using a volumetric pipet and transfer it to a second 10-mL volumetric flask. Dilute this solution with fresh chloroform to the 10-mL mark.
4. Place 1–2 mL of this final solution in a cuvette and record the spectrum after first calibrating the instrument with pure chloroform.

It is very important that the cells used for the analysis be absolutely clean. In addition, the sample cell should be rinsed several times with the solution to be analyzed before being filled and placed in the instrument. A spectrum of the solvent used to prepare the analyte solution should be run as a blank to determine whether any interfering absorption might be present. This check should be repeated between successive determinations.

Typical ultraviolet–visible spectra are shown in Experiments 6, 21D, 41, and 49. As part of the characterization data, λ_{max} is also given in Experiments 17, 20, 21A, 29, 40A, and 44.

Experiment 6

Photochemical Isomerization of an Olefin: *cis*-1,2-Dibenzoylethylene

(*cis*-1,4-diphenyl-2-butene-1,4-dione)

The isomerization of a trans olefin to the corresponding cis isomer under photochemical conditions is demonstrated in this experiment.

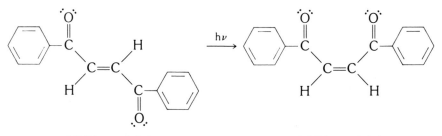

trans-1,2-Dibenzoylethylene   *cis*-1,2-Dibenzoylethylene

DISCUSSION The photochemical geometric cis–trans isomerization reaction is demonstrated in this experiment by the conversion of trans-1,4-diphenyl-2-butene-1,4-dione to the corresponding cis isomer. The reaction proceeds by way of an excited state of the trans isomer.

The bonding between two carbon atoms of a C=C consists of a sigma bond and a pi bond created by overlap of the carbon *p* orbitals. This *p*-orbital overlap forming the pi bond imparts a certain rigidity to the C—C and, thus, free rotation of the parts of the molecule connected through the C=C is prohibited. The

---

[6] The authors are indebted to D. F. Dedolph and C. N. Sukenik of Case-Western Reserve University for permission to use the analysis employed in their laboratories.

isomers do not interconvert unless enough energy (60–65 kcal/mol) is supplied to break the pi bond.

The nonrotation situation about the C=C gives rise to the possibility of stereoisomerism. Cis and trans geometrical isomers are not mirror images of one another and are thus not optical isomers, but diastereoisomers. The isomers differ only in the arrangement of the atoms in space. Cis–trans isomerization of this type is not possible if either carbon carries two identical groups. Geometrical isomers have different physical properties such as melting point, boiling point, dipole moment, density, and solubility. Because of the differences in physical properties, the possibility exists that the isomers may be separated, as in this experiment.

The course of the isomerization may be followed using *thin-layer chromatography* (Part B) or *NMR* analysis (Part C).

PART A: Purification of *trans*-1,2-Dibenzoylethylene

Estimated time to complete the reaction: 2.0 hours of actual laboratory time. The reaction requires approximately 1 hour of irradiation.

Physical Properties of Reactants and Products

Compound	MW	Wt/Vol	mmol	mp(°C)	bp(°C)
trans-1,2-Dibenzoylethylene	236.27	150 mg	0.62	111	
Ethanol (95%)		6.0 mL			78.5
Methylene chloride		4.0 mL			40
cis-1,2-Dibenzoylethylene	236.27			134	

The starting olefin is purified by recrystallization.

To a 10-mL Erlenmeyer flask are added 150 mg (0.62 mmol) of *trans*-1,2-dibenzoylethylene and 3.0 mL of methylene chloride. Decolorizing charcoal pellets (10 mg) are added, and after swirling (or addition of a magnetic stirring bar and use of a magnetic stirring hot plate ■) for several minutes, the solution is filtered by gravity through a fast-grade filter paper (■). The filtrate is collected in a 10-mL Erlenmeyer flask containing a boiling stone. The filter paper is rinsed with an additional 1 mL of methylene chloride and this rinse is collected in the

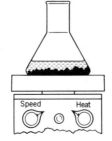

trans–1, 2-Dibenzoylethylene,
150 mg +CH$_2$Cl$_2$, 3.0 mL,
charcoal pellets

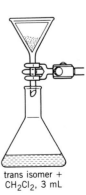

trans isomer +
CH$_2$Cl$_2$, 3 mL

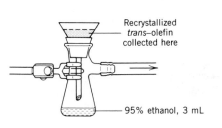

Recrystallized
trans–olefin
collected here

95% ethanol, 3 mL

HOOD

same flask. The solution is now concentrated to dryness in a sand bath under a slow stream of nitrogen gas in the **hood**.

To the flask are now added 3 mL of 95% ethanol and the yellow solid dissolved by warming on a sand bath with stirring until a homogeneous solution is obtained.

The solution is now allowed to cool slowly to room temperature over a period of 15 minutes and then placed in an ice bath for an additional 10 minutes. The yellow needles are collected by vacuum filtration using a Hirsch funnel (■) and then air-dried on a porous clay plate or on filter paper.

Weigh the olefin and calculate the percentage recovery. Determine the evacuated melting point and compare your result with the literature value. Also compare the melting point with the sample obtained from the distillation residue of Experiment 2A, in which a simple crystallization was performed without the aid of decolorizing charcoal.

PART B: Isomerization of the Olefin: TLC Analysis

EXPERIMENTAL

Reagents and Equipment

To a 13 × 100-mm test tube are added 25 mg (0.21 mmol) of recrystallized *trans*-1,2-dibenzoylethylene and 3.0 mL of 95% ethanol.

Reaction Conditions

GENTLY

A sand bath is used to warm the mixture (*gently*) until a homogeneous solution is obtained. The test tube is *loosely* stoppered or covered with filter paper and then placed approximately 2–4 in. from a 275-watt sunlamp. The solution is irradiated for approximately one hour. (■)

Note. *If a lower-wattage lamp is used, longer irradiation times will be necessary. In either case solvent evaporation can be significantly reduced by directing a current of cool air (fan) over the reaction tube. An alternative is to allow the tube to stand in sunlight for several days.*

The course of the isomerization may be followed by TLC analysis.

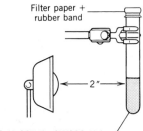

Filter paper +
rubber band

2″

$(C_6H_5CO)HC=CH(COC_6H_5)$,
25 mg + 95% ethanol, 3 mL

Information. *The TLC analysis is carried out with Eastman Kodak silica gel–polyethylene terephthalate plates having a fluorescent indicator. The plates are activated at an oven temperature of 100°C for 30 minutes and then placed in a desiccator to be cooled and stored until use. Development of the plates, after spotting, is carried out with pure methylene chloride solvent. Visualization is done using a UV lamp. As the course of the reaction is followed, a small sample (2–3 drops) of solution is removed from the test tube (**hot**) using a Pasteur pipet and placed in a $\frac{1}{2}$-dram vial. See Technique 8 for the method of TLC analysis and the determination of R_f values.*

R_f values: trans = 0.67; cis = 0.58.

HOT

Isolation of Product

The test tube is now removed from the light source and allowed to cool to room temperature.

Note. *Be careful when removing the test tube from the light source. IT IS HOT!*

The resulting mixture is placed in an ice bath to complete crystallization of the *colorless cis*-1,2-dibenzoylethylene product. The solid is collected by vacuum filtration using a Hirsch funnel, washed with 0.5 mL of cold 95% ethanol, and dried on a porous clay plate or on filter paper.

Purification and Characterization

The purity of the isolated product may be determined by TLC analysis (if not used previously). A portion of the isolated product may be further purified by recrystallization from 95% ethanol using a Craig tube.

Weigh the dried product and calculate the percentage yield. Determine the evacuated melting point and compare your result with the literature value. Obtain an IR spectrum (KBr pellet technique) and compare it with Figure 5.45 and also with the trans isomer (Fig. 5.46).

This experiment involves the photochemical isomerization of a trans double bond to a cis configuration. The reaction can be followed by a number of spectroscopic techniques.

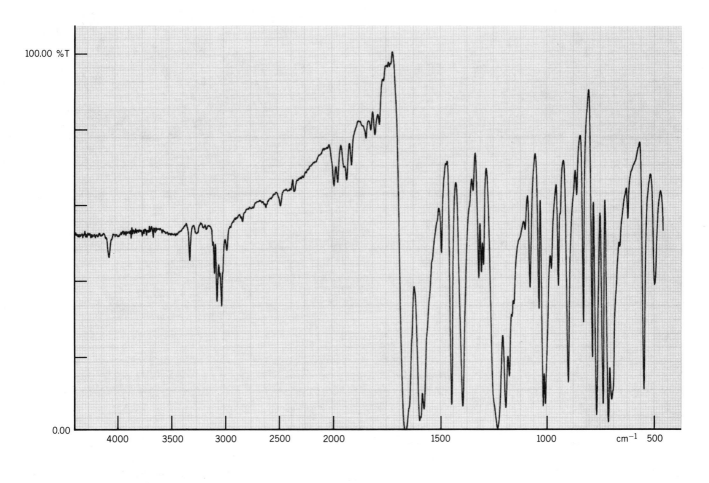

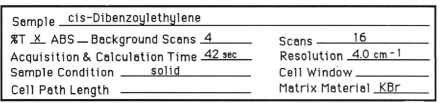

Fig. 5.45 *IR spectrum: cis-dibenzoylethylene.*

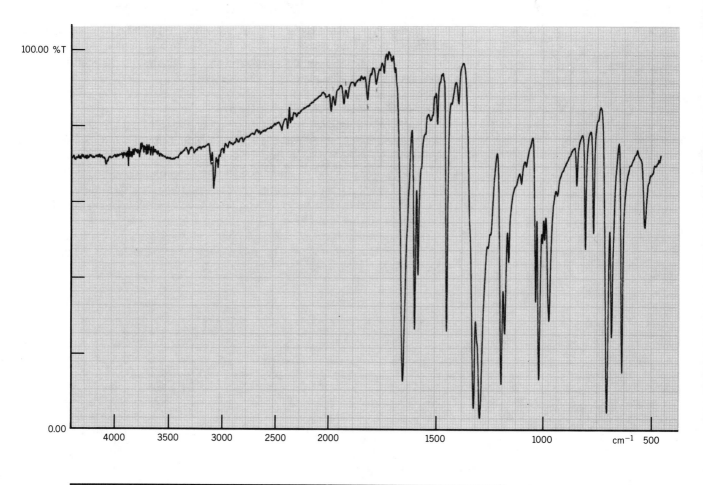

Sample __trans-Dibenzoylethylene__	
%T _X_ ABS __ Background Scans _4_	Scans _____ _16_
Acquisition & Calculation Time _42 sec_	Resolution _4.0 cm⁻¹_
Sample Condition _____ solid _____	Cell Window _____
Cell Path Length _____	Matrix Material _KBr_

Fig. 5.46 *IR spectrum: trans-dibenzoylethylene.*

Ultraviolet–Visible Spectrum

The bright yellow color of the *trans*-dibenzoylethylene rapidly fades as the conversion to the white cis compound progresses under irradiation. This visual observation is supported by an examination of the absorption spectra of the isomers in methanol solution over the region 225–400 nm (see Figs. 5.47 and 5.48). The λ_{max} of the trans isomer drops from 268 to 259 nm in the cis compound. This shift to shorter wavelengths is just enough to move the long-wavelength wing of the absorption band out of the visible region (thus, the cis compound does not possess absorption to which the eye is sensitive and the compound appears white). This observation is consistent with the contraction of the extended π system which takes place during the trans-to-cis isomerization.

Infrared Analysis

The infrared spectral changes are consistent with the proposed reaction product. Consider the spectrum of the trans starting material (Fig. 5.46). It contains the following.

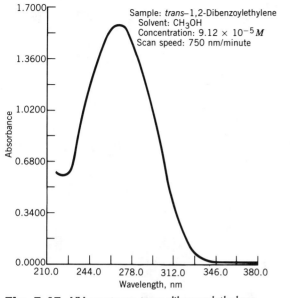

Fig. 5.47 *UV spectrum: trans-dibenzoylethylene.*

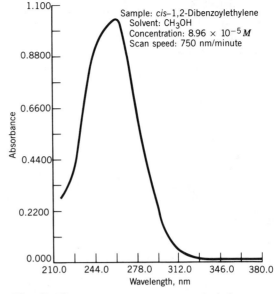

Fig. 5.48 *UV spectrum: cis-dibenzoylethylene.*

a. The macro group frequency train for conjugated aromatic ketones: 3080 & 3030 (C—H, aromatic), 1652 (doubly conjugated carbonyl), 1599 & 1581 (v_{8a}, v_{8b} degenerate ring stretch, strong intensity of 1581 cm^{-1} peak confirms ring conjugation), 1495 & 1450 (v_{19a}, v_{19b} degenerate ring stretch, v_{19a} weak) cm^{-1}.

b. The monosubstituted phenyl ring macro group frequency train: 1980(d), 1920(d), 1820, 1780 (mono combination-band pattern), 708 (C—H out-of-plane bend, C=O conjugated), 686 (ring puckering) cm^{-1}.

The presence of the trans double bond is indicated only by the relatively weak 970 cm^{-1} band, as the C—H stretch is over lapped by the aromatic ring systems.

The spectrum of the cis-photo product (Fig. 5.45) possesses the same macro group frequencies as the starting material. The conjugated aromatic ketone frequency train is as follows: 3335 (overtone of C=O stretch), 3080 & 3040, 1667, 1601, 1581, 1498 (weak), and 1450 cm^{-1}. The monosubstituted phenyl ring macro frequency train is assigned the following peaks: 1990, 1920, 1830, 1795, 710, and 695 cm^{-1}.

The cis double bond is clearly present and utilizes the following macro frequency train: 3030 (overlapped by aromatic C—H stretch), 1650 (C=C stretch, shoulder on the low wavenumber side of conjugated carbonyl; resolved in spectra run on thin samples), 1403 (=C—H out-of-phase, in-plane bending mode, strong band not present in trans compound), 970 (trans, in-phase, out-of-plane bend missing), 820 (cis, in-phase out-of-plane bend; band not found in spectrum of trans isomer) cm^{-1}.

Examine the spectrum of the reaction product that you have obtained in a potassium bromide matrix. Discuss the similarities and differences of the experimentally derived spectral data to the reference spectra (Figs. 5.45 and 5.46).

If the photoreaction exposure is continued, the cis isomer slowly undergoes conversion to a new product, ethyl 4-phenyl-4-phenoxy-3-butenoate. The conversion may be followed conveniently by thin-layer chromatography and maximum yields are obtained over approximately 24 hours under the above conditions. This rearrangement product has $R_f = 0.83$. Evaporation of the solvent yields a yellow oil.

The infrared spectrum of this new material (Fig. 5.49) indicates that significant changes have occurred in the structure of the material. The proposed structure involves a rather spectacular molecular rearrangement of the unsaturated ketone to an ethyl ester containing a phenoxy substituted double bond. Can you

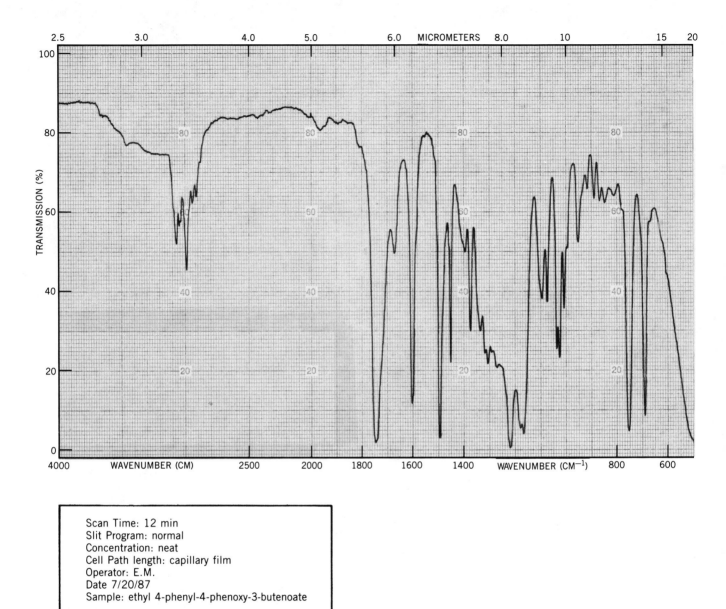

Scan Time: 12 min
Slit Program: normal
Concentration: neat
Cell Path length: capillary film
Operator: E.M.
Date 7/20/87
Sample: ethyl 4-phenyl-4-phenoxy-3-butenoate

Fig. 5.49 *IR spectrum: ethyl 4-phenyl-4-phenoxy-3-butenoate.*

rationalize the spectral data to fit this structure? Suggest possible macro group frequencies that are operating in the molecular environment of the rearranged product.

It is also of interest to take a series of mixture melting points (see Chapter 4) with mixtures (75 : 25, 50 : 50, 25 : 75) of the isomers to determine the eutectic temperature of this system. This technique is used to advantage in Experiment 34A to aid in identifying the isolated product.

PART C: Isomerization of the Olefin: NMR Analysis

EXPERIMENTAL   **Reagents and Equipment**

Prepare a sample of 4–5 mg of recrystallized *trans*-1,2-dibenzoylethylene in 500 μL of CDCl$_3$ in an NMR tube. Using a fine capillary, spot a silica gel TLC plate (see Part B) with a sample of this solution. TLC is used to track the results of the NMR experiment.

Obtain an NMR spectrum of this solution.

Reaction Conditions

Clamp the NMR tube 3–4 in. from a 275-watt sunlamp for 15–20 minutes (see Part B).

Note. *A 15-minute exposure is adequate, but total isomerization might not occur in this time frame as evidenced by the TLC analysis.*

Analysis of the Results

After irradiation, spot the original TLC plate with this irradiated solution and develop the plate in methylene chloride solvent (see Part B). Obtain an NMR spectrum of the irradiated solution.

Does the TLC analysis correlate with the NMR data? Is there evidence that the isomerization did occur?

The data in the following table were taken on a high-field (300 MHz) NMR instrument.

Chemical Shift Data

trans isomer (ppm)	cis isomer (ppm)
8.06 doublet	7.9 doublet
8.01 singlet^a	7.55 triplet
7.63 triplet	7.43 triplet
7.52 triplet	7.14 singlet^a
7.24 singlet ($CHCL_3$)	7.24 singlet ($CHCl_3$)

^{a} Most intense band in the spectrum.

ENVIRONMENTAL DATA

Substance	Amount	TLV (mg/m^3)	Emission (mg)	Volume (m^3)
Ethanol, 95% (Max. filter hot soln., vac. filtr.)	3.0 mL	1900	2250	1.2
Ethanol, 95% (Max., vac. filtr.)	5.5 mL	1900	4120	2.2
Methylene chloride (Max., less in hood)	4.0 mL	350	5300	15

TXDS: 1,2-Dibenzoylethylene—ipr-mus LD50: 25 mg/kg

QUESTIONS

5-27. List the properties that an ideal solvent should have to perform the purification of an organic compound by the crystallization technique.

5-28. What is meant by the term *solvent pair* as it relates to crystallization of solid materials?

5-29. What is the purpose of adding a decolorizing agent such as powdered charcoal to the solvent system during a recrystallization sequence?

5-30. Why do crystals form better if a solution of an organic compound is allowed to cool slowly?

5-31. List several advantages of using the Craig tube to separate crystalline products in contrast to using the Hirsch funnel filtration method.

5-32. The stereochemistry of the more highly substituted alkenes is difficult to define using the cis and trans designations. Therefore, chemists have developed a more systematic manner of indicating stereochemistry in these systems using the symbols E and Z.

    a. What do these letters signify?

    b. Draw the structures of the E and Z stereoisomers of 1,4-diphenyl-2-butene-1,4-dione used in this experiment.

5-33. The cis-H—C=C—H in-phase out-of-plane bending frequency normally comes in the range 740–680 cm^{-1}. This mode can be quite variable in intensity. On conjugation with C=O groups, the frequency rises to near 820 cm^{-1} and stabilizes as a medium to strong band with a narrow wavenumber range. Can you explain the underlying cause of this observation? (*Hint:* See the discussion of cis double-bond group frequencies in Chapter 7.)

5-34. In trans-1,2-dibenzoylethylene, even though the conjugated C=O group frequency coincides directly with the expected C=C stretching frequency, we would not have expected to observe this mode in the infrared. Why?

REFERENCES The preceding reaction was adapted from the following references.

1. Pasto, D. J.; Ducan, J. A.; Silversmith, E. F. *J. Chem. Educ.* **1974,** *51,* 277.
2. Silversmith, E. F.; Dunsun, F. C. *J. Chem. Educ.* **1973,** *50,* 568.

Reviews on photochemical isomerization reactions may be found in the following references:

1. Crombie, L. *Quart. Rev.* **1952,** *6,* 101.
2. DeMayo, P. *Adv. Org. Chem.,* **1960,** *2,* 367.
3. Fonken, G. L. In *Organic Photochemistry;* Chapman, O. L., Ed.; Marcel Dekker: New York, 1967; Vol. 1, p 197.

TECHNIQUE 7: EXTRACTION: ACID–BASE TECHNIQUE

Extraction

The present experiment demonstrates the separation of a carboxylic acid from a neutral alcohol using the "acid–base" extraction technique. For a discussion of this extraction procedure, see Experiment 4C.

Experiments 7A, 7B

The Cannizzaro Reaction with 4-Chlorobenzaldehyde or 4-Bromobenzaldehyde: 4-Chlorobenzoic Acid and 4-Chlorobenzyl Alcohol; 4-Bromobenzoic Acid and 4-Bromobenzyl Alcohol[7]

(benzoic acid, p-chloro- and benzyl alcohol, p-chloro- or benzoic acid, p-bromo- and benzyl alcohol, p-bromo-)

This experiment illustrates the simultaneous oxidation and reduction of an aromatic aldehyde to form the corresponding benzoic acid and benzyl alcohol.

---

[7] Portions of this experiment were previously published. Mayo, D. W.; Butcher, S. S.; Pike, R. M.; Foote, C. M.; Hotham, J. R.; Page, D. S. *J. Chem. Educ.* **1985,** *62,* 149.

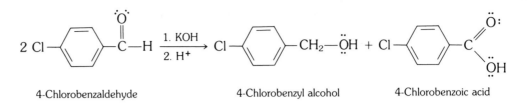

4-Chlorobenzaldehyde   4-Chlorobenzyl alcohol   4-Chlorobenzoic acid

DISCUSSION   The carbonyl group of an aldehyde represents the intermediate stage of oxidation between an alcohol and a carboxylic acid.

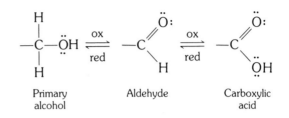

Primary   Aldehyde   Carboxylic
alcohol   acid

It is not surprising then, to find a reaction in which an aldehyde is oxidized and reduced in a disproportionation sequence to form the corresponding alcohol and carboxylic acid. Such a reaction is the Cannizzaro reaction. In the presence of strong base, aldehydes that lack α-hydrogen atoms undergo a self oxidation–reduction reaction. One molecule of the aldehyde oxidizes a second aldehyde molecule to the acid anion and is itself reduced to the corresponding primary alcohol. Aldehydes with α-hydrogen atoms do not give the reaction because in the presence of base they undergo the aldol reaction (see Experiments 22 and 49).

The first step in the mechanistic sequence is the nucleophilic attack of the hydroxide anion on the carbonyl group of the aldehyde. This is followed by the key step in the reaction: the transfer of a hydrogen atom with its pair of electrons (a hydride species) to a carbonyl group of a second molecule of aldehyde. This sequence is diagrammed here.

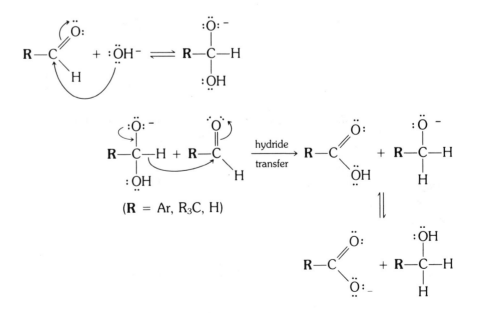

(R = Ar, R_3C, H)

The strong electron-donating character of the negatively charged oxygen atom in the anion greatly facilitates the ability of the aldehydic hydrogen to transfer, with its pair of electrons, to a second molecule of aldehyde. As seen in the preceding diagram, this nucleophilic addition to a carbonyl group leads to the formation of a carboxylic acid and an alkoxide anion. The final stage, involving a fast acid–base equilibrium to yield the alcohol and the carboxylic acid anion, lies far to the right. Thus, even though the Cannizzaro reaction is an equilibrium reaction, it proceeds nearly to completion. The mechanism is supported by evidence obtained by running the reaction in D_2O. It was found that the product alcohol did not contain α-deuterium substitution, indicating that the transferred hydride ion must come from a molecule of aldehyde and not from the solvent.

High yields of alcohol can be obtained from almost any aromatic aldehyde by running the reaction in the presence of excess formaldehyde. The formaldehyde, acting as the reducing agent, is in turn oxidized to formic acid.

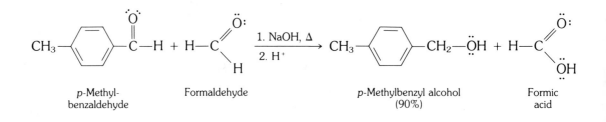

| p-Methyl- | Formaldehyde | p-Methylbenzyl alcohol | Formic |
| benzaldehyde | | (90%) | acid |

This procedure is known as a *crossed* Cannizzaro reaction.

α-Keto aldehydes undergo an internal Cannizzaro reaction to yield α-hydroxycarboxylic acids.

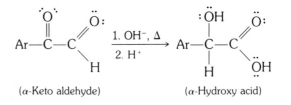

(α-Keto aldehyde)   (α-Hydroxy acid)

PART A: Reaction with 4-Chlorobenzaldehyde

EXPERIMENTAL   Estimated time to complete the experiment: two laboratory periods.

Physical Properties of Reactants and Products

Compound	MW	Wt/Vol	mmol	mp(°C)	bp(°C)	n_D
4-Chlorobenzaldehyde	140.57	150 mg	1.1	47.5		
Methanol	32.04	400 μL			65	1.3288
Potassium hydroxide (11 M)	56.11	400 μL				
4-Chlorobenzoic acid	156.57			243		
4-Chlorobenzyl alcohol	142.59			75		

This reaction may be run in a centrifuge tube containing a boiling stone. The tube should be loosely fitted with a cotton plug.

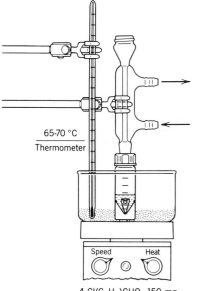

65-70 °C
Thermometer

Speed   Heat

4-Cl(C₆H₄)CHO, 150 mg
+ KOH/CH₃OH, 0.4 mL

Reagents and Equipment

To a 5.0-mL conical vial containing a magnetic spin vane and equipped with a reflux condenser are added 150 mg (1.1 mmol) of 4-chlorobenzaldehyde and 0.4 mL of methanol.(■) With gentle swirling, 0.4 mL of an 11 M aqueous solution of potassium hydroxide is then added.

Note. *It is convenient to dispense the methanol and KOH solution using automatic delivery pipets. It is recommended that the glass equipment NOT be rinsed or cleaned with acetone followed by air-drying. The residual acetone undergoes the aldol reaction, which causes problems in isolation of the desired products.*

> **CAUTION: *The concentrated KOH solution is very caustic. Do not allow it to come into contact with the skin or eyes.***

Reaction Conditions

The reaction mixture is heated with stirring in a sand bath at 65–75°C for a period of one hour.

Isolation of Products

The reaction mixture is allowed to cool and 2.0 mL of chilled distilled water is added. The resulting solution is now extracted with three 0.5-mL portions of methylene chloride using a Pasteur filter pipet to transfer the extracts to a 3.0-mL conical vial. On each addition of the methylene chloride, the vial is capped, shaken gently, and then carefully vented by loosening the cap. A Vortex mixer may be used for this extraction step. After separation of the layers, the lower methylene chloride layer is removed using a Pasteur filter pipet.

IMPORTANT. *Save the alkaline phase for further workup.*

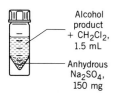

Alcohol
product
+ CH₂Cl₂,
1.5 mL

Anhydrous
Na₂SO₄,
150 mg

1. *4-Chlorobenzyl alcohol.* The combined methylene chloride extracts are washed with two 0.25-mL portions of saturated sodium bicarbonate solution followed by one 0.5-mL. portion of distilled water. The aqueous upper phase is removed (Pasteur filter pipet) and discarded. The methylene chloride layer is now dried over 150 mg of granular anhydrous sodium sulfate.(■) The solution, after drying, is transferred by use of a Pasteur filter pipet to a tared 3.0-mL conical vial. The sodium sulfate drying agent is rinsed with 0.3 mL of fresh methylene chloride, and the rinse is combined with the dried organic phase. The methylene chloride solvent is now evaporated using a stream of dry nitrogen gas in a warm sand bath in the **hood** to yield the crude 4-chlorobenzyl alcohol.

HOOD

2. *4-Chlorobenzoic acid.* The alkaline phase remaining from the original extraction procedure is diluted by the addition of 2.0 mL of water and acidified with the addition of 0.4 mL of concentrated hydrochloric acid. The voluminous white precipitate of the product is collected under reduced pressure by use of a Hirsch funnel, and the filter cake is rinsed with 2.0 mL of distilled water.(■) Air-drying on a porous clay plate gives crude 4-chlorobenzoic acid.

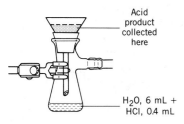

Acid
product
collected
here

H₂O, 6 mL +
HCl, 0.4 mL

This point in the procedure is a logical place to divide the experiment into two laboratory sessions if this seems appropriate.

Purification and Characterization

1. *4-Chlorobenzyl alcohol.* The crude alcohol is purified by recrystallization from a solution of 4% acetone in hexane (0.25 mL). Collection of the product

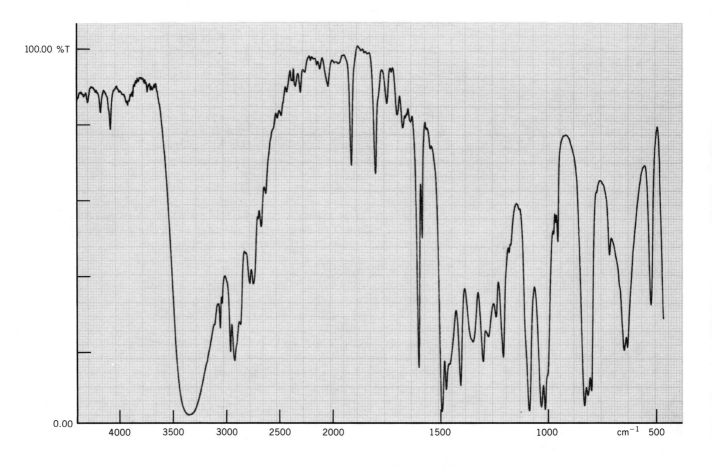

Sample	4-Chlorobenzyl alcohol		
%T _X_ ABS __ Background Scans _4_		Scans ___16___	
Acquisition & Calculation Time _42 sec_		Resolution _4.0 cm⁻¹_	
Sample Condition ___solid – melt___		Cell Window ___KBr___	
Cell Path Length ___capillary film___		Matrix Material ___	

Fig. 5.50 *IR spectrum: 4-chlorobenzyl alcohol.*

under reduced pressure using a Hirsch funnel, followed by washing of the filter cake with 0.2 mL of cold hexane, gives the desired 4-chlorobenzyl alcohol. The product is air-dried on a porous clay plate or on filter paper.

Weigh the 4-chlorobenzyl alcohol and calculate the percentage yield. Determine the melting point and compare your value with that found in the literature. Obtain the IR spectrum and compare it with that in Figure 5.50.

2. *4-Chlorobenzoic acid.* The crude acid may be purified by recrystallization from methanol by use of a Craig tube.

Weigh the dried material and calculate the percentage yield. Determine the melting point and compare your value to that found in the literature. Obtain the IR and compare it with that in Figure 5.51.

Infrared Analysis

The substrate molecule is 4-chlorobenzaldehyde. The infrared spectrum of this aromatic aldehyde (Fig. 5.52) is rich and interesting. The aromatic aldehyde macro group fre-

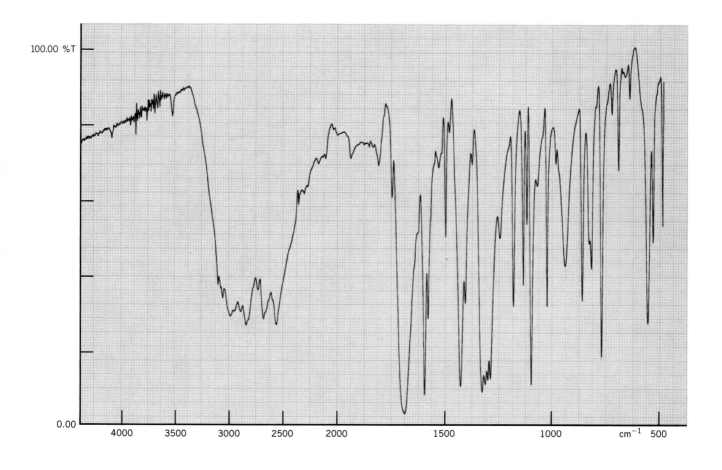

Fig. 5.51 *IR spectrum: 4-chlorobenzoic acid.*

Sample ___4-Chlorobenzoic acid___

%T _X_ ABS __ Background Scans _4_   Scans ___16___

Acquisition & Calculation Time _42 sec_   Resolution _4.0 cm⁻¹_

Sample Condition ___solid___   Cell Window _____

Cell Path Length _____   Matrix Material _KBr_

quency train consists of peaks near 3090, 3070, 2830 & 2750, 1706, 1602, 1589, and 1396 cm⁻¹.

a. 3070 cm⁻¹: C—H stretch on sp^2 carbon.

b. 2840 & 2740 cm⁻¹: this pair of bands is a famous example of powerful Fermi coupling (see Appendix B). The unperturbed C—H stretching of the aldehyde C—H group would be expected to occur near 2790 cm⁻¹. The in-plane bending mode of this C—H group occurs at 1390 cm⁻¹. Thus, the first harmonic should fall near 2780 cm⁻¹, very close (10 cm⁻¹) to the stretching frequency of this oscillator. The basic conditions are met for strong Fermi coupling, which occurs and gives rise to the split peaks 2840 and 2740 cm⁻¹. The latter band is moved well away from the normal sp^3 C—H symmetric stretching modes. Thus, the lower wavenumber component of the Fermi coupled aldehyde C—H mode leads to easy identification even when it represents a very small fraction of an aliphatic system. In the present case where no aliphatic C—H oscillators are present, both components are observed.

c. 1704 cm⁻¹: the carbonyl stretch of the aldehyde group. The frequency observed in aliphatic aldehydes falls in the range 1735–1720 cm⁻¹, but when conjugated, the

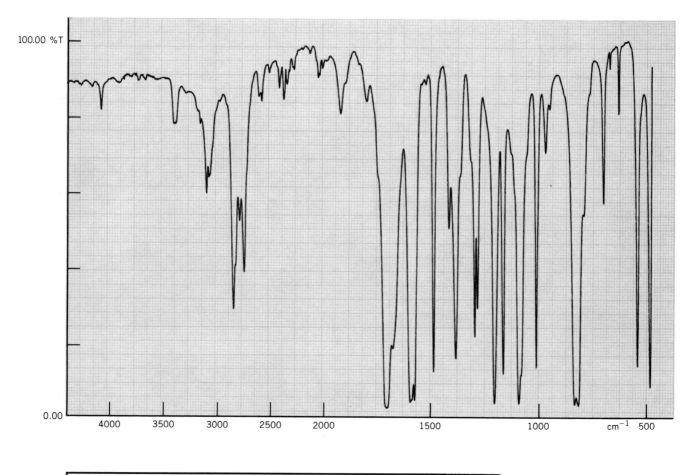

Sample	4-Chlorobenzaldehyde		
%T _X_ ABS — Background Scans _4_		Scans	_16_
Acquisition & Calculation Time _42 sec_		Resolution	_4.0 cm⁻¹_
Sample Condition _solid – melt_		Cell Window	_KBr_
Cell Path Length _capillary film_		Matrix Material	_____

Fig. 5.52 *IR spectrum: 4-chlorobenzaldehyde.*

value drops 15–25 cm^{-1} (see Chapter 7) and is found in the range 1720–1700 cm^{-1}.

d. 1601 & 1578 cm^{-1}: this pair of bands is related to the degenerate ring stretching vibrations, ν_{8a} and ν_{8b}, of benzene (also see infrared discussions in Experiments 16B and Chapter 7).

e. 1390 cm^{-1}: the aldehyde C—H in-plane bending vibration. The first harmonic of this vibration is Fermi coupled to the aldehyde C—H stretching mode as discussed earlier.

This compound also possesses a second powerful macro group frequency train which assesses the substitution pattern of the aromatic ring system (see Chapter 7). The para-disubstituted benzene ring macro group frequency train requires peaks in the following regions: 1950, 1880, 1800, 1730, 750, and 690 cm^{-1}.

a. 1905, 1795 cm^{-1}: this pair of two weak bands, with the higher wavenumber band more intense than the lower member, arises from combination bands (see Appendix B) which involve the out-of-plane bending frequencies of the ring C—H bonds (see below). The exact wavenumber positions are not very important, but the overall shape of the pattern can be used to determine the ring substitution pattern (see Chapter 7).

b. 832 cm^{-1}: this strong band is very characteristic of para-disubstituted benzene rings. The 832 cm^{-1} peak arises from the in-phase out-of plane bending vibration of the two pairs of C—H groups on opposite sides of the six-membered ring (see Chapter 7).

The C—Cl stretching mode when substituted on an aromatic ring drops below the range of the instrument. The presence of this group must be determined by other methods such as a Beilstein or sodium fusion test (see Chapter 7). The presence of a hidden group (very likely halogen), however, is strongly indicted. Because the system must be para-substituted (see earlier) one of the substituents on the ring must exhibit very little absorption from 4000 to 500 cm^{-1}.

The reaction involves the formation of two products, one neutral and one acidic, both of which incorporate large portions of the substrate molecule. These materials can be characterized by their infrared spectra. The neutral product is proposed to be 4-chlorobenzyl alcohol. An examination of the spectrum (Fig. 5.50) supports the presence of two macro group frequency trains. One (a) is the same one found in the starting aldehyde, the para-disubstituted benzene ring frequency train. The other (b) is a primary aliphatic alcohol macro frequency train. Macro (a) has been expanded to include all aromatic ring specific group frequencies, as the alcoholic side chain is decoupled from the ring and the carbon—chlorine frequencies fall outside the range of the instrumentation normally used in these measurements.

The macro group frequencies for the neutral product are as follows.

a. 3055 (aromatic C—H stretch), 1906 & 1793 (para combination band pattern), 1596 & 1583 (ring stretch degenerate pair, ν_{8a} & ν_{8b}; 1583 intensity indicates weak conjugation with the ring), 1495 & 1475 (ring stretch degenerate pair, ν_{19a} & ν_{19b}), 834 (C—H, out-of-plane bend) cm^{-1}.

b. 3340 (broad, O—H stretch), 2965–2925 (C—H, aliphatic—sp^3), 1450–1300 (broad, O—H bend, associated), 1015 (C—O, stretch, primary alcohol), 630 (broad, weak O—H bend, associated) cm^{-1}.

The acidic product generated in the reaction is assumed to be 4-chlorobenzoic acid. This material also is a para-substituted substance; thus, the spectrum of this compound (Fig. 5.51) possesses a macro frequency train (a) similar to those of the aldehyde and alcohol. In addition, this benzoic acid derivative exhibits an extended aromatic acid macro group frequency train (b). The macro frequencies are as follows.

a. 1935 & 1795 (para combination band pattern), 852 (ring C—H, in-phase, out-of-plane bend) cm^{-1}.

b. 3400–2200 (very broad, very strong, O—H stretch, associated acid), 1683 (C=O, stretch, out-of-phase, associated acid dimer), 1596 & 1578 (degenerate ring stretch), 1500 & 1432 (degenerate ring stretch), 1320–1280 (C—O, stretch), 885 (O—H, out-of-plane, ring dimer bend) cm^{-1}.

The carbon chlorine stretch is not observed.

Examine the spectra of the reaction products you have obtained in a potassium bromide matrix. Discuss the similarities and differences of the experimentally derived spectral data to the reference spectra (Figs. 5.50–5.52).

It is interesting to compare the results of chemical tests used to classify the preceding compounds with the spectral data available from the IR technique.

Chemical Tests

Perform each of the following tests (see Chapter 7). Do the results confirm that you have isolated an aromatic carboxylic acid and an aromatic alcohol?

a. The ignition test.

b. The Beilstein or the sodium fusion test for the halogen present.

c. The ceric nitrate and/or the Jones oxidation test for the alcohol.

d. The solubility of the acid in sodium bicarbonate and sodium hydroxide solutions. Is carbon dioxide evolved in the bicarbonate test?

If you were to prepare a derivative for the alcohol and acid products, which one would you choose? See Chapter 7, Preparation of Derivatives.

PART B: Reaction with 4-Bromobenzaldehyde

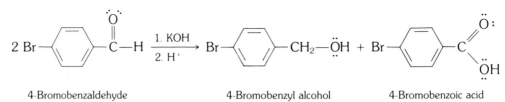

4-Bromobenzaldehyde   4-Bromobenzyl alcohol   4-Bromobenzoic acid

EXPERIMENTAL   Estimated time to complete the experiment: two laboratory periods.

Physical Properties of Reactants and Products

Compound	MW	Wt/Vol	mmol	mp(°C)	bp(°C)
4-Bromobenzaldehyde	185.05	150 mg	0.81	67	
Methanol	32.04	400 μL			65
Potassium hydroxide (11 M)	56.11	400 μL			
4-Bromobenzoic acid	201.03			254.5	
4-Bromobenzyl alcohol	187.04			77	

Amounts of reagents (on a weight basis), reaction conditions, isolation technique, purification procedures, and characterization methods used in this experiment are identical to those employed for 4-chlorobenzaldehyde.

The chemical tests outlined in Part A may also be carried out on these products.

ENVIRONMENTAL DATA

Substance	Amount (mL)	TLV (mg/m³)	Emissions (mg)	Volume (m³)
Experiment 7A				
4-Chlorobenzaldehyde (Max. from yield, transfer)	150 mg	—	54	
Methanol (SKIN) (Max., Craig tube)	1.4	260	1120	4.3
Methylene chloride (Max., extract, evap.)	1.8	350	2400	6.9
Hexane (Max., recryst.)	0.45	180	300	1.7
Experiment 7B				
4-Bromobenzaldehyde (Max. from yield, transfer)	150 mg	—	54	
Methanol (SKIN) (Max., Craig tube)	1.4	260	1120	4.3
Methylene chloride (Max., extract, evap.)	1.8	350	2400	6.9
Hexane (Max., recryst.)	0.45	180	300	1.7

TLV: Potassium hydroxide—2 mg/m³
        Hydrochloric acid—5 ppm
TXDS: Sodium bicarbonate—orl-rat LD50: 4220 mg/kg
        Sodium sulfate—orl-mus LD50: 5989 mg/kg

QUESTIONS   **5-35.** The discussion mentions that the crossed Cannizzaro reaction can be realized when one of the components is formaldehyde. Predict the products of the reaction below and give suitable names to the reactants and products.

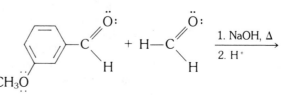

5-36. One group of investigators has suggested that a dianion

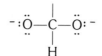

might be the source of hydride in the Cannizzaro reaction. Explain why this species would be a better source of hydride than the species depicted in the mechanism in the discussion section.

5-37. The Cannizzaro reaction is an oxidation–reduction sequence. In reference to Question 5-35, formaldehyde is acting as what type of reagent?

5-38. Sketch the NMR spectrum you would expect to observe for each of the following compounds. (*Hint:* See Experiment 5.)

- **a.** Acetone
- **b.** 1,1,2-Tribromoethane
- **c.** Ethyl chloride
- **d.** 2-Iodopropane
- **e.** 1-Bromo-4-methoxybenzene

5-39. Propose a mechanism for the internal Cannizzaro reaction depicted here.

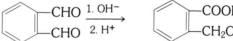

o-Phthaldehyde   o-(Hydroxymethyl)benzoic acid

REFERENCES

1. Reviews on the Cannizzaro reaction:
   a. Geissman, T. A. *Org. React.* **1944,** *2,* 94.
   b. Swain, C. G.; Powell, A. L.; Sheppard, L. A.; Morgan, C. R. *J. Am. Chem. Soc.* **1979,** *101,* 3576.
2. Examples of the Cannizzaro reaction:
   a. Wilson, W. C. *Organic Syntheses*; Wiley: New York, 1941; Collect. Vol. 1, p 256.
   b. Davidson, D.; Weiss, M. *Organic Syntheses*; Wiley: New York, 1943; Collect. Vol. II, p 590.

TECHNIQUE 8:
CHROMATOGRAPHY: COLUMN
AND THIN-LAYER;
CONCENTRATION OF
SOLUTIONS

The technique of chromatography is defined in Experiment 1 during the development of gas-phase separations. The term is derived from the Greek word for color, *chromatos*. Tswett discovered the technique (1903) during studies centered on the separation of mixtures of natural plant pigments.[8] The chromatographic zones were detected simply by observing the visual absorption bands (see Experiment 32, 34D). Thus, as originally applied, the name was not an inconsistent use of terminology. Today, however, the very large majority of mixtures that are chromatographed are colorless materials. The separated zones in these cases are established by other methods. Interestingly, Tswett means "color" in Russian (the discoverer's nationality).

In this experiment two additional chromatographic techniques are explored. Both of these procedures depend on adsorption and distribution between a stationary solid phase and a moving liquid phase. The first to be discussed is "column chromatography," a very powerful technique used extensively throughout microscale experiments. It was one of the earliest of the modern chromatographic methods to be applied to the separation of organic mixtures by

[8] Tswett, M. *Ber. Deut. Botan. Ges.* **1906,** *24,* 235.

Tswett. The second procedure, "thin-layer chromatography," was developed in the late 1950s. TLC is particularly effective in rapid assays of sample purity. It is also effective when employed as a preparative technique for obtaining high-quality material for analytical data.

COLUMN CHROMATOGRAPHY

The term *column chromatography* is derived from the use of a glass column packed with a solid stationary phase, a relationship similar to the liquid phase of gas chromatography. The moving liquid phase descends by gravity through the column.

A wide variety of substances have been employed as the stationary phase in this technique. In practice, however, two materials have become dominant in this type of separation chemistry. Finely ground (100–200 mesh) alumina (aluminum oxide, Al_2O_3) and silicic acid (silica gel, SiO_2) are by far the most useful of the known adsorbents. The liquids, which act as the moving phase and elute (wash) sample materials through the column, are many of the common organic solvents. Table 5.6 lists the better known column packings and elution solvents.

Silica gel impregnated with silver nitrate (usually 5–10% $AgNO_3$) is an attractive solid-phase adsorbent. The silver salt selectively binds to unsaturated sites via a silver ion π complex. Traces of olefinic materials are easily removed from saturated reaction products by chromatography with this system (Experiment 12). This adsorbent, however, must be protected from light until used or the mixture will rapidly darken and become ineffective.

The procedure is usually carried out as follows.

Packing the Column

The quantity of stationary phase required is determined by the sample size. A common rule of thumb is to use 30–100 times the weight of packing to the amount of sample to be chromatographed. The size of the column is chosen to give roughly a $10:1$ ratio of height to diameter for the amount of adsorbent required (Fig. 5.53).

In the microscale laboratory program two standard chromatographic columns are employed.

1. A Pasteur pipet modified by shortening the capillary tip is used for the separation of smaller mixtures (10–100 mg). Approximately, 0.5–2.0 g of packing is used in the pipet column.

2. A 50-mL titration buret modified by reducing the length of the column (10 cm beyond the stopcock) is used for the larger sample mixtures (50–200 mg) and for the difficult-to-separate mixtures. Approximately 5–20 g of packing is employed in the buret column (Fig. 5.53).

Table 5.6 Column Chromatography Materials

Stationary Phase		Moving Phase	
Alumina	↑ Increasing	Water	↑ Increasing
Silicic acid	adsorption	Methanol	solvation
Magnesium sulfate	of polar	Ethanol	of polar
Cellulose-Paper	materials	Acetone	materials
		Ethyl acetate	
		Diethyl ether	
		Methylene chloride	
		Cyclohexane	
		Pentane	

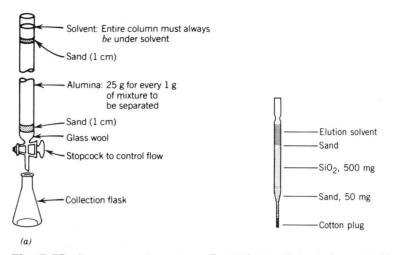

(a)

Fig. 5.53 *Chromatographic columns. Fig. 5.53a from* Zubrick, James W. *The Organic Chem Lab Survival Manual,* 2nd ed.; Wiley: New York, 1988. (Reprinted by permission of John Wiley & Sons, New York.)

Both columns are prepared by first clamping the empty column in a vertical position and then seating a small cotton or glass wool plug at the bottom. The cotton is covered with a thin layer of sand in the case of the buret. The Pasteur pipets are loaded by adding the adsorbent with gentle tapping, "dry packing." The column is then premoistened just prior to use. The burets are packed by a slurry technique. In this procedure the column is filled part way with solvent; then the stopcock is opened slightly, and as the solvent slowly drains from the column a slurry of the adsorbent–solvent is poured into the top of the column. The column should be gently tapped while the slurry is added. The solvent is then drained to the top of the adsorbent level and held at that level until use.

Sample Application
The sample is applied in a minimum amount of solvent (usually the least polar solvent in which the material is readily soluble) to the top of the column by Pasteur pipet. The pipet is rinsed and the rinsings are added to the column just as the sample solution drains to the top of the adsorbent layer.

Elution of the Column
The critical step in resolving the sample mixture is eluting the column. Once the sample has been applied to the top of column, the elution begins (a small layer of sand can be added to the top of the buret column after addition of the first charge of elution solvent). *It is very important not to let the column run dry.* The Pasteur pipet is free flowing (the flow rate is controlled by the size of the capillary tip), and once the sample is on the column, the chromatogram will require constant attention. Buret column flow is controlled by the stopcock. The flow rate should be set to allow time for equilibrium to be established between the two phases. The choice of solvent is dictated by a number of factors. A balance between the adsorption power of the stationary phase and the solvation power of the elution solvent will govern the rate of travel of the material descending through the stationary phase. If the material travels rapidly down the column, then too few adsorption–elution cycles will occur and the materials will elute together in one fraction. If the sample travels too slowly, diffusion broadening takes over and resolution is degraded. In the latter case, samples then elute over

many fractions with overlapping broad bands. The ideal solvent and elution rate strike a balance between these two situations and maximizes the separation. It can take considerabie time to develop a solvent or mixture of solvents that produces a satisfactory separation of a particular mixture. The optimum elution parameters for the experiments in the microscale laboratory have been established to conserve time during the laboratory period.

Fraction Collection

As the solvent elutes from the column, it is collected in a series of "fractions" using small Erlenmeyer flasks or vials. Under ideal conditions, as the mixture of material travels down the column, it will separate into several individual bands of pure substances. By careful collection of the fractions, these bands can be separated as they sequentially elute from the column (similar to the collection of GC fractions in Experiment 1). The bands of material being eluted can be detected by a number of techniques (weighing fraction residues, visible absorption bands, thin-layer chromatography, etc.). The collection protocol is usually established in the microscale experiments.

CONCENTRATION OF SOLUTIONS   The solvent can be removed from the chromatographic fractions by a number of different methods.

Distillation

Concentration of solvent by distillation is straightforward, and the standard routine is described in Experiment 2. This allows for high recovery of volatile solvents and often can be done outside a hood. The Hickman still head and the 5-mL round-bottom flask are useful in this purpose. Distillation should be used primarily for concentration of the solution, followed by transfer of the concentrate with a Pasteur filter pipet to a vial for final isolation.

Evaporation with Nitrogen Gas

A very convenient method for removal of final solvent traces is concentration of the last 0.5 mL of solution by evaporation with a gentle stream of nitrogen gas while the sample is warmed in a sand bath. This is usually done at a hood station where several Pasteur pipets can be attached to a manifold leading to a tank of compressed gas. Gas flow to the individual pipets is controlled by needle valves. *Always test the gas flow with a blank vial of solvent.* This is a hand-held operation. The sample vial will cool rapidly on evaporation of the solvent, and gentle warming of the vial with agitation will thus aid removal of the last traces of the volatile material. This procedure avoids possible moisture condensation on the sample residue. *Do not leave the heated vial in the gas flow after the solvent is removed!* This is particularly important in the isolation of liquids. Remember to tare the vial before loading the solution to be concentrated, because achievement of constant weight is the best indication of total solvent removal.

Removal of Solvent Under Reduced Pressure

Concentration of solvent under reduced pressure is very efficient. It reduces the time of solvent removal in microscale experiments to a few seconds or at most a few minutes. In contrast, distillation or evaporation procedures require several minutes to tens of minutes for even relatively small volumes. Vacuum concentration, however, is tricky and should be practiced prior to committing hard-won reaction product to this test. The procedure is most beneficial when applied to fairly large chromatographic fractions (5–10 mL).

The sequence of operations is as follows (see also Fig. 5.54):

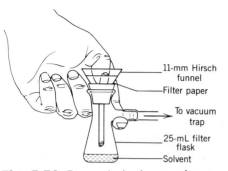

Fig. 5.54 *Removal of solvent under reduced pressure.*

1. Transfer the chromatographic fraction to the 25-mL filter flask.
2. Insert the 11-mm Hirsch funnel and rubber adapter into the flask.
3. Turn on the water pump (with trap) and connect the vacuum tubing to the pressure flask side arm while holding the flask in one hand (this is another hand-held operation).
4. Place the thumb of the hand holding the filter flask over the Hirsch funnel filter bed to shut off the air flow through the system (Fig. 5.54). This will result in an immediate drop in pressure. The volatile solvent will rapidly come to a boil at room temperature. Thumb pressure adjusts air leakage through the Hirsch funnel and thereby controls the pressure in the system. It is also good practice to learn to manipulate the pressure so that the liquid does not foam up into the side arm of the filter flask.

It is essential that the filter flask be warmed by the sand bath during this operation, for rapid evaporation of the solvent will quickly cool the solution. The air leak used to control the pressure results in a stream of moist laboratory air being rapidly drawn over the surface of the solution. If the evaporating liquid becomes cold, water will condense over the interior of the filter flask and eventually contaminate the isolated residue. Warming the flask while the evaporation process is being carried out will avoid this problem and help to speed solvent removal. The temperature of the flask should be checked from time to time by touching it with the palm of the free hand. The flask is kept slightly above room temperature by adjusting the heating and evaporation rates. It is best to practice this operation a few times with solvent blanks to see whether you can avoid boilovers and the accumulation of water residue in the flask.

Column chromatography is a powerful technique for the purification of organic materials. In general, it is significantly more efficient than crystallization procedures. Thus, this technique is used extensively in microscale laboratory experiments. Recrystallization is avoided until the last stages of purification, where it will be most efficient, but chromatography does the real dirty work. *One major advantage of working with small amounts of product is that the chromatographic times are shortened dramatically.* Column chromatography of a few milligrams of product usually takes no more than half an hour, but to chromatograph 10 g of product might take a whole afternoon or the better part of a day. Large-scale chromatograms (50–100 g) may take several days or even several weeks to complete.

THIN-LAYER CHROMATOGRAPHY

Thin-layer chromatography is another solid–liquid partition technique of more recent development. It is a close relative to column chromatography in that the phase materials used in both techniques are essentially identical. That is, alumina and silica gel are used as stationary phases and the moving phases are the usual solvents. There are, however, some distinct operational differences between thin-layer and column chromatography. While the moving phase descends in column chromatography, in thin-layer chromatography the solvent front ascends. The column of stationary-phase material used in column chromatography is replaced, in thin-layer chromatography, by a very thin layer (100 μm) of the material spread over a flat surface. The technique has some distinct advantages at the microscale level. It is very rapid (10–20 minutes), and it employs *very* small quantities of material (2–20 μg). The chief disadvantage of this type of chromatography is that it is not very amenable to preparative scale work. Even when large surfaces and thicker layers are used, separations are most often restricted to the 5–10 mg level unless sophisticated research equipment is available.

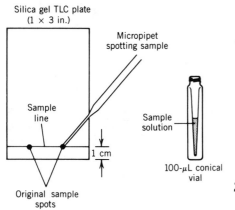

Silica gel TLC plate
(1 × 3 in.)

Micropipet
spotting sample

Sample
line

Sample
solution

1 cm

100-μL conical
vial

Original sample
spots

Fig. 5.55 *Sample application to a TLC plate.*

The sequence of operations is as follows.

1. A piece of window glass, a microscope slide, or a sheet of plastic can be used as a support for a thin layer of adsorbent spread over the surface. It is possible to prepare locally the glass surfaces, but plastic-supported thin-layer systems are only commercially available. The plastic-supported layers are particularly attractive because they possess very uniform coatings and are highly reproducible in operation. Another convenient feature of the plastic-backed plates is that they can be cut with scissors into very economical 1 × 3-in. strips. The latter style is used exclusively in the microscale laboratory.

2. A pencil line is drawn parallel to a short side of the plate 1.0 cm from the edge. One or two points evenly spaced are marked on the line. The sample (1 mg) to be analyzed is placed in a 100-μl conical vial and a few drops of solvent are added. A micropipet (prepared by the same technique used for constructing the capillary insert in the ultramicro boiling point determination; see Chapter 4) is used to apply a small fraction of the solution from the vial to the plate (Fig. 5.55).

3. The chromatogram is carried out by placing the spotted thin-layer plate in a screw-capped wide-mouth jar or a beaker with a watch glass cover containing a small amount of developing solvent (Fig. 5.56). The material spot on the TLC plate must initially be positioned above the solvent line. The jar is quickly recapped or the watch glass replaced to maintain an atmosphere saturated with the developing solvent. The elution solvent rapidly ascends the plate by capillary action. The choice of solvent will be similar to that used in column chromatography, but need not be identical. The spotted material is eluted vertically up the plate. Resolution of mixtures into individual spots along the vertical axis occurs by precisely the same mechanism as in column chromatography. Development is interrupted when the solvent line nears the top of the plate.

4. Visualization of colorless separated components is achieved by placing the plate in an iodine vapor chamber for a few seconds. Iodine forms a reversible complex with most organic substances. Thus, dark spots will develop in those areas containing sample material. On removal from the iodine chamber, the spots are marked by pencil, because they will fade rather rapidly. Sample quenching of UV activated fluorescent indicator coated TLC plates is an alternative mode of detection. The elution characteristics are reported as R_f values. The R_f value is a measure of the travel of a substance up the plate during the chromatogram relative to the solvent movement. This value is defined as the length of migration by the substance divided by the distance

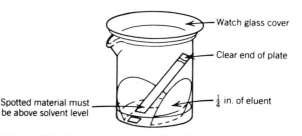

Watch glass cover

Clear end of plate

$\frac{1}{4}$ in. of eluent

Spotted material must be above solvent level

Fig. 5.56 *Development of a TLC plate.* From Zubrick, James W. *The Organic Chem Lab Survival Manual,* 2nd ed.; Wiley: New York, 1988. (Reprinted by permission of John Wiley & Sons, New York.)

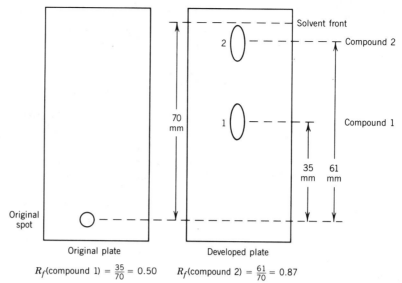

Fig. 5.57 *A sample calculation of R_f values.*

R_f(compound 1) $= \frac{35}{70} = 0.50$ R_f(compound 2) $= \frac{61}{70} = 0.87$

traversed by the solvent front (the position of the solvent front should be quickly marked on the plate when the chromatogram is terminated; see Fig. 5.57).

Thin-layer chromatography is used in a number of applications. The speed of the technique makes it quite useful for monitoring large-scale column chromatograms. Analysis of fractions can guide decisions on the solvent elution sequence. TLC analysis of column-derived fractions can also give an indication of how best to combine collected fractions. Following the progress of a reaction by periodically removing small aliquots for thin-layer analysis is another useful application of thin-layer chromatography. The technique is utilized in several microscale experiments, for example, Experiments 6, 17, 21A, 32, 34D, 40A, 41, and 45.

Experiments 8A, 8B, 8C, 8D, 8E

The Esterification Reaction: Ethyl Laurate; Octyl Acetate; Isopentyl Acetate; 4-*tert*-Butylcyclohexyl Acetate; Benzoin Acetate

(lauric acid, ethyl ester; acetic acid, octyl ester; acetic acid, isopentyl ester; cyclohexanol, 4-*t*-butyl acetate; ethanone, 2-(acetyloxy)-1,2-diphenyl-)

The experiments to be described demonstrate the preparation of organic esters by treatment of a carboxylic acid or anhydride with an alcohol in the presence of an acid catalyst.

DISCUSSION   Esterification is one of the important reactions in organic chemistry and has been studied extensively.

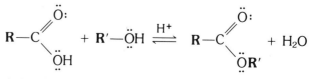

A wide range of esters are important as chemical intermediates in organic synthesis and find extensive use in industrial product applications from fingernail polish remover and pain-relieving drugs to polymeric fiber compositions and surfactants. They are also distributed widely in nature in the form of fatty acid esters, waxes, and oils.

The majority of esters are prepared by one of four basic routes.

1. Direct esterification of a carboxylic acid with an alcohol in the presence of an acid catalyst.
2. Alcoholysis of acid chlorides, anhydrides, or nitriles.
3. Reaction of a carboxylic acid salt with an alkyl halide or sulfate.
4. The transesterification reaction.

Direct esterification, known as Fischer esterification, is the method used for the preparation of ethyl laurate in Part A and of isopentyl acetate in Part B of this series. The reaction proceeds by nucleophilic attack of the alcohol on the protonated carbonyl group of the carboxylic acid to form a tetrahedral intermediate. Regeneration of the carbonyl group produces the ester and water. The overall sequence is outlined here.

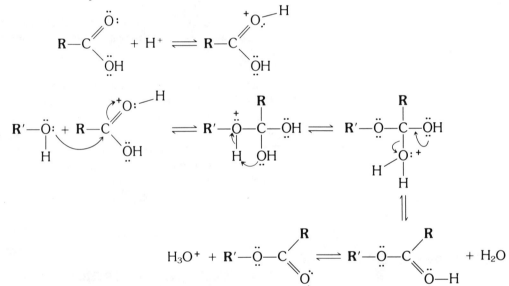

The Fischer method of preparing esters is an equilibrium reaction and, therefore, to obtain substantial yields of the ester product, the reaction must be shifted to the side of the ester. This is accomplished in several ways. If the alcohol is relatively cheap, such as methanol or ethanol, an excess of the alcohol is used to shift the position of equilibrium. This technique is used in the preparation of ethyl laurate (Part A). An alternative is to use an excess of the carboxylic acid. A third option is removal of the ester or water (preparation of isopentyl acetate, Part C) formed in the reaction. The acid catalyst employed is generally dry hydrogen chloride, concentrated sulfuric acid, or p-toluenesulfonic acid.

When the carboxyl and hydroxyl groups are present in the same molecule a cyclic ester (called a lactone) may be formed. This is especially prevalent in the formation of five- or six-membered ring systems.

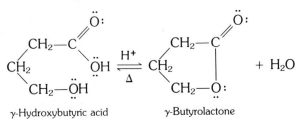

γ-Hydroxybutyric acid   γ-Butyrolactone

As noted earlier, the Fischer method of preparing esters is an equilibrium reaction. Heating an ester in the presence of an acid catalyst with water regenerates the corresponding carboxylic acid and alcohol. This reaction is called *acid hydrolysis* of an ester. The rate-determining step in both the esterification and hydrolysis reaction is the formation of the tetrahedral intermediate. It is therefore evident that the rate will be determined by the ease with which the nucleophile (alcohol on esterification and water on hydrolysis) approaches the carbonyl group. Steric and electrical factors have a large impact on the rate. The increase of bulky substituents on the α and β positions of the carboxylic acid or ester decreases the rate. Electron-withdrawing groups near the carbonyl group increase the rate because they increase the electrophilicity of the carbonyl carbon atom. Conversely, electron-donating groups retard the rate.

Another of the methods for the preparation of esters is also illustrated in this series. Octyl acetate, 4-*t*-butylcyclohexyl acetate, and benzoin acetate are prepared by the reaction of an alcohol with acetic anhydride, usually in the presence of an acid catalyst. This is a common technique for the preparation of acetate esters. One portion of the anhydride acylates the alcohol; the other forms a carboxylic acid.

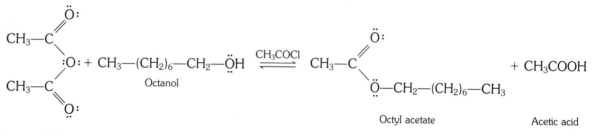

Of the preparations for the acetate esters listed earlier, two use acetyl chloride as a catalyst source. Acid chlorides are more reactive than anhydrides. Why? The acetyl chloride added to the reaction mixture reacts with a molecule of the alcohol to produce HCl *in situ*. The formation of the HCl is shown here; notice that the other product of this reaction is a molecule of the desired ester.

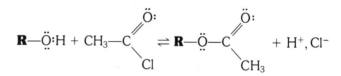

The hydrogen chloride generated acts as the catalyst for the reaction of the anhydride and alcohol to form the desired ester.

If the hydrolysis of an ester is carried out in the presence of base, the reaction is called *saponification*:

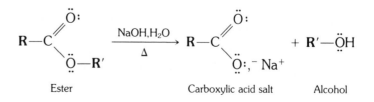

Saponification is essentially an irreversible reaction in which one mole of base is consumed per mole of ester to generate a carboxylic acid anion (as its salt). It is possible to carry out the reaction in a quantitative manner so that a value called the *saponification equivalent* can be obtained. This value is the molecular weight of the ester divided by the number of ester groups in the species being analyzed.

A weighed amount of the *ester* is heated with an excess (known volume) of standard alkaline hydroxide solution, and the excess base is then determined by titration with standard acid.

When the saponification reaction is carried out on a fat or oil belonging to the naturally occurring class of compounds called *lipids*, a soap is obtained:

$$CH_3-(CH_2)_{16}-C \begin{matrix} \ddot{O}: \\ \end{matrix} \ddot{O}-CH_2$$

Stearic acid ester of glycerol
(a triglyceride)

Sodium stearate
(a soap)

Glycerol

Fats and oils are esters of glycerol with long-chain carboxylic acids. We should note that waxes are esters of long-chain carboxylic acids and long-chain primary alcohols.

Cetyl stearate

Triacontyl palmitate

The condensation of difunctional carboxylic esters with difunctional alcohols using a reaction called *transesterification* has led to the synthesis of high-molecular-weight polyesters. Many of these polymeric materials have achieved industrial importance. The reaction used to prepare the textile fiber named Dacron is

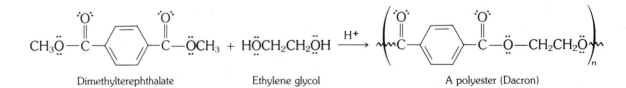

Dimethylterephthalate   Ethylene glycol   A polyester (Dacron)

The transesterification reaction is acid-catalyzed and is used to convert one ester into another by heating with an excess of an alcohol or carboxylic acid. In the following example, the equilibrium is shifted to the right by removal of the ethanol as it is formed.

PART A: Ethyl Laurate

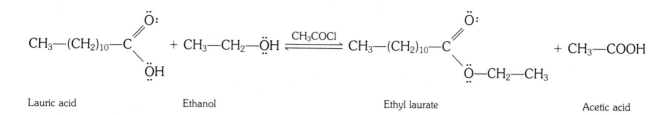

| Lauric acid | Ethanol | Ethyl laurate | Acetic acid |

DISCUSSION   See experimental introductory discussion.

EXPERIMENTAL   Estimated time to complete the experiment: 3.0 hours.

Physical Properties of Reactants and Products

Compound	MW	Wt/Vol		mmol	mp(°C)	bp(°C)	Density	n_D
Lauric acid	200.33	70	mg	0.35	44			
Ethanol	46.07	1.0 mL		17		78.5	0.7	1.3611
Acetyl chloride	78.50	30	μL	0.42		50.9	1.11	1.3898
Ethyl laurate	228.36					273	1.43	1.4311

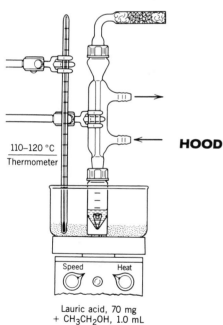

110–120 °C
Thermometer

Speed   Heat

Lauric acid, 70 mg
+ CH₃CH₂OH, 1.0 mL
+ CH₃COCl, 30 μL

HOOD

Reagents and Equipment
To a 3.0-mL conical vial containing a magnetic spin vane and equipped with a reflux condenser protected by a calcium chloride drying tube, add 70 mg (0.35 mmol) of lauric acid.(■) Using a graduated 1.0-mL pipet, add 1.0 mL of absolute ethanol followed by 30 μL (0.43 mmol) of acetyl chloride.

> **WARNING: Acetyl chloride is an irritant and should be dispensed in the hood using an automatic delivery pipet. The vial should be capped immediately as this reagent is moisture sensitive.**

Reaction Conditions
The reaction mixture is now heated with stirring at reflux using a sand bath temperature of 110–120°C for a period of one hour. The resulting mixture is then cooled to room temperature and the spin vane removed with forceps.

Isolation of Product
A boiling stone is added to the vial and the reaction solution concentrated to a volume of ~0.25 mL by warming in a sand bath in the **hood**.

To the resulting product mixture are added 0.5 mL of diethyl ether and 0.25 ml of 5% sodium bicarbonate solution using a 1.0-mL graduated pipet. The conical vial is capped and shaken gently (or mixed on a Vortex mixer) and the cap carefully loosened to vent the two-phase mixture. The aqueous layer is removed with a Pasteur filter pipet and discarded. The organic phase is then extracted with three additional 0.25-mL portions of 5% sodium bicarbonate solution.

HOOD

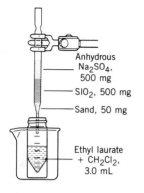

Anhydrous
Na₂SO₄,
500 mg

SIO₂, 500 mg

Sand, 50 mg

Ethyl laurate
+ CH₂Cl₂,
3.0 mL

This extraction procedure also ensures removal of traces of ethyl acetate that might remain in the reaction mixture after the initial concentration of the reaction mixture. Ethyl acetate has a boiling point (76.5–77.5°C) in the same range as that of ethanol; thus, the major portion of this ester is removed by evaporation.

The wet, crude ethyl laurate is dried and purified by column chromatography. In a Pasteur filter pipet is placed 500 mg of activated silica gel followed by 500 mg of anhydrous sodium sulfate.(■) The column is first wet with 0.5 mL of methylene chloride and the crude ethyl laurate placed on the column using a Pasteur pipet. A tared 5-mL conical vial containing a boiling stone is used as a collection flask. The reaction vial is rinsed with two 0.5-mL portions of methylene chloride and each rinse is also transferred to the column using the same pipet. An additional 1.0 mL of methylene chloride is then added directly to the column to ensure complete elution of the product.

HOOD   The methylene chloride is removed by evaporation in the **hood** by using a stream of nitrogen gas or by gentle warming in a sand bath.

Purification and Characterization

Ethyl laurate, a clear, viscous, pleasant-smelling ester, is sufficiently pure as isolated for characterization. Weigh the material and calculate the percentage yield. Determine the density, refractive index (optional), and boiling point and compare the results with the literature values.

Obtain an IR spectrum of the ester and compare it with that recorded in the literature.

Chemical Tests

Does this ester give a positive hydroxamate test (see Chapter 7)?

Check the solubility of this material in water. Would you have predicted the result? Is the ester soluble in 85% phosphoric acid or in concentrated sulfuric acid?

PART B: Octyl acetate

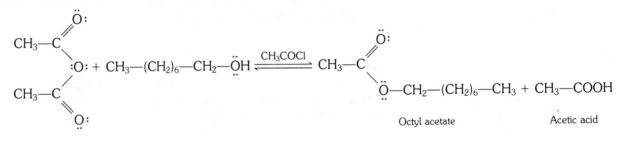

Acetic
anhydride

1-Octanol

Octyl acetate

Acetic acid

DISCUSSION   See experimental introductory discussion. The starting octanol may be prepared using the procedure in Experiment 13.

EXPERIMENTAL   Estimated time to complete the experiment: 3.0 hours.

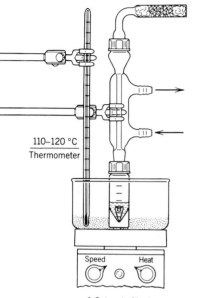

110–120 °C
Thermometer

Speed   Heat

1-Octanol, 61 µL
+ (CH₃CO)₂O, 121 µL
+CH₃COCl, 24 µL

Physical Properties of Reactants and Product

Compound	MW	Wt/Vol	mmol	bp(°C)	Density	n_D
1-Octanol	130.23	61 µL	0.38	194.4	0.83	1.4295
Acetyl chloride	78.50	24 µL	0.38	50.9	1.11	1.3898
Acetic anhydride	102.09	121 µL	1.28	139.6	1.08	1.3901
Octyl acetate	172.27			210	0.87	1.4150

Reagents and Equipment

In a 1.0-mL conical vial containing a magnetic spin vane and equipped with a reflux condenser protected by a calcium chloride drying tube is placed 61 µL (50 mg, 0.38 mmol) of 1-octanol followed by 121 µL (1.28 mmol) of acetic anhydride and 24 µL (0.34 mmol) of acetyl chloride.(■)

HOOD

> **WARNING: Acetic anhydride and acetyl chloride are irritants. They are dispensed in the hood using automatic delivery pipets. The vial is capped after addition of each reagent, as they are moisture sensitive.**

Reaction Conditions

The reaction mixture is now heated in a sand bath maintained at a temperature of 110–120°C with stirring for a period of one hour. The resulting mixture is then cooled to room temperature and the spin vane removed with forceps.

Isolation of Product

The product is isolated using an extraction procedure identical to that given in Part A.(■)

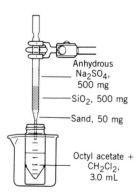

Anhydrous
Na₂SO₄,
500 mg

SiO₂, 500 mg

Sand, 50 mg

Octyl acetate +
CH₂Cl₂,
3.0 mL

Purification and Characterization

The clear, viscous, pleasant-smelling ester, octyl acetate, is sufficiently pure for characterization.

Weigh the ester and calculate the percentage yield. Determine the density, refractive index (optional), and boiling point and compare your values with those in the literature.

Obtain an IR spectrum of your octyl acetate product and compare it with that shown in Figure 5.58.

Infrared Analysis

The conversion of a straight-chain aliphatic primary alcohol (see infrared discussion of secondary alicyclic alcohols in Experiments 5A and 5B) to an acetate ester significantly modifies the spectrum of the substrate. The infrared spectrum of 1-octanol (Fig. 5.59) possesses key bands at 3350 (broad), 3000–2850, 1470, 1420 (broad), 1382, 1060, and 690 (broad) cm⁻¹ which make up the "macro group frequency train for straight-chain aliphatic alcohols." The bands are individually assigned as follows.

a. 3350 cm⁻¹ (very intense, broad): O—H stretching.

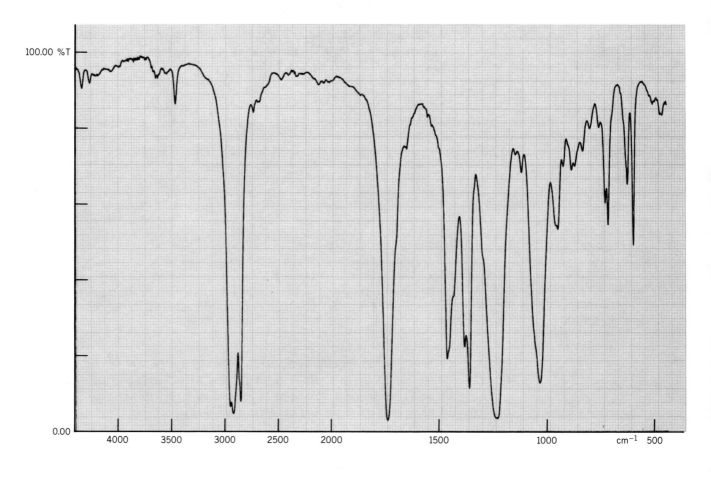

Sample ___Octyl acetate_____

%T _X_ ABS __ Background Scans _4_____   Scans _____16_____
Acquisition & Calculation Time _42 sec__   Resolution _4.0 cm⁻¹____
Sample Condition __liquid, neat_____   Cell Window ___KBr_____
Cell Path Length __capillary film_____   Matrix Material _____

Fig. 5.58 *IR spectrum: octyl acetate.*

b. 3000–2850 cm⁻¹ (strong): C—H stretch of aliphatic section.

c. 1420 cm⁻¹ (broad): this to a large extent is associated with an O—H in-plane bending mode.

d. 1382 cm⁻¹ (sharp): C—H symmetric deformation of a methyl group at the end of the aliphatic chain.

d. 1060 cm⁻¹ (strong): C—O stretch.

e. 690 cm⁻¹ (broad): O—H out-of-plane bending.

After acetylation of the hydroxyl group, the spectrum of the purified ester product (Fig. 5.58) exhibits the following modifications.

a. The 3350 cm⁻¹ band has vanished along with the 1420 and 690 cm⁻¹ bands. These three broad bands are related to the three fundamental vibrations directly identified with the hydroxyl group of alcohols.

b. The bands at 3000–2850, 1382, and 1090 cm⁻¹ survive the transformation and are identified at 3000–2850, 1387, and 1047 cm⁻¹ in the ester. The first two arise from the aliphatic portion of the molecule which remains unchanged. The band at 1090 cm⁻¹ has been somewhat more displaced to lower wavenumbers, but still

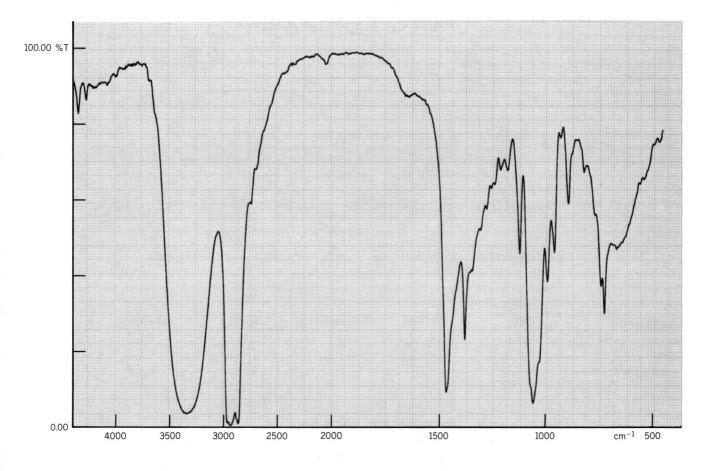

Sample	1-Octanol	
%T _X_ ABS __Background Scans _4_		Scans _____16_____
Acquisition & Calculation Time _42 sec_		Resolution _4.0 cm⁻¹_
Sample Condition __liquid, neat__		Cell Window ___KBr___
Cell Path Length __capillary film__		Matrix Material _____

Fig. 5.59 *IR spectrum: 1-octanol.*

represents predominantly C—O stretching of the alcohol-derived section of the ester group.

c. New bands are observed at 3475 (weak, sharp), 1743 (very strong), 1365 (sharp), and 1246 (strong) cm⁻¹. These absorptions are related principally to the newly formed ester linkage. The 3475 cm⁻¹ peak is the first harmonic of the very intense carbonyl mode (see discussion of C=O overtones in Experiment 5A) located at 1743 cm⁻¹. The 1365 cm⁻¹ band is associated with the newly introduced acetate methyl group. The intense absorption found at 1246 cm⁻¹ also involves C—O stretching.

Thus, the macro group frequency train derived for aliphatic acetates requires bands near 3475, 3000–2850, 1745, 1380, 1365, 1250, and 1050 cm⁻¹.

Chemical Tests

Does this ester give a positive hydroxamate test (see Chapter 7)?

Check the solubility of the ester in water, ether, concentrated sulfuric acid, and 85% phosphoric acid. To what solubility group does this material belong (see Chapter 7)? Would you have predicted the results you observed and why?

PART C: Isopentyl Acetate

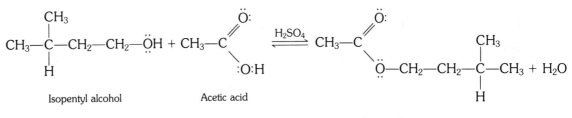

Isopentyl alcohol   Acetic acid

Isopentyl acetate

DISCUSSION   See experimental introductory discussion.

EXPERIMENTAL   Estimated time to complete the experiment: 3.0 hours.

Physical Properties of Reactants and Products

Compound	MW	Wt/Vol	mmol	bp(°C)	Density	n_D
Isopentyl alcohol	88.15	1 mL	9.2	132	0.81	1.4053
Acetic acid	102.09	550 μL	1.59	139.6	1.08	1.3901
Isopentyl acetate	130.19			142	0.87	1.4003

Reagents and Equipment

In a 5.0-mL conical vial containing a magnetic spin vane and equipped with a reflux condenser protected by a calcium chloride drying tube are placed 1 mL (810 mg, 9.2 mmol) of isopentyl alcohol, 550 μL (576 mg, 10 mmol) of glacial acetic acid, 4 drops (Pasteur pipet) of concentrated sulfuric acid, and approximately 100 mg of silica gel beads. (■)

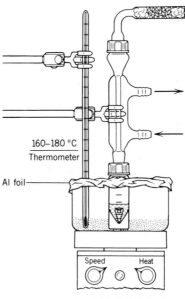

160–180 °C
Thermometer

Al foil

Speed   Heat

Isopentyl alcohol, 1 mL
+ CH_3CO_2H, 0.55 mL
+ 4 drops conc. H_2SO_4
+ silica gel beads, 100 mg

HOOD

> **CAUTION:** *The vial should be capped immediately after addition of each reagent. The reagents should be dispensed in the hood using automatic delivery pipets. Acetic and sulfuric acids are corrosive reagents.*

The silica gel beads (the Desiccant) are used to absorb the water as it is generated in the system.

Reaction Conditions

The reaction mixture is heated with stirring in a sand bath at a temperature of 160–180°C for a period of one hour. The sand bath is covered with aluminum foil during this heating operation. The resulting mixture is then cooled to room temperature and the spin vane removed with forceps.

Isolation of Product

The solution is transferred using a Pasteur filter pipet to a 5-mL conical vial. The crude organic product is then extracted with three 2-mL portions of 5% sodium bicarbonate solution followed by 1 mL of water. During each extraction, the vial is capped, shaken gently, vented, and then allowed to stand so the layers may separate. A Vortex mixer can be used to good advantage in this sequence. The aqueous layer is removed and discarded after each extraction step.

After the water extraction, the organic fraction is dried by addition of anhydrous sodium sulfate to the vial. The dried material is then transferred using a Pasteur filter pipet to a clean, dry 3-mL conical vial containing a boiling stone. The vial is then attached to a Hickman still equipped with an air condenser and arranged in a sand bath for distillation. (■) The sand bath is covered with aluminum foil during the distillation procedure.

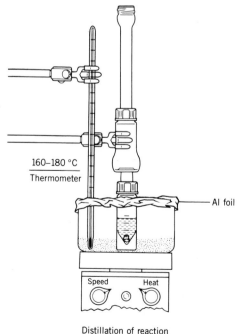

160–180 °C
Thermometer

Al foil

Speed   Heat

Distillation of reaction products

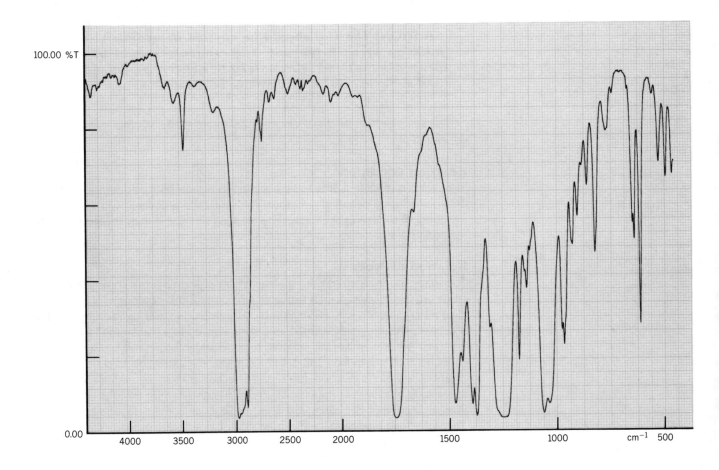

Sample	Isopentyl acetate		
%T _X_ ABS __ Background Scans _4_		Scans ___16___	
Acquisition & Calculation Time _42 sec_		Resolution _4.0 cm⁻¹_	
Sample Condition _liquid, neat_		Cell Window ___KBr___	
Cell Path Length _capillary film_		Matrix Material _____	

Fig. 5.60 *IR spectrum: isopentyl acetate.*

Purification and Characterization

The isopentyl acetate product is distilled at a sand bath temperature of 160–180°C. As the material collects in the collar of the still, it is transferred by Pasteur pipet (9 in.) to a tared ½-dram shell vial.

The clear, viscous, pleasant-smelling ester, isopentyl acetate is weighed and the percentage yield calculated. Determine the refractive index (optional) and boiling point and compare your values with those found in the literature.

Obtain an IR spectrum of the ester using the capillary film technique and compare it with that shown in Figure 5.60.

Infrared Analysis

The conversion of a branched-chain aliphatic primary alcohol to an acetate ester results in significant changes in the infrared spectrum of the starting material compared with the product. These changes are similar to those observed in straight-chain systems (see

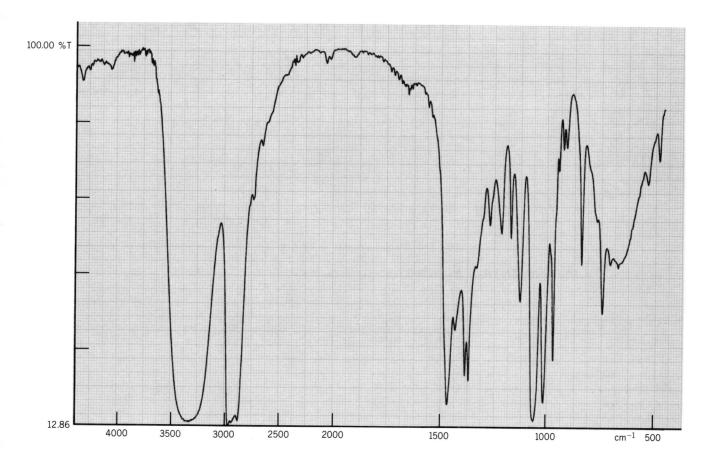

Sample	Isopentyl alcohol

%T X ABS — Background Scans 4 　　　Scans 16
Acquisition & Calculation Time 42 sec 　Resolution 4.0 cm⁻¹
Sample Condition liquid, neat 　　　Cell Window KBr
Cell Path Length capillary film 　　　Matrix Material _____

Fig. 5.61 *IR spectrum: isopentyl alcohol.*

infrared discussion in the formation of straight-chain acetates, Experiment 8B). The two macro group frequency trains in the present example are

a. Isopentyl alcohol (Fig. 5.61): 3350 (broad), 3000–2850, 1460–1300, 1060, and 660 cm⁻¹.

b. Isopentyl acetate (Fig. 5.60): 3490 (weak), 3000–2850, 1746 (strong), 1367 (broad), 1250, and 1040 cm⁻¹.

If the macro group frequencies are further narrowed to include only those systems in which the chain branching involves the presence of an isopropyl group, two additional bands near 1385 and 1365 cm⁻¹ are required. These peaks, which are present in both the alcohol and the ester, arise from the spatially coupled and split symmetric methyl bending vibrations of the aliphatic backbone. In both isopentyl alcohol and isopentyl acetate, this pair of bands is found at identical locations, 1386 and 1367 cm⁻¹.

Chemical Tests

Does the product give a positive hydroxamate test for an ester (Chapter 7)?

Check the solubility of this ester in water, ether, and 85% phosphoric acid. In which solubility group does isopentyl acetate fall?

PART D: 4-*tert*-Butylcyclohexyl Acetate

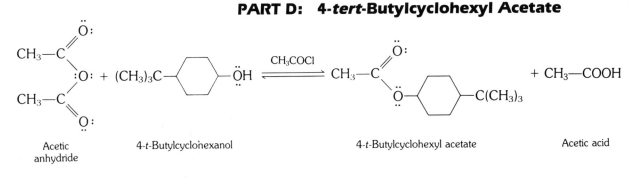

Acetic   4-*t*-Butylcyclohexanol   4-*t*-Butylcyclohexyl acetate   Acetic acid
anhydride

DISCUSSION   See experimental introductory discussion. The starting 4-*t*-butylcyclohexanol may be prepared using the procedure described in Experiment 5B.

EXPERIMENTAL   Estimated time to complete experiment: 3.0 hours.

Physical Properties of Reactants and Products

Compound	MW	Wt/Vol	mmol	mp(°C)	bp(°C)	Density	n_D
4-*t*-Butylcyclohexanol (mixture of isomers)	156.27	80 mg	0.51	62–70			
Acetic anhydride	102.09	100 µL	1.06		139.6	1.08	1.3901
Acetyl chloride	78.50	60 µL	0.85		50.9	1.11	1.3898
4-*t*-Butylcyclohexyl acetate	198.30				100–105$^{(8 mm)}$	0.94	1.4507

Reagents and Equipment

To a 3.0-mL conical vial containing a magnetic spin vane and equipped with a reflux condenser protected by a calcium chloride drying tube, add 80 mg (0.51 mmol) of 4-*t*-butylcyclohexanol, 100 µL (1.06 mmol) of acetic anhydride, and 60 µL (0.85 mmol) of acetyl chloride.(■)

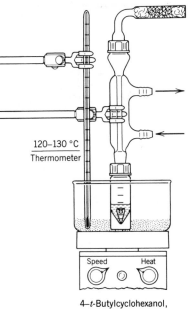

120–130 °C
Thermometer

4-*t*-Butylcyclohexanol,
80-mg +
(CH₃CO)₂0, 100 µL +
CH₃COCl, 60 µL

HOOD

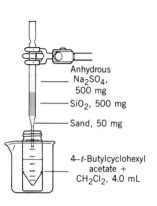

Anhydrous
Na$_2$SO$_4$,
500 mg

SiO$_2$, 500 mg

Sand, 50 mg

4-*t*-Butylcyclohexyl
acetate +
CH$_2$Cl$_2$, 4.0 mL

Reaction Conditions

The reaction mixture is heated with stirring in a sand bath at a temperature of 120–130°C for a period of one hour. It is then cooled to room temperature and the spin vane removed with forceps.

Isolation of Product

The extraction procedure is identical to that in Part A. The only exception is the last step in the chromatographic sequence; 2.0 mL of methylene chloride is used in the final elution step in place of 1.0 mL of this solvent.(■)

Purification and Characterization

The clear, pleasant-smelling ester, 4-*t*-butylcyclohexyl acetate, is sufficiently pure for characterization. Weigh the ester and calculate the percentage yield. Determine the refractive index (optional) and boiling point and compare your values with those reported in the literature. Obtain an IR spectrum of the ester and compare it with that shown in Figure 5.62.

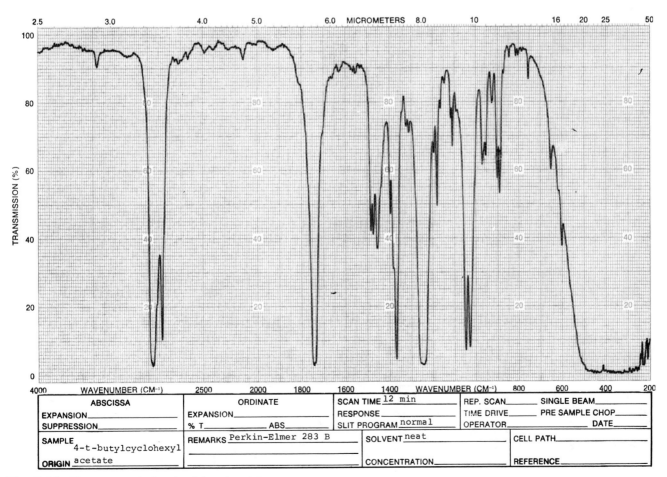

Fig. 5.62 *IR spectrum: 4-t-butylcyclohexyl acetate.*

Infrared Analysis

The acetylation of alicyclic secondary alcohols gives rise to modifications of the infrared spectrum of the starting alcohol not unlike those found in both straight- and branched-chain alcohols (see infrared discussions in Experiments 8B and 8C). In the specific case in hand, the spectra of 4-t-butylcyclohexanol (mixed cis and trans isomers; see Fig. 5.37) and 4-t-butylcyclohexyl acetate (mixed cis and trans isomers; see Fig. 5.62, not derived from the same mixture of isomers as present in Fig. 5.37) possess macro group frequencies based on the following sets of peaks.

a. 4-t-Butylcyclohexanol: 3300, 1450–1300 (weak, broad), 1075, 1053, ~700 (weak, broad) cm^{-1}.

b. 4-t-Butylcyclohexyl acetate: 3465 (weak), 1745, 1367, 1245, 1047 (equatorial) and 1027 (axial) cm^{-1}.

For macro group frequencies restricted to t-butylcyclohexyl derivatives, the coupled symmetric methyl deformation modes occur near 1395 and 1365 cm^{-1}. In the specific case of 4-t-butylcyclohexanol, the bands occur at 1396 (weak) and 1376 (medium) cm^{-1}. In the case of the acetate derivative they are found at 1396 (weak) and 1378 (medium) cm^{-1} (see also discussions of this pair of bands in Experiment 5B and similar effects in Experiment 8C).

Chemical Tests

Does the ester give a positive hydroxamate test (see Chapter 7)?

Check the solubility of the ester in water, ether, concentrated sulfuric acid, and 85% phosphoric acid. To what solubility class does this ester belong (see Chapter 7)?

PART E: Benzoin Acetate

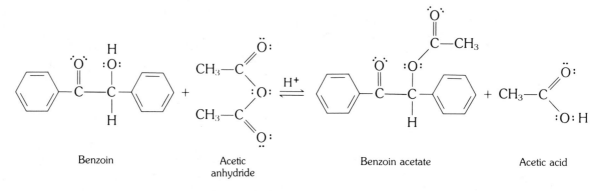

| Benzoin | Acetic anhydride | Benzoin acetate | Acetic acid |

The formation of a solid acetate derivative of an alcohol by the esterification reaction is demonstrated in this experiment. Benzoin, used as the starting material, is prepared in Experiment 20.

DISCUSSION   See experimental introductory discussion.

EXPERIMENTAL   Estimated time to complete the experiment: 0.5 hour.

Physical Properties of Reactants and Products

Compound	MW	Wt/Vol	mmol	mp(°C)	bp(°C)	Density
Benzoin	212.25	20 mg	0.09	137		
Acetic anhydride	102.09	48 μL	0.51		139.6	1.08
Glacial acetic acid	60.05	48 μL	0.84		117.9	
Conc. sulfuric acid	98.08	2 μL	0.04			
Benzoin acetate	254.29			83		

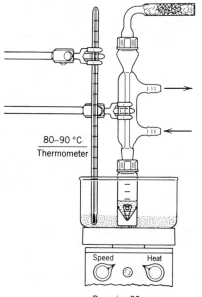

Benzoin, 20 mg +
acylation solution, 100 μL

Reagents and Equipment

In a 3.0-mL conical vial containing a magnetic spin vane and equipped with a reflux condenser protected by a calcium chloride drying tube are placed 20 mg of benzoin followed by 100 μL of acylation solution.(■)

Purification of the benzoin used in the experiment is not necessary if the melting point is above 132°C.

Instructor Preparation

The acylation solution is prepared by mixing 5.0 mL of acetic anhydride, 5.0 mL of glacial acetic acid, and 0.5 mL of concentrated sulfuric acid. It should be

HOOD   dispensed in the **hood** by use of an automatic delivery pipet.

Reaction Conditions

The reaction mixture is warmed in a sand bath at a temperature of 80–90°C for 5 minutes.

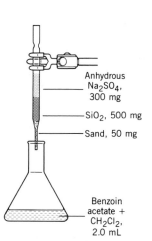

Anhydrous
Na₂SO₄,
300 mg

SiO₂, 500 mg

Sand, 50 mg

Benzoin
acetate +
CH₂Cl₂,
2.0 mL

Isolation of Product

The solution is cooled to room temperature and 0.25 mL of water added using a calibrated Pasteur pipet. The resulting mixture is cooled in an ice bath for a period of 5 minutes and then extracted with three 0.5-mL portions of methylene chloride. On each addition of methylene chloride, the vial is capped, shaken, and vented and the layers are allowed to separate. The use of a Vortex mixer is a viable alternative in this extraction step. Using a Pasteur filter pipet, the bottom methylene chloride layer is separated and transferred to a Pasteur filter pipet containing 0.5 g of activated silica gel (100 mesh) and 0.3 g of anhydrous sodium sulfate. This chromatographic column should be previously wet with methylene chloride solvent. After transfer of the third portion of the methylene chloride, an additional 0.5 mL of solvent is added to the column. The eluate is collected in a 10-mL Erlenmeyer flask containing a boiling stone.(■)

HOOD   The eluate is concentrated in the **hood** by warming in a sand bath under a slow stream of nitrogen gas. The resulting white, solid product is removed from the flask and air-dried on a porous clay plate or on filter paper.

Purification and Characterization

The product is generally pure enough for characterization. However, it may be recrystallized from methanol/water, if desired, using the Craig tube.

Note. *In the recrystallization sequence, the material is dissolved in warm methanol. Water is then added dropwise until the solution becomes cloudy. The mixture is again warmed until a clear solution is observed and then allowed to cool. It is not necessary to wash the collected crystals before drying.*

Weigh the benzoin acetate and calculate the percentage yield. Determine the melting point and compare your value with that found in the literature. Obtain an IR spectrum of the ester and compare it with that of the starting material and also with that of an authentic sample.

Chemical Tests

Several chemical tests also assist in classifying this product. Does the ignition test confirm the presence of the aromatic rings? Are positive results obtained for the 2,4-dinitrophenyl-hydrazine test and for the hydroxamate test, demonstrating the presence of the ketone unit and the ester group, respectively?

ENVIRONMENTAL DATA

Substance	Amount		TLV (mg/m^3)	Emissions (mg)	Volume (m^3)
Experiment 8A					
Ethanol (reflux only up to Max.)	1.0	mL	1900	70–800	0.4
Methylene chloride (Max., col. chromatog.)	2.5	mL	350	3338	9.5
Acetyl chloride (Max.)	30	μL	10[a]	33	3.3
Experiment 8B					
1-Octanol (Max. from yield, transfer)	50	mg	—	30	—
Acetic anhydride (reflux, liquid transfer)	121	μL	20	12	0.6
Acetyl chloride (Max.)	24	μL	10[a]	26	2.6
Methylene chloride (Max., col. chromatog., evap.)	2.5	mL	350	3338	9.5
Experiment 8C					
Isopentyl alcohol (Max. from yield, transfer)	1	mL	—	81	—
Acetic acid	0.55	mL	25	580	23
Experiment 8D					
4-*t*-Butylcyclohexanol (Max. from yield, transfer)	80	mg	—	28	
Acetic anhydride (reflux, liquid transfer)	100	μL	20	12	0.6
Acetyl chloride (Max.)	60	μL	10[a]	66	6.6
Methylene chloride (Max., col. chromatog., evap.)	2.5	mL	350	3338	9.5
Experiment 8E					
Acetic anhydride (Max., heat in open)	100	μL	20	108	5.4
Methanol (SKIN) (Max., Craig tube)	1.0	mL	260	800	3.1

[a] This TLV value is calculated assuming the generation of equivalent amounts of acetic acid and hydrochloric acid.

TLV: Acetic acid—10 ppm
  Isopentyl acetate—100 ppm
TXDS: Lauric acid—orl-rat LD50: 12 g/kg
  Acetyl chloride—inl-hum TCLo: 2 ppm/1 m TFX:IRR
  Sodium sulfate—orl-mus LD50: 5989 mg/kg
  1-Octanol—orl-mus LD50: 1790 mg/kg
  Octyl acetate—orl-rat LD50: 3000 mg/kg
  4-*t*-Butylcyclohexanol—orl-rat LD50: 4200 mg/kg
  4-*t*-Butylcyclohexyl acetate—orl-rat LD50: 5000 mg/kg
  Benzoin—orl-rat LDLo: 5460 mg/kg
  Sulfuric acid—orl-rat LD50: 2140 mg/kg

QUESTIONS **5-40.** A series of dyes is separated by TLC. Calculate the R_f value for each dye based on the following data:

Dye	Distance moved (mm)
Solvent	66
Methyl red	56
Rhodamine B	38
Bismark brown	16
Congo red	0.5

5-41. The technique of high-performance liquid chromatography (HPLC) was developed to combat the problems of slow mass transfer and of relatively slow flow rates in achieving a large number of theoretical plates. This technique involves the use of a high pressure (~5000 psi) at the top of the column, which causes the mobile phase to be driven through the column at a faster rate. Suitable packings and pressure systems have been developed to accommodate the high pressures. The method is particularly useful for biochemical compounds that would decompose under gas chromatographic conditions.

Trace amounts of phenol in water can be detected by HPLC. If a 50-ng phenol standard gives a peak area of 140 units and a 20-μL water sample gives a peak area for phenol of 26 units, calculate the phenol content of the water sample as parts per million (ppm or μg/mL).

5-42. If the straight-chain alcohol involved in the esterification contains at least four methylene groups in a row, a further specific infrared group frequency becomes part of the macro group frequencies for both the alcohol and the ester. What is this mode and where does it occur in the spectrum? (*Hint:* It appears in the spectra of 1-octanol and octyl acetate.)

5-43. In the infrared spectrum of acetates two intense bands are usually observed in the $1270-1000$ cm^{-1} region. These peaks are related to the stretching vibrations of the two C—O bonds of the ester group. Should we expect to be able to assign the C—O bonds to individual peaks, and if so, which mode is associated with which band? Why? (*Hint:* The peak that occurs at higher wavenumbers is consistently close to 1250 cm^{-1}.)

5-44. In the case of octyl acetate, the symmetric deformation vibration of the methyl group attached to the acetate carbonyl occurs 22 cm^{-1} below that of the methyl group positioned at the end of the aliphatic chain. The band is also more intense. Explain these observations.

5-45. In the spectrum of isopentyl acetate, the 1367 cm^{-1} peak assigned to the coupled symmetric methyl deformation frequency appears to be measurably broadened compared with the higher wavenumber component found at 1386 cm^{-1}. Suggest a reason for the increased bandwidth.

5-46. Fatty acids are long-chain carboxylic acids, usually of more than 12 carbon atoms, isolated from saponification of fats (esters of glycerol). Draw the structure of each of the fatty acids named below and also give its common name.

hexadecanoic acid
octadecanoic acid
(Z)-9-octadecenoic acid
($9Z$, $12Z$)-9,12-octadecadienoic acid

5-47. It has been found that the rate constant for the saponification of ethyl trifluoroacetate is more than one million times greater than the rate constant for ethyl acetate under similar conditions. Explain.

5-48. Explain why acetyl chloride is more reactive than acetic anhydride toward alcohols.

5-49. What values for saponification equivalent would be obtained from the following compounds: benzaldehyde, ethyl chloroacetate, diethyl succinate?

5-50. Write a mechanism for the acid-catalyzed transesterification reaction of ethyl acetate with butyl alcohol to prepare butyl acetate.

REFERENCES These references are selected from the large number of examples of esterification given in *Organic Syntheses*.

1. Fuson, R. C.; Wojcik, B. H. *Organic Syntheses*; Wiley: New York, 1943; Collect. Vol. II, p 260.
2. Mic'ovic', V. M. *Ibid.*, p 264.
3. Bowden, E. *Ibid.*, p 414.
4. McCutcheon, J. W. *Organic Syntheses*; Wiley: New York, 1955; Collect. Vol. III, p 526.
5. Weissberger, A.; Kibler, C. J. *Ibid.*, p 610.
6. Eliel, E. L.; Fisk, M. T. *Organic Syntheses*; Wiley: New York, 1963; Collect. Vol. IV, p 169.
7. Emerson, W. S.; Longely, R. I., Jr. *Ibid.*, p 302.

TECHNIQUE 9: PREPARATION AND ANALYSIS OF A GASEOUS PRODUCT

Experiment 9

The Dehydration of a Secondary Alcohol, 2-Butanol: 1-Butene; *trans*-2-Butene; *cis*-2-Butene

(1-butene; 2-butene, *E*-; 2-butene, *Z*-)

This experiment illustrates the acid-catalyzed elimination reaction of the secondary (2°) alcohol 2-butanol. The gaseous products formed in the reaction are separated and analyzed by the use of gas chromatography.

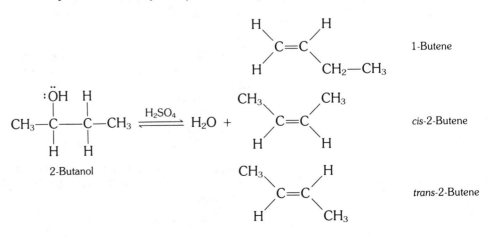

DISCUSSION The introduction of unsaturation into an aliphatic hydrocarbon generally involves the loss of a proton and a nucleophile from adjacent carbon atoms in a chain or ring.

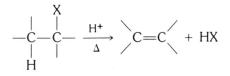

One of the common methods of introducing unsaturation into an organic compound is by dehydration of the corresponding alcohol. The dehydration reaction (loss of water) belongs to the important class of organic reactions called *elimination reactions*, which are often used in synthesis and in the determination of molecular structure.

Dehydration of alcohols is an acid-catalyzed elimination reaction. Experimental evidence shows that alcohols react in the order $3° > 2° > 1°$, which relates to the stability of the carbocation intermediate formed in the reaction. Generally, sulfuric or phosphoric acid is used as a catalyst in the laboratory. Lewis acids such as aluminum oxide or silica gel are frequently used at fairly high temperatures as catalysts on the industrial scale.

The E_1 acid-catalyzed elimination dehydration reaction proceeds in several steps, as outlined:

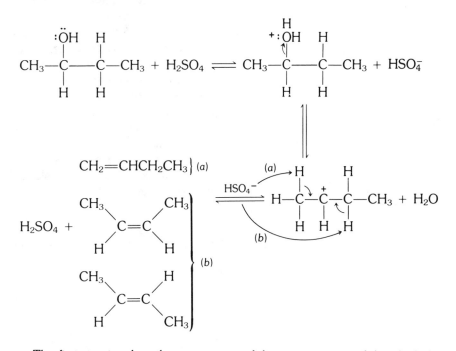

The first step involves the protonation of the oxygen atom of the alcohol to form an oxonium ion in a fast equilibrium reaction. This protonization step is important because it turns a poor leaving group, —OH, into a good leaving group, H_2O (a small neutral molecule). The second step of the reaction is the decomposition of this oxonium ion to yield a carbocation intermediate and water. This is the rate-determining step of the reaction. In the third step, the carbocation stabilizes itself by transferring a proton located on an adjacent carbon atom to a molecule of water or other base present in the system, to regenerate the catalyst and form an alkene.

The E_1 elimination reactions involve equilibrium conditions and thus, to maximize the yield, the alkene is generally removed from the reaction site. Distillation is often used since the alkene *always* has a lower boiling point than the corresponding alcohol. In the present reaction, the alkenes are gases and therefore are easily removed and collected as described.

It is important to realize that many primary alcohols undergo the dehydration reaction by the E_2 mechanism, mainly because the primary carbocation that would be generated in an E_1 process is relatively unstable.

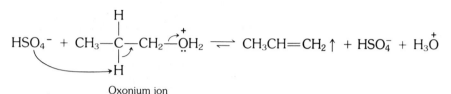

Oxonium ion

The E_1 elimination is usually accompanied by a competing S_N1 substitution reaction. This competing reaction involves the same carbocation intermediate; however, since the reactions are reversible, and the alkene is easily removed from the reaction site, the substitution reaction is not too troublesome.

Certain alcohols dehydrate to yield more than one alkene. In the present reaction involving the dehydration of 2-butanol, three alkenes are formed. As seen in the mechanism scheme, this is due to the fact that the hydrogen atom removed to form the alkene can come from different adjacent carbon atoms. According to the Saytzeff rule, the alkene formed in the larger amount is that having the highest degree of substitution. Moreover, trans alkenes are thermodynamically more stable than their cis counterparts. Knowing these general observations, one is often able to predict the ratio of alkenes to expect from a given elimination reaction.

Rearrangement of alkyl groups and hydrogen is often observed in the dehydration of alcohols, especially in the presence of strong acid. The dehydration of 3,3-dimethyl-2-butanol with sulfuric acid catalyst results in the formation of a mixture of alkenes as shown below. Can you predict which olefin is formed in the greatest amount?

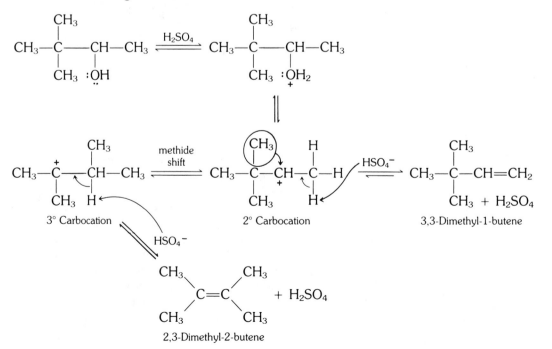

Reactions of this type are very susceptible to rearrangement because a more stable carbocation can be formed in the system. In the preceding example, a secondary carbocation rearranges to a more stable tertiary carbocation. The rearrangement involves the movement of an alkyl group (in this case a methyl) *with its pair of electrons* to the adjacent carbon atom. This type of shift is referred to as a 1,2 shift and is very common in aliphatic systems involving carbocation intermediates. Hydride shifts are also frequently observed.

EXPERIMENTAL Estimated time to complete the experiment: 2 hours.

Note. *It is suggested that the starting times of the reaction be staggered to allow access to the GC instrument by the students when the product gases are analyzed.*

Physical Properties of Reactants and Products

Compound	MW	Wt/Vol	mmol	bp(°C)	Density	n_D
2-Butanol	74.12	100 μL	1.1	99.5	0.81	1.3978
Conc. sulfuric acid	98.08	50 μL				
1-Butene	56.12			−6.3	0.60	
trans-2-Butene	56.12			0.9	0.61	
cis-2-Butene	56.12			3.7	0.62	

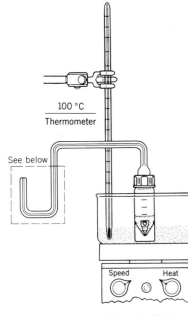

100 °C
Thermometer

See below

Speed   Heat

2-Butanol, 100 μL, +
conc. H$_2$SO$_4$, 50 μL

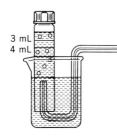

3 mL
4 mL

Reagents and Equipment
Assemble the gas collection apparatus shown in the figure *before* the reactants are mixed.(■)

Determine the capacity of the gas collection reservoir by capping the tube, inverting it, and then adding 3.0 mL of water. Mark the level at 3.0 mL. Add an additional 1.0 mL and mark the 4.0-mL level.

A Teflon-lined septum should be used. This is to prevent loss of the collected butene gases by permeation. A GC septum is even less likely to leak under repeated use.

To position the collection reservoir, fill it with water, place a finger over the open end, invert it, and place the reservoir open end down into a beaker (250 mL) filled with water. When the finger is removed, the column of water should remain in the reservoir.

Place 100 μL (81 mg, 1.1 mmol) of 2-butanol and 50 μL of concentrated sulfuric acid in a clean, dry 1.0-mL conical vial containing a magnetic spin vane. The vial is then attached to the gas collection apparatus. Position the delivery tube under water into the open end of the collection reservoir and clamp the reservoir in place.

> **CAUTION: Sulfuric acid is a strong, corrosive material. Contact with the skin or eyes can cause severe burns. It is convenient to dispense the reactants using automatic delivery pipets.**

Reaction Conditions
With stirring, the reaction mixture is heated using a sand bath until the evolution of gas takes place (~110–120°C sand bath temperature). The mixture should be warmed slowly to prevent foaming.

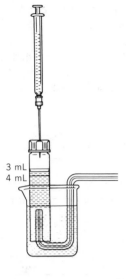

3 mL
4 mL

Isolation of Product

Collect about 3–4 mL of gas in the collection reservoir and then, using a hypodermic syringe, withdraw a 0.5 mL sample through the rubber septum for GC analysis.(■)

Remove the delivery tube from the collecting reservoir and then from the water before discontinuing the heat on the reaction vial. This order prevents water from being sucked back into the reaction flask.

Purification and Characterization

The collected gas is analyzed by gas chromatography without further purification.

> *Gas Chromatography Conditions*
>
> Column: $\frac{1}{4}$ in. × 8 ft packed with 20% silicone DC-710
>
> Room temperature
>
> Flow rate (He gas): 20 mL/minute
>
> Sample size: 0.5 mL of collected gas.

Assuming that the amount of each substance in the gas is proportional to the area of its corresponding peak, determine the percentage ratio of the three components in the gas sample.

Area Under a Curve. *Several techniques may be used. The following method gives reproducible results of ±3–4%: peak height (mm) × width at half-height (mm), measured from the baseline of the curve.*

The order of elution of the butenes is 1-butene, *trans*-2-butene, and *cis*-2-butene. *If the reaction mixture is heated strongly, rearrangement can occur and isobutene is also formed.*

ENVIRONMENTAL DATA

Substance	Amount	TLV (mg/m³)	Emissions (mg)	Volume (m³)
2-Butanol	100 µL	305	10	<0.1
Sulfuric acid	50 µL	1		

Losses unknown, but assumed small because of gas trapping.

QUESTIONS

5-51. Gas chromatographic analysis of a mixture of organic compounds gave the following peak areas (cm²): hexane, 2.7, heptane, 1.6; hexanol, 1.8; toluene, 0.5.

    a. Calculate the mole percent composition of the mixture. Assume that the response of the detector (area/mole) is the same for each component.

    b. Calculate the weight percent composition of the mixture under the same assumptions used in a.

5-52. It is noted at the end of the experiment that if the mixture is heated strongly, rearrangement can occur and isobutene is also formed. Suggest a mechanism to account for the formation of this compound.

5-53. When *t*-pentyl bromide is treated with 80% ethanol, the following amounts of olefinic products are detected on analysis.

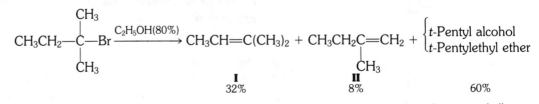

Explain why compound **I** is formed in far greater amount than the terminal alkene.

5-54. The —SR_2^+ group is easily removed in elimination reactions, but the —**SR** group is not. Explain.

5-55. Why is sulfuric acid used to catalyze the dehydration of alcohols rather then hydrochloric acid?

REFERENCES

1. Several dehydration reactions of secondary alcohols using sulfuric acid as the catalyst are given in *Organic Syntheses*.
   a. Coleman, G. H.; Johnstone, H. F. *Organic Syntheses*; Wiley: New York, 1941; Collect. Vol. I, p 183.
   b. Norris J. F. *Ibid.*, p 430.
   c. Bruce, W. F. *Organic Syntheses*; Wiley: New York, 1943; Collect. Vol. II, p 12.
   d. Adkins, H.; Zartman, W. *Ibid.*, p 606.
   e. Wiley, R. H.; Waddey, W. E. *Organic Syntheses*; Wiley: New York, 1955; Collect. Vol. III, p 560.
   f. Grummitt, O.; Becker, E. I. *Organic Syntheses*; Wiley: New York, 1963; Collect. Vol. IV, p 771.
2. The present experiment is adapted from the method given by Helmkamp, G. K.; Johnson, H. W. Jr. *Selected Experiments in Organic Chemistry*, 3rd ed.; Freeman: New York, 1983; p 99.

TECHNIQUE 10: MEASUREMENT OF SPECIFIC ROTATION

Solutions of optically active substances, when placed in the path of a beam of polarized light, may rotate the plane of the polarized light clockwise or counterclockwise. The observed optical rotation is measured using a *polarimeter*. This technique is one of the oldest instrumental procedures used to characterize chemical compounds. The results obtained from the measurement of the observed angle of rotation α are generally expressed in terms of *specific rotation* $[\alpha]$.

Theory

Natural light behaves as though it is composed of electromagnetic waves in which the oscillating electric field vectors may take up all possible orientations around the direction of propagation (see Fig. 5.63).

Fig. 5.63 *Oscillation of the electrical field of ordinary light occurs in all possible planes perpendicular to the direction of propagation. From* Solomons, T. W. Graham *Organic Chemistry*, 3rd ed.; Wiley: New York, 1984. (Reprinted by permission of John Wiley & Sons, New York.)

Note. *A beam of light behaves as though it is composed of two mutually perpendicular oscillating fields: electrical and magnetic. The oscillating magnetic field is not considered in the following discussion.*

The planes in which the electrical waves oscillate are perpendicular to the direction of propagation of the light beam. If one separates one particular plane of oscillation from all other planes by passing the beam of light through a polarizer, the resulting radiation is said to be *plane polarized* (see Fig. 5.64). In the interaction of light with matter, this plane-polarized radiation is represented as the vector sum of two circularly polarized waves. The electric vector of one of the waves moves in a clockwise direction, whereas the other moves in a counterclockwise direction, both waves having the same amplitude (see Fig. 5.65). These two components add vectorially to produce plane-polarized light.

If the passage of plane-polarized light through a material results in the velocity of one of the circularly polarized components being decreased more than the other by interaction with bonding and nonbonding electrons, the transmitted beam of radiation has the plane of polarization rotated from its *original* position (Figs. 5.66 and 5.67).

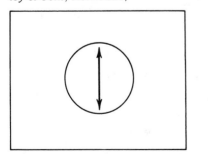

Fig. 5.64 *The plane of oscillation of the electrical field of plane polarized light. In this example the plane of polarization is vertical. From* Solomons, T. W. Graham *Organic Chemistry*, 3rd ed.; Wiley: New York, 1984. (Reprinted by permission of John Wiley & Sons, New York.)

The Polarimeter

The *polarimeter* measures the amount of rotation by an optically active compound of the plane-polarized light. The principal parts of the instrument are diagrammed in Figure 5.66. Two Nicol prisms are used in the instrument. The

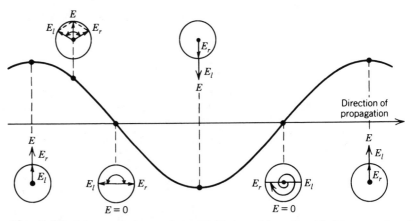

Fig. 5.65 *A beam of plane-polarized light viewed from the side (sine wave) and along the direction of propagation at specific times (circles) where the resultant vector* E *and the circularly polarized components* E$_l$ *and* E$_r$ *are shown. From* Douglas, Bodie; McDaniel, Darl H.; Alexander, John J. *Concepts and Models of Inorganic Chemistry,* 2nd ed.; Wiley: New York, 1983. (Reprinted by permission of John Wiley & Sons, New York.)

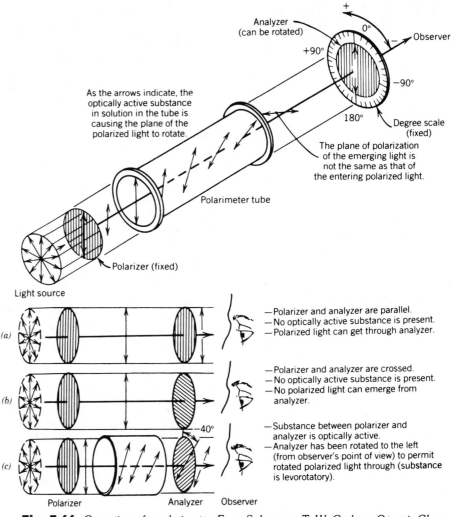

Fig. 5.66 *Operation of a polarimeter. From* Solomons, T. W. Graham *Organic Chemistry,* 3rd ed.; Wiley: New York, 1984. (Reprinted by permission of John Wiley & Sons, New York.)

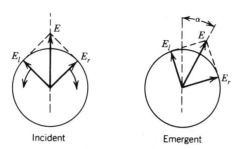

Incident   Emergent

Fig. 5.67 *Plane-polarized light before entering and after emerging from an optically active substance. From* Douglas, Bodie; McDaniel, Darl H.; Alexander, John J. *Concepts and Models of Inorganic Chemistry,* 2nd ed.; Wiley: New York, 1983. (Reprinted by permission of John Wiley & Sons, New York.)

prism by which the original light source is polarized is called the polarizer. The second prism, called the analyzer, is used to examine the polarized light after it passes through a solution of the optically active species.

When the axes of the analyzer and polarizer prisms are parallel and no optically active substance is present, the maximum amount of light is passed and the instrument dial is set to zero degrees. However, if the axes of the analyzer and polarizer are at right angles to each other, no transmission of light is observed and the field is dark. The introduction of a solution of an enantiomer in the path of the plane-polarized light causes one of the circularly polarized components, through dissymmetric interaction, to be slowed more than the other. The refractive indices are, therefore, different in the two circularly polarized beams. Figure 5.66 represents a case in which the left-hand component has been affected the most.

Note. *In this simplified figure the effect on only one of the circularly polarized waves is diagrammed. See Figure 5.67 for a more accurate description (view from behind the figure).*

As seen, this results in a tilt of the plane of polarization. The analyzer prism must be rotated to the left to maximize the transmission of radiation. If rotation is counterclockwise, the angle of rotation is defined as $(-)$ and the enantiomer that caused the effect is termed *levorotatory* (*l*). Conversely, clockwise rotation is $(+)$ and the enantiomer is *dextrorotatory* (*d*). It is important to note that if a solution of equal amounts of a *d* and an *l* enantiomorphic pair is placed in the beam path of the polarimeter, no rotation is observed. Such a solution is called a *racemic* mixture if, as in this case, it is a mixture of enantiomers.

The magnitude of optical rotation depends on several factors: (1) the nature of the substance, (2) the path length through which the light passes, (3) the wavelength of light used as a source, (4) the temperature. It also depends on the concentration of the solution of the optically active material. Optical rotation is usually expressed as *specific rotation* $[\alpha]$:

$$[\alpha]_D^t = \alpha/lc$$

where $[\alpha]$ = specific rotation
α = angle of rotation
l = path length (dm)
c = concentration (g/mL)
t = temperature (°C)
D = the D line of sodium

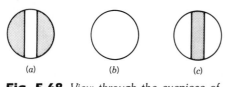

Fig. 5.68 *View through the eyepiece of the polarimeter. The analyzer should be set so that the intensity in all parts of the field is the same (b). When the analyzer is displaced to one side or the other, the field will appear as in (a) or (c).*

The observed angle of rotation α must be less than 360°. However, the *specific rotation* [α] can be very large because of the concentration effect. Indeed, this is the case with usnic acid, which has a specific rotation value near 460°.

For increased sensitivity, most polarimeters are equipped with an optical device that divides the viewed field into three adjacent parts (triple-shadow polarimeter; Fig. 5.68). A very slight rotation of the analyzer will cause one portion to become dimmer, the other lighter. The angle of rotation reading is taken when the sections of the fields all have the same intensity. An accuracy of ±0.01° is easily obtained using this technique.

Instructions should be provided on the use of the particular model of polarimeter available in each laboratory.[9]

Experiment 10

Isolation and Characterization of an Optically Active Natural Product: Usnic Acid

[1,3(2*H*, 9*bH*)-Dibenzofurandione, 2,6-diacetyl-7,9-dihydroxy-8,9*b*-dimethyl-]

This experiment illustrates the extraction of a natural product from its native source, a lichen. It also introduces the method used to measure the specific rotation of an optically active compound.

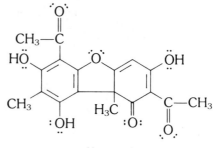

Usnic acid

DISCUSSION   The common method of extracting chemical constituents from natural sources is presented in this experiment. In the present case, only one main chemical species, usnic acid, is soluble in the extraction solvent. Therefore, the purification and characterization sequence is straightforward.

Usnic acid has a chiral carbon atom (see structure) and therefore exists as optically active isomers (enantiomers). Generally, in a given lichen, only one of the isomers (**R** or **S**) is present. Usnic acid has a very high specific rotation (~460°) and for this reason is an ideal candidate to measure rotation on the microscale level.

PART A: Isolation of Usnic Acid

EXPERIMENTAL   Estimated time for completion of the experiment: 2.5 hours.

---

[9] For further discussion of optical activity and the use of the polarimeter see Solomons, T. W. G. *Organic Chemistry*, 4th ed.; Wiley: New York, 1988; p 163; and Willard, H. H.; Merritt, Jr., L. L.; Dean, J. A.; Settle, F. A. *Instrumental Methods of Analysis*, 6th ed.; Van Nostrand: New York, 1981; p 421.

Physical Properties of Reactants and Products

Compound	MW	Wt/Vol	mp(°C)	bp(°C)
Lichen		1.0 g		
Acetone	58.08	15 mL		56.2
Usnic acid	344.31		204	

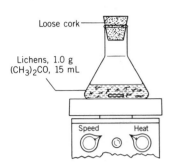

Loose cork

Lichens, 1.0 g
$(CH_3)_2CO$, 15 mL

Speed   Heat

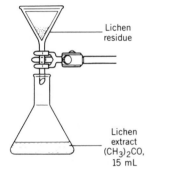

Lichen
residue

Lichen
extract
$(CH_3)_2CO$,
15 mL

HOOD

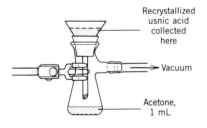

Recrystallized
usnic acid
collected
here

→ Vacuum

Acetone,
1 mL

Reagents and Equipment

In a 50-mL Erlenmeyer flask containing a magnetic stirrer and loosely capped with a cork stopper are placed 1.0 g of oven-dried (40°C), crushed or cut up lichens, and 15.0 mL of acetone.(■) *The lichens used in this experiment are Usnea sp (commonly called "Old Man's Beard").*

Reaction Conditions

The mixture is stirred for 0.5 hour at room temperature. It may be necessary periodically to push the lichens below the surface of the acetone solvent using a glass rod.

Isolation of Product

The resulting mixture is now filtered by gravity and the filtrate collected in a 25-mL Erlenmeyer flask.(■) The acetone solvent is evaporated under a stream of air in the **hood**, yielding the crude usnic acid.

Purification and Characterization

The crude material is recrystallized from acetone–95% ethanol (10 : 1). Dissolve the crystals in the minimum amount of hot acetone and add the ethanol. The yellow crystals are collected by vacuum filtration,(■) washed with *cold* acetone, and dried on a porous clay plate or a sheet of filter paper. As an alternative, the crude material may be recrystallized using the Craig tube, thus avoiding the filtration step with the Hirsch funnel.

Weigh the yellow needles of usnic acid and calculate the percentage of the acid in the lichen. Determine the melting point (use the evacuated mp technique) and compare your value with that found in the literature. Obtain an IR spectrum and compare it with that of an authentic sample.

Chemical Tests

Chemical tests can assist in establishing the nature of the functional groups in usnic acid. Perform the 2,4-dinitrophenylhydrazine test and the ferric chloride test (see Chapter 7). Are the results significant?

PART B: Determination of the Specific Rotation

Usnic acid is an optically active compound with a very high specific rotation. For determination of its rotation on the microscale, a low-volume, long-path-length cell must be used.

Usnic acid (80 mg) is dissolved in 4.0 mL of tetrahydrofuran (THF) solvent and transferred to the polarimetry cell using a Pasteur pipet.

This amount of usnic acid is obtained by having eight or nine students pool their crystals. The THF used is spectral grade.

The cell is placed in the polarimeter and the angle of rotation determined. The specific rotation is then calculated using the equation given in the discussion section.

ENVIRONMENTAL DATA

Substance	Amount	TLV (mg/m^3)	Emissions (mg)	Volume (m^3)
Acetone (Max., extract, evap.)	15 mL	1780	11875	6.7
Acetone (Max., recryst.)	3.0 mL	1780	2375	1.3
Tetrahydrofuran (Max.)	0.5 mL	590	438	0.7.

TXDS: Usnic acid—inv-mus LD50: 25 mg/kg

QUESTIONS   **5-56.** Give an *R* or *S* designation to each of the following molecules.

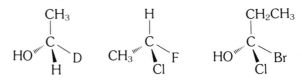

5-57. The structure originally proposed for cordycepic acid, for which $[\alpha] = +40.3°$, is

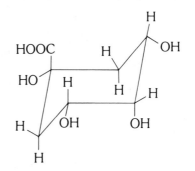

Do you agree that this a plausible structure? If not, why not?

5-58. A sample of 150 mg of an organic compound was dissolved in 7.5 mL of water. The solution was placed in a 20-cm polarimeter tube and the rotation read in a polarimeter. The value obtained was +2.676°. Distilled water, in the same tube at the same temperature, gave a reading of +0.016°. Calculate the specific rotation for the compound.

5-59. Compound A is optically active and has the molecular formula $C_5H_{10}O$. On catalytic hydrogenation (addition of molecular hydrogen, H_2) of A, compound B is obtained. B has the molecular formula $C_5H_{12}O$ and is optically inactive. Give the structure for compounds A and B.

5-60. Which of the following substances has a *meso* form?
2,3-Dibromopentane
2,4-Dibromopentane
2,3-Dibromobutane

REFERENCES   **1.** This experiment is adapted from that given by Todd, D. *Experimental Organic Chemistry*; Prentice–Hall: Englewood Cliffs, NJ, **1979;** p 57.
2. Usnic acid has been synthesized:
   a. Barton, D. H. R.; DeFlorin, A. M.; Edwards, O. E. *J. Chem. Soc.* **1956,** 530.
   b. Penttila, A.; Fales, H. M. *Chem. Commun.* **1966,** 656.
3. A large-scale method of isolation of usnic acid has been reported: Stark, J. B.; Walter, E. D.; Owens, H. S. *J. Am. Chem. Soc.* **1950,** *72,* 1819.
4. Optical, crystallographic, and X-ray diffraction data have been reported for usnic acid: Jones, F. T.; Palmer, K. J. *J. Am. Chem. Soc.* **1950,** *72,* 1820.

TECHNIQUE 11: SUBLIMATION

Sublimation is a technique that is especially suitable for the purification of solid substances at the microscale level. It is particularly advantageous when the impurities present in the sample are nonvolatile under the conditions employed. Sublimation is a relatively straightforward method in that the impure solid need only be heated and mechanical losses can be kept to a minimum.

Materials sublime most easily when heated below their melting points under reduced pressure. Substances that can be purified by sublimation are those that do not have strong intermolecular attractive forces. Caffeine, isolated in Experiment 11A, and ferrocene, used as a reactant in Experiment 32, meet these requirements.

Theory

The processes of *sublimation* and *distillation* are closely related. Crystals of a solid substance, in an evacuated container, will gradually generate molecules in the vapor state by the process of *evaporation* (that is, the solid exhibits a vapor pressure). Occasionally, one of these gas molecules will strike the crystal surface and be held by attractive forces of the crystalline molecules. This process is termed *condensation*.

Sublimation is the complete process of *evaporation* from the solid phase to *condensation* from the gas phase to form crystals *directly* without passing through the liquid state.

A typical single-component phase diagram is shown in Fig. 5.69 relating the solid, liquid, and vapor states of a substance with temperature and pressure. Where two of the areas (solid, liquid, or vapor) touch, there is a line, and along each line the two phases exist in *equilibrium*. *BO* is the sublimation vapor pressure curve of the substance in question and *only* along line *BO* can solid and vapor exist together in equilibrium. At temperatures and pressure along the *BO* curve the liquid state is thermodynamically unstable. Where the three lines representing pairs of phases intersect, all three phases exist together in equilibrium. This point is called the triple point.

Many solid substances have a sufficiently high vapor pressure near their melting point that they can be sublimed easily *under reduced pressure* in the laboratory. Sublimation occurs when the vapor pressure of the solid equals the applied pressure. Simple apparatus suitable for sublimation of small quantities of material are used for the purification of caffeine. Two such devices are shown in Figure 5.70.

Heating the sample with a microburner or a sand bath to just below the melting point of the solid causes sublimation to occur. The caffeine vapors condense on the cold finger surface, whereas any nonvolatile material remains at the bottom of the flask. Detailed instructions of the procedure are outlined in Experiment 11A. Sublimators suitable for the sublimation of small quantities are also commercially available (Fig. 5.71).

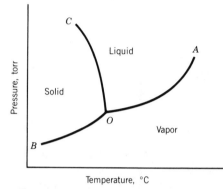

Fig. 5.69 *Single-component phase diagram.*

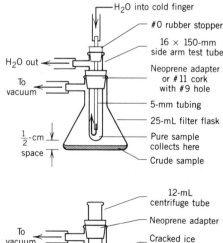

Fig. 5.70 *Sublimation apparatus.*

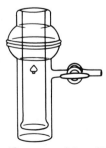

Fig. 5.71 *Vacuum sublimator.* (Courtesy of Ace Glass Inc., Vineland, NJ.)

Experiments 11A, 11B

Isolation and Characterization of a Natural Product: Caffeine; Caffeine 5-Nitrosalicylate

(1,3,7-trimethyl-2,6-dioxopurine)

The extraction of a natural product, caffeine, from its native source, tea, is demonstrated in this experiment. The preparation of a caffeine derivative, 5-nitrosalicylate, is also described.

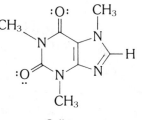

Caffeine

DISCUSSION

This experiment illustrates (as does Experiment 10) the common technique of isolating natural products from their common sources. In the present case, the alkaloid (meaning alkali-like, a basic characteristic imparted by the presence of nitrogen) caffeine is readily soluble in hot water and is thus easily separated from the tea source. However, other substances, mainly tannic acids, present in the tea leaves are also water soluble. Sodium carbonate, a base, is added to the aqueous extracting medium to remove these tannic acids (phenols) as water-soluble sodium salts. Subsequent extraction of the aqueous extract with methylene chloride, in which caffeine has moderate solubility, allows separation of the caffeine from these water-soluble sodium salt impurities.

Extraction of the tea leaves directly with methylene chloride to remove the caffeine gives very poor results. Water is a superior extraction solvent because it swells the tea leaves and allows the caffeine to be liberated.

The caffeine molecule contains the purine ring system which plays an important role in living systems. Caffeine, the most widely used of all stimulants, is present in tea leaves at the 2–3% level and is the major stimulant in coffee.

PART A: Isolation of Caffeine

EXPERIMENTAL

Estimated time to complete the experiment: 2.5 hours.

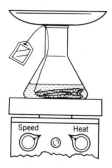

Anhydrous Na₂CO₃, 1.1 g
+ H₂O, 10 mL + tea bag
with tea leaves, 1.0 g

Physical Properties of Reactants and Products

Compound	MW	Wt/Vol	mmoles	mp(°C)
Tea		1.0 g		
Water		10 mL		
Sodium carbonate	105.99	1.1 g	10	851
Caffeine	194.20			238

Reagents and Equipment

To a 50-mL Erlenmeyer flask are added 1.1 g (0.01 mol) of anhydrous sodium carbonate and 10 mL of water.(■) The mixture is heated on a hot plate to dissolve the solid. To this solution is added 1.00 g of tea leaves contained in a tea bag.

Carefully open a commercial tea bag (2.0–2.5 g of tea leaves) and empty the contents. Weigh out 1 g of tea leaves and place them back in the empty tea bag. Close and secure the bag with staples. Place the bag in the flask so that it lies flat across the bottom.

Reaction Conditions

GENTLY

A small watch glass is placed on the Erlenmeyer flask and the tea mixture then heated at boiling [*gently*] for 30 minutes.

Isolation of Product

The cooled, aqueous liquid extract is removed from the Erlenmeyer flask using a Pasteur filter pipet and transferred to a 15-mL centrifuge tube. In addition, the tea bag is gently squeezed by pressing it against the side of the Erlenmeyer flask to recover as much of the basic extract as possible. Discard the tea bag and contents.

The aqueous solution is now extracted with 2.0 mL of methylene chloride (see Experiment 3).

The tea solution contains some constituents that cause emulsions. If on mixing of the aqueous and organic solvent layers (shaking or using a Vortex mixer) an emulsion is obtained, it can readily be broken by centrifugation.

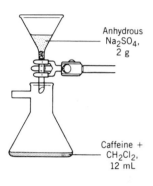

Anhydrous
Na$_2$SO$_4$,
2 g

Caffeine +
CH$_2$Cl$_2$,
12 mL

The lower methylene chloride layer is separated using a 9-in. Pasteur pipet and the wet solution passed through a filter containing a small plug of cotton that is covered with ~2.0 g of anhydrous sodium sulfate, previously moistened with a small amount of methylene chloride.(■)

The dried filtrate is collected in a 25-mL filter flask. The aqueous solution is now extracted with four 2.0-mL portions of methylene chloride. Each extract is separated, dried as before, and collected in the same filter flask. Finally, the sodium sulfate is rinsed with an additional 2.0 mL of methylene chloride.

HOOD

A boiling stone is added to the flask and the solution concentrated to dryness by warming the flask in a sand bath under reduced pressure in the **hood**. The crude caffeine crystallizes as an off-white solid.

Purification and Characterization

The crude caffeine is purified by the process of sublimation.

Assemble a sublimation apparatus as shown in Figure 5.70.(■); either arrangement is satisfactory. Apply a vacuum to the system through the filter flask, using a water-trap bottle between the flask and the aspirator. After the system is evacuated, cold water is run through the cold finger or ice is added to the centrifuge tube. This procedure will minimize the amount of atmospheric water condensation that collects on the cold finger surface.

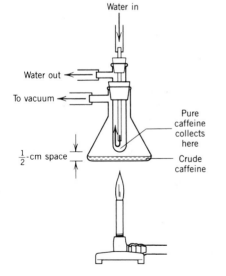

Water in

Water out

To vacuum

$\frac{1}{2}$-cm space

Pure
caffeine
collects
here

Crude
caffeine

The sublimation is started by heating the flask gently with a microburner (or sand bath), moving the flame back and forth around the bottom and sides of the flask.

Be Careful. *Do not MELT the caffeine. If the sample does begin to melt, remove the flame for a few seconds before heating is resumed. This aids in avoiding decomposition as higher temperatures are not necessary since the sublimation temperature of caffeine is below its melting point.*

CAREFULLY

When all the caffeine has completely sublimed onto the cold finger, remove the heat and allow the apparatus to cool to room temperature under reduced pressure. Carefully remove the vacuum tubing from the aspirator, returning the system to atmospheric pressure. Shut off the water to the cold finger and *carefully* remove the cold finger from the sublimation apparatus.

If this operation is done carelessly, the sublimed crystals may be dislodged from the tube and fall back into the impure residue.

Scrape the caffeine from the cold finger onto weighing paper using a micro-spatula and sample brush. Weigh the caffeine product and calculate the percent by weight of caffeine in the tea leaves. Determine the melting point and compare your value with that in the literature.

If the Thomas–Hoover melting point apparatus is used to determine the melting point, an evacuated sealed tube is necessary since caffeine sublimes, the melting point being above the sublimation temperature (see Chapter 4). The melting point may be obtained using the Fischer–Johns apparatus without this precaution.

Obtain an IR spectrum and compare it with that of an authentic sample.

Chemical Test

Does the soda lime or the sodium fusion test (see Chapter 7) confirm the presence of nitrogen in your caffeine product?

PART B: Caffeine 5-Nitrosalicylate

The caffeine product may be further characterized by preparation of its 5-nitro-salicylate derivative.

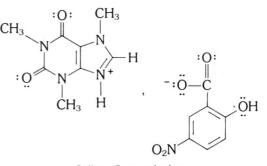

Caffeine 5-nitrosalicylate

EXPERIMENTAL   Estimated time to complete the experiment: 0.5 hours.

Physical Properties of Reactants and Products

Compound	MW	Wt/Vol		mmol	mp(°C)	bp(°C)
Caffeine	194.20	11	mg	0.06	238	
5-Nitrosalicylic acid	183.12	10	mg	0.06	229–230	
Petroleum ether (60–80°C)		0.5 mL				
Ethyl acetate	88.12	0.7 mL				77
Caffeine 5-nitrosalicylate	377.32				180	

Reagents and Equipment

In a 3.0-mL conical vial containing a magnetic spin vane and equipped with an air condenser are placed 11 mg (0.06 mmol) of caffeine, 10.0 mg (0.06 mmol) of 5-nitrosalicyclic acid (prepared in Experiment 34C) and 0.7 mL of ethyl acetate.

Reaction Conditions

GENTLY *Gently* warm the mixture on a hot plate, with stirring, to dissolve the solids. Add 0.5 mL of petroleum ether (60–80°C) to the warm ethyl acetate solution, mix, and warm for several seconds. Remove the spin vane using forceps.

Isolation of Product

Now cool the mixture to room temperature and place it in an ice bath for 10–15 min. Collect the crystals under reduced pressure using a Hirsch funnel and wash the filter cake with 0.5 mL of cold ethyl acetate. Dry the product on a porous clay plate.

Purification and Characterization

The product is sufficiently pure for characterization. Weigh the caffeine 5-nitro-salicylate and calculate the percentage yield. Determine the melting point and compare your value with that reported above.

Obtain an IR spectrum by the KBr pellet technique and compare it with that shown in Fig. 5.72. The infrared spectrum reveals some of the details of derivative formation.

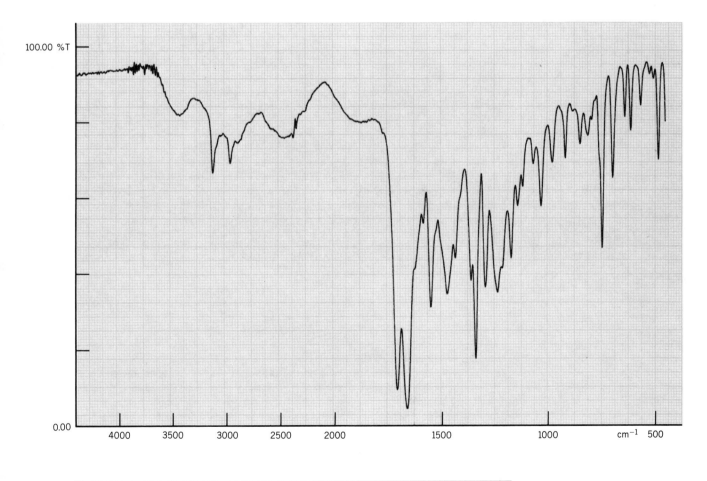

Sample	Caffeine 5-nitrosalicylate		
%T X ABS — Background Scans 4		Scans 16	
Acquisition & Calculation Time 42 sec		Resolution 4.0 cm⁻¹	
Sample Condition solid		Cell Window	
Cell Path Length		Matrix Material KBr	

Fig. 5.72 *IR spectrum: caffeine 5-nitrosalicylate.*

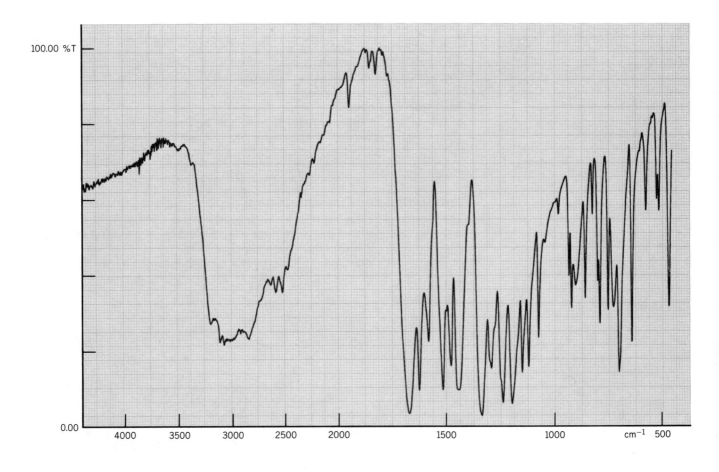

Sample	5-Nitrosalicylic acid		
%T _X_ ABS — Background Scans _4_		Scans _16_	
Acquisition & Calculation Time _42 sec_		Resolution _4.0 cm⁻¹_	
Sample Condition _solid_		Cell Window _____	
Cell Path Length _____		Matrix Material _KBr_	

Fig. 5.73 *IR spectrum: 5-nitrosalicyclic acid.*

Infrared Analysis

The infrared spectrum of 5-nitrosalicylic acid (Fig. 5.73) is characteristic of aromatic carboxylic acids (see Discussion, Experiment 7A). Three points should be noted in the spectrum:

a. The substitution of the ring is revealed by the presence of the 1,2,4 combination band pattern with peaks at 1940, 1860, 1815 cm⁻¹.

b. The strongest band in the spectrum below 1750 cm⁻¹ is assigned to the symmetric stretch of the —NO_2 group found at 1339 cm⁻¹.

c. the conjugated carboxyl C═O stretch is located at 1675 cm⁻¹.

In the spectrum of the complex (Fig. 5.72) we do not find evidence for ionized carboxyl. This group should give rise to two very strong broad bands 1600–1550 & 1400–1330 cm⁻¹. What is observed is the carboxyl C═O stretch at 1665 cm⁻¹ overlapped with a caffeine band. Evidence for very strong H bonding, however, is indicated by the series of very broad bands extending from 3550 to 2000 cm⁻¹. The complex association does not, therefore, very likely involve a complete proton transfer as indicated in the preceding chemical structures. The unambiguous presence of the

5-nitrosalicylic acid in the complex is best ascertained by the N–O symmetric stretch of the nitro group (peak located at 1345 cm^{-1}). Caffeine does not possess a band in this region.

Examine the spectrum that you have obtained in a potassium bromide matrix of the derivative of the material isolated from tea leaves. Discuss the similarities and differences of the experimentally derived spectral data to the reference spectra (Figs. 5.72 and 5.73).

ENVIRONMENTAL DATA

Substance	Amount		TLV (mg/m^3)	Emissions (mg)	Volume (m^3)
Experiment 11A					
Methylene chloride (Max., extract, filter, evap.)	12	mL	350	16,020	46
Experiment 11B					
Ethyl acetate	1.2	mL	1400	1080	0.8
Petroleum ether (Max., treated as hexane)	0.5	mL	180	320	1.8

TXDS: Caffeine—orl-hum LDLo: 192 mg/kg
        Sodium carbonate—orl-rat LDLo: 4000 mg/kg

QUESTIONS

5-61. Compounds such as naphthalene and 1,4-dichlorobenzene find use as mothballs because they sublime at a slow rate at atmospheric pressure. Explain this behavior in terms of the structure of the molecules.

5-62. How many peaks would you expect to find in the proton NMR spectrum of caffeine?

5-63. The vapor pressures of 1,2-diphenylethane, p-dichlorobenzene, and 1,3,5-trichlorobenzene are 0.06, 11.2, and 1.4 torr, respectively, at their melting points (52–54°C). Which compound can be sublimed most rapidly at a reduced pressure of 15 torr and a temperature of 40°C?

5-64. Spots on TLC plates are often developed in an iodine chamber. Explain how the iodine solid–vapor equilibrium operates in this instance.

5-65. Caffeine is rather soluble in ethyl acetate solvent. Do you think that the purity of your product could be checked by TLC using ethyl acetate as a developing solvent? Explain.

5-66. The infrared spectrum of 5-nitrosalicylic acid (Fig. 5.73) possesses the typical broad medium band found in acid dimers (908 cm^{-1}). In the caffeine 5-nitrosalicylate complex, however, this band is missing. Suggest a reason why the 908 cm^{-1} peak vanishes.

REFERENCES

Several extraction procedures for isolating caffeine from tea, designed for the introductory organic laboratory, have been reported.

1. Mitchell, R. H.; Scott, W. A.; West, P. R. *J. Chem. Educ.* **1974,** *51,* 69.
2. Ault, A.; Kraig, R. *J. Chem. Educ.* **1969,** *46,* 767.
3. Laswick, J. A.; Laswick, P. H. *J. Chem. Educ.* **1972,** *49,* 708.

Chapter 6

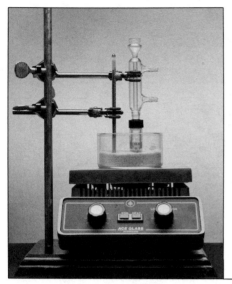

Preparative Organic Chemistry

Experiment 12   **Reductive Hydrogenation of an Olefin: Octane**

(octane)

In this experiment the addition of hydrogen to an unsaturated hydrocarbon to produce a saturated alkane is demonstrated. This addition reaction has great utility in laboratory and industrial syntheses.

$$CH_3-(CH_2)_5-CH=CH_2 \xrightarrow[\text{NaBH}_4]{\text{H}_2\text{PtCl}_6} CH_3-(CH_2)_5-CH_2-CH_3$$

1-Octene C_2H_5OH Octane

HCl(6 *M*)

DISCUSSION   The addition of hydrogen to an alkene is a very important reaction in organic chemistry because by use of the reaction, an alkene is converted into an alkane. The reaction is exothermic; the *heat of hydrogenation* is approximately 125 kJ/mol for most alkenes. Generally, the reaction takes place at room temperature with hydrogen gas pressure in the range of 1–4 atm, in the presence of a finely divided metal catalyst. For this reason, the reaction is referred to as *catalytic hydrogenation*. It represents a class of organic reactions known as addition reactions; one hydrogen atom of the hydrogen molecule adds to each carbon of the C=C. The metals most often used as catalysts in low-pressure hydrogenation are nickel, platinum, rhodium, ruthenium, and palladium. The addition of hydrogen to an alkene does not take place at room temperature and atmospheric pressure in the absence of a metal catalyst.

In the present experiment, the metal catalyst, platinum, is generated in situ by the treatment of chloroplatinic acid with sodium borohydride. The hydrogen gas necessary for the reduction is generated by the addition of excess sodium borohydride in the presence of hydrochloric acid.

$$4\,NaBH_4 + 2\,HCl + 7\,H_2O \rightarrow Na_2B_4O_7 + 2\,NaCl + 16\,H_2$$

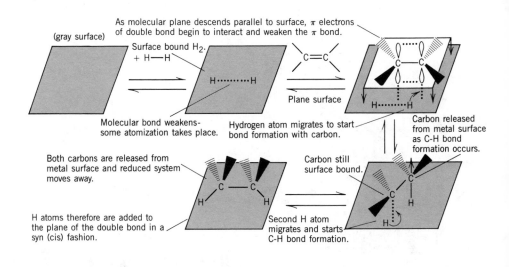

This reduction technique is useful for easily reducible groups such as unhindered olefins, but a major limitation is that other functional groups that are normally reduced by sodium borohydride cannot be present in the molecule.

The platinum catalyst adsorbs both the molecular hydrogen and the alkene on its surface. The transfer of the adsorbed hydrogen to the alkene molecule on the metal surface results in reduction, producing an alkane. The addition is syn (or cis) since both hydrogen atoms add to the same side of the C=C of the alkene. The mechanistic sequence is partially outlined above.

An additional aspect of this reaction is important in regard to purification of the product using column chromatography. A silver nitrate/silica gel column is used to remove any unreacted olefin from the final product mixture. The ability of olefins to form coordination complexes with certain metal ions having nearly filled d orbitals has long been known. In the case of the silver ion complex with olefins, the nature of the bonding has been pictured to involve a sigma bond formed by overlap of the filled π orbital of the olefin with the free s orbital of the silver ion and a π bond formed by overlap of the vacant antibonding π orbitals of the olefin with the filled d orbitals of the silver. The 1-octene is removed from the product mixture by forming such a metal ion complex, the pure pentane being eluted very rapidly.

EXPERIMENTAL   Estimated time to complete the experiment: 2.5 hours.

Physical Properties of Reactants and Products

Compound	MW	Wt/Vol	mmol	bp(°C)	Density	n_D
1-Octene	112.22	120 μL	0.76	121.3	0.72	1.4087
Ethanol	46.07	1.0 mL		78.5		
Chloroplatinic acid (0.2 M)	517.92	50 μL				
Sodium borohydride (1 M)	37.83	125 μL				
Dilute HCl (6 M)		100 μL				
Pentane	72.15	6.5 mL		36.1		
Octane	114.23			125.7	0.70	1.3974

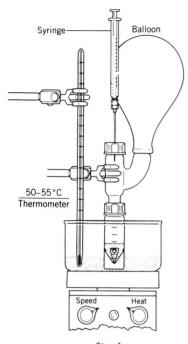

Syringe — Balloon

50–55°C
Thermometer

Speed   Heat

Step 1:
0.2 M H$_2$PtCl$_6$ solution, 50 μL
+ CH$_3$CH$_2$OH, 1.0 mL
+ NaBH$_4$ solution, 125 μL
Step 2:
6 M HCl, 100 μL
+ 1-octene, 120 μL
+ NaBH$_4$ solution, 1.0 mL

Reagents and Equipment

A 5.0-mL conical vial containing a magnetic spin vane is equipped with a Claisen head fitted with a rubber balloon and Teflon-lined rubber septum.(■) Add 50 μL of a 0.2 M solution of chloroplatinic acid, H$_2$PtCl$_6$, dispensed from an automatic delivery pipet and 1.0 mL of absolute ethanol (calibrated Pasteur pipet).

Note. *No residual acetone (perhaps from cleaning the equipment) can be present since it destroys the catalyst. If the balloon does not fit snugly, secure it with copper wire or a rubber band.*

Instructor Preparation

The 0.2 M H$_2$PtCl$_6$ solution is prepared by adding 41 mg (0.1 mmol) of the acid to 0.5 mL of deionized water.

A solution (125 μL) of sodium borohydride reagent (see next Instructor Preparation) is now added by automatic delivery pipet with vigorous stirring. The solution should turn black immediately as the catalyst, finely divided platinum, is formed.

Important. *The rubber septum is put in place IMMEDIATELY after the NaBH$_4$ solution is added.*

Instructor Preparation

The sodium borohydride reagent is prepared by adding 0.38 g (0.01 mol) of NaBH$_4$ to a solution previously obtained by mixing 0.5 mL of 2.0 M aqueous NaOH and 9.5 mL of absolute ethanol.

After one minute, 100 μL of 6 M HCl solution is added through the septum, using a syringe. With a fresh syringe, a solution of 120 μL (86 mg, 0.76 mmol) of 1-octene (delivered from an automatic pipet to a 1-mL conical vial) dissolved in 250 μL of absolute ethanol is added immediately to the acid solution.

The addition of 1-octene is followed by the dropwise addition (clean syringe) of 1.0 mL of the NaBH$_4$ solution over a two-minute time interval.

Important. *At this point the balloon should inflate and remain inflated for at least 30 minutes. If it does not, the procedure must be repeated.*

Reaction Conditions

The reaction mixture is vigorously stirred in a sand bath at 50°C for 45 minutes.

Isolation of Product

Water (1 mL) is added dropwise to the cooled reaction medium and the resulting mixture is extracted with three 1.0-mL portions of pentane. Each pentane extract is transferred to a stoppered 10-mL Erlenmeyer flask containing 0.5 g of anhydrous sodium sulfate.

On addition of each portion of pentane, the vial is capped, shaken, and vented and the layers are allowed to separate. A Vortex mixer may be used if available. The transfers are made using a Pasteur filter pipet.

Using a pasteur filter pipet, the dried solution is transferred to a second 10-mL Erlenmeyer flask. The drying agent is rinsed with an additional 0.5 mL of pentane (calibrated Pasteur pipet) and the rinse also added to the second flask. The solution is concentrated using a gentle stream of nitrogen gas or by warming very gently in a sand bath in the **hood** to a volume of ~1.0 mL.

HOOD

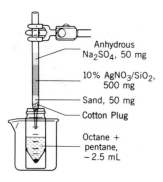

Anhydrous
Na$_2$SO$_4$, 50 mg

10% AgNO$_3$/SiO$_2$,
500 mg

Sand, 50 mg

Cotton Plug

Octane +
pentane,
~2.5 mL

Purification and Characterization

The product, octane, is purified by column chromatography.

In a Pasteur filter pipet are placed 50 mg of sand, 500 mg of 10% silver nitrate on activated silica gel (200 mesh), and 50 mg of anhydrous sodium sulfate.(■)

Instructor Preparation

The ~10% silver nitrate/silica gel used in this separation may be prepared using a mixture of 3% by weight AgNO$_3$ to SiO$_2$. A typical run is as follows: 3 g of the 3% mixture of AgNO$_3$/SiO$_2$ and 30 mL of water are placed in a 100-mL round-bottom flask. The flask is stoppered and shaken for several minutes. The flask is then attached to a rotovap and the water removed under reduced pressure while warming on a hot water bath. After removal of the majority of the water, 10 mL of ethanol are added, the flask again placed on the rotovap, and the gel taken to dryness. The resulting mixture is removed and, if necessary, stored in a brown bottle in the dark.

> **CAUTION: *Silver nitrate causes stains on the skin. Protective gloves should be worn during this operation.***

The column is wetted with 0.5 mL of pentane (calibrated Pasteur pipet) and the crude product, obtained earlier, is transferred to the column by Pasteur filter pipet. The octane is eluted using 1.5 mL of pentane and the eluate collected in a tared 5.0-mL conical vial containing a boiling stone.

The vial is fitted with an air condenser and then placed in a sand bath maintained at a temperature of 90–100°C to evaporate the pentane solvent. *The evaporation is continued until a constant weight of product is obtained.*

Record the weight of product and calculate the percentage yield. Determine the boiling point, and refractive index (optional) of your material and obtain an IR spectrum. Compare your results with those reported in the literature. Also, compare your IR spectrum with that of the 1-octene starting material.

ENVIRONMENTAL DATA

Substance	Amount	TLV (mg/m^3)	Emissions (mg)	Volume (m^3)
Ethanol (solv., water soluble)	1.0 mL	1900	140	<0.1
Ethanol (NaBH$_4$ solution)	1.13 mL	1900	140	<0.1
Hydrochloric acid, 6 M (Max., transfer)	100 μL	7	22	3.1
1-Octene (Max., react.)	86 mg	—	85.6	
Pentane (Max., evap.)	5.5 mL	1800	5320	3.0

TLV: NaOH—2 μg/m^3
  Octane—300 ppm
TXDS: Sodium borohydride—orl-rat LD50: 160 mg/kg
  Chloroplatinic acid—inv-rat LD50: 49 mg/kg
  Silver nitrate—unk-man LDLo: 29 mg/kg
  Silica gel—inv-mus LDLo: 234 mg/kg
  Anhydrous sodium sulfate—LD50: 5989 mg/kg

QUESTIONS

6-1. Squalene, first isolated from shark oil and a biological precursor of cholesterol, is a long-chain aliphatic alkene, C$_{30}$H$_{50}$. The compound undergoes catalytic hydrogenation to yield a compound of molecular formula C$_{30}$H$_{62}$. How many double bonds does a molecule of squalene have?

6-2. An optically active carboxylic acid A (C$_5$H$_6$O$_2$) reacts with one mole of hydrogen gas on catalytic hydrogenation. The product is an optically inactive carboxylic acid B (C$_5$H$_8$O$_2$). What are the structures of compounds A and B?

6-3. Two hydrocarbons, A and B, each contain six carbon atoms and one C=C. Compound A exhibits geometrical isomerism but compound B does not. However, on catalytic hydrogenation both A and B give only 3-methylpentane. Draw the structures and give suitable names for compounds A and B.

6-4. What chemical test would you use to distinguish between the 1-octene starting material and the octane product?

6-5. Give the structures and names of five alkenes having the molecular formula C_6H_{12} that produce hexane on catalytic hydrogenation.

REFERENCES   **1.** Selected reviews on catalytic hydrogenation:
   a. Adkins, H.; Shriner, R. L. in *Advanced Organic Chemistry,* 2nd ed.; Gilman, H., Ed; Wiley: New York, 1943; Vol. I, p 779.
   b. Carruthers, W. *Some Modern Methods of Organic Synthesis,* 2nd ed.; Cambridge University Press: London, 1978; p 437.
   c. March, J. *Advanced Organic Chemistry* 3rd ed., Wiley: New York, 1985, p. 691.
2. Selected examples of catalytic hydrogenation of olefins in *Organic Syntheses:*
   a. Adams, R.; Kern, J. W.; Shriner, R. L. *Organic Syntheses;* Wiley: New York, 1941; Collect. Vol. I, p 101.
   b. Bruce, W. F.; Ralls, J. O. *Organic Syntheses;* Wiley: New York, 1943; Collect. Vol. II, p 191.
   c. Herbst, R. M.; Shemin, D. *Ibid.,* p 491.
   d. Cope, A. C.; Herrick, E. C. *Organic Syntheses;* Wiley: New York, 1963; Collect. Vol. IV, p 304.

Experiment 13

Hydroboration–Oxidation of an Olefin: Octyl Alcohol

(1-octanol)

The hydroboration–oxidation of 1-octene to prepare the anti-Markownikoff addition product, 1-octanol, is performed in this experiment.

$$3\ CH_3-(CH_2)_5-CH=CH_2 \xrightarrow{THF-BH_3} [CH_3(CH_2)_7]_3B \xrightarrow[OH^-]{H_2O_2} 3\ CH_3-(CH_2)_7-OH$$

1-Octene   Trioctylborane   1-Octanol

DISCUSSION   This reaction demonstrates the addition of diborane as a BH_3–THF complex to an alkene to form an intermediate trialkylborane, which on oxidation with alkaline hydrogen peroxide yields the corresponding alcohol.

   The addition reaction is known as "hydroboration" and is a general reaction for all classes of acyclic and cyclic alkenes as well as alkynes. As depicted in the following mechanism, the boron hydride reagent adds successively to three molecules of the alkene to form a trialkylborane.

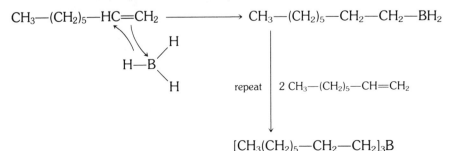

It is important to note that with the 1-octene component, the boron atom becomes attached to the less substituted carbon atom of the double bond. As the alkene donates the π electron pair to the boron atom, which possesses a vacant p orbital, a partial positive charge is developed on the more highly substituted carbon. This occurs because of the electron-releasing effect of the alkyl groups, which enables the carbon to better accommodate the positive charge. The boron atom in turn develops a partial negative charge, which aids cleavage of a hydrogen–boron bond. The hydride ion is thus transferred from the boron atom to the carbon atom of the alkene, which has developed the partial positive charge. The reaction is said to have a four-centered transition state wherein the four atoms involved undergo changes in bonding at the same time.

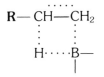

When internal alkenes with different degrees of substitution undergo hydroboration, the boron atom also attaches itself to the less substituted carbon atom. This direction of addition appears to result in part from steric factors. That is, a bulky boron-containing group can approach the least substituted carbon atom more easily. Thus, we see that due to electronic and steric factors, the addition of the B—H elements to the C=C is highly selective with respect to orientation (*regioselectivity*). Accumulated evidence demonstrates that the reaction occurs by syn addition, a consequence of the four-centered transition state. Therefore, as a result of the mechanism operating in this reaction, the new C—B and C—H bonds are necessarily formed on the same face of the C=C bond as shown in the following example.

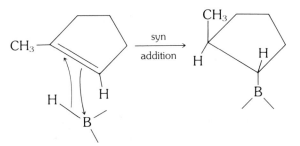

The organoboranes are important in organic synthesis because of the reactions they undergo. Procedures have been developed by which the boron atom may be replaced by a wide variety of groups, such as —H, —OH, —NH$_2$, —Br, and —I. The present experiment demonstrates the conversion of an organoborane to an alcohol by oxidation with alkaline hydrogen peroxide solution. As observed in the experimental section, it is not necessary to isolate the organoborane. This is a great advantage because most alkylboranes are pyrophoric (flammable).

There is conclusive evidence that oxidation of the C—B bond proceeds with retention of configuration of the carbon atom. That is, the hydroxyl group that replaces the boron atom has the identical orientation in the molecule as that of the boron atom.

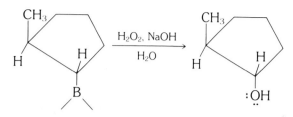

As a result of the hydroboration–oxidation sequence of reactions, the elements of H—OH are added to the original C=C in an anti-Markownikoff manner when the alkene is unsymmetrical.

In the oxidation reaction a hydroperoxide anion is generated in the alkaline medium. This species makes a nucleophilic attack on the boron atom. Migration of an alkyl group from boron to oxygen then occurs, forming a boron ester (a borate). Hydrolysis of the final boron ester, generated by rearrangement of the three alkyl groups, produces the desired alcohol. The mechanism of the oxidation sequence is given below.

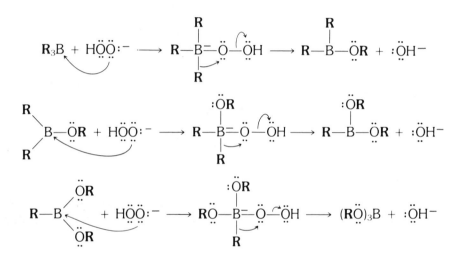

In the final step, alkaline hydrolysis of the trialkyl borate ester, (RÖ)$_3$B, yields the alcohol. Three moles of alcohol is obtained per mole of borate ester.

$$B(OR)_3 + 3\ OH^- \xrightarrow{H_2O} 3\ ROH + BO_3^{3-}$$

EXPERIMENTAL   Estimated time to complete the experiment: 4.0 hours.

Physical Properties of Reactants and Products

Compound	MW	Wt/Vol	mmol	bp(°C)	Density	n_D
1-Octene	112.22	210 μL	1.34	121	0.72	1.4087
Borane–THF (1 *M*)		500 μL				
Sodium hydroxide (3 *M*)	40.00	300 μL				
Hydrogen peroxide (30%)	34.01	300 μL				
Octanol	130.23			194	0.83	1.4295

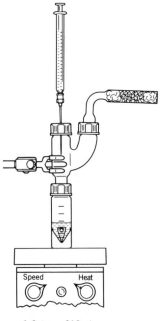

1-Octene, 210 μL,
+1 *M* BH₃–THF, 500 μL

Reagents and Equipment

A 5.0-mL conical vial containing a spin vane is equipped with a Claisen head fitted with a rubber septum and calcium chloride drying tube.(■) Through the rubber septum is added 210 μL (150 mg, 1.34 mmol) of 1-octene (in one portion) with a 1.0-cm³ syringe.

Important. *The glassware and syringe are dried in a 100°C oven for at least 0.5 hour before use. An alternate method is to "flame-out" the glassware with a micro burner and then add the drying tube, caps, and syringe.*

The reaction vessel is cooled in an ice bath and, using the same syringe, 500 μL (0.5 mmol) of the borane–THF solution are added through the septum over a 5-minute period.

> **CAUTION:** *The BH₃–THF reagent reacts violently with water.*

Reaction Conditions

The reactants are allowed to warm to room temperature and then stirred for a period of 45 minutes. Using a Pasteur pipet, two drops of water are now added to hydrolyze any unreacted borane complex.

At this stage of the procedure the vial may be removed from the Claisen head, capped, and allowed to stand until the next laboratory period.

The reaction vial is removed from the Claisen head and, using a graduated 1.0-mL pipet, 300 μL of 3 *M* sodium hydroxide solution is added, followed by the dropwise addition of 300 μL of 30.0% hydrogen peroxide solution over a 10-minute period. The hydrogen peroxide solution is also added with a graduated 1.0-mL pipet. The reaction vial is stirred gently after each addition.

> **CAUTION:** *Hydrogen peroxide blisters the skin. Concentrated solutions can explode!*

The vial is fitted with a reflux condenser and the reaction mixture heated at reflux with stirring for a period of one hour at a sand bath temperature of 100–110°C.(■)

The resulting two-phase mixture is cooled to room temperature and the spin vane removed by the use of forceps.

Isolation of Product

With a Pasteur filter pipet, the bottom layer is separated and transferred to a **SAVE** 3.0-mL reaction vial. *Save the organic phase.*

This separated water layer is extracted with two 1.0-mL portions of diethyl ether and the ether extracts are combined with the original organic phase. On the addition of each portion of ether, the vial is capped, shaken, and vented, and the layers are allowed to separate. The top ether layer is then separated using a Pasteur filter pipet. *If a solid forms during the extraction, a few drops of 0.1 N HCl are added.*

The combined organic phase and ether extracts are now extracted with 750 μL of 0.1 N hydrochloric acid solution followed by extraction with 0.5-mL portions of water until the aqueous extract is neutral to pH paper.

A boiling stone is added to the vial and the organic phase concentrated by **HOOD** warming in a sand bath (60–65°C) in the **hood.**

Gas Chromatographic Analysis

This crude product is easily analyzed by gas chromatography. The procedure involves the injection of 10 μL of the liquid material onto a ¼ in. × 8 ft. steel column packed with

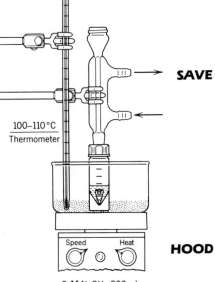

3 *M* NaOH, 300 μL
+ 30% H₂O₂, 300 μL
+ [CH₃(CH₂)₇]₃B product

100–110°C
Thermometer

10% Carbowax 80/100 20M PAW-DMS. Experimental conditions are He flow rate, 50 mL/minute; current, 150 mA; attenuator setting, 8; chart speed, 1 cm/minute; temperature, 190°C.

The liquid effluents elute in the order unreacted 1-octene, a small amount of 2-octanol, and the major product 1-octanol. Retention times are 1.3, 3.1, and 4.2 minutes, respectively.

The eluted material may be collected in an uncooled 4-mm-diameter collection tube and transferred to a 0.1-mL conical vial (see Chapter 5). This material may then be included in the purification step outlined next.

Purification and Characterization

A Pasteur filter pipet is packed with 400 mg of 10% silver nitrate-treated activated silica gel (see Experiment 12 for preparative method) followed by 100 mg anhydrous sodium sulfate.

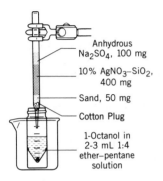

Anhydrous
Na₂SO₄, 100 mg

10% AgNO₃–SiO₂,
400 mg

Sand, 50 mg

Cotton Plug

1-Octanol in
2-3 mL 1:4
ether–pentane
solution

HOOD

The organic residue isolated earlier is dissolved in 500 μL of pentane (spectral or HPLC grade) and transferred by Pasteur filter pipet to the column.(■) The material is eluted from the column with 3 mL of a 1 : 4 diethyl ether/pentane solution. The eluate is collected in a tared 5.0-mL conical vial containing a boiling stone.

The eluate is concentrated to a constant weight by warming on a sand bath (60–65°C) in the **hood.** Weigh the octanol product and calculate the percentage yield.

This product may be analyzed again by gas chromatography. The procedure and experimental conditions outlined earlier are followed. The 1-octene impurity should have been removed in the purification step and the small percentage of 2-octanol still present detected. The actual percentage composition of the mixture can be calculated by determination of the areas under the chromatographic peaks (see Chapter 5).

Obtain an IR spectrum of the alcohol and compare your result with that recorded in the literature (also see Fig. 5.59).

Chemical Tests

A positive ceric nitrate test (Chapter 7) should confirm the presence of the alcohol group. The ignition test may be used to establish that the material is an aliphatic species. The phenyl or α-naphthylurethane derivative may also be prepared to further characterize the alcohol (Chapter 7).

It might also be of interest to determine the solubility characteristics of this C_8 alcohol in water, ether, concentrated sulfuric acid, and 85% phosphoric acid (Chapter 7). Do your results agree with what you would predict for this alcohol?

What chemical tests would you perform to determine the difference between the starting olefin and the alcohol product?

ENVIRONMENTAL DATA

Substance	Amount	TLV (mg/m³)	Emissions (mg)	Volume (m³)
1-Octene (Max. from yield, transfer)	150 mg	—	48	
Tetrahydrofuran (Max., solv.)	500 μL	590	444	0.8
Hydrogen peroxide (30%; Max.)	300 μL	1.5	90	65
Ether (Max., extract to col. chromatog.)	2.6 mL	1200	1860	1.5
Pentane (Max., col. chromatog.)	2.9 mL	1800	1815	1.0
Diborane (Max.)	7 mg	0.1	7	70

TLV: Hydrochloric acid—5.0 ppm
   Sodium hydroxide—2 μg/m³
TXDS: Silica gel—inv-mus LDLo: 234 mg/kg
   Sodium sulfate—orl-mus LD50: 5989 mg/kg
   Silver nitrate—unk-man LDLo: 29 mg/kg
   1-Octanol—orl-mus LD50: 1790 mg/kg

6-6. Using the hydroboration reaction, outline a reaction sequence for each of the following conversions.

    a. 1-Pentene to 1-pentanol

    b. 1-Methylcyclopentene to *trans*-2-methylcyclopentanol

    c. 2-Phenylpropene to 2-phenyl-1-propanol

6-7. When diborane, B_2H_6, dissociates in ether solvents such as tetrahydrofuran (THF), a complex between BH_3 (Borane) and the ether is formed. For example,

$$B_2H_6 \ + \ 2 \ \overset{..}{O}\!\!\bigcirc \ \longrightarrow \ 2 \ \bigcirc\!\!\overset{+}{O}\!:\!-\!\bar{B}H_3$$

    a. In the Lewis sense, what is the function of BH_3 as it forms the complex? Explain.

    b. Write the Lewis structure for BH_3. Diagram its expected structure, indicating the bond angles in the molecule.

6-8. In reference to Question 6-7a, explain why borane (BH_3) reacts readily with the π electron system of an olefin.

6-9. In an unsymmetrical alkene, the boron atom adds predominantly to the least substituted carbon atom. For example, 2-methyl-2-butene gives the products indicated below.

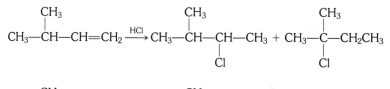

(Diglyme: $CH_3\overset{..}{O}\!-\!CH_2CH_2\!-\!\overset{..}{O}\!-\!CH_2CH_2\!-\!\overset{..}{O}CH_3$, diethyleneglycol dimethyl ether)

Offer a reasonable explanation to account for the ratio obtained.

6-10. An advantage of the hydroboration reaction is that rearrangement of the carbon skeleton does not occur. This contrasts with results obtained on the addition of hydrogen chloride to the double bond. For example,

$$CH_3\!-\!\underset{\underset{\textstyle CH_3}{|}}{CH}\!-\!CH\!=\!CH_2 \ \xrightarrow{\text{HCl}} \ CH_3\!-\!\underset{\underset{\textstyle CH_3}{|}}{CH}\!-\!\underset{\underset{\textstyle Cl}{|}}{CH}\!-\!CH_3 \ + \ CH_3\!-\!\underset{\underset{\textstyle Cl}{|}}{\overset{\overset{\textstyle CH_3}{|}}{C}}\!-\!CH_2CH_3$$

$$CH_3\!-\!\underset{\underset{\textstyle CH_3}{|}}{CH}\!-\!CH\!=\!CH_2 \ \xrightarrow{\text{BH}_3} \ CH_3\!-\!\underset{\underset{\textstyle CH_3}{|}}{CH}\!-\!CH_2CH_2BH_2$$

Offer an explanation for the difference in these results.

 1. For references relating to diborane as a hydroboration agent, see Experiment 16.

2. This experiment was based on that reported by Kabalka, G. W. *J. Chem. Educ.* **1975**, *52*, 745.

EXPERIMENT 14

Diels–Alder Reaction:
4-Cyclohexene-*cis*-1,2-dicarboxylic Acid Anhydride

(*cis*-4-cyclohexene-1,2-dicarboxylic anhydride)

This experiment illustrates the Diels–Alder reaction of 1,3-butadiene (generated in situ) with maleic anhydride to form a carbocyclic six-membered ring.

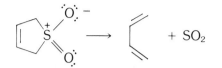

3-Sulfolene   s-cis-1,2-Butadiene

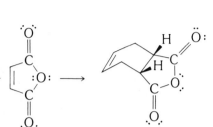

s-cis-1,3-Butadiene Maleic anhydride   Diels-Alder adduct
4-cyclohexene-cis-1,2-dicarboxylic acid anhydride

DISCUSSION   The Diels–Alder reaction is one of the most useful synthetic conversions in organic chemistry. It is an example of a [4 + 2] cycloaddition reaction between a conjugated diene and a dienophile (an -ene or -yne), which leads to the formation of six-membered cyclic rings.

The reaction proceeds well if the dienophile bears electron-attracting groups and the diene, electron-donating groups. Thus, α,β-unsaturated esters, ketones, nitriles, and so on, make excellent dienophiles and are often used in the Diels–Alder reaction. A very large number of compounds can be prepared by varying the nature of the diene and dienophile. Unsubstituted alkenes, such as ethylene, are poor dienophiles and react with 1,3-butadiene only at elevated temperatures and pressures. Since the Diels–Alder reaction is reversible, the lowest possible temperature is generally used.

The reaction is highly stereospecific; the diene component must be in the s-cis configuration to yield a cyclic product having a cis C=C. For this reason, cyclic dienes react more readily than acyclic species. The configuration of the groups on the diene and dienophile are retained in the adduct, and thus the mode of addition must be in the suprafacial–suprafacial manner.

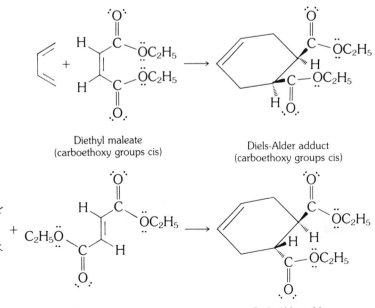

Diethyl maleate
(carboethoxy groups cis)

Diels-Alder adduct
(carboethoxy groups cis)

Diethyl fumarate
(carboethoxy groups trans)

Diels-Alder adduct
(carboethoxy groups trans)

The reaction of cyclopentadiene with maleic anhydride demonstrates a further consequence of the mode of addition. In this situation, there are two possible ways in which the reactants may bond. This leads to the formation of two products, designated the *endo* and *exo* forms.

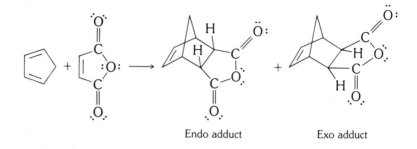

Endo adduct   Exo adduct

Generally, the *endo* form of the molecule predominates, but *endo/exo* ratios may vary depending on reaction conditions.

The reaction is a thermal cycloaddition sequence, and the mechanism, based on orbital symmetry considerations, shows no formation of an intermediate. A simplified electron flow sequence is given below.

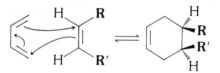

In the present reaction the 3-sulfolene decomposes at a moderate temperature to yield sulfur dioxide and 1,3-butadiene. This is an example of a retro-Diels–Alder reaction. By using this reagent, the diene component is generated in situ.

EXPERIMENTAL   Estimated time to complete the experiment: 1.5 hours.

Physical Properties of Reactants and Products

Compound	MW	Wt/Vol	mmol	mp(°C)	bp(°C)
3-Sulfolene	118.15	85 mg	0.72	66	
Maleic anhydride	98.06	45 mg	0.46	60	
Xylene		40 μL			137–140
cis-4-Cyclohexene-1,2-dicarboxylic acid anhydride	152.15			104	

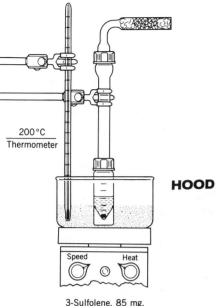

200°C
Thermometer

HOOD

3-Sulfolene, 85 mg,
+ maleic anhydride, 45 mg
+ xylene, 40 μL

Reagents and Equipment

In a 3.0-mL conical vial equipped with an air condenser protected with a calcium chloride drying tube and containing a boiling stone are placed 85 mg (0.72 mmol) of 3-sulfolene, 45 mg (0.46 mmol) of maleic anhydride, and 40 μL of xylene.(■)

Note. *The maleic anhydride should be finely ground and protected from moisture to prevent hydrolysis to the corresponding acid. A mixture of xylenes in the boiling point range 137–140°C will suffice. The solvent should be freshly distilled or dried over molecular sieves before use. The xylene is dispensed in the* **hood** *using an automatic delivery pipet.*

Reaction Conditions

The reaction mixture is now heated at reflux using a sand bath at a temperature of 200°C for a period of 20 minutes.

> **CAUTION: *The reaction is exothermic. Overheating should be avoided. Sulfur dioxide is evolved in the process; adequate ventilation should be provided.***

The conical vial is removed and the contents are allowed to cool to room temperature.

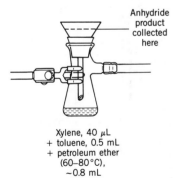

Anhydride
product
collected
here

Xylene, 40 μL
+ toluene, 0.5 mL
+ petroleum ether
(60–80°C),
~0.8 mL

Isolation of Product

To the resulting solution is added 0.5 mL of toluene followed by the dropwise addition of petroleum ether until a slight cloudiness persists.

The amount of petroleum ether required is in the range of 0.25–0.35 mL. During the recrystallization step, the sides of the vial may have to be scratched with a glass rod to induce crystallization.

The solution is reheated until it becomes clear and is then allowed to cool in an ice bath. The crystalline product is collected by vacuum filtration and washed with 0.5 mL of *cold* petroleum ether.(■)

Be careful. *Do not use an excess of the petroleum ether wash. Loss of product results if an excess is used.*

Purification and Characterization

The product is of sufficient purity for characterization. Weigh the product and calculate the percentage yield. Determine the melting point and compare your result with the literature value.

Obtain an IR spectrum of the material utilizing the KBr pellet technique. The reaction involves two reactants, both of which contribute carbon sections to the product. The infrared spectrum of the adduct reflects these contributions.

Infrared Analysis

The spectrum of one of the starting materials, 3-sulfolene (Fig. 6.1), is representative of an olefinic sulfone. The macro group frequency train for an unconjugated five-membered ring olefin fits the data reasonably well: 3090 (=C—H, stretch), 1635 (C=C, stretch), and 657 (H—C=C—H, cis, out-of-plane bend) cm^{-1}. The presence of the sulfone group is convincingly identified by the very strong bands at 1287 and 1127 cm^{-1}, which are assigned to the coupled in-phase and out-of-phase S—O stretching vibrations. The antisymmetric and symmetric C—H stretching vibrations of the methylene groups are observed at 2975 and 2910 cm^{-1}. Note that the influence of the heterocyclic five-membered ring system has raised all of the C—H stretching modes by 40–50 cm^{-1}.

The spectrum of the other reactant, maleic anhydride (Fig. 6.2), possesses all the peaks representative of a conjugated five-membered ring anhydride: 3110 (=C—H, stretch), 1858 (C=O, in-phase stretch, weak), 1777 (C=O, out-of-phase stretch, strong), 1595 (C=C, stretch, weak), 1060 (C—O, antisymmetric stretch), 899 (C—O, symmetric stretch), 835 (H—C=C—H, out-of-plane bend, normally near 700 cm^{-1}, but raised by conjugation to carbonyls). The in-phase stretch of the anhydride carbonyls is particularly weak as the five-membered ring system forces the two oscillators into nearly opposed positions.

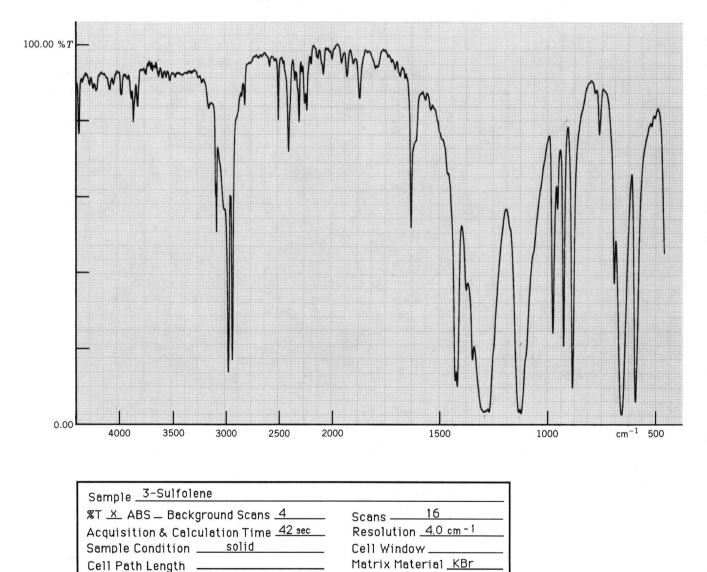

Fig. 6.1 *IR spectrum: 3-sulfolene.*

The Diels–Alder adduct exhibits all of the spectral properties of an unconjugated olefinic five-membered anhydride (Fig. 6.3). The macro group frequency train for the anhydride portion of the molecule involves the following peaks: 1844, 1772, 998, and 935 cm^{-1}.

a. 1844 cm^{-1}: This weak band is similar to the 1858 cm^{-1} band found in maleic anhydride and involves the in-phase stretch of the two coupled carbonyl groups. The band is somewhat more intense in the case of this saturated five-membered ring example, as the ring system is more easily distorted from planarity. This band often exhibits some secondary splitting on the high-wavenumber side (see discussion of the 935 cm^{-1} band below).

b. 1772 cm^{-1}: This band arises from the out-of-phase stretch of the anhydride carbonyls. It is the most intense band in the spectrum. The separation or splitting between the two carbonyl modes of anhydrides is relatively consistent and falls in the range of 60–90 cm^{-1}.

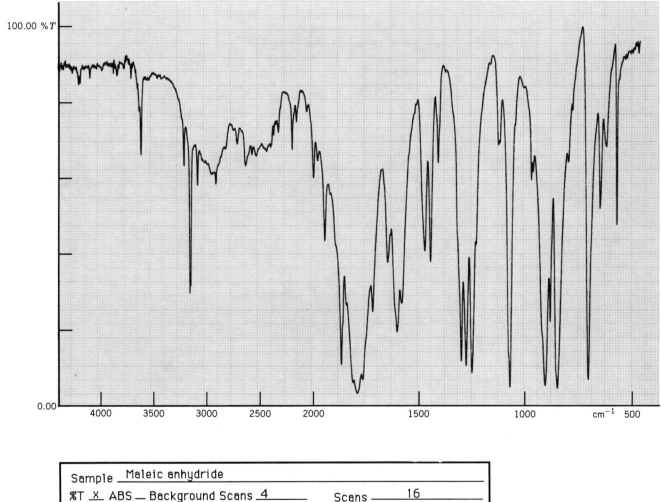

Fig. 6.2 *IR spectrum: maleic anhydride.*

c. 998 and 935 cm^{-1}: These two bands are identified with the antisymmetric and symmetric C—O—C stretching modes. It is the overtone of the symmetric stretch that is Fermi coupled (see Chapter 7) to the symmetric carbonyl stretch (935 × 2 = 1870 cm^{-1}). The frequency match is variable, with the harmonic generally falling on the high wavenumber side of the fundamental.

The olefinic section of the adduct possesses the following macro group frequency bands: 3065 (=C—H stretch), 1635 (C=C stretch, unsubstituted C=C, five-membered ring), and 685 (H—C=C—H, cis, out-of-plane bend) cm^{-1}.

The saturated C—H section of the cyclohexene ring possesses two identical methylene groups that contain the short macro frequency train 2972, 2929 & 2860, 1447 cm^{-1}.

a. 2972 cm^{-1}: antisymmetric C—H stretch of the —CH$_2$— group.

b. 2929 & 2860 cm^{-1}: split symmetric C—H stretch of the —CH$_2$— group. This mode is split by Fermi coupling (see Chapter 7) with the overtone of the symmetric methylene scissoring vibration found at 1447 cm^{-1} (1447 × 2 = 2894). The uncoupled fundamental mode would be expected to occur near 2890 cm^{-1} and, thus, falls very close to the predicted location of the overtone. As both levels involved in the interaction depend on methylene C—H motion, strong Fermi coupling is not unexpected.

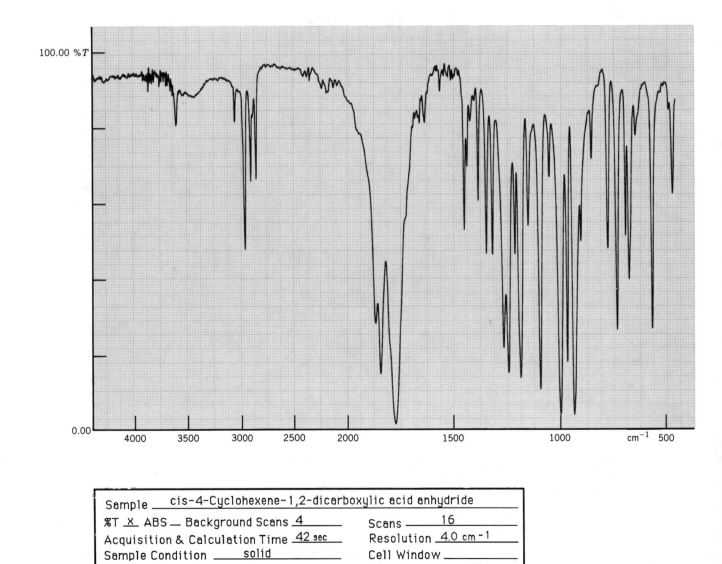

Sample	cis-4-Cyclohexene-1,2-dicarboxylic acid anhydride	
%T _X_ ABS __ Background Scans __4__		Scans __16__
Acquisition & Calculation Time _42 sec_		Resolution _4.0 cm^{-1}_
Sample Condition __solid__		Cell Window __
Cell Path Length __		Matrix Material _KBr_

Fig. 6.3 *IR spectrum: 4-cyclohexene-cis-1,2-dicarboxylic acid anhydride.*

c. 1447 cm^{-1}: symmetric deformation vibration (scissoring motion) of the methylene groups present in the cyclohexene system.

Examine the spectrum of the reaction product you have obtained in a potassium bromide matrix. Discuss the similarities and differences of the experimentally derived spectral data to the reference spectra (Figs. 6.1–6.3).

Nuclear Magnetic Resonance Analysis

The 4-cyclohexene-*cis*-1,2-dicarboxylic acid anhydride provides an ideal example of the utility of ^{13}C NMR when only limited information is available from the ^{1}H NMR spectrum. The 300-MHz ^{1}H spectrum is shown in Fig. 6.4. Due in part to the presence of two chiral centers as well as long-range coupling through the π system of the olefin, the entire ^{1}H spectrum is second order and no information is available from the coupling constants because the spectrum is too complex. Limited assignments to peaks could be made on the basis of chemical shift, but it would be difficult to make any statements regarding the purity of your sample based on the ^{1}H NMR spectrum, as an impurity could well be hidden beneath any of the complex signals.

On the other hand, the ^{13}C spectrum (Fig. 6.5) of 4-cyclohexene-*cis*-1,2-dicarboxylic acid anhydride is much less complex. Because of the mirror plane of symmetry in the

300 MHz NMR OF 4-CYCLOHEXENE-CIS-1,2-DICARBOXYLIC ACID ANHYDRIDE

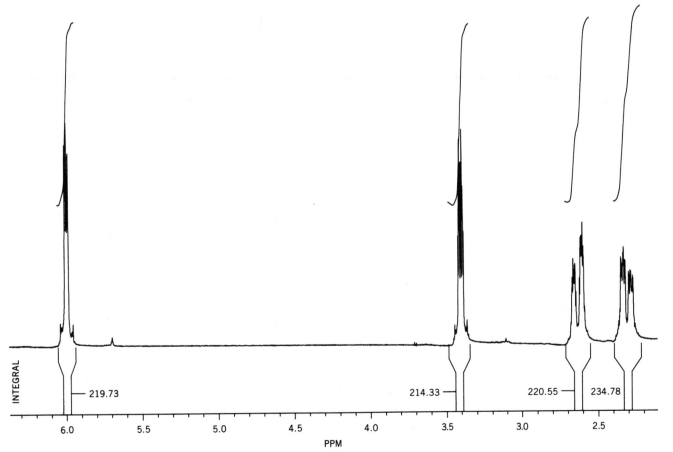

219.73   214.33   220.55   234.78

Fig. 6.4 ^{1}H-NMR spectrum: 4-cyclohexene-cis-1,2-dicarboxylic acid anhydride.

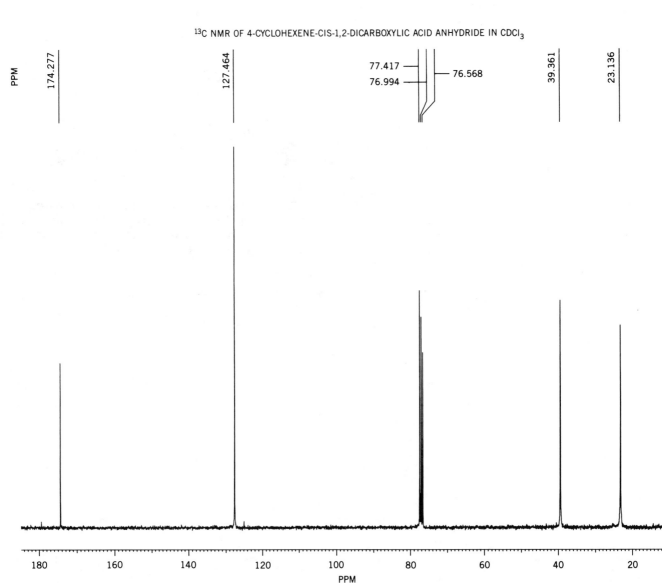

¹³C NMR OF 4-CYCLOHEXENE-CIS-1,2-DICARBOXYLIC ACID ANHYDRIDE IN CDCl₃

Fig. 6.5 *¹³C-NMR spectrum: 4-cyclohexene-cis-1,2-dicarboxylic acid anhydride.*

compound, there are only four different carbons and thus only four lines are seen in the ¹³C NMR spectrum. The 1 : 1 : 1 triplet centered at 77 ppm is due to the solvent, CDCl₃. Although no ¹H–¹³C coupling is seen as a result of the ¹H broad-band decoupling, ²H–¹³C coupling is observed because ¹H and ²H resonate at drastically different frequencies.

Sample preparation for ¹³C NMR is essentially the same as for ¹H NMR spectroscopy except that significantly more material is required. Generally, decent signal-to-noise levels can be obtained on 40 mg of material in about 10 or 15 minutes. Although TMS can be added as a reference, it is often more convenient to reference the spectrum relative to the known chemical shift of the solvent signal. Do not hesitate to use all of your material for the ¹³C spectrum; it can be easily recovered later by emptying the NMR tube into a small vial, rinsing once with solvent, and evaporating the solvent in a hood under a gentle stream of dry nitrogen.

Selected chemical classification tests can also be used to aid in characterization of this compound.

Chemical Tests

Is the compound soluble in water? If so, does the aqueous solution turn blue litmus paper red? Is the compound soluble in 5% NaOH and 5% $NaHCO_3$? Is there evidence of CO_2 evolution with the bicarbonate solution? If so, what does this test indicate? Give the structure of the product formed when the material is added to the sodium hydroxide solution.

Perform the Baeyer test for unsaturation (Chapter 7). Is there evidence of the presence of a C=C bond?

OPTIONAL SCALEUP This preparation may be scaled up by a factor of 10. The procedure is similar to that outlined earlier in the micro preparation with the exceptions as noted below.

a. A 10-mL round-bottom flask containing a magnetic stirrer is used. It is equipped with a reflux condenser protected by a calcium chloride drying tube. An insulating aluminum foil guard is placed between the bottom of the reflux condenser and the round-bottom flask.(■)

HOOD

> **CAUTION: At this scale, due to the generation of a larger quantity of sulfur dioxide the reaction must be run in the hood or provisions made to trap the gas.[1]**

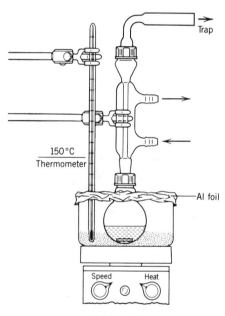

150°C
Thermometer

Reflux conditions
10-mL RB flask

b. The amounts of reagent and solvent are increased tenfold:

Compound	MW	Wt/Vol	mmol	mp(°C)	bp(°C)
3-Sulfolene	118.15	850 mg	7.2	66	
Maleic anhydride	98.06	450 mg	4.6	60	
Xylene		400 μL			137–140

c. The reaction mixture is heated at 150–155°C for 20 minutes. Overheating should be avoided.
d. After the mixture cools, 5 mL of toluene is added and the resulting clear solution transferred to a 25-mL Erlenmeyer flask. The petroleum ether (2.5–3.5 mL) is then added. The toluene solution may require heating and stirring to dissolve any crystals that may have formed.
e. After isolation by vacuum filtration, the product is washed with 5 mL of cold petroleum ether.
f. The product may be characterized as outlined earlier in the micro procedure.

---

[1] For two common methods used to trap water-soluble gases see Horodniak, J.W.; Indicator, N. *J. Chem. Educ.* **1970,** *47,* 568).

ENVIRONMENTAL DATA

Substance	Amount	TLV (mg/m³)	Emissions (mg)	Volume (m³)
3-Sulfolene (treated as SO_2)	85 mg			
Sulfur dioxide (Max.)	46 mg	5	46	9.2
Xylene (SKIN; Max., solv.)	40 μL	435	35	<0.1
Toluene (SKIN) (Max., vac. filtr.)	0.5 mL	375	434	1.2
Petroleum ether (cryst., wash) (treated as hexane)	0.85 mL	180	544	3.0

TXDS: 3-Sulfolene—ins-rbt LD50: 25 mg/kg
　　　　4-Cyclohexene-*cis*-1,2-dicarboxylic acid anhydride—orl-rat LDLo: 4590 mg/kg

QUESTIONS　**6-11.** Predict the product in each of the following Diels–Alder reactions.

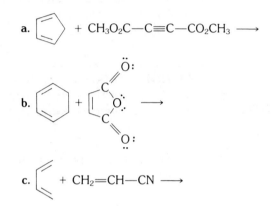

6-12. Cyclopentadiene reacts as a diene in the Diels–Alder reaction a great deal more readily than does 1,3-butadiene. Explain.

6-13. Predict the diene and dienophile that would lead to each of the following products.

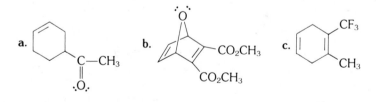

6-14. The dissociation of 3-sulfolene to form 1,3-butadiene generates one mole of sulfur dioxide gas per mole of 1,3-butadiene. Substantial quantities of SO_2 would be generated if this decomposition were carried out on a large scale. Suggest a method for trapping the gas to prevent its escape into the environment.

6-15. Two structural isomers are formed when 2-methyl-1,3-butadiene reacts with ethyl acrylate. Draw structures for these isomers.

6-16. In general, on going from the unsaturated fully conjugated anhydrides to the saturated systems, the C—O stretching modes begin to collapse, and in many cases only a single band is observed. In maleic anhydride, these modes occur at 1060 and 899 cm^{-1}, whereas in 4-cyclohexene-*cis*-1,2-dicarboxylic acid anhydride they are found at 998 and 935 cm^{-1}. Explain this observation.

6-17. Open-chain anhydrides exhibit a reversal of the intensity relationship of the carbonyl stretching vibrations found in the cyclic anhydrides. That is, the lower wavenumber band is now the weaker member of the pair. Explain why this intensity exchange occurs.

6-18. If you had prepared the other possible diastereomer of the compound, 4-cyclohexene-*trans*-1,2-dicarboxylic acid anhydride, how many lines would you expect to see in the ^{13}C NMR spectrum? This isomer does not have a mirror plane of symmetry.

6-19. Does the ^{13}C NMR spectrum unambiguously demonstrate the position of the carbon–carbon double bond? How?

6-20. In ^1H spectra, the relative intensities of lines within a triplet are $1:2:1$. The signal for $CDCl_3$ is a triplet with relative intensities of $1:1:1$. Why?

REFERENCES   1. The preparation of 4-cyclohexene-*cis*-1,2-dicarboxylic acid anhydride by the reaction of 1,3-butadiene with maleic anhydride is recorded: Cope, A. C.; Herrick, E. C. *Organic Syntheses*; Wiley: New York, 1963; Collect. Vol. IV, p 890.

2. The conditions of this reaction were adapted from those reported by Sample, T. E., Jr.; Hatch, L. F. *J. Chem. Educ.* **1968,** *45*, 55.

Experiment 15

Diels–Alder Reaction: 9,10-Dihydroanthracene-9,10-α,β-succinic Acid Anhydride

(9,10-ethanoanthracene-11,12-dicarboxylic anhydride,9,10-dihydro-)

In this experiment the Diels–Alder reaction is demonstrated in which an aromatic ring system constitutes the diene component.

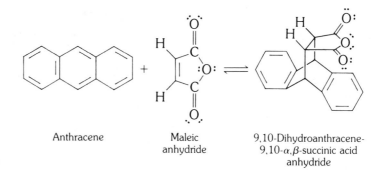

Anthracene   Maleic anhydride   9,10-Dihydroanthracene-9,10-α,β-succinic acid anhydride

DISCUSSION   This experiment is a further example of the Diels–Alder reaction. For a discussion of the Diels–Alder reaction, see Experiment 14. The central ring of anthracene has a characteristic diene system and thus reacts to form stable adducts with many dienophiles at the 9,10 position. The reaction of anthracene and maleic anhydride has been studied extensively and is a reversible reaction.

Higher hydrocarbons containing this anthracene nucleus also react with maleic anhydride, but differ widely in the rates at which they react. 1,2,5,6-Dibenzanthracene (**a**), 2,3,6,7-dibenzanthracene (**b**, pentacene), and 9,10-diphenyl-anthracene (**c**), are examples of aromatic hydrocarbons that undergo the Diels–Alder reaction.

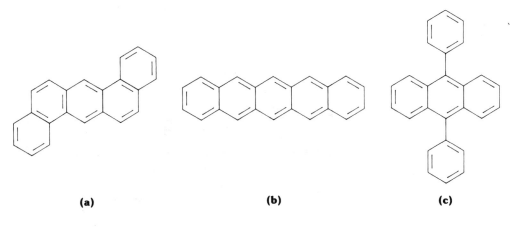

(a)   (b)   (c)

EXPERIMENTAL   Estimated time to complete the experiment: 1.5 hours.

Physical Properties of Reactants and Products

Compound	MW	Wt/Vol	mmol	mp(°C)	bp(°C)
Anthracene	178.24	40 mg	0.22	216	
Maleic anhydride	98.06	20 mg	0.20	60	
Xylene		0.5 mL			137–140
9,10-Dihydroanthracene-9,10-α,β-succinic acid anhydride	276			261–262	

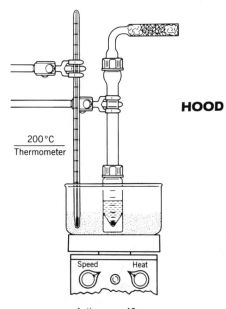

200 °C
Thermometer

Speed   Heat

Anthracene, 40 mg
+ maleic anhydride, 20 mg
+ xylene, 0.5 mL

HOOD

Reagents and Equipment

In a 3.0-mL conical vial containing a boiling stone and equipped with an air condenser protected by a calcium chloride drying tube are placed 40 mg (0.22 mmol) of anthracene and 20 mg (0.20 mmol) of maleic anhydride followed by 500 μL of xylene **[hood]** using an automatic delivery pipet.(■)

Note. *A high-purity grade of anthracene and maleic anhydride is recommended. Anthracene may be recrystallized from 95% ethanol; maleic anhydride may be purified by sublimation. A mixture of xylenes with a boiling point range of 137–140°C is sufficient and should be dried over molecular sieves before use.*

Reaction Conditions

The reaction mixture is now heated under fairly vigorous reflux using a sand bath temperature of ~200°C for a period of 30 minutes. During this time the initial yellow color of the mixture gradually disappears. The resulting solution is allowed to cool to room temperature and is then placed in an ice bath for 10 minutes to complete crystallization of the product.

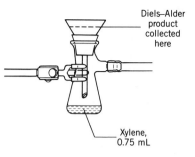

Diels–Alder product collected here

Xylene, 0.75 mL

Isolation of Product

The crystals are collected by vacuum filtration using a Hirsch funnel and the filter cake is washed with 250 μL of ice-cold xylene. (■) The material is now placed on a porous clay plate under an inverted watch glass overnight along with paraffin shavings or film to remove the xylene. *The high boiling point of the xylenes make it difficult to remove the last traces of solvent simply by air-drying. Xylene, as well as other hydrocarbons used for recrystallizations, is absorbed by paraffin. Keep the crystals and paraffin separated.*

Purification and Characterization

The product is of sufficient purity for characterization. However, it may be further purified by recrystallization from ethyl acetate using the Craig tube. Weigh the dried Diels–Alder product and calculate the percentage yield. Determine the melting point and compare your result with the literature value. *It is suggested that a hot-stage melting point apparatus, such as the Fischer–Johns, be used because of the high melting point of the product.*

Chemical Tests

The ignition test is often used as a preliminary method to categorize hydrocarbon materials (see Chapter 7, Table 7.1). Aromatic compounds give a yellow, sooty flame. Perform this test on anthracene. Based on your results can anthracene be classified as aromatic?

Carry out the ignition test on the following compounds and determine whether they should be classified as aromatic materials.

toluene   octane   isopropyl alcohol   nitrobenzene   *trans*-stilbene

OPTIONAL SCALEUP   This preparation may be scaled up by a tenfold or a hundredfold increase in reagent amounts. The procedure is similar in each case to that given for the microscale preparation with the following exceptions.

Tenfold Scaleup

a. A 10-mL round-bottom flask containing a magnetic stirrer is used. It is fitted with a water-jacketed reflux condenser protected by a calcium chloride drying tube. (■)

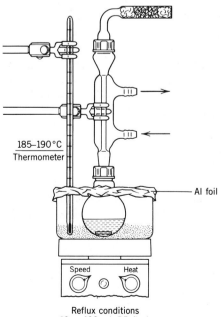

185–190 °C Thermometer

Al foil

Speed   Heat

Reflux conditions
10- or 100-mL RB flask

b. The amounts of reagents and solvent are increased tenfold.

Compound	MW	Wt/Vol	mmol	mp(°C)	bp(°C)
Anthracene	178.24	400 mg	2.2	216	
Maleic anhydride	98.06	200 mg	2.0	60	
Xylene		5.0 mL			137–140

Note. *The product is washed with 2.5 mL of ice-cold xylene.*

c. The reaction flask is **heated at 185–190°C** for a period of 30 minutes to optimize yields.

Hundredfold Scaleup

a. A 100-mL round-bottom flask is used. It is equipped with a water reflux condenser fitted with a calcium chloride drying tube.

b. The amounts of reagents and solvent are increased as given in the following table.

Compound	MW	Wt/Vol	mol	mp(°C)	bp(°C)
Anthracene	178.24	4 g	0.2	216	
Maleic anhydride	98.06	2 g	0.2	60	
Xylene		50 mL			137–140

c. The reaction mixture is heated at vigorous reflux for a period of 2 hours.
d. The product is recrystallized from ethyl acetate.

ENVIRONMENTAL DATA

Substance	Amount	TLV (mg/m³)	Emissions (mg)	Volume (m³)
Xylene—o,m, or p (SKIN) (Max., recryst.)	0.75 mL	435	652	1.5

TLV: Maleic anhydride—0.25 ppm

TXDS: Anthracene—orl-rat TDLo: 20 g/kg/79W-1 TFX:ETA
Anthracene, 11,12-dicarboxylic acid anhydride,9,10-dihyro—ipr-mus LDLo: 200 mg/kg

QUESTIONS **6-21.** Given the data tabulated below for the rate of reaction of maleic anhydride with a series of substituted 1,3-butadienes, offer a reasonable explanation to account for the trend in the rates.

R	k (rel) at 25°C
—H	1
—CH₃	4.2
—C(CH₃)₃	<0.05

6-22. Predict the structure of the product formed in the reactions given below.

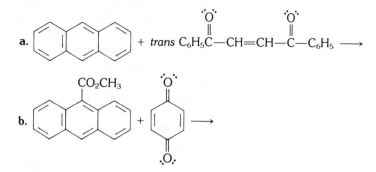

a.

b.

6-23. Offer an explanation of why anthracene preferentially forms a Diels–Alder adduct at the 9,10 position.

6-24. Experiment 38 demonstrates the monobromination of anthracene to yield one specific monosubstituted product. Other conditions may be used to form other mono-brominated anthracenes. Draw the structures and name the possible monobromo substituted anthracenes that could be prepared in the laboratory.

6-25. There are four resonance structures for anthracene. Draw them.

6-26. A large number of polycyclic benzenoid aromatic hydrocarbons are known. One of these named benz[a]pyrene, a powerful carcinogen, is found to be present in tobacco smoke. Locate and then draw the structure for this hydrocarbon. Can you name other sources of this material?

6-27. Anthracene undergoes a suprafacial [4 + 4] cycloaddition reaction at the 9,10 position when subjected to photochemical irradiation to form a cyclic dimer. Draw the structure of this product.

REFERENCES   For references on the Diels–Alder reaction, refer to Experiment 14.

Experiments 16A; 16B   **Diborane Reductions: Thioxanthene or Xanthene**

(thioxanthene or xanthene)

The reduction of an aldehyde or ketone with diborane normally yields the corresponding alcohol. In the selected reactions described below, the carbonyl group is reduced to a methylene unit.

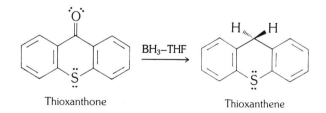

Thioxanthone   Thioxanthene

DISCUSSION   Diborane is a useful selective reducing agent and is prepared by reaction of boron trifluoride etherate with sodium borohydride.

$$3\ NaBH_4 + 4\ BF_3 \rightarrow 3\ NaBF_4 + 2\ B_2H_6$$

The diborane produced may be distilled into tetrahydrofuran to form the BH_3–THF complex, which is commercially available as a 1.0 M solution. Diborane reacts rapidly with water, and for this reason procedures using the BH_3–THF complex must be run under anhydrous conditions.

Diborane is a Lewis acid that attacks electron-rich centers. Thus, when aldehydes or ketones are treated with the BH_3–THF complex, the borate ester is formed, which, on hydrolysis, gives the corresponding alcohol. Reduction of the carbonyl group is believed to take place by addition of the electron-deficient boron atom to the oxygen atom, followed by irreversible transfer of hydride ion from boron to the carbon.

In the case of the xanthone ring system, the corresponding xanthydrol is not formed. It was found that the intermediate borate ester undergoes an elimination reaction, forming a borate ion and a resonance-stabilized xanthonium ion. The second stage of the reaction involves the addition of the BH_3 to the onium ion, which is a special example of the well-known hydroboration reaction. In the addition process the hydride group of the borane adds to the C-9 carbon atom, forming the —CH_2— group. The overall result is the reduction of a carbonyl to a methylene unit.

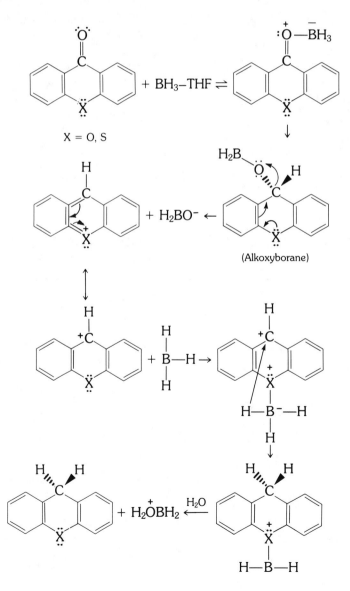

Conventional methods for the reduction of a carbonyl to a methylene unit are the well-known Clemmensen reduction [Zn(Hg), HCl], the Wolff–Kishner reaction (H_2NNH_2, KOH), and the desulfurization of the corresponding thioacetal with Raney nickel.

PART A: Thioxanthene

The reaction is shown above.

EXPERIMENTAL   Estimated time to complete the experiment: 1.5 hours.

Physical Properties of Reactants and Products

Compound	MW	Wt/Vol	mmol	mp(°C)	bp(°C)
Thioxanthone	212.28	50 mg	0.24	209	
Tetrahydrofuran	72.12	1.7 mL			67
Borane–THF, 1 M		1.0 mL			
Thioxanthene	198.19			128–129	

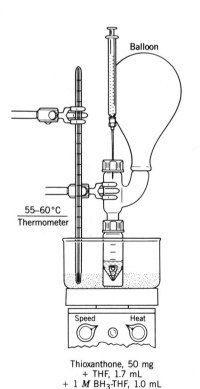

Balloon

55–60°C
Thermometer

Speed   Heat

Thioxanthone, 50 mg
+ THF, 1.7 mL
+ 1 M BH$_3$-THF, 1.0 mL

Reagents and Equipment

To a 5.0-mL conical vial containing a magnetic spin vane and equipped with a Claisen head, which in turn is fitted with a nitrogen inlet tube (prepared from a syringe) and a rubber septum, are added 50 mg (0.24 mmol) of thioxanthone and 1.7 mL of dry tetrahydrofuran.(■)

Note. *The tetrahydrofuran must be absolutely dry. It is recommended that HPLC-grade reagent be used and that it be distilled once from calcium hydride and stored over molecular sieves. It may then be used for up to a week without adversely affecting the yield of product.*

The reaction vial is flushed with a gentle stream of nitrogen gas for several minutes, and then a small balloon is placed over the Claisen head outlet so as to maintain a dry atmosphere in the system.

The mixture is heated with stirring at 55–60°C until the thioxanthone dissolves, yielding a yellow solution. Then with stirring, 1.0 mL of a 1.0 M solution of BH$_3$–THF is added in one portion through the rubber septum with a 1.0-mL syringe.

> **CAUTION: The BH$_3$–THF solution reacts violently with water.**

Reaction Conditions

The solution is heated at 55–60°C in a sand bath with stirring for a period of 5 minutes. The solution becomes colorless during this time.

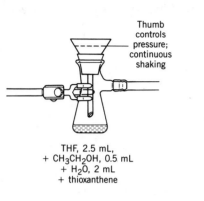

CAREFUL

Thumb
controls
pressure;
continuous
shaking

THF, 2.5 mL,
+ CH₃CH₂OH, 0.5 mL
+ H₂O, 2 mL
+ thioxanthene

Isolation of Product

The reaction is quenched by the *careful,* dropwise addition of approximately 10 drops of 95% ethanol (Pasteur pipet), with stirring, until the observed foaming subsides. *Alcohol is added to decompose any unreacted BH₃–THF reagent.*

After the solution has cooled to room temperature, the mixture is transferred by Pasteur pipet to a 25-mL filter flask containing a boiling stone. Using a calibrated Pasteur pipet, two 1.0-mL portions of water are added to the reaction mixture. The THF/water mixture is carefully removed under reduced pressure with frequent swirling of the flask.(■) As the solvent evaporates, white crystals of thioxanthene appear. The product crystals are collected under reduced pressure by use of a Hirsch funnel, and the filter cake is washed with two 1.0-mL portions of water. The crystals are then air-dried on a porous clay plate.

Purification and Characterization

The thioxanthene product is essentially pure. It may be recrystallized from an ethanol–chloroform mixture, if necessary. Weigh the product and calculate the percentage yield. Determine the melting point of the material and compare it with the value reported in the literature. Obtain IR spectra of thioxanthone and thioxanthene and compare them with each other as well as with those given in the literature.

PART B: Xanthene

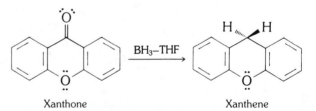

Xanthone $\xrightarrow{\text{BH}_3\text{–THF}}$ Xanthene

EXPERIMENTAL   Estimated time to complete the experiment: 3.0 hours.

Physical Properties of Reactants and Products

Compound	MW	Wt/Vol	mmol	mp(°C)	bp(°C)
Xanthone	196.22	50 mg	0.26	174	
Tetrahydrofuran	72.12	0.7 mL			67
Borane–THF, 1 *M*		0.75 mL			
Xanthene	182.23			100.5	

Reagents and Equipment

Using the experimental apparatus described in Part A, 50 mg (0.26 mmol) of xanthone followed by 0.7 mL of dry tetrahydrofuran (see Part A, Reagents and Equipment) are placed in the reaction flask.(■) A dry nitrogen atmosphere is maintained in the system and the reaction mixture is heated with stirring to 55–60°C using a sand bath. After the xanthone dissolves, 0.75 mL of a 1.0 *M*

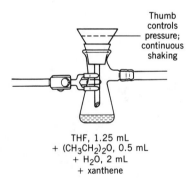

Xanthone, 50 mg
+ THF, 0.7 mL
+ 1 *M* BH₃-THF, 0.75 mL

Thumb
controls
pressure;
continuous
shaking

THF, 1.25 mL
+ (CH₃CH₂)₂O, 0.5 mL
+ H₂O, 2 mL
+ xanthene

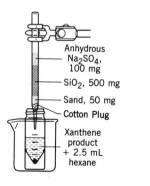

Anhydrous
Na₂SO₄,
100 mg

SiO₂, 500 mg

Sand, 50 mg

Cotton Plug

Xanthene
product
+ 2.5 mL
hexane

solution of BH₃–THF is added through the rubber septum retained on the threaded port of the Claisen head in one portion using a 1.0-mL syringe.

> **CAUTION: *The BH₃–THF solution reacts violently with water.***

Reaction Conditions

The reaction solution is stirred at 55–60°C in a sand bath for a period of one hour.

Isolation of Product

The reaction solution is now quenched by the careful addition (Pasteur pipet) with stirring of approximately 10 drops of 95% ethanol to the warm solution or until the observed foaming subsides. *Alcohol is added to decompose any unreacted BH₃–THF reagent.*

The solution is transferred by Pasteur pipet to a 25-mL filter flask containing a boiling stone. Two 1.0-mL portions of water are added (calibrated Pasteur pipet) to the solution. The tetrahydrofuran solvent is carefully removed under reduced pressure with frequent swirling of the flask.(■) As the solution becomes more concentrated, white crystals of xanthene appear. The crystals are collected under reduced pressure using a Hirsch funnel.

Purification and Characterization

The crude xanthene is purified by column chromatography. Place 0.5 g of activated silica gel followed by 0.1 g of anhydrous sodium sulfate in a Pasteur filter pipet.(■) The column is first wetted with a small amount of hexane and then a solution of the crude xanthene in 0.25 mL of methylene chloride is placed on the column using a Pasteur pipet. Xanthene is eluted by adding more hexane (approximately 2.5 mL). The eluate is collected in a tared 5-mL conical vial containing a boiling stone. The hexane solvent is removed by evaporation **HOOD** **[hood]** by warming the solution on a sand bath to yield pure xanthene. This compound may be recrystallized from ethanol if further purification is necessary for characterization.

Weigh the solid and calculate the percentage yield of xanthene. Determine the melting point and compare it with the value found in the literature. Obtain an IR spectrum of the xanthene formed in the reaction by the KBr pellet technique.

In this experiment, an aromatic ketone has been reduced to a methylene group. Let us examine the infrared spectra of the starting material and the reduction product to see what evidence is present to substantiate that what we expected to happen in the reaction has actually taken place.

Infrared Analysis

We will first consider the spectrum of xanthone (Fig. 6.6). The macro group frequency associated with six-membered carbocyclic aromatic ring systems applies in this instance (peaks at 3100–3000, 1600, 1585, 1500, and 1450 cm⁻¹). This frequency train involves the bands located at 3085–3020, 1610–1570, 1484, and 1460 cm⁻¹. These peaks are assigned as follows:

a. 3085–3020 cm^{-1}: C—H stretch on sp^2 hybridized carbon. The breadth and complexity of this set of absorption bands indicate the presence of a fairly complex aromatic system.

b. 1610–1570 cm^{-1}: The peaks observed in this region are related to the two degenerate fundamental stretching motions, ν_{8a} and ν_{8b}, of the simple aromatic ring system. Normally, ν_{8a} is found near 1600 cm^{-1} and is considerably more intense than ν_{8b} which is located near 1580 cm^{-1}. In xanthone the situation is more complicated, as the central γ-pyrone ring introduces a pseudo-aromatic six-membered ring system. Thus, this ring system might be expected to possess fundamental frequencies somewhat shifted from those of the carbon rings. It is not unexpected then that we observe a band system of four major components in this region.

c. 1484 and 1460 cm^{-1}: aromatic ring stretch related to ν_{19a} and ν_{19b}. These frequencies are less disturbed by the presence of the pyrone system and occur near their expected locations of 1500 and 1450 cm^{-1}.

The very strong band at 756 cm^{-1} in the absence of strong absorption near 700 cm^{-1} is supporting evidence for the presence of four adjacent ring C—H groups, which implies ortho substitution.

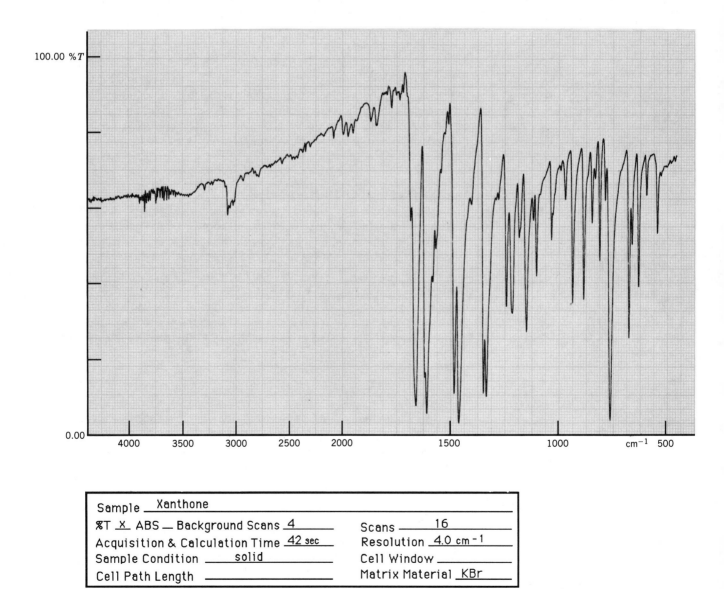

Sample	Xanthone		
%T _X_ ABS __ Background Scans _4_		Scans ____16____	
Acquisition & Calculation Time _42 sec_		Resolution _4.0 cm⁻¹_	
Sample Condition ____solid____		Cell Window _____	
Cell Path Length _____		Matrix Material _KBr_	

Fig. 6.6 *IR spectrum: Xanthone.*

The intense band observed at 1656 cm^{-1} is strong evidence for the presence of a highly conjugated carbonyl group which, of course, is consistent with the structure of the starting material.

The infrared spectrum of the product (Fig. 6.7) supports the complete reduction to the methylene system. The macro group frequency train defined for six-membered carbocyclic aromatic systems still applies. The expected frequencies are very close to the observed values:

a. 3075–3025, 1603, 1589, 1490, and 1457 cm^{-1}.

b. The carbonyl band has vanished and in its place two weak bands have arisen near 2902 and 2840 cm^{-1}. These latter peaks are assigned to the antisymmetric and symmetric stretching modes of the newly formed methylene group.

c. The most intense band in the spectrum occurs at 750 cm^{-1} (four in a row, C—H all-in-phase, out-of-plane bend) and indicates that the basic substitution pattern has not changed on the ring system during the reaction.

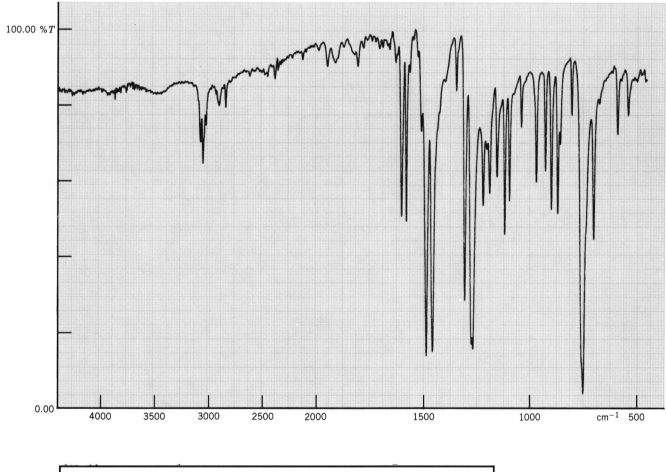

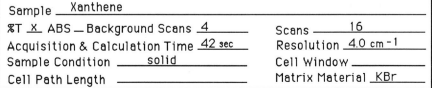

Fig. 6.7 *IR spectrum: Xanthene.*

ENVIRONMENTAL DATA

Substance	Amount	TLV (mg/m³)	Emissions (mg)	Volume (m³)
Tetrahydrofuran (Max., evap.)	2 mL	590	1776	3.0
Ethanol (Max., react., evap.)	0.5 mL	1900	357	0.2
Diborane (Max.)	13.8 mg	0.1	13.8	138
Hexane (Max., col. chromatog.)	3 mL	180	1980	11
Methylene chloride (Max.)	0.25 mL	350	335	1

TXDS: Xanthone—inv-mus LD50: 180 mg/kg
Xanthene—suc-mus LD50: 690 mg/kg
Sodium sulfate—orl-mus LD50: 5989 mg/kg
Silica gel—inv-mus LDLo: 234 mg/kg

QUESTIONS

6-28. In the reaction performed in this experiment, assume that the first stage of the reaction is the rate-controlling step. If so, would you predict that the relative rate of reduction of the carbonyl group to the methylene unit in compound A would be faster or slower than that of xanthone under the conditions of this experiment? Explain.

Compound A: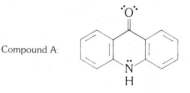

6-29. The reduction of aldehydes or ketones to the methylene group occurs with hydride reagents only when some special feature of the substrate promotes cleavage of the C—OH linkage. Suggest a suitable mechanism by which the reduction given below might occur.

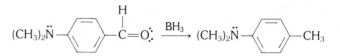

6-30. Diborane, on dissociation to borane, also forms complexes with sulfides and amines. Draw a suitable structure to represent the complex formed between BH_3 and dimethyl sulfide and that between BH_3 and triethylamine.

6-31. Using your lecture textbook as a reference source, show three different methods for the conversion of 4-methylcyclohexanone to methylcyclohexane.

6-32. Using the hydroboration–hydrogen peroxide oxidation reaction sequence, carry out the following conversions.
a. 1-Pentene to 1-pentanol
b. 1-Methylcyclopentene to *trans*-2-methylcyclopentanol

6-33. The infrared spectrum of the xanthene reduction product contains evidence that demonstrates that conjugation of the rings is still maintained after removal of the carbonyl group. What is this evidence?

REFERENCES

1. The reactions carried out in the experiments outlined in Parts A and B were adapted from the work of Wechter, W. J. *J. Org. Chem.* **1963,** *28,* 2935.

2. Diborane as a reducing agent:
   a. Carruthers, W. *Some Modern Methods of Organic Synthesis,* 2nd ed.; Cambridge University Press: New York, 1978; Chap. 5.
   b. Fieser, L. F.; Fieser, M. *Reagents for Organic Synthesis*; Wiley: New York, 1967; Vol. 1, p 199. (Subsequent volumes of this series have further examples of diborane as a reducing agent.)
   c. Pelter, A. *Chem. Ind. (London)* **1976,** 888.
   d. Lane, C. F. *Aldrichim. Acta* **1973,** 6, 36.
3. Diborane as a hydroborating agent:
   a. Brown, H. C. *Boranes in Organic Chemistry*; Cornell University Press: New York, 1972.
   b. Brown, H. C. *Organic Synthesis via Boranes*; Wiley: New York, 1975.
   c. Brown, H. C. *Hydroboration*; Benjamin: New York, 1962.
   d. Zweifel, G.; Brown, H. C. *Org. React.* **1963,** 13, 1.

Experiment 17

Grignard Reaction with a Ketone: Triphenylmethanol

(methanol, triphenyl-)

This is the first of a number of Grignard reactions presented in the text. The present experiment involves the addition of a Grignard reagent to a ketone. It is a classic method for the synthesis of *tert*-alcohols.

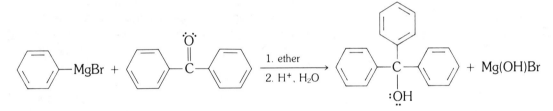

| Phenylmagnesium bromide | Benzophenone | Triphenylmethanol |

DISCUSSION   Addition of nucleophilic Grignard reagents to electrophilic carbonyl carbon centers is an important method for the formation of C—C bonds. This experiment involves the addition of phenylmagnesium bromide to benzophenone to yield the corresponding tertiary (3°) alcohol. Because it is possible to vary the structure of the Grignard reagent and also that of the carbonyl species, a wide variety of 3° alcohols are prepared by this method.

Reaction of Grignard reagents with other carbonyl derivatives yields a number of important classes of compounds. For example, reaction with formaldehyde yields 1° alcohols; with higher aldehydes, 2° alcohols; with esters, 3° alcohols or ketones; with carbon dioxide, carboxylic acids.

The mechanism involves a nucleophilic attack of the Grignard reagent at the carbon atom center of the carbonyl group. The alkoxide ion intermediate is then hydrolyzed with dilute acid to yield the desired alcohol. The sequence is outlined below.

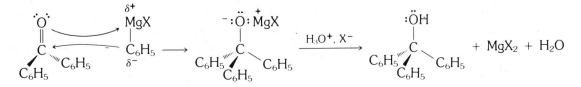

EXPERIMENTAL Estimated time to complete the experiment: two laboratory periods.

> **CAUTION: *Ether is a flammable liquid and also a narcotic. All flames should be extinguished during the time of this experiment.***

Physical Properties of Reactants and Products

Compound	MW	Wt/Vol	mmol	mp(°C)	bp(°C)	Density	n_D
Bromobenzene	157.02	76 μL	0.72	156		1.50	1.5597
Diethyl ether	74.12	1.4 mL			34.5	0.73	
Magnesium	24.3	18 mg	0.73				
Iodine		1 crystal					
Benzophenone	182.21	105 mg	0.58	48			
Triphenylmethanol	260.34			164			

Reagents and Equipment

Preparation of Phenylmagnesium Bromide

Note. *All the glassware used in the experiment should be cleaned, dried in an oven at 110°C for at least 30 minutes, and then cooled in a desiccator before use. Flame-drying of the apparatus with a micro burner may also be used, preferably before assembly (care must be taken when flaming the reaction vial; too rapid heating may cause cracking). If done after assembly, be careful of the plastic caps.*

In a 3.0-mL conical vial containing a magnetic spin vane and equipped with a Claisen head fitted with a calcium chloride drying tube and a rubber septum are placed 18 mg (0.73 mmol) of magnesium, a small crystal of iodine, and 100 μL of anhydrous ether.(■)

Note. *A 2- to 3-in. piece of magnesium ribbon is scraped clean of oxide coating and cut into 1-mm-long sections. This freshly cut material is handled only with forceps.*

Alternate Procedure
The magnesium metal and the iodine crystal are placed in the vial and the apparatus assembled. The mixture is warmed (micro burner or hot plate) *gently* until evidence of a purple vapor from the iodine is seen. *If a micro burner is used in this step, the flame should be extinguished before proceeding.* The 100 μL of anhydrous ether is now added by use of a 1.0-mL syringe inserted through the septum.

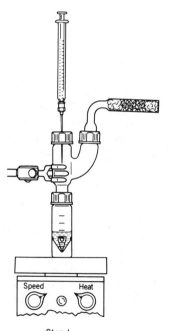

Step I
BrC₆H₅, 76μL
+ Mg, 18 mg
+ I₂, 1 crystal
+ (CH₃CH₂)₂O, 700 μL
Step II
(C₆H₅)₂CO, 105 mg
+ (CH₃CH₂)₂O, 600 μL

HOOD

A solution of 76 μL (113 mg, 0.72 mmol) of bromobenzene in 400 μL of anhydrous diethyl ether is prepared in a dry, screw-capped vial. *Automatic delivery pipets are used to deliver these reagents in the* **hood**.

The bromobenzene–ether solution is drawn into a 1.0-mL syringe, and the syringe then inserted through the rubber septum on the Claisen head.

An additional 300 μL of diethyl ether is placed in the empty vial and the vial is capped and set aside for later use.

While the mixture is stirred, 6–8 drops of the bromobenzene solution are added to initiate the formation of the Grignard reagent. Evidence of reaction is the evolution of tiny bubbles from the surface of the magnesium.

SLOWLY When the reaction has started, the remainder of the bromobenzene is added dropwise *slowly* over a 3- to 5-minute period and the reactants are warmed slightly.

On completion of this addition, the rinse in the capped vial is drawn into the syringe and added through the septum in a single portion.

The resulting solution is now heated gently with stirring for a period of 15 minutes.

> **CAUTION:** *DO NOT OVERHEAT! This will cause loss of ether solvent and promote formation of by-products. Small fragments of magnesium may remain at the end of this time.*

The gray-brown mixture of the Grignard reagent, phenylmagnesium bromide, is cooled to room temperature.

The Benzophenone Reagent

A solution of 105 mg (0.58 mmol) of benzophenone in 300 μL of anhydrous diethyl ether is prepared in a dry shell vial. *The ether is measured using an* **HOOD** *automatic delivery pipet and is dispensed in the* **hood**.

The solution is immediately drawn into a 1.0-mL syringe, and the syringe then inserted into the rubber septum on the Claisen head.

An additional 300 μL of the anhydrous diethyl ether is placed in the empty vial and the vial is capped and set aside for later use.

Reaction Conditions

CAREFULLY The benzophenone solution is added *carefully*, with stirring, to the Grignard reagent over a period of 30 seconds at such a rate as to maintain the ether solvent at a steady reflux.

On completion of this addition, the rinse in the capped vial is added, in like manner, in one portion.

After stirring for 2–3 minutes, the reaction mixture is allowed to cool to room temperature and the reaction vial removed from the Claisen head and capped. During this cooling period the reaction mixture generally solidifies. *It is recommended that the vial be stood in a beaker to prevent loss of product by tipping.*

Note. *If the laboratory is done in two periods, one may stop at this point or after the hydrolysis sequence in the next step.*

Isolation of Product

CAREFUL The magnesium alkoxide salt is hydrolyzed by the *careful*, dropwise addition of 3 *M* HCl from a Pasteur pipet using a small stirring rod to break up the solid. The addition is continued until the aqueous phase is acid to litmus paper. A two-layer reaction solution forms (ether–water) as the solid dissolves.

> **CAUTION:** *The addition of the acid is accompanied by the evolution of heat and foaming. An ice bath should be handy to cool the solution if necessary. Additional ether may be added if required to maintain the volume. Check the acidity of the mixture periodically. The total solution must be acidic; either insufficient or excess amounts of HCl will be detrimental in the subsequent workup.*

The magnetic spin vane is now removed with forceps and set aside to be rinsed with an ether wash. The vial is capped tightly, shaken, and vented, and the layers are allowed to separate. Using a Pasteur filter pipet, the lower aqueous layer is transferred to a clean 5.0-mL conical vial.

Important. *The ether layer is saved since it contains the crude reaction product.*

After transfer to the 5.0-mL vial, the aqueous layer is now washed with three 0.5-mL portions of diethyl ether using a calibrated Pasteur pipet. Rinse the spin vane with the first portion of ether as it is added to the vial. The vial is capped, shaken (or a Vortex mixer is used, if available), and vented and the layers are allowed to separate. After each extraction, the ether layer is combined with the ether solution retained before. The bottom aqueous layer is finally discarded.

The combined ether layers are now extracted with 0.5 mL of cold water to remove any acidic residue. The aqueous rinse is discarded. The ether solution is dried over 250–300 mg of anhydrous sodium sulfate for approximately 10 minutes. The drying agent is stirred intermittently with a glass rod. A larger amount of the drying agent may be used if necessary.

HOOD Using a dry Pasteur filter pipet, the ether solution is transferred to a previously tared Craig tube containing a boiling stone. The ether solution is transferred in 0.5-mL portions. Between transfers, the ether is evaporated by being warmed in a sand bath in the **hood** to concentrate the solution. The vial and drying agent are rinsed with an additional 0.5 mL of ether, the rinse is added to the crude product in the Craig tube, and the solution is concentrated to dryness.

Determine the weight of the crude triphenylmethanol product.

Purification and Characterization

The major impurity in the crude triphenylmethanol is biphenyl, formed by a coupling reaction.

$$2\,\mathbf{R}X + Mg \xrightarrow{\text{ether}} \mathbf{R}\!-\!\mathbf{R} + MgX_2$$

The purity of your material may be determined by *thin-layer chromatography.*

TLC Conditions. *Eastman Kodak fluorescent silica gel sheets (1.0 × 7 cm) are used. They are developed with methylene chloride and visualized by UV light. Reference R_f values for triphenylmethanol and biphenyl are 0.58 and 0.90, respectively.*

These two materials may be separated by taking advantage of their differing solubility in ligroin. Ligroin, a nonpolar solvent, easily dissolves the nonpolar biphenyl molecule, but much less readily dissolves the more polar triphenylmethanol.

To the crude product in the Craig tube is added 0.5 mL of cold ligroin and the solid material scraped into suspension with a small stirring rod. Swirl the solid product with the solvent for several minutes. *Ligroin, bp 30–60°C, is used. Petroleum ether may be substituted if desired.*

SAVE Isolate the solid triphenylmethanol using the Craig tube in the usual manner. *Save the ligroin solution containing any biphenyl by transferring it, using a Pasteur filter pipet, to a tared 10-mL Erlenmeyer flask.*

HOOD Repeat the above extraction with a second 0.5-mL portion of ligroin, again stirring the solid suspension and combining the recovered ligroin solution with that saved before. The open Erlenmeyer flask is now placed in the **hood** overnight, or warmed in a sand bath, to allow the ligroin to evaporate. Determine the amount of biphenyl (and other impurities) produced in the reaction.

The Craig tube containing the solid triphenylmethanol is heated in a 100°C oven for a period of 5 minutes and then the crystals are placed on a clay plate to complete the drying process.

The purity of this material may be checked using TLC as described above.

Weigh the product and calculate the percentage yield. Determine the melting point of the triphenylmethanol and compare your result with the literature value. If desired, the product may be purified further by recrystallization from isopropyl alcohol using the Craig tube.

Characterization of the triphenylmethanol is best done by obtaining the IR and NMR spectra and comparing them with the spectra of an authentic sample or with the published spectra in the literature.

Triphenylmethanol (0.1 g/l in methanol) and biphenyl (0.05 g/l in methanol) also have markedly different UV spectra. These data can further establish the identity of the products formed in this Grignard reaction.

ENVIRONMENTAL DATA

Substance	Amount	TLV (mg/m³)	Emissions (mg)	Volume (m³)
Benzophenone	105 mg			
Bromobenzene (Max. from yield, transfer)	133 mg	—	45	
Diethyl ether (Max., evap.)	3.4 mL	1200	2428	2
Ligroin (Recrystal., treated as hexane)	1.0 mL	180	860	4.8
(Preparation of Grignard also included here.)				

TLV: Iodine—0.1 ppm
    Hydrochloric acid—5 ppm

TXDS: Benzophenone—orl-mus LD50: 2895 mg/kg
    Magnesium—orl-dog LDLo: 230 mg/kg
    Sodium sulfate—orl-mus LD50: 5989 mg/kg

QUESTIONS   **6-34.** Predict the product formed in each of the reactions below and give each reactant and product a suitable name.

a. $CH_3CH_2MgBr + CH_2O \xrightarrow[\text{2. H}^+]{\text{1. ether}}$

b. $p\text{-}CH_3C_6H_4MgBr + CH_3CH_2CHO \xrightarrow[\text{2. H}^+]{\text{1. ether}}$

c. $C_6H_5MgBr + D_2O \xrightarrow{\text{ether}}$

d. $CH_3 - \underset{\underset{CH_3}{|}}{\overset{\overset{CH_3}{|}}{\bigcirc}} - MgBr + CO_2 \xrightarrow[\text{2. H}^+]{\text{1. ether}}$

6-35. Using the Grignard reaction, carry out the following transformations. Any necessary organic or inorganic reagents may be used. Name all reactants and products.

a.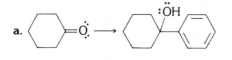

b.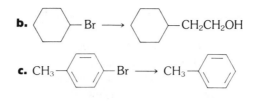

c. CH_3—⬡—Br ⟶ CH_3—⬡

6-36. Outline a balanced stepwise reaction scheme for the preparation of triphenylmethanol by the reaction of phenylmagnesium bromide with (a) methyl benzoate and (b) diethyl carbonate.

6-37. In the experiment, ligroin may be used as a solvent for the separation of the product from biphenyl.
>   **a.** What is the composition of ligroin?
>   **b.** Can you suggest an alternative solvent that might be used in this step?

6-38. Give the reaction scheme, showing the products formed (before hydrolysis), when one equivalent of ethylmagnesium bromide is treated with one equivalent of 5-hydroxy-2-pentanone. Does addition of two equivalents of the Grignard reagent to one equivalent of this same ketone produce a different material(s)? If so, give the structure(s).

REFERENCES

1. The alternate method of preparing the Grignard reagent is adapted from Eckert, T. S. *J. Chem. Educ.* **1987,** *64,* 179.
2. General references on the Grignard reagent:
   a. Grignard, V. *Compt. rend.* **1900,** *130,* 1322.
   b. Jones, R. G.; Gliman, H. *Chem. Rev.* **1954,** *54,* 835.
   c. Wakefield, B. *J. Organometal. Chem. Rev.* **1966,** *1,* 131.
   d. Ashby, E. C. *Quart. Rev.* **1967,** *21,* 259.
   e. Coates, G. E.; Green, M. L. H.; Wade, K. *Organometallic Compounds,* 3rd ed.; Methuen: London, 1968; Vol. II.
   f. Yoffe, S. T.; Nesmeyanov, A. N. *Handbook of Magnesium—Organic Compounds*; Pergamon: London, 1957; Vols. I–III.
   g. Kharasch, M. S.; Reinmuth, O. *Grignard Reactions of Non-metallic Substances*; Prentice–Hall: New York, 1954.
3. A synthesis of triphenylmethanol (triphenylcarbinol) is reported in Bachmann, W. E.; Hetzner, H. P. *Organic Syntheses*; Wiley: New York, 1955; Collect. Vol. III, p 839.
4. The preparation of a series of tertiary alcohols by the addition of Grignard reagents to diethyl carbonate is reported in Moyer, W. W.; Marvel, C. S. *Organic Syntheses*; Wiley: New York, 1943; Collect. Vol. II, p 602–603.

Experiment 18

Grignard Reaction with an Aldehyde: 4-Methyl-3-heptanol

(3-heptanol, 4-methyl-)

The addition of a Grignard reagent to an aldehyde is described in this experiment and constitutes a classic method for the synthesis of *secondary* alcohols.

$$CH_3CH_2CH_2-\overset{\overset{\displaystyle CH_3}{|}}{\underset{\underset{\displaystyle H}{|}}{C}}-MgBr \;+\; CH_3CH_2-\overset{\overset{\displaystyle \ddot{O}}{\|}}{C}-H \quad\xrightarrow[\text{2. H}^+,\text{ H}_2\text{O}]{\text{1. ether}}\quad CH_3CH_2CH_2-\overset{\overset{\displaystyle CH_3}{|}}{\underset{\underset{\displaystyle H}{|}}{C}}-\overset{\overset{\displaystyle H}{|}}{\underset{\underset{\displaystyle \ddot{O}H}{|}}{C}}-CH_2CH_3$$

1-Methylbutylmagnesium bromide   Propanal   4-Methyl-3-heptanol

DISCUSSION In this experiment the addition of a nucleophilic Grignard reagent, 1-methyl-butylmagnesium bromide, to the electrophilic carbonyl carbon atom of an aldehyde (propanal) is described. The product is a 2° alcohol, 4-methyl-3-heptanol. Because it is possible to vary the structure of both the Grignard reagent and the aldehyde, a wide variety of 2° alcohols are prepared by this route. The reaction of ethyl formate with Grignard reagents is also used as a method of preparing 2° alcohols. Primary alcohols are obtained when formaldehyde is used as the aldehyde component. The mechanistic sequence is outlined below.

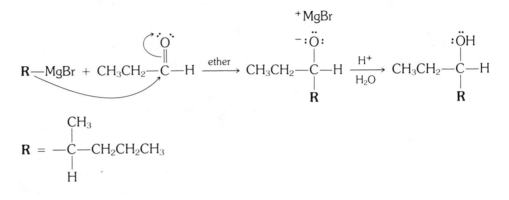

EXPERIMENTAL Estimated time to complete the experiment: 3.0 hours. The chromatographic separation requires approximately an additional 15 minutes per student.

> **CAUTION: *Ether is a flammable liquid and also a narcotic. All flames should be extinguished during the time of this experiment.***

Physical Properties of Reactants and Products

Compound	MW	Wt/Vol	mmol	bp(°C)	Density	n_D
2-Bromopentane	151.05	125 μL	1.0	117	1.21	1.4413
Diethyl ether	74.12	600 mL		34.5		
Magnesium	24.31	36 mg	1.48			
Iodine	253.81	1 crystal				
Propanal	58.08	50 μL	0.69	49	0.81	1.3636
4-Methyl-3-heptanol	130.23			160–161		1.4310

Reagents and Equipment

Preparation of 1-Methylbutylmagnesium Bromide

Note. *All the glassware used in the experiment should be cleaned, dried in an oven at 110°C for at least 30 minutes, and then cooled in a desiccator before use. Flame-drying of the apparatus with a micro burner may also be used (care must be taken when flaming the reaction vial; too rapid heating may cause cracking). If done after assembly, be careful of the plastic caps.*

In a 3.0-mL conical vial containing a magnetic spin vane and equipped with a Claisen head fitted with a calcium chloride drying tube and a rubber septum are

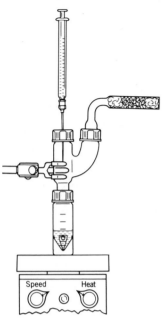

Step I
Mg, 36 mg + I$_2$, 1 crystal
+ (CH$_3$CH$_2$)$_2$O, 400 μL
+ CH$_3$CHBr(CH$_2$)$_2$CH$_3$, 125 μL
Step II
CH$_3$CH$_2$CHO, 50 μL
+ (CH$_3$CH$_2$)$_2$O, 200 μL

placed 36 mg (1.5 mmol) of magnesium, a small crystal of iodine, and 100 μL of anhydrous ether.(■)

Alternate Procedure

The magnesium metal and the iodine crystal are placed in the vial and the apparatus assembled. The mixture is warmed (micro burner or hot plate) *gently* until evidence of purple vapor from the iodine is seen. *If a micro burner is used in this step, the flame should be extinguished before proceeding.* The 100 μL of anhydrous ether is now added by use of a 1.0-mL syringe inserted through the septum.

Note. *A 2- to 3-in. piece of magnesium ribbon is scraped clean of oxide coating and cut into 1-mm-long sections. This freshly cut material is handled only with forceps.*

A solution of 125 μL (153 mg, 1.0 mmol) of 2-bromopentane in 300 μL of anhydrous diethyl ether is prepared in a dry, screw-capped vial. *An automatic delivery pipet is used to deliver these reagents.*

The 2-bromopentane solution is drawn into a 1.0-mL syringe and the syringe then inserted through the rubber septum on the Claisen head. An additional 100 μL of diethyl ether is placed in the empty vial, and the vial is capped and set aside for later use. The equipment is allowed to cool to room temperature before the syringe is inserted into the septum. While the mixture is stirred, 6–8 drops of the 2-bromopentane–ether solution are added to initiate the formation of the Grignard reagent. Evidence of reaction is the evolution of tiny bubbles from the surface of the magnesium.

SLOWLY When the reaction starts, the remainder of the 2-bromopentane-ether solution is added dropwise **slowly** over a 3- to 5-minute period and the reactants are warmed slightly. On completion of this addition, the rinse in the capped vial is drawn into the syringe and added through the septum in a single portion. The resulting solution is now warmed slightly for a period of 15 minutes.

> **CAUTION: *DO NOT OVERHEAT. This will cause loss of ether solvent. Small fragments of magnesium may remain at the end of this time.***

The gray-colored Grignard reagent is cooled to room temperature.

The Propanal Reagent

A solution of the aldehyde is prepared by weighing 50 μL (40 mg, 0.7 mmol) of propanal into a tared, oven-dried, capped vial followed by the addition of 100 μL of anhydrous diethyl ether. The propanal is the limiting reagent and therefore an accurate weight should be recorded for the percentage yield calculations. *The aldehyde and diethyl ether are measured by automatic delivery* **HOOD** *pipets and dispensed in the* **hood.**

The aldehyde solution is immediately drawn into a 1.0-mL syringe and the syringe then inserted into the rubber septum on the Claisen head.

An additional 100 μL of the anhydrous diethyl ether is placed in the empty vial, and the vial is capped and set aside for later use.

Reaction Conditions

CAREFULLY The propanal solution is now added *carefully* with stirring to the Grignard reagent over a period of 30 seconds at such a rate as to keep the ether solvent at a steady reflux.

Following this addition, the rinse in the capped vial is added, in one portion, in a similar manner.

The reaction mixture is stirred for a period of 5 minutes and allowed to cool to room temperature, and the conical vial removed and capped. *It is recommended that the vial be placed in a beaker to prevent loss of product by tipping.*

Note. *If the laboratory is done in two periods, one may stop at this point or after the hydrolysis sequence in the next step.*

Isolation of Product

The alkoxide magnesium salt is hydrolyzed by the *careful*, dropwise addition of 2–3 drops of water from a Pasteur pipet, and the resulting solution stirred for a period of 5 minutes.

> **CAUTION:** *The addition of water causes the evolution of heat. An ice bath should be handy to cool the solution if necessary.*

Two to three drops of 3 M HCl are now added and the capped vial is allowed to stand at room temperature for a period of 5 minutes. A two-phase reaction mixture develops (ether–water) as the magnesium salt is hydrolyzed. Test the aqueous layer with litmus paper. The solution should be slightly acidic. *Insufficient HCl or a large excess of this reagent will cause problems in the subsequent work up.*

The magnetic spin vane is removed with forceps and set aside to be rinsed with an ether wash. The vial is capped tightly, shaken (or a Vortex mixer is used), and vented, and the layers are allowed to separate.

SAVE Using a Pasteur filter pipet, the bottom aqueous layer is transferred to a clean 5.0-mL conical vial. *Save* the ether layer since it contains the crude reaction product.

The aqueous layer previously transferred to the 5.0-mL vial is now washed with three 0.5-mL portions of diethyl ether, and the magnetic spin vane rinsed with the first portion as it is added to the vial. On addition of each portion of ether (calibrated Pasteur pipet), the vial is capped, shaken (or a Vortex mixer is used), and vented and the layers are allowed to separate. With the aid of a Pasteur filter pipet, each ether layer is removed and combined with the ether solution retained earlier. After the final extraction the lower aqueous layer is discarded. The combined ether fractions are now extracted with 0.5 mL of cold water to remove any acidic material.

The ether solution is dried by transferring it by aid of a Pasteur filter pipet to a shortened Pasteur filter pipet containing 500 mg of anhydrous sodium sulfate. The eluate is collected in a tared 10 × 75-mm test tube. The ether solvent is **HOOD** removed from the eluate by warming the solution in a sand bath in the **hood** to concentrate the solution to a weight less than 90 mg.

Cotton packing

Alcohol product
collects here

Purification and Characterization

The product, 4-methyl-3-heptanol, is isolated and purified using gas chromatography.

The entire sample of crude material obtained above is injected onto the GC column using a 100-µL syringe, and the components of interest are collected as they elute, employing the technique described in Experiment 1. (■)

Gas Chromatographic Conditions

Gow Mac Series 150, Thermal Conductivity Detector

20% Carbowax 20 *M* column, $\frac{1}{4}$ in. × 8 ft

Temperature: 145°C

Flow rate: 50 mL/min (He gas)

The retention time for 4-methyl-3-heptanol under these conditions is 7–8 minutes (2 minutes after any other peak). An alternate method of purification is distillation using a Hickman still.

Determine the weight of 4-methyl-3-heptanol collected and calculate the percentage yield. Determine the boiling point and refractive index (optional) of the alcohol and compare them with the values reported in the literature.

Obtain an IR spectrum of the product and compare it with that recorded in the literature. If enough of the hydrocarbon by-product is collected, obtain an IR spectrum to verify its identity.

Chemical Tests

The ignition test should indicate that this compound is an aliphatic species. Does your result confirm this fact?

Perform the ceric nitrate test to demonstrate the presence of the OH group and the Lucas test to substantiate that a secondary alcohol has been prepared.

If you were required to prepare a solid derivative of this alcohol, which one would you select?

It may be of interest to determine the solubility of this product in water, ether, concentrated sulfuric acid, and 85% phosphoric acid. Do your results agree with those you would predict?

What test(s) would you perform to establish that one of the starting reagents was an aldehyde?

ENVIRONMENTAL DATA

Substance	Amount	TLV (mg/m³)	Emissions (mg)	Volume (m³)
2-Bromopentane (Max. from yield, transfer)	125 µL	—	53	
Propanal (Max. from yield, transfer)	40 mg	—	20	
Diethyl ether (Max., evap.)	2.6 mL	1200	1860	1.6
(Preparation of Grignard also included here.)				

TLV: Iodine—0.1 ppm
   Hydrochloric acid—5 ppm

TXDS: Propanal—orl-rat LDLo: 800 mg/kg
   2-Bromopentane—ipr-mus LD50: 150 mg/kg
   4-Methyl-3-heptanol—ivn-mus LD50: 180 mg/kg
   Magnesium—orl-dog LDLo: 230 mg/kg
   Sodium sulfate—orl-mus LD50: 5989 mg/kg

QUESTIONS   **6-39.** Carry out each of the following transformations using the Grignard reaction. Any necessary organic or inorganic reagents may be used. Name each reactant and product.

a. $CH_3(CH_2)_2CH_2Br \longrightarrow (CH_3CH_2CH_2CH_2)_2CH\overset{..}{O}H$

b. —CH=CH—CH$\overset{..}{O}$: $\longrightarrow$ H$_2$C=CH—CH—CH=CH—
   |
   :$\overset{..}{O}$H

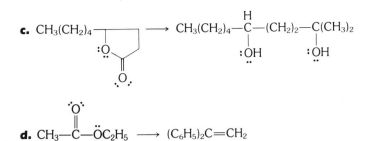

c. $CH_3(CH_2)_4 \longrightarrow CH_3(CH_2)_4-\overset{\overset{H}{|}}{C}-(CH_2)_2-C(CH_3)_2$

d. $CH_3-\overset{\overset{\cdot\cdot O\cdot}{\|}}{C}-\ddot{O}C_2H_5 \longrightarrow (C_6H_5)_2C=CH_2$

6-40. Explain why Grignard reagents cannot be prepared from an organic halide that contains a hydroxyl (—OH), acid carbonyl (—COOH), thiol (—SH), or amino (—NH$_2$) group.

6-41. What would be the final product of the reaction between methyl benzoate and two equivalents of ethylmagnesium bromide?

6-42. Consider the reaction in Question 6-41, but carry out the reaction with ethyl benzoate. What product would be expected in this case?

6-43. Grignard reagents may be used to prepare other organometallic reagents; for example, ethylmagnesium bromide on reaction with cadmium chloride yields diethyl cadmium.

$$2\ CH_3CH_2MgCl + CdCl_2 \rightarrow (CH_3CH_2)_2Cd + 2\ MgCl_2$$

Complete each of the following reactions and name the organometallic product.

$$4\ CH_3MgCl + SiCl_4 \rightarrow$$

$$2\ C_6H_5MgCl + HgCl_2 \rightarrow$$

REFERENCES

1. For references relating to the preparation of Grignard reagents, see Experiment 17.
2. Summary of secondary alcohol preparations presented in *Organic Syntheses*:
   a. Coleman, G. H.; Craig, D. *Organic Syntheses*; Wiley: New York, 1943; Collect. Vol. II, p 179.
   b. Drake, N. L., *Ibid.*, p 406.
   c. Overberger, C. G.; Saunders, J. H.; Allen, R. E.; Gander, R. *Organic Syntheses*; Wiley: New York, 1955; Collect, Vol. III, p 200.
   d. Coburn, E. R. *Ibid.*, p 696.
   e. Skattebol, L.; Jones, E. R. H.; Whiting, M. C. *Organic Syntheses*; Wiley: New York, 1963; Collect. Vol. IV, p 792.

Experiment 19

The Perkin Reaction: Condensation of Rhodanine with an Aromatic Aldehyde: *o*-Chlorobenzylidene Rhodanine

(rhodanine, 5-(*o*-chlorobenzylidene)-)

In this experiment the base-catalyzed condensation of an aromatic aldehyde with an active methylene component is described. This condensation constitutes an example of the well-known Perkin reaction.

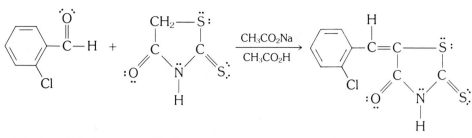

o-Chlorobenzaldehyde   Rhodanine   o-Chlorobenzylidene rhodanine

DISCUSSION The original Perkin reaction, the base-catalyzed condensation of an aromatic aldehyde with a carboxylic acid anhydride, was used extensively to prepare α,β-unsaturated carboxylic acids. In general, the condensation is an aldol-type reaction.

One variation of the Perkin reaction is the condensation of rhodanine with aromatic aldehydes. The resulting compounds have been shown to exhibit antibacterial, antitubercular, antimalarial, antifungal, and antiparasitic activity. Rhodanine has an active methylene group that can be deprotonated with a relatively mild base to generate a nucleophilic agent.

The reaction of rhodanine with the aldehyde group is an example of a nucleophilic attack on a trigonal carbon atom. Under the present conditions, elimination follows the initial nucleophilic addition to yield the benzylidene derivative. The mechanism is shown below.

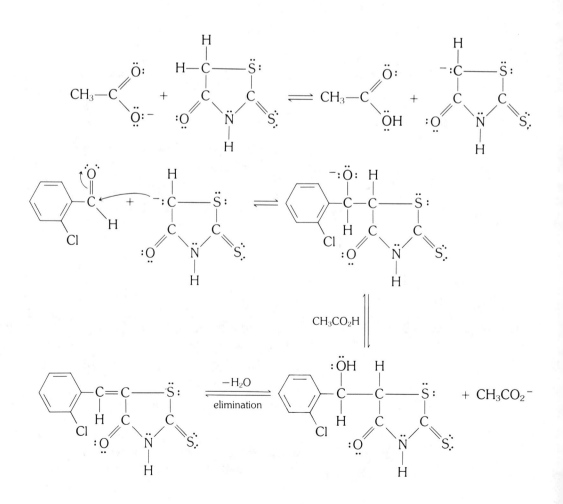

EXPERIMENTAL   Estimated time of the experiment: 1.5 hours.

Physical Properties of Reactants and Products

Compound	MW	Wt/Vol	mmol	mp(°C)	bp(°C)	Density	n_D
Rhodanine	133.19	30 mg	0.23	170			
Sodium acetate	82.03	52 mg	0.63	324			
Glacial acetic acid	60.05	1.0 mL			118		
o-Chlorobenzaldehyde	140.57	58 mg	0.41		212	1.25	1.5662
o-Chlorobenzylidene rhodanine	259.76			191			

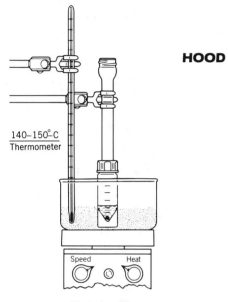

140–150°C
Thermometer

Rhodanine, 30 mg
+ NaOAc, 52 mg
+ CH₃COOH, 1.0 mL
+ o–ClC₆H₄CHO, 58 mg
<u>NOTE</u>: It is important that
the vial be immersed in
sand to the level of the
reaction mixture.

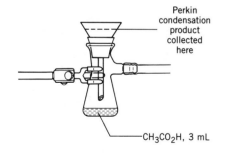

Perkin
condensation
product
collected
here

CH₃CO₂H, 3 mL

Reagents and Equipment

HOOD

In a 3.0-mL conical vial containing a boiling stone and equipped with an air condenser are placed 30 mg (0.23 mmol) of rhodanine, 52 mg (0.63 mmol) of anhydrous sodium acetate, and 0.8 mL of glacial acetic acid **[hood]** dispensed from a graduated pipet.(■) To this mixture is added 58 mg (0.41 mmol) of o-chlorobenzaldehyde delivered from a Pasteur pipet (use 0.2 mL AcOH rinse).

Note. *The sodium acetate is dried in the oven for one hour before use. The aldehyde must be free from the corresponding acid or lower yields of product will result. It is recommended that the purity be checked by IR analysis. The reaction vial may be weighed before and after the addition to obtain an accurate weight of aldehyde.*

Reaction Conditions

The reaction mixture is heated at a sand bath temperature of 140–150°C for a period of 30 minutes.

Important. *The vial is immersed in the sand to the level of the reaction mixture.*

During this time, the reaction mixture becomes homogeneous and turns yellow. The resulting solution is allowed to cool to room temperature and is then placed in an ice bath to complete crystallization of the product.

Isolation of Product

The yellow crystals are collected by vacuum filtration using a Hirsch funnel and the reaction vial and filter cake are washed with two 1.0-mL portions of cold glacial acetic acid delivered from a calibrated Pasteur pipet.(■) The crystals are dried on a porous clay plate or on filter paper.

Purification and Characterization

Weigh the dried product and calculate the percentage yield of the crude material.

Fine, bright yellow needles are obtained on recrystallization (Craig tube) of 5 mg of the crude o-chlorobenzylidene rhodanine from 0.5 mL of glacial acetic acid.

Determine the melting point and compare your results with the literature value.

Chemical Tests

This product is an interesting one since the molecule contains chlorine, nitrogen, and sulfur atoms. The sodium fusion test (see Chapter 7) may be used to substantiate this fact.

It may also be of interest to determine if the Beilstein test can be used to detect chlorine in the presence of sulfur and nitrogen and whether the soda–lime test for nitrogen will give a positive result in the presence of the chlorine and sulfur (see Chapter 7).

Does ignition of the material (see Chapter 7, Table 7.1) indicate that the compound contains an aromatic ring?

OPTIONAL SCALEUP   This experiment can be scaled up to be carried out with five or ten times the amounts used in the preceding microscale preparation. The following data are for the tenfold scaleup. The procedure is identical to the microscale method with the following exceptions:

a. A 10-mL round-bottom flask containing a spin bar and fitted with a water jacketed condenser is used.(■)
b. The reagent and solvent amounts are summarized below.

Compound	MW	Wt/Vol	mmol	mp(°C)	bp(°C)	Density	n_D
Rhodanine	133.19	300 mg	2.3	170			
Sodium acetate	82.03	520 mg	6.3	324			
Glacial acetic acid	60.05	5.0 mL			118		
o-Chlorobenzaldehyde	140.57	580 mg	4.1		212	1.25	1.5662

c. The reaction mixture is heated at 140–150°C for a period of 30 minutes.
d. After the product is air-dried on a clay plate or on filter paper, it is placed in a 100°C oven to dry overnight. It may also be dried in a vacuum drying oven.

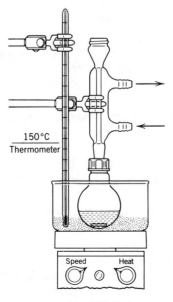

150°C
Thermometer

Speed   Heat

10-mL RB flask

ENVIRONMENTAL DATA

Substance	Amount	TLV (mg/m³)	Emissions (mg)	Volume (m³)
o-Chlorobenzaldehyde (Max. from yield)	58 mg	—	7	
Acetic acid (Max., vac. filtr.)	1.0 mL	25	1049	42
Acetic acid (Max., recryst., wash)	2.0 mL	25	2098	84
Acetic acid (recryst., Craig)	0.5 mL	25	524	21

(Note that the acetic acid is handled in the open except for the Craig tube step.)

TXDS: Rhodanine—ipr-mus LD50: 200 mg/kg
       2-Chlorobenzaldehyde—ipr-mus LDLo: 10 mg/kg

QUESTIONS

6-44. In which of the two species represented below is the underlined hydrogen atom more acidic? Explain.

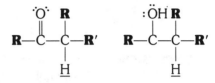

6-45. A number of other active methylene compounds similar to rhodanine have been used in the Perkin condensation reaction. Two are shown in the margin.

Draw the structure of the product that would be formed if each underwent the Perkin condensation with p-chlorobenzaldehyde.

6-46. p-Nitrobenzaldehyde reacts at a faster rate than benzaldehyde in the Perkin reaction, whereas p-N,N-dimethylaminobenzaldehyde is much less active toward the same substrate. Explain.

6-47. Explain the fact that the C=O group in —C—CH— is effective in increasing the reactivity of the α-hydrogen.

6-48. A large number of condensations are closely related to the Perkin reaction. Among these are the aldol, Knoevenagel, Claisen, and Dieckmann condensations.

    a. Give an example of each of these reactions.
    b. Indicate how the reactions are similar to the Perkin reaction in terms of the mechanism that operates in each case and how they differ in terms of reagents.
    c. What general class of compounds can be prepared using each of these well-known reactions?

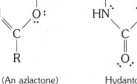

(An azlactone)   Hydantoin

REFERENCES

1. For a review on the general Perkin reaction, see Johnson, J. R. *Org. React.* **1942,** *1,* 210.
2. For examples of the general Perkin reaction, see
   a. Thayer, F. K. *Organic Syntheses*; Wiley: New York, 1941; Collect. Vol. I, p 398.
   b. Herbst, R. M.; Shemin, D. *Organic Syntheses*; Wiley: New York, 1943; Collect. Vol. II, p 1.
   c. Weiss, R. *Ibid.*, p 61.
   d. Corson, B. B. *Ibid.*, p 229.
   e. Johnson, J. R. *Organic Syntheses*; Wiley: New York, 1955; Collect. Vol. III, p 426.
3. For references using rhodanine in the Perkin reaction, see
   a. Brown, F. C. *Chem. Rev.* **1961,** *61,* 463.
   b. Andreasch, R. *Monatsh. Chem.* **1928,** *49,* 122.
   c. Campbell, N.; McKail, J. E., *J. Chem. Soc.* **1948,** 1215.
   d. Julian, P. L.; Sturgis, B. M. *J. Am. Chem. Soc.* **1935,** *57,* 1126.
   e. Foye, W. O.; Tovivich, P. *J. Pharm. Sci.* **1977,** *66,* 1607.

Experiment 20

The Benzoin Condensation with Benzaldehyde: Benzoin

(ethanone, 2-hydroxy-1,2-diphenyl-)

The condensation of an aromatic aldehyde in the presence of a cyanide catalyst to yield an α-hydroxy ketone is known as the benzoin condensation.

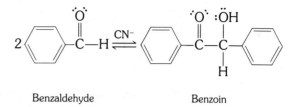

Benzaldehyde   Benzoin

DISCUSSION   Aromatic aldehydes in the presence of cyanide ion catalyst dimerize to form the corresponding α-hydroxy ketone (acyloin). The reaction, which is reversible, is known as the benzoin condensation. Cyanide ion is a *specific* catalyst for the reaction and can function in this capacity because it is a good nucleophile, it stabilizes the anion intermediate, and it is also a good leaving group. In the mechanism outlined below, it is observed that the cyanide ion makes a nucleophilic attack on one molecule of the aromatic aldehyde to form the conjugate base of a cyanohydrin. The effect of the CN group is to increase the acidity of the aldehydic hydrogen atom, thus allowing the formation of the anion **(I)**

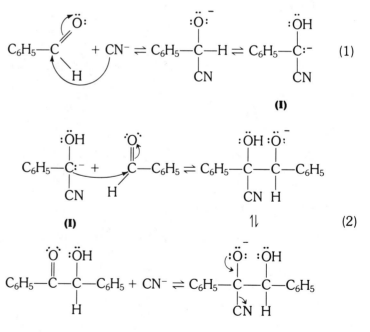

(An acyloin)

Once generated, the nucleophilic carbanion **(I)** attacks a second molecule of the aromatic aldehyde to yield a substituted cyanohydrin. This species can be stabilized by loss of cyanide ion to form the acyloin product.

The electrical effects of various substituents on the aromatic ring have been investigated. If a strongly electron-donating group is in the para position of the ring, the reaction fails. As seen for the *N,N*-dimethylamino group, the effect is to render the carbonyl carbon atom less electrophilic.

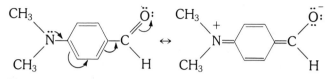

Thus, the nucleophilic attack on the carbonyl carbon by the cyanide ion is retarded. Of interest is the fact that a mixed benzoin condensation with benzaldehyde and *p-N,N*-dimethylaminobenzaldehyde leads to the formation of compound **II**, but not **III**. This result indicates that the cyanohydrin anion **IV** formed from *p-N,N*-dimethylaminobenzaldehyde may add to benzaldehyde, but that the anion **I** formed from benzaldehyde will not add to *p-N,N*-dimethylaminobenzaldehyde. This is due to the increase in electron density at the carbonyl carbon brought about by the presence of the *p-N,N*-dimethylamino substituent.

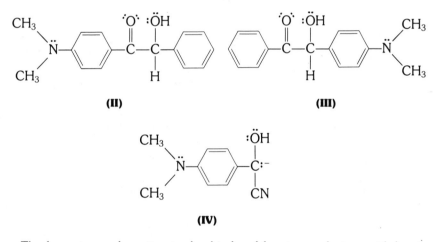

The benzoin condensation is also hindered by strong electron-withdrawing groups on the ring. The presence of a *p*-nitro group decreases the electron density on the carbonyl carbon atom in the cyanohydrin anion, making it less nucleophilic. This effect retards the addition of the anion to the second molecule of aldehyde.

Benzoin, the compound prepared in this experiment, is the starting material for one of the **Sequential Experiments** (Sequence **C**). Benzoin is converted by oxidation (Experiment 41) to benzil and then to tetraphenylcyclopentadienone (Experiment 49). The latter compound is reacted with diphenylacetylene (Experiment 50B) to obtain hexaphenylbenzene (Experiment 51). If the sequence is to be carried out, it is recommended that the **Optional Scaleup** procedure be followed so that sufficient material will be available for the subsequent steps.

EXPERIMENTAL   Estimated time to complete the experiment: 2.0 hours.

Physical Properties of Reactants and Products

Compound	MW	Wt/Vol	mmol	mp(°C)	bp(°C)	Density	n_D
Benzaldehyde	106.13	200 μL	1.96		178	1.04	1.5463
Sodium cyanide (0.54 *M*)		1 mL					
Ethanol (95%)		1 mL					
Benzoin	212.25			137			

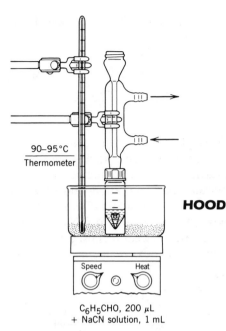

90–95°C
Thermometer

HOOD

Speed   Heat

C_6H_5CHO, 200 μL
+ NaCN solution, 1 mL

Reagents and Equipment

In a tared 5.0-mL conical vial containing a magnetic spin vane and equipped with a reflux condenser are placed 200 μL (208 mg, 1.96 mmol) of fresh acid-free benzaldehyde and 1 mL of a 0.54 M sodium cyanide solution.(■)

The 0.54 M sodium cyanide solution should be prepared by the instructor.

WARNING: *Sodium cyanide is extremely toxic.*

Note. *The conical vial is tared so that the exact weight of benzaldehyde used can be determined. The vial is capped during this operation. It is recommended that both reagents be dispensed in the **hood** using automatic delivery pipets.*

Reaction Conditions

The mixture is heated with stirring at a sand bath temperature of 90–95°C. It is maintained at this temperature for a period of 30 minutes. The reaction solution turns yellow and in approximately 5 minutes may become cloudy.

CAUTION: *Do not overheat the reaction mixture. If the solution begins to darken, remove the vial from the heat source immediately.*

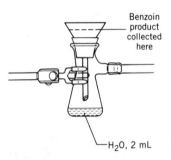

Benzoin
product
collected
here

—H_2O, 2 mL

Isolation of Product

At the end of the reflux period the solution is cooled to room temperature and then placed in an ice bath for an additional 10 minutes. The benzoin product is collected by filtration under reduced pressure using a Hirsch funnel.(■) The filter cake is washed with two 0.5-mL portions of cold water and air-dried under suction for 5 minutes. The crude material is further dried on a porous clay plate or on filter paper.

Purification and Characterization

This crude material is sufficiently pure to use in the preparation of benzoin acetate (see Experiment 8E). It may be purified by recrystallization from methanol or ethanol (95%) by use of a Craig tube.

Weigh the benzoin product and calculate the percentage yield. Determine the melting point and compare your value with that found in the literature.

Obtain the IR spectrum of the compound by KBr pressed disk. Compare your spectrum with that recorded in the literature.

This compound also has a characteristic ultraviolet spectrum showing a peak of maximum absorption at 247 nm (ε_{max} = 13,200, ethanol) characteristic of

$$\text{the } C_6H_5 - \overset{\displaystyle \overset{\cdot\cdot}{O}\cdot}{\overset{\|}{C}} - \text{ group.}$$

Chemical Tests

Benzoin contains the aromatic ring system. Confirm this fact by performing the ignition test (Chapter 7, Table 7.1). To confirm the presence of the alcohol and ketone functions in benzoin carry out the chromic anhydride test for the OH group and the 2,4-dinitrophenylhydrazine test for the C=O group. Isolate the solid 2,4-dinitrophenylhydrazine derivative and compare its melting point with the literature value.

There is a specific test for the presence of benzoin. Place a few crystals of your material in 800 μL of 95% ethanol. The addition of a few drops of 10% sodium hydroxide solution produces a purple coloration. The color fades when shaken in air but reappears if the solution is allowed to stand.

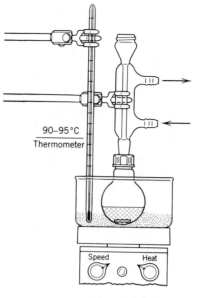

90–95°C
Thermometer

Speed   Heat

10-mL RB flask

OPTIONAL SCALEUP   The procedure is identical to that just given except that a 10-mL round-bottom flask is used(■).

The scale of the reaction can be increased by a factor of 2 or 5.

Twofold Scaleup

Compound	MW	Wt/Vol	mmol	mp(°C)	bp(°C)	Density
Benzaldehyde	106.13	400 µL	3.99		178	1.04
Sodium cyanide (0.54 M)		2 mL				
Ethanol (95%)		2 mL				

Note. *The product is washed with two 1-mL portions of cold water.*

Fivefold Scaleup

Compound	MW	Wt/Vol	mmol	mp(°C)	bp(°C)	Density
Benzaldehyde	106.13	1.0 mL	9.8		178	1.04
Sodium cyanide (0.54 M)		5 mL				
Ethanol (95%)		5 mL				

Note. *The product is washed with two 2-mL portions of cold water*

ENVIRONMENTAL DATA

Substance	Amount	TLV (mg/m³)	Emissions (mg)	Volume (m³)
Benzaldehyde (Max. from yield, transfer)	200 µL	—	74	
Methanol (SKIN) (Max., Craig)	1.0 mL	260	800	3.1
Ethanol (Max., Craig)	1.0 mL	1900	800	0.4

TXDS: Sodium cyanide—orl-hmn LDLo: 2857 µg/kg TWA 5 mg (CN)/m³ (SKIN)
    Benzoin—orl-rat TDLo: 5460 mg/kg
    Benzaldehyde—orl-rat LD50: 1300 mg/kg

QUESTIONS

6-49. The benzoin product produced in this experiment contains a chiral carbon atom, but the product itself is not optically active. Explain.

6-50. The cyanide ion is a highly specific catalyst for the benzoin condensation. Can you list three functions this ion performs in this catalytic role?

6-51. Can you suggest a reason why *p*-cyanobenzaldehyde does not undergo the benzoin condensation to yield a symmetrical benzoin product?

6-52. Tollen's regent is used as a qualitative test for the presence of the aldehyde functional group (see Chapter 7, Classification Tests). However, benzoin, which does not have an aldehyde unit present, gives a positive test with this reagent. Explain.

6-53. Outline equations showing how one might accomplish the conversion of *p*-methylbenzaldehyde to each of the following compounds.

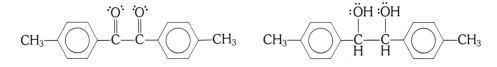

REFERENCES

1. Ide, W. S.; Buck, J. S. *Org. React.* **1948**, *4*, 269.

2. Kuebrich, I. P.; Schowen, R. L.; Wang, M.; Lupes, M. E. *J. Am. Chem. Soc.* **1971**, *93*, 1214.

3. Adams, R.; Marvel, C. S. *Organic Syntheses*; Wiley: New York, 1941; Collect. Vol. I, p 94.

Experiments 21A, 21B, 21C, 21D

Olefin Preparation by the Wittig Reaction: *E*-Stilbene; 1-Methylene-4-*tert*-butylcyclohexane; *trans*-9-(2-Phenylethenyl)anthracene

(*E*-benzene, 1,1′-(1,2-ethenediyl)bis-; cyclohexane, 1-methylene-4-(*t*-butyl)-; anthracene, 9-(2-phenylethenyl)-*trans*)

These preparations illustrate the Wittig reaction, which is used extensively in organic synthesis to prepare olefins. The reaction involves the condensation of a *phosphorus ylid* with a selected carbonyl-containing compound.

$$C_6H_5CHO + [(C_6H_5)_3PCH_2C_6H_5^+, Br^- + NaNH_2] \xrightarrow{\text{THF}} C_6H_5CH{=}CHC_6H_5 + (C_6H_5)_3PO$$

Benzaldehyde "Instant ylid" benzyltriphenylphosphonium bromide–sodium amide *E*-Stilbene (70%) + *Z*-Stilbene (30%) Triphenylphosphine oxide

DISCUSSION

The Wittig reaction constitutes a valuable method for the preparation of alkenes. The major advantage of this approach is that there is no ambiguity in the location of the C=C generated by the reaction.

The reaction involves the formation of a phosphorus ylid obtained by treatment of a phosphonium salt with a strong base. Phosphorus ylids are stable, highly reactive species for which resonance structures may be written; the phosphorus atom accommodates the electron donation into its 3*d* orbitals.

$$[\overset{+}{R_3}P{-}\overset{-}{C}H_2{:} \leftrightarrow R_3P{=}CH_2]$$

The phosphonium salts are available through a nucleophilic displacement reaction with various alkyl halides.

$$(C_6H_5)_3P + \mathbf{R}CH_2X \rightarrow (C_6H_5)_3\overset{+}{P}CH_2\mathbf{R}, X^-$$
$$(X = I, Br, Cl)$$

Treatment of the phosphonium salt with a strong base such as butyl lithium in THF or sodium hydride in dimethyl sulfoxide (DMSO) produces the ylid.

$$(C_6H_5)_3\overset{+}{P}CH_2\mathbf{R}, X^- + C_4H_9{:}Li \xrightarrow{\text{THF}} (C_6H_5)_3\overset{+}{P}{-}\overset{-}{\ddot{C}}H{-}\mathbf{R} + C_4H_{10} + LiX$$

The ylids are generally not isolated, but treated directly with the carbonyl compound.

The "instant ylids" are solid-phase mixtures of a phosphonium salt with sodium amide and are now commercially available. Thus, the mixture need only be suspended in a suitable solvent to generate the desired ylid. This is a marked advantage over the usual methods employed to obtain these species. The ylid is then reacted with a suitable carbonyl compound to form an intermediate *betaine*, which on alkaline hydrolysis gives the olefin product. The mechanistic sequence is outlined below.

Generation of the Ylid

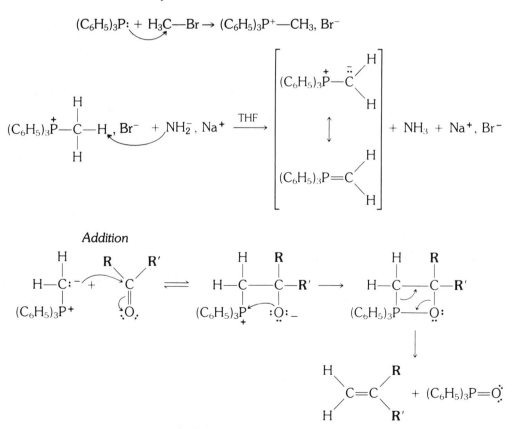

Addition

The Wittig reaction is very general. The carbonyl entity may be aliphatic or aromatic and contain C=C bonds and functional groups such as OH, O$\mathbf{R}$, NO$_2$, N$\mathbf{R}_2$, and X. It is the only general method for the preparation of exocyclic carbon–carbon double bonds. The one drawback is that it is not stereoselective, although the trans isomer is usually dominant. This is most likely due to the relative stabilities of the *cis* and *trans* alkenes.

An important modification (often called the Horner–Emmons reaction) of the Wittig sequence makes use of phosphate esters. It has the advantage of being *highly stereoselective*. The reaction depicted here is for the preparation of E-stilbene.

$$C_6H_5CHO + (C_2H_5O)_2\overset{\overset{\displaystyle :\!O\!:}{\|}}{P}{-}CH_2C_6H_5 \rightarrow C_6H_5CH{=\!=}CHC_6H_5 + (C_2H_5O)_2\overset{\overset{\displaystyle :\!O\!:}{\|}}{P}{-}\ddot{\underset{..}{O}}{:}^-, Na^+$$

Benzaldehyde   Diethylbenzyl phosphonate   E-Stilbene   Sodium diethyl phosphate

The use of the phosphonate ester reactant allows much easier separation of the product olefin since the sodium phosphate also produced in the reaction is soluble in water. This modification has been used generally in cases where the organo group (such as phenyl or acyl) can assist in the stabilization of the carbanion entity. This modification (Part B) may be used as an alternative to Part A for the preparation of E-stilbene. *The "instant ylid" technique yields predominantly the E-product (~70%). The Horner–Emmons reaction yields exclusively the E isomer. Both procedures are given below.* The experiment given in Part D also employes the Horner–Emmons modification.

PART A: *E*-Stilbene by the "Instant Ylid" Method

> **CAUTION:** *Tetrahydrofuran is a flammable liquid. All flames should be extinguished in the laboratory when this solvent is used.*

The reaction for this sequence was shown above.

EXPERIMENTAL   Estimated time to complete the reaction: two 3-hour laboratory periods.

Physical Properties of Reactants and Products

Compound	MW	Wt/Vol	mmol	mp(°C)	bp(°C)	Density	n_D
Benzaldehyde	106.13	100 μL	0.95		178	1.0	1.5463
"Instant ylid" benzyltriphenylphosphonium bromide–sodium amide		600 mg	~1.2				
Tetrahydrofuran	72.12	1.0 mL			67		
E-Stilbene	180.25			124–125			

Reagents and Equipment

In a dry 5.0-mL conical vial containing a magnetic spin vane and equipped with an air condenser protected by a calcium chloride drying tube is placed 600 mg (~1.2 mmol) of benzyltriphenylphosphonium bromide–sodium amide ("*instant*

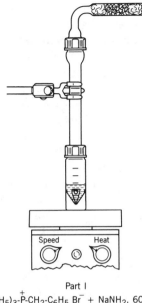

Part I
(C$_6$H$_5$)$_3$-$\overset{+}{P}$-CH$_2$-C$_6$H$_5$ Br$^-$ + NaNH$_2$, 600 mg
+ THF, 1.0 mL
Part II
C$_6$H$_5$CHO, 100 μL

ylid") mixture.(■) *Freshly distilled* tetrahydrofuran (1.0 mL) is now added using a calibrated Pasteur pipet and the mixture stirred for 15 minutes at room temperature. During this time the mixture turns orange.

After generation of the ylid, 100 μL (0.95 mmol) of benzaldehyde (*freshly distilled*) is added to the reaction flask using an automatic delivery pipet.

Reaction Conditions

The resulting heterogeneous mixture is stirred at room temperature for an additional 15 minutes. The system develops a light brown color during this time.

Isolation of Product

The reaction is quenched by adding 1.0 mL of a 25% aqueous NaOH solution (calibrated Pasteur pipet), and the resulting mixture transferred to a 12-mL centrifuge tube using a Pasteur pipet. The reaction vial is rinsed with three 2.0-mL portions of diethyl ether, each of which is also transferred to the centrifuge tube (Pasteur pipet). The resulting two-phase mixture is now partially neutralized by careful addition of 3.0 mL of 0.1 N HCl. The ether layer (top) is transferred by Pasteur filter pipet to a short microcolumn prepared from a Pasteur filter pipet containing 1.5 g of anhydrous sodium sulfate.(■)

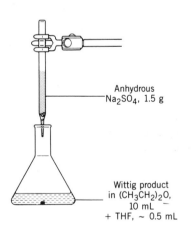

Anhydrous Na$_2$SO$_4$, 1.5 g

Wittig product in (CH$_3$CH$_2$)$_2$O, 10 mL + THF, ~ 0.5 mL

The dried eluate is collected in a 25-mL Erlenmeyer flask containing a boiling stone. The remaining aqueous layer is extracted with two additional 2.0-mL portions of ether and these ether extracts are transferred as before to the microcolumn containing anhydrous sodium sulfate. The eluate solution (~10 mL) is now concentrated to dryness using a gentle stream of nitrogen and warming it in **HOOD** a sand bath **[hood]**, to give a white solid.

Purification and Characterization

The triphenylphosphine oxide by-product is first separated from the mixture of stilbenes by extracting the crude solid product with hexane; the triphenylphosphine oxide is relatively insoluble in the hexane solvent.

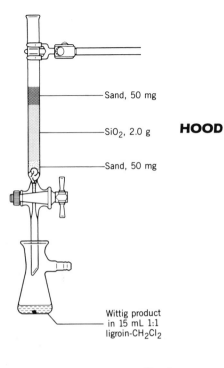

Sand, 50 mg

SiO₂, 2.0 g

HOOD

Sand, 50 mg

Wittig product
in 15 mL 1:1
ligroin-CH₂Cl₂

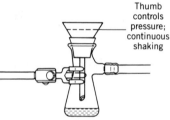

Thumb
controls
pressure;
continuous
shaking

Ligroin (60–80 °C)/CH₂Cl₂, 15 mL
+ Wittig product

To the 25-mL Erlenmeyer flask containing the crude product is added 2.0 mL of hexane, and the solution is agitated with swirling. Some breakup of the material with a microspatula may be necessary. The hexane solution is then transferred by Pasteur filter pipet to a 10-ml Erlenmeyer flask containing a boiling stone. The remaining crude solid is extracted with two additional 2.0-mL portions of hexane, and these extracts are transferred to the 10-mL Erlenmeyer flask as before. The hexane solution (~6 mL) is concentrated to ~1.5 mL using a gentle stream of nitrogen gas and warming it on a sand bath **[hood]**.

A short (1 × 10 cm) chromatography column is packed with 2.0 g of activated silica gel and premoistened with ligroin (60–80°C).(■) The hexane solution is added directly to the column. The column is eluted, in a single fraction, with 15 mL of 1 : 1 ligroin (60–80°C)–methylene chloride solution. The eluate is collected in a tared 25-mL filter flask containing a boiling stone. The solution is concentrated to dryness under reduced pressure in a warm sand bath to yield the pure product mixture.(■)

Note. *A 25-mL side arm filter flask equipped with a Hirsch funnel and filter paper disks to control the pressure is a convenient system for the removal of a small volume (5–20 mL) of solvent.*

Weigh the product residue and calculate the percentage yield.

The isolated product mixture may be analyzed by gas chromatography and/ or thin-layer chromatography. *This constitutes the second week of laboratory for this experiment.*

Gas Chromatographic Analysis

The product mixture isolated above is dissolved in the minimum amount of 1 : 1 ligroin (60–80°C)–methylene chloride solution (~0.5–0.75 mL). A 10-μL sample is injected into a gas chromatograph set at the following conditions:

Column: ¼ in. × 8 ft, 20% Carbowax 20*M* on Chromosorb P80/100 mesh

Temperature: 220°C

Flow rate: 30 mL/min (He gas)

Current: 150 mA

Attenuator setting: 2

Chart speed: 1 cm/min

The compounds elute in the order of *Z*-stilbene, followed by *E*-stilbene. Measure the ratio of peak heights and compare it with that of a standard *E/Z*-stilbene mixture. Calculate the percentage composition of isomers in the mixture.

Thin-Layer Chromatographic Analysis

Spot a TLC plate with a sample from the product solution used for GC analysis and also with the standard solution.

Hexane is used as the developing solvent, silica gel (containing a fluorescent indicator) as the stationary phase, and UV light for visualization. *R*f values: *E*-stilbene, 0.21; *Z*-stilbene, 0.27.

The product mixture is now concentrated as before and the *E*-stilbene is separated and purified by recrystallization from a minimum amount of 95% ethanol using the Craig tube.

Weigh the dried purified *E*-stilbene product and calculate the percentage yield. The purity of this material may be determined by TLC using the conditions outlined above.

Obtain a melting point and IR spectrum of the material and compare your results with those reported in the literature.

E- and *Z*-stilbene exhibit different absorption in the ultraviolet region. The data are summarized below.

Z-Stilbene (1-mm cell)

λ_{max} 223 nm (ε_{max} = 20,600, methanol, 0.05 g/L)

λ_{max} 276 nm (ε_{max} = 10,900, methanol, 0.1 g/L)

E-Stilbene: (0.05 g/L, 1-mm cell)

λ_{max} 229 nm (ε_{max} = 21,000, methanol)

λ_{max} 294 nm (ε_{max} = 33,200, methanol)

λ_{max} 307 nm (ε_{max} = 32,100, methanol)

These ultraviolet data illustrate an interesting example of steric effects on the absorption pattern exhibited by geometrical isomers. Steric interference is present in the *Z* isomer, causing destruction of the coplanarity in the molecule. This is reflected in the lower intensity of the 276-nm band as compared with the 294-nm band in the *E* isomer.

Chemical Tests

Further characterization may be accomplished by performing the Br_2-CH_2Cl_2 test for unsaturation. Note that the dibromo compound is prepared in Experiment 50A. It may be used here as a derivative to characterize the *E*-stilbene. The ignition test (Chapter 7) may be used to confirm the presence of the aromatic portion of the molecule.

PART B: *E*-Stilbene by the Horner–Emmons Reaction

$$C_6H_5CHO + (C_2H_5O)_2\overset{\displaystyle :O:}{\overset{\|}{P}}—CH_2C_6H_5 \xrightarrow[\substack{\text{Hexane}\\\text{Aliquat 336}}]{\substack{40\%\\\text{NaOH}}} C_6H_5CH{=}CHC_6H_5 + (C_2H_5O)_2\overset{\displaystyle :O:}{\overset{\|}{P}}—O^-, Na^+$$

Benzaldehyde   Diethylbenzyl phosphonate   *E*-Stilbene   Sodium diethyl phosphate

Note. *This experiment makes use of a phase transfer catalyst (Aliquat 336). For a discussion on the use of phase transfer catalysts see the Discussion for Experiment 25.*

EXPERIMENTAL   Estimated time to complete the experiment: 2.5 hours.

Physical Properties of Reactants and Products

Compound	MW	Wt/Vol	mmol	mp(°C)	bp(°C)	Density	n_D
Benzaldehyde	106.13	100 μL	0.95		178	1.0	1.5463
Diethylbenzyl phosphonate	228.23	200 μL	0.93		106–8 (1 mm)	1.0	1.4970
Aliquat 336 (tricaprylyl-methylammonium chloride)	404.17	88 mg	0.22				
Hexane	86.18	2.0 mL					
40% Sodium hydroxide		2.0 mL					
E-Stilbene	180.25			124–125			

Reagents and Equipment

In a 10-mL round-bottom flask containing a magnetic stirrer and fitted with a reflux condenser are placed 88 mg (100 μL) of tricaprylylmethylammonium chloride, 100 μL (0.95 mmol) of benzaldehyde, 200 μL (0.93 mmol) of diethyl-benzyl phosphonate, 2.0 mL of hexane, and 2 mL of 40% sodium hydroxide solution.(■)

HOOD

*The benzaldehyde, diethylbenzyl phosphonate, hexane, and NaOH solution are dispensed in the **hood** using automatic delivery pipets. Aliquat 336 is very viscous and is best measured by weighing. A medicine dropper is used to dispense this material.*

It is advisable to lightly grease the bottom joint of the condenser since strong base is being used.

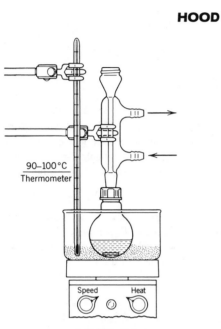

90–100°C
Thermometer

Speed   Heat

C₆H₅CHO, 100 μL
+ diethylbenzyl phosphonate, 200 μL
+ Aliquat 336, 88 mg
+ 40% NaOH, 2 mL, + hexane, 2 mL
10-mL RB flask

Reaction Conditions

The two-phase mixture is now heated at reflux on a sand bath (temperature ~90–100°C) for a period of one hour. The reaction mixture is stirred vigorously during this period. The resulting solution is then allowed to cool nearly to room temperature. Crystals of product may appear as cooling occurs.

Isolation of Product

Methylene chloride (700 μL) is added to dissolve any crystalline material that may have formed and then, using a Pasteur pipet, the contents of the round-bottom flask are transferred to a 15-mL centrifuge tube. The flask is rinsed with an additional 300 μL of methylene chloride and the rinse also transferred to the same centrifuge tube. The aqueous layer is carefully removed using a Pasteur filter pipet. The organic layer is then washed with two portions of water (1 mL each). The mixture is stirred with a small glass rod after each addition (or a Vortex mixer may be used), and then the water layer is removed. The methylene chloride solution is dried by addition of a small amount of anhydrous sodium sulfate. Using a Pasteur filter pipet, the dried solution is transferred to a 10-mL Erlenmeyer flask. The sodium sulfate remaining in the centrifuge tube is washed with two 1-mL portions of methylene chloride. These washings are removed using the Pasteur filter pipet and also transferred to the Erlenmeyer flask.

The solution containing the desired product is concentrated to dryness on a warm sand bath under a slow stream of nitrogen.

Purification and Characterization

The *E*-stilbene obtained is in most cases sufficiently pure for characterization. However, this should be confirmed by thin-layer chromatography as outlined earlier (see Purification and Characterization, Part A). If only a trace of the *Z* isomer is detected, the product may be recrystallized directly using 95% ethanol. The recrystallized material is collected by vacuum filtration using a Hirsch funnel. The vacuum is maintained for an additional 10 minutes to partially dry the crystalline product. It is then placed on a clay plate or filter paper and allowed to dry thoroughly. As an alternative, the product may be dried in a vacuum drying oven or pistol for 10–15 minutes at 30°C (1–2 mm).

Weigh the dried *E*-stilbene product and calculate the percentage yield. The *E*-stilbene may be characterized as outlined in Part A.

PART C: Methylene-4-*tert*-butylcyclohexane

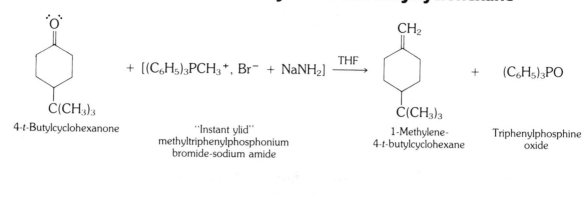

4-*t*-Butylcyclohexanone

"Instant ylid"
methyltriphenylphosphonium
bromide–sodium amide

1-Methylene-
4-*t*-butylcyclohexane

Triphenylphosphine
oxide

EXPERIMENTAL Estimated time to complete the experiment: 4.0 hours.

Physical Properties of Reactants and Products

Compound	MW	Wt/Vol	mmol	mp(°C)	bp(°C)	n_D
4-*t*-Butylcyclohexanone	154.26	100 mg	0.64	50		
"Instant ylid"						
Methyltriphenylphosphonium bromide–sodium amide		320 mg	~0.72			
Tetrahydrofuran		1.0 mL			67	
Methylene-4-*t*-butylcyclohexane	151.27				185	1.4630

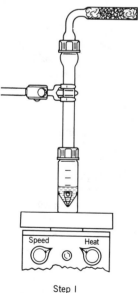

Step I
$(C_6H_5)_3 - \overset{+}{P} - CH_3, \overset{-}{Br} + NaNH_2$, 320 mg
+ THF, 1.0 mL
Step II
4–*t*–Butylcyclohexanone,
100 mg

Reagents and Equipment

In a dry 5.0-mL conical vial containing a magnetic spin vane and equipped with an air condenser protected by a calcium chloride drying tube is placed 320 mg (0.72 mmol) of methyltriphenylphosphonium bromide–sodium amide ("instant ylid") mixture. (■) Freshly distilled tetrahydrofuran (1.0 mL) is added using a calibrated Pasteur pipet and the mixture stirred for 15 minutes at room temperature. During this period it turns a bright yellow.

After generation of the ylid, 100 mg (0.64 mmol) of 4-*t*-butylcyclohexanone is added to the reaction flask. *The THF must be freshly distilled before use.*

Reaction Conditions

The resulting heterogeneous mixture is stirred at room temperature for an additional 90 minutes. The system develops a light tan color over this period of time.

Isolation of Product

The reaction is quenched by adding 1.0 mL of a 25% aqueous NaOH solution (calibrated Pasteur pipet) and the resulting mixture transferred to a 12-mL centrifuge tube using a Pasteur filter pipet. The reaction flask is rinsed with three 2.0-mL portions of diethyl ether, which are also transferred to the centrifuge tube. The resulting two-phase system is partially neutralized by the careful addition of 2.0 mL of 0.1 N HCl.

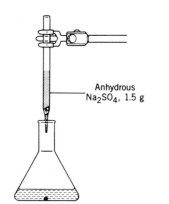

Anhydrous
Na₂SO₄, 1.5 g

Wittig product in
(CH₃CH₂)₂O, ~ 21 mL
+ THF, ~ 0.5 mL

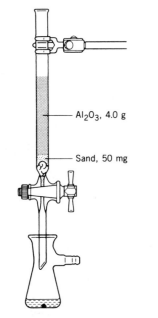

Al₂O₃, 4.0 g

Sand, 50 mg

Wittig product in 8 mL
9:1 ligroin–CH₂Cl₂

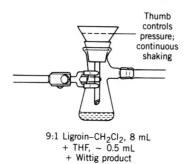

Thumb
controls
pressure;
continuous
shaking

9:1 Ligroin–CH₂Cl₂, 8 mL
+ THF, ~ 0.5 mL
+ Wittig product

The ether layer is separated by Pasteur filter pipet and placed in a 25-mL Erlenmeyer flask. The remaining aqueous layer is extracted with three additional 5-mL portions of diethyl ether, and these extracts are separated as before and combined with the original ether layer.

The combined ether fractions are transferred by Pasteur filter pipet to a short microcolumn prepared from a Pasteur filter pipet containing 1.5 g of anhydrous sodium sulfate.(■) The dried eluate is collected in a 50-ml Erlenmeyer flask containing a boiling stone. This solution is concentrated to dryness using a gentle **HOOD** stream of nitrogen and warming it on a sand bath **[hood],** yielding a colorless liquid residue.

Purification and Characterization

The crude product isolated above is purified by chromatography on an alumina column.

A short (1 × 10 cm) buret column is packed with 4.0 g of activated basic alumina and premoistened with ligroin.(■) The crude material is dissolved in 0.5 mL of 9:1 ligroin (60–80°C)–methylene chloride solvent and the solution transferred to the column using a Pasteur filter pipet. The product is eluted in a single fraction using 8.0 mL of 9:1 ligroin (60–80°C)–methylene chloride. The eluate is collected in a tared 25-mL filter flask containing a boiling stone. The **HOOD** solvent is evaporated **[hood]** under vacuum, with swirling, in a warm sand bath leaving a liquid residue. Product loss may occur if the concentrated residue is heated to excess.(■)

The product is of sufficient purity for characterization. Weigh the methylene-4-t-butylcyclohexane and calculate the percentage yield. Determine the boiling point and compare your result with the literature value.

Obtain an IR spectrum of the olefinic product using the capillary film technique.

Infrared Analysis

The spectrum (Fig. 5.36) of the ketone was discussed in Experiment 5B. The major change observed in the spectrum of the product (Fig. 6.8), when compared with the starting material, is the loss of the carbonyl absorption band (1717 cm⁻¹) and its replacement by a less intense new band at 1654 cm⁻¹. This latter absorption is associated with the stretching motion of the C=C system exocyclic to the six-membered ring. The key bands associated with group frequencies present in the reaction product are identified at 3075, 3000–2850, 1783, 1654, 1394, 1368, and 883 cm⁻¹.

The presence of these bands confirms the projected structure of the Wittig reaction product. The data may be interpreted as follows: the bands at 3075 (sharp spike), 1783 (weak), 1654 (sharp, medium), and 883 (strong) cm⁻¹ form a frequency train that is defined as a "terminal olefin macro frequency." The overall interpretation requires the presence of all four bands in the spectrum if the specific assignment of any one of them is to be correct. Thus, these four data points lead to a structural interpretation with very high confidence limits. The 3075 cm⁻¹ band arises from the coupled antisymmetric stretch of the two C—H oscillators on the terminal olefinic methylene group. The weak 1783 cm⁻¹ peak is an overtone of the strong band observed near 883 cm⁻¹ (883 × 2 = 1766 cm⁻¹) and is unusually intense for an isolated harmonic. The fundamental can be assigned as the =CH₂ out-of-plane wag (C—H deformation). In this fairly rare example, the observed harmonic occurs at a wavenumber value higher than twice the fundamental frequency. A situation of this type is termed negative anharmonicity (see Appendix B). Finally, the C=C stretching mode is assigned to the 1654 cm⁻¹ band, which is consistent with the requirements of this frequency train as it is found below 1660 cm⁻¹ (see Chapter 7—cis, terminal, or vinyl). The bands identified at 1396 and 1368 cm⁻¹ indicate that the tertiary butyl group, as expected, has been preserved during the conversion of the ketone to a terminal olefin (see Discussion in Experiment 5B).

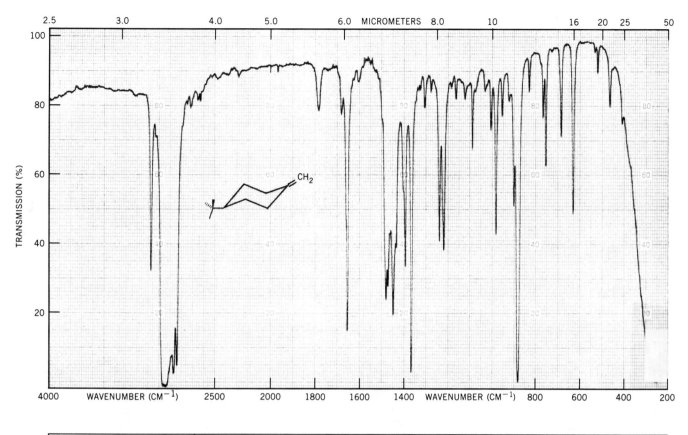

ABSCISSA	ORDINATE	SCAN TIME **12 min**	REP. SCAN_____ SINGLE BEAM_____
EXPANSION_____	EXPANSION_____	RESPONSE_____	TIME DRIVE_____ PRE SAMPLE CHOP_____
SUPPRESSION_____	% T __√__ ABS_____	SLIT PROGRAM **N**	OPERATOR **CD** DATE **8/15/86**
SAMPLE **4-t-Butylmethylene**	REMARKS **283B**	SOLVENT **neat**	CELL PATH **KBr 0.015mm**
cyclohexane			
ORIGIN_____		CONCENTRATION_____	REFERENCE_____

Fig. 6.8 *IR spectrum: methylene-4-t-butylcyclohexane.*

PART D: *trans*-9-(2-Phenylethenyl)anthracene

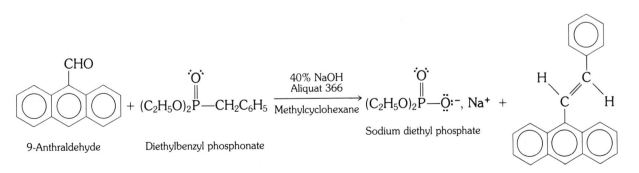

9-Anthraldehyde   Diethylbenzyl phosphonate   Sodium diethyl phosphate

trans-9-(2-Phenylethenyl)anthracene

EXPERIMENTAL   Estimated time to complete the reaction: 2.5 hours.

Physical Properties of Reactants and Products

Compound	MW	Wt/Vol	mmol	mp(°C)	bp(°C)	Density	n_D
9-Anthraldehyde	206.24	106 mg	0.52	104–105			
Diethylbenzyl phosphonate	228.23	120 μL	0.52		106–108 (1 mm)	1.0	1.4970
Aliquat 336 (tricaprylylmethyl-ammonium chloride)	404.17	2 drop					
Methylcyclohexane	98.19	2.0 mL			101	0.77	1.4215
40% sodium hydroxide		2.0 mL					
trans-9-(2-Phenylethenyl)anthracene	280			130–132			

Reagents and Equipment

In a 10-mL round-bottom flask containing a magnetic stirrer and fitted with a reflux condenser are placed 2 drops of tricaprylylmethylammonium chloride, 106 mg (0.52 mmol) of 9-anthraldehyde, 120 μL (0.52 mmol) of diethylbenzyl phosphonate, 2.0 mL of methylcyclohexane, and 2 mL of 40% sodium hydroxide solution.(■)

HOOD   *The diethylbenzyl phosphonate, methylcyclohexane, and NaOH solution are dispensed in the **hood** using automatic delivery pipets. Aliquat 336 is very viscous. A medicine dropper is used to dispense this material.*

It is advisable to lightly grease the bottom joint of the condenser since strong base is being used.

Reaction Conditions

The two-phase mixture is heated on a sand bath at a temperature of ~125–130°C for 45 minutes. The reaction mixture is stirred vigorously during this period; the upper organic layer turns deep red. The resulting solution is then allowed to cool to room temperature.

Isolation of Product

The two-phase solution is transferred to a 15-mL glass centrifuge tube using a Pasteur pipet. The flask is rinsed with 1 mL of methylene chloride and this rinse also transferred to the centrifuge tube. The aqueous layer is now carefully removed using a Pasteur filter pipet. The organic layer is washed with two portions of water (2 mL each). The mixture is stirred with a small glass rod after each addition (or a Vortex mixer may be used) and then the water layer is removed. The methylene chloride solution is dried by addition of sodium sulfate to the tube. Using a Pasteur filter pipet, the dried solution is transferred to a tared 10-mL Erlenmeyer flask containing a boiling stone. The sodium sulfate remaining in the centrifuge tube is washed with two 1-mL portions of methylene chloride. These washings are also transferred to the Erlenmeyer flask.

HOOD   The solution containing the desired product is concentrated almost to dryness on a sand bath in the **hood** under a slow stream of nitrogen gas. Approximately

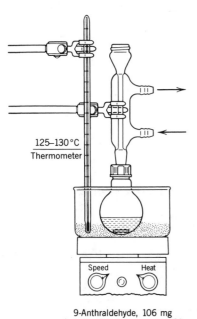

125–130°C
Thermometer

Speed   Heat

9-Anthraldehyde, 106 mg
+ diethylbenzyl phosphonate, 120 μL
+Aliquat 336, 2 drops, + 40% NaOH, 2 mL
+ methylcyclohexane, 2 mL

1–2 mL of 2-propanol is added to the flask and the resulting solution allowed to stand at room temperature and subsequently placed in an ice bath to complete the crystallization of the product.

Purification and Characterization

The yellow crystals are collected by vacuum filtration using a Hirsch funnel and the filter cake is washed with two 1-mL portions of *cold* methanol. The vacuum is maintained for an additional 10 minutes to partially dry the crystalline product. It is then placed on a clay plate or on filter paper and allowed to dry thoroughly. As an alternative, the product may be dried in a vacuum drying oven or pistol for 10–15 minutes at 30°C (1–2 mm).

The purity of the product may be checked using *thin-layer chromatography* (see Chapter 5, Technique 8). A small amount of the starting aldehyde and the product are dissolved separately in ethanol and applied to the TLC plate. Toluene is used as the developing solvent, silica gel (containing a fluorescent indicator) as the stationary phase, and UV light for visualization. R_f values: *trans-9-(2-phenylethenyl)anthracene*, 0.92; 9-anthraldehyde, 0.50.

Weigh the product and calculate the percentage yield. A portion of the material may be further purified by recrystallization from 2-propanol using the Craig tube.

Determine the melting point and compare it with the literature value. Obtain the IR spectrum of the olefinic product (KBr pellet technique). The product should have a strong absorption band at 962 cm^{-1}, which confirms the presence of a trans double bond.

This material has a characteristic UV spectrum in methanol. The following data were obtained at a concentration of $3.1 \times 10^{-6} M$ (see Fig. 6.9):

λ_{max} 255 nm (ε_{max} = 35,207, methanol)

λ_{max} 383 nm (ε_{max} = 3618, methanol)

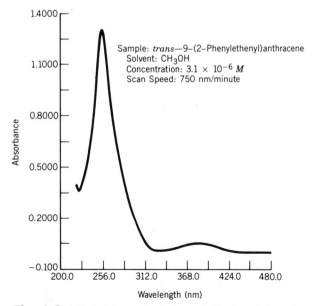

Fig. 6.9 *UV-visible spectrum:* trans-9-(2-phenylethenyl)anthracene.

ENVIRONMENTAL DATA

Substance	Amount	TLV (mg/m³)	Emissions (mg)	Volume (m³)
Experiment 21A				
"Instant ylid"	600 mg			
Benzaldehyde	100 µL			
Tetrahydrofuran (Max., water soluble)	1.0 mL	590	890	1.5
Diethyl ether (Max., extractant, dried, evap.)	10 mL	1200	7,140	6.0
Ligroin (Max., col. chromatog., treated as hexane)	7.5 mL	180	6450	35
Methylene chloride (Max., col. chromatog., evap.)	7.5 mL	350	10,000	28
Ethanol, 95% (Max., recryst.)	1.0 mL	1900	750	0.4
Experiment 21B				
Benzaldehyde	100 µL			
Diethylbenzyl phosphonate	200 µL			
Hexane	2 mL	180	1320	7
Methylene chloride (Max., rinse, extract, evap.)	3 mL	350	4000	11.4
Experiment 21C				
"Instant ylid"	320 mg			
4-t-Butylcyclohexanone (react.)	100 mg			
Tetrahydrofuran (Max., water soluble)	1.0 mL	590	890	1.5
Diethyl ether (Max., rinse, extract, dry, evap.)	21 mL	1200	15,000	12.5
Methylene chloride (Max., col. chromatog., evap.)	0.85 mL	350	1,135	3.2
Ligroin (Max., col. chromatog. evap., treated as hexane)	8.65 mL	180	7,440	41
Experiment 21D				
9-Anthraldehyde	106 mg			
Diethylbenzyl phosphonate	120 µL			
Methylcyclohexane (Max.)	2 mL	1600	1540	1.0
Methylene chloride (Max., rinse, dry, evap.)	3 mL	350	3990	11.4
2-Propanol (Max., recryst.)	2 mL	980	1600	1.6

TLV: Hydrochloric acid—5 ppm
   Sodium hydroxide—2 µg/m³

TXDS: Benzaldehyde—orl-rat LD50: 1300 mg/kg
   Diethylbenzyl phosphonate—ivn-mus LD50: 180 mg/kg
   Tricaprylylmethylammonium chloride—orl-rat LD50: 223 mg/kg
   Sodium sulfate—ori-mus LD50: 5989 mg/kg
   Stilbene—ipr-mus LD50: 1150 mg/kg
   t-Butylcyclohexanone—orl-rat LD50: 5000 mg/kg

QUESTIONS   **6-54.** Complete each of the following reactions by giving a suitable structure for the species represented by the letters. Give a suitable name for compound B in each reaction.

a. $(C_6H_5)_3\overset{+}{P}-CH_2C_6H_5$, $Cl^- \xrightarrow[C_2H_5OH]{NaOC_2H_5}$ **A**

$$\mathbf{A} + C_6H_5CH{=}CH-\overset{\overset{\displaystyle H}{\displaystyle |}}{C}{=}O \rightarrow \mathbf{B} + (C_6H_5)_3PO$$

b. $(C_6H_5)_3\overset{+}{P}-CH_3$, $Br^- \xrightarrow[DMSO]{NaH}$ **A**

$$\mathbf{A} + \left\langle\bigcirc\hspace{-1.2em}\right\rangle{=}\ddot{O}{:} \longrightarrow \mathbf{B} + (C_6H_5)_3P\ddot{O}{:}$$

6-55. Why is it important that any aldehyde used in the Wittig reaction be free of carboxylic acid impurity?

6-56. Reaction of triphenylphosphine with benzyl bromide produces the corresponding phosphonium salt. Suggest a suitable mechanism for this reaction.

6-57. Heteroatoms other than P are also capable of stabilizing the negative charge on C to yield ylids. For example, nitrogen is capable of forming such a system.

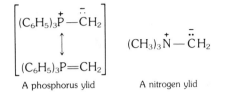

A phosphorus ylid   A nitrogen ylid

Why are resonance structures not drawn for the nitrogen ylid as in the phosphorus system?

6-58. Would you expect that the sulfonium salt, $(C_6H_5)_2S^+$—CH_3, Br^-, is capable of forming an ylid when reacted with a strong base? If so, would its structure be represented in a manner resembling the phosphorus or nitrogen ylid? Explain.

6-59. Explain why the C=C stretching mode gives rise to a rather weak band in 1-methylcyclohexene, whereas in its isomer, methylene–cyclohexane, the band is of medium to strong intensity.

6-60. Predict the C=C stretching frequencies of the olefins formed when cyclopentanone, cyclobutanone, and cyclopropanone undergo the Wittig reaction with methyltriphenylphosphonium bromide–sodium amide reagent.

6-61. What concentration of a given compound (moles/liter) in methanol, having a molecular weight of 165, would have been used if the solution gave an absorbance of 0.68 with a calculated $\varepsilon_{max} = 14,800$?

REFERENCES   **1.** Three of the many reviews on the Wittig reaction:
   a. Maerker, A. *Org. React.* **1965,** *14,* 270.
   b. Trippett, S. *Quart. Rev.* **1963,** *17,* 406.
   c. Schoellkopf, U. In *Newer Methods of Preparative Organic Chemistry;* Foerst, W., Ed.; Academic Press: New York, 1964; Vol. 3, p 111.
2. Selected references pertaining to the Horner–Emmons modification of the Wittig reaction:
   a. Boutgay, J.; Thomas, R. *Chem. Rev.* **1974,** *74,* 87.
   b. Wardsworth, W. S., Jr.; Emmons, W. D. *J. Am. Chem. Soc.* **1961,** *83,* 1733.
   c. Silversmith, E. F. *J. Chem. Educ.* **1968,** *63,* 645.
3. Selected examples where the Wittig reaction is applied in *Organic Syntheses:*
   a. McDonald, R. N.; Campbell, T. W. *Organic Syntheses;* Wiley: New York, 1973; Collect. Vol. V, p 499.
   b. Jorgenson, M. J.; Thacher, A. F. *Ibid.,* p 509.
   c. Wadsworth, W. G., Jr.; Emmons, W. D. *Ibid.,* p 547.
   d. Witting, G., Schoellkopf, U. *Ibid.,* p 751.
   e. Campbell, T. W.; McDonald, R. N. *Ibid.,* p 985.

Experiment 22   Aldol Condensation: Dibenzalacetone

(1,4-pentadien-3-one, 1,5-diphenyl-)

The aldol condensation is used extensively in organic synthesis to form carbon–carbon bonds. The example presented here illustrates the utility of the method. Experiment 49 is another example of the aldol condensation.

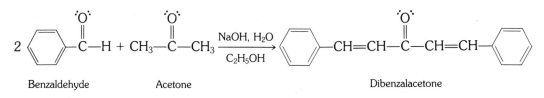

| Benzaldehyde | Acetone | | Dibenzalacetone |

DISCUSSION   The aldol condensation is one of the fundamental reactions of organic chemistry. The reaction is wide in scope and may be used to condense various combinations of aldehydes and ketones leading to formation of new carbon–carbon bonds. The mixed condensation of an aldehyde having no α-hydrogen atom with a ketone is known as the Claisen–Schmidt reaction. This variation of the aldol condensation is illustrated by the synthesis of dibenzalacetone.

The reaction conditions of this experiment favor the formation of dibenzalacetone. This product is insoluble in the aqueous alcohol solvent and precipitates from the reaction medium as it is formed, whereas the starting materials and the intermediate, benzalacetone, are soluble in aqueous alcohol. These experimental conditions assist in driving the reaction to completion.

The aldol reaction involves the addition of a deprotonated α-carbon atom of an aldehyde or ketone to a carbonyl carbon of an aldehyde or ketone. The reaction is generally base catalyzed and involves several mechanistic steps.

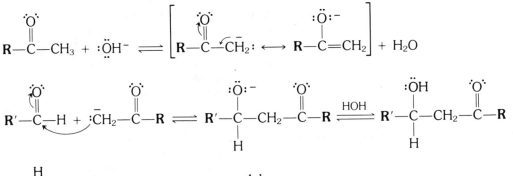

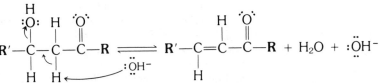

(not concerted rx, carbanion forms)

An additional example of the aldol reaction is shown in Experiment 49, where tetraphenylcyclopentadienone is prepared.

Note. *This reaction in Experiment 22 may be carried out in a 10 × 75-mm test tube. However, the reagents must be stirred efficiently with a glass rod at frequent intervals. If a larger test tube is used, a small magnetic stirring bar or vane is more efficient as an agitator.*

EXPERIMENTAL   Estimated time to complete the experiment: 2 hours.

Physical Properties of Reactants and Products

Compound	MW	Wt/Vol	mmol	mp(°C)	bp(°C)	Density	n_D
Benzaldehyde	106.13	80 μL	0.79		178	1.04	1.5463
Acetone	58.08	29 μL	0.40		56	0.79	1.3588
NaOH catalyst solution		1 mL					
Dibenzalacetone	234.30			111			

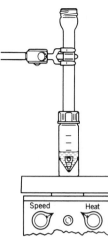

C$_6$H$_5$CHO, 80 μL,
+ acetone, 29 μL
+ ethanolic NaOH, 1.0 mL

Reagents and Equipment

In a 3.0-mL conical vial containing a magnetic spin vane are placed 80 μL (84 mg, 0.79 mmol) of benzaldehyde and 29.0 μL (23 mg, 0.40 mmol) of acetone.(■)

Important. *It is recommended that the purity of the benzaldehyde be checked by IR. The presence of benzoic acid in the benzaldehyde can substantially lower the yield of product. It may be purified by distillation under reduced pressure (bp 178–179 °C; 57–59 °C/8 torr).*

The benzaldehyde may be added to the vial by weight, or by volume using an automatic delivery pipet. To prevent loss of acetone by evaporation, the vial, fitted with a septum cap, is cooled in an ice bath and the acetone added by volume through the septum using a syringe.

The stoichiometric quantities of reagents are used. An excess of benzaldehyde results in a more intractable product; excess acetone favors the formation of benzalacetone.

To this reaction mixture is added 1.0 mL of aqueous ethanolic sodium hydroxide catalyst solution delivered from a calibrated Pasteur pipet.

Instructor Preparation

CAUSTIC The catalyst solution is prepared by dissolving 0.4 g of sodium hydroxide **(caustic)** in 4.0 mL of water. To this solution is added 3.0 mL of 95% ethanol.

Reaction Conditions

The reaction mixture is stirred at room temperature for a period of 30 minutes. During this time a yellow solid product precipitates from solution.

Isolation of Product

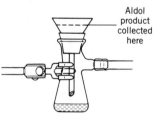

Aldol product collected here

H$_2$O/CH$_3$CH$_2$OH, ~ 5 mL

The crude, yellow dibenzalacetone is collected by vacuum filtration using a Hirsch funnel.(■) The magnetic spin vane is now removed from the reaction vial with forceps. *Some of the product adheres to the magnetic spin vane. This material should be removed by carefully scraping the vane with a microspatula. The material is added to the product collected by filtration.*

The collected filter cake is washed with three 1.0-mL portions of water. The filtrate should be nearly at the neutral point as indicated by pH test paper. If not, repeat the washing until the test indicates the filtrate is at the neutral point.

Important. *It is essential to remove the NaOH completely. If it is not removed, the recrystallization step proves difficult.*

The product is air-dried by maintaining the suction on the Hirsch funnel for approximately 10 minutes. During this operation a piece of filter paper may be placed over the mouth of the Hirsch funnel to prevent dust contamination.

Note. *An alternate procedure may be used to isolate and purify the dibenzalacetone. Transfer the reaction product directly to a large Craig tube. The washings and the ethanol recrystallization step (see next section) may then be carried out with no transfer of the material. In this way loss of product is minimized.*

Purification and Characterization

The crude dibenzalacetone may be purified by recrystallization from 95% ethanol using the Craig tube.

Weigh the dried product and calculate the percentage yield. Determine the melting point and compare your result with the literature value.

Obtain an infrared spectrum of the material using the KBr pellet technique.

In this reaction a double condensation takes place between the two principal reactants in which both contribute to the carbon skeleton of the product. Acetone forms the central portion of the dibenzal derivative and the original ketone carbonyl survives two aldol reactions in the process. A comparison of the infrared spectra of the starting reagents and the product is given below.

Infrared Analysis

The simple aliphatic ketone possesses the short macro group frequency train 3415 (overtone of C=O stretch, $2 \times 1712 = 3424$ cm^{-1}), 3000–2850 (sp^3 C—H stretch), 1712 (C=O stretch), and 1360 (symmetric methyl bend α to a carbonyl; see discusssion of methyl bends in acetates Experiments 8B,C,D) cm^{-1}. The substrate molecule is benzaldehyde. The infrared spectrum of this aromatic aldehyde (Fig. 6.10) is rich and interest-

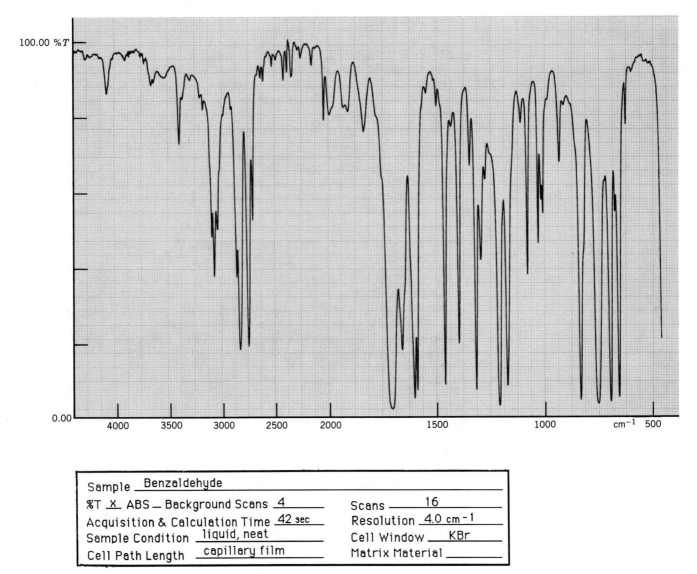

Sample	Benzaldehyde		
%T _X_ ABS _ Background Scans _4_		Scans ___16___	
Acquisition & Calculation Time _42 sec_		Resolution _4.0 cm^{-1}_	
Sample Condition _liquid, neat_		Cell Window ___KBr___	
Cell Path Length _capillary film_		Matrix Material _____	

Fig. 6.10 *IR spectrum: benzaldehyde.*

ing. The aromatic aldehyde macro group frequency consists of the following peaks: 3070, 2830 and 2750, 1706, 1602, 1589, and 1396 cm^{-1}.

a. 3070 cm^{-1}: C—H stretch on sp^2 carbon.

b. 2830 & 2750 cm^{-1}: This pair of bands is another example of Fermi coupling (see discussion in Experiment 7A and Appendix B).

c. 1706 cm^{-1}: The carbonyl stretch of the aldehyde group. The frequency observed in aliphatic aldehydes falls in the range 1735–1720 cm^{-1}, but when conjugated the value drops 15–25 cm^{-1} (see Chapter 7) and is found in the range 1720–1700 cm^{-1}.

d. 1602 & 1589 cm^{-1}: This pair of bands is related to the degenerate ring stretching vibrations ν_{8a} and ν_{8b} of benzene (see Infrared Analyses in Experiment 16B and Chapter 7).

e. 1396 cm^{-1}: The aldehyde C—H in-plane bending vibration. The first harmonic of this vibration is Fermi coupled to the aldehyde C—H stretching mode.

This compound also possesses a second powerful macro group frequency train that assesses the substitution pattern of the aromatic ring system (see Chapter 7). The monosubstituted benzene ring macro group frequency train requires peaks in the following regions: 1950, 1880, 1800, 1730, 750, and 690 cm^{-1}.

a. 1990, 1920, 1830, 1770 cm^{-1}: This series of four weak bands with generally decreasing intensity from high to low wavenumber values (the third peak near 1800 cm^{-1} may be intensified if the ring is conjugated as in this case) arise from combination bands (see Appendix B), which involve the out-of-plane bending frequencies of the ring C—H bonds (see below). The exact wavenumber positions are not very important, but the overall shape of the pattern can be used to determine the ring substitution pattern (see Chapter 7).

b. 750 and 690 cm^{-1}: This pair of bands is very characteristic of monosubstituted phenyl groups. The 750 cm^{-1} peak arises from the all-in-phase out-of-plane bending vibration of the five C—H groups adjacent to each other on the ring (see Chapter 7). The 690 cm^{-1} companion band involves an out-of-plane displacement of the ring carbon atoms (see Chapter 7).

The infrared spectrum of the dibenzalacetone product (Fig. 6.11) contains many features of the starting materials plus new and shifted bands unique to the newly formed structure.

a. The monosubstituted aromatic macro group frequency train remains 1960, 1890, 1815, 1770, 760, and 690 cm^{-1}. In addition the two pairs of degenerate ring stretching bands are present: 1598, 1580 and 1500, 1450 cm^{-1}. The 1500 cm^{-1} band is rather weak in benzaldehyde but intensifies in the product.

b. The ketone carbonyl remains the most intense band in the spectrum but is shifted to 1653 cm^{-1} by double conjugation. The aldehyde carbonyl mode at 1706 cm^{-1} has vanished.

c. The methyl C—H stretching bands between 3000 and 2850 cm^{-1} and the coupled aldehyde C—H stretching peaks at 2830 and 2750 cm^{-1} have disappeared.

d. New bands appear at 3058 and 3035 cm^{-1} (olefinic =C—H, stretch) that overlap with the aromatic ring (C—H stretching) peaks (3075 cm^{-1}). In addition, bands appear at 1627 cm^{-1} (conjugated C=C) and 985 cm^{-1} (trans substituted C=C, C—H in-phase out-of-plane bend). The latter band occurs slightly above its usual location near 965 cm^{-1}. This rise in frequency is the result of conjugation of the double bond to the carbonyl group. For an example of a much more dramatic rise in frequency, refer to the discussion of the cis-C—H bending modes in the spectrum of maleic anhydride (Experiment 14).

Examine the spectrum of the reaction product you have obtained in a potassium bromide matrix. Discuss the similarities and differences of the experimentally derived spectral data to the reference spectra (Figs. 6.10–6.11).

Dibenzalacetone is a classic compound that can be characterized by chemical tests.

Chemical Tests

A preliminary ignition test (see Chapter 7, Table 7.1) should indicate that it contains an aromatic ring system. Perform the test to confirm this fact.

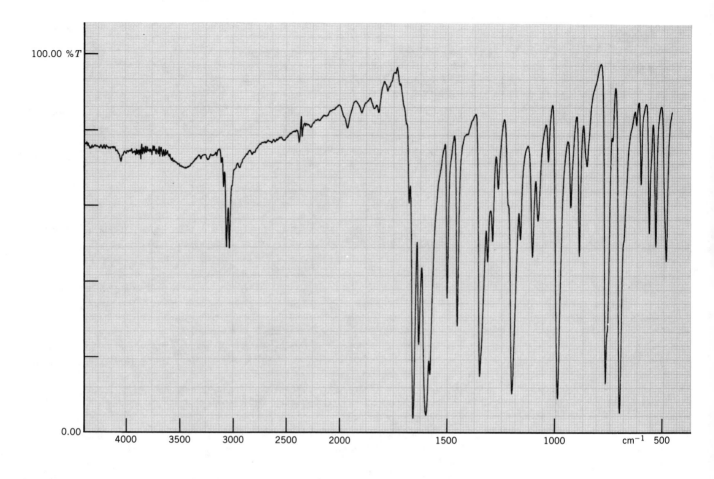

Sample	Dibenzalacetone		
%T _X_ ABS __ Background Scans _4_		Scans _____16_____	
Acquisition & Calculation Time _42 sec_		Resolution _4.0 cm⁻¹_	
Sample Condition _____solid_____		Cell Window _____	
Cell Path Length _____		Matrix Material _KBr_	

Fig. 6.11 *IR spectrum: dibenzalacetone.*

Several classification tests might also be of assistance in classifying this compound. Does the 2,4-dinitrophenylhydrazine test for an aldehyde or ketone give a positive result? Isolate the 2,4-dinitrophenylhydrazone derivative and determine the melting point. Does it correspond to the literature value of 180°C? What further test should be run to determine whether the carbonyl is present as an aldehyde or a ketone?

A test for unsaturation should be enlightening. Would you perform the Br_2–methylene chloride or the Baeyer test (check Chapter 7, Classification Tests)? Did the correct test give a positive result?

OPTIONAL SCALEUP   This experiment can be scaled up to be carried out with five or ten times the amounts used in the microscale preparation. The data summarized here are for the tenfold scaleup.

The procedure is identical to that for the microscale preparation with the following exceptions:

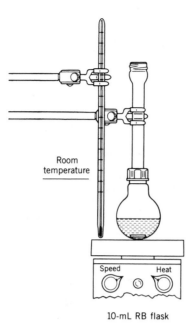

Room temperature

Speed   Heat

10-mL RB flask

a. A 10-mL round-bottom flask fitted with an air condenser is used. (■)

b. The amounts of reagents and solvents are increased tenfold.

Compound	MW	Wt/Vol	mmol	bp(°C)	Density	n_D
Benzaldehyde	106.13	800 μL	7.9	178	1.04	1.5463
Acetone	58.08	300 μL	4.0	56	0.79	1.3588
NaOH catalyst solution		10 mL				

c. The collected filter cake is transferred to a 10-mL beaker, stirred with 5.0 mL of water, and then recollected by vacuum filtration. This process is repeated, usually about three times, until the filtrate is neutral to litmus paper.

ENVIRONMENTAL DATA

Substance	Amount	TLV (mg/m^3)	Emissions (mg)	Volume (m^3)
Benzaldehyde (Max. from yield, transfer)	84 mg	—	37	
Acetone (Max. from yield, transfer)	23 mg	1780	20	<0.1
Ethanol (Max., solv., recryst., Craig)	1.43 mL	1900	1140	0.6

TLV: Sodium hydroxide—$2.0\ mg/m^3$

TXDS: Benzaldehyde—orl-mus LD50: 28 mg/kg

QUESTIONS **6-62.** A key step in the total synthesis of the hydrocarbon azulene is shown below.

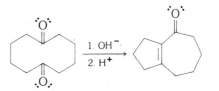

Outline a suitable mechanism to account for the reaction.

6-63. The aldol reaction has been utilized extensively for the generation of five- and six-membered rings. Suggest a suitable mechanism for the cyclization reactions shown below.

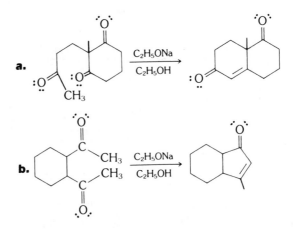

6-64. Predict the major organic product formed in each of the following reactions.

a. $CH_3CH_2NO_2 + CH_2O \xrightarrow{\text{NaOH}}$

b. $C_6H_5-CH=CH-CHO + CH_3-\overset{\overset{\displaystyle \cdot\cdot O\cdot}{\|}}{C}-C_6H_5 \xrightarrow[\text{C}_2\text{H}_5\text{OH}]{\text{C}_2\text{H}_5\text{ONa}}$

c. $C_6H_5CHO + C_6H_5CH_2CN \xrightarrow[\text{C}_2\text{H}_5\text{OH}]{\text{C}_2\text{H}_5\text{ONa}}$

6-65. "Crossed" or "mixed" aldol condensations are practical for synthesis, if one of the aldehydes (or ketones) has no α-hydrogen atoms. Explain.

6-66. Give several examples of aldehydes or ketones that could be used in the "crossed" aldol condensation with propanal. Assign structures and names to the products that could be formed and point out any side reactions that might occur.

6-67. In the aldol condensation using the conditions of this experiment, why is it essential that the aldehyde component contain none of the corresponding acid?

REFERENCES   **1.** Review articles:
    a. Neilsen, A. T.; Houlihan, W. *J. Org. React.* **1968,** *16,* 1.
    b. Mukaiyama, T. *Ibid.* **1982,** *28,* 203.
  2. Examples of the aldol and Claisen–Schmidt reactions:
    a. Hill, G. A.; Bramann, G. M. *Organic Syntheses*; Wiley: New York, 1941; Collect. Vol. I, p 81.
    b. Leuck, G. J.; Cejka, L. *Ibid.,* p 283.
    c. Kohler, E. P.; Chadwell, H. M. *Ibid.,* p 78.
    d. Conrad, C. R.; Dolliver, M. A. *Organic Syntheses*; Wiley: New York, 1943; Collect. Vol. II, p 167.
    e. Russel, A.; Kenyon, R. L. *Organic Syntheses*; Wiley: New York, 1955; Collect. Vol. III, p 747.

Experiment 23

Quantitative Analysis of a Grignard Reagent: 1-Methylbutylmagnesium Bromide

(magnesium, bromo(1-methylbutyl-))

This experiment demonstrates the technique by which a Grignard reagent may be analyzed. Also see Experiment 35.

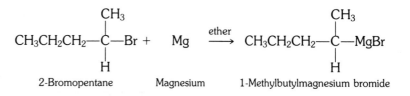

2-Bromopentane   Magnesium   1-Methylbutylmagnesium bromide

DISCUSSION   This experiment demonstrates the formation of a Grignard reagent and a titration method by which the amount of the reagent prepared can be analyzed.

The discovery by Victor Grignard in 1900 that organic halides react with magnesium metal to give organomagnesium compounds was a landmark in organic chemistry. It is one of the most useful and versatile reagents in organic synthesis.

The reaction of the Grignard reagent with water is the basis of the analytical method used in this experiment.

$$\mathbf{R}MgX + HOH \rightarrow \mathbf{R}H + Mg(OH)X$$

In the Grignard reagent, the carbon atom bound to the electropositive magnesium atom has a high charge density, which is responsible for the strong nucleophilic and basic character exhibited by this organometallic reagent. The carbon, acting as a base, can abstract a proton from protic reagents such as water, carboxylic acids, alcohols, etc. In this process, the corresponding hydrocarbon (the conjugate acid of the **R** group anion) and the basic magnesium halide species are produced. This reaction sequence can be used in the laboratory as a synthetic method to convert organohalides to hydrocarbons.

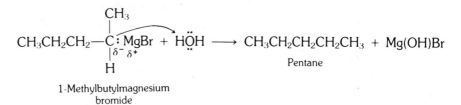

1-Methylbutylmagnesium
bromide

Pentane

Titration of the Mg(OH)X species with standardized acid solution allows one to determine the equivalents of Grignard reagent originally present in the solution.

$$2\,Mg(OH)X + H_2SO_4 \rightarrow 2\,HOH + MgSO_4 + MgX_2$$

An excess of the sulfuric acid is generally added to ensure that the Mg(OH)X is completely reacted. The excess acid is then neutralized with the standard sodium hydroxide solution. The difference between the total amount of sulfuric acid used and the amount of sodium hydroxide required corresponds to the number of equivalents of the acid actually used to neutralize the Mg(OH)X species. This value is then directly related to the equivalents of Grignard reagent by the equations given above.

PART A: 1-Methylbutylmagnesium Bromide

The reaction is shown above.

> **CAUTION:** *Ether is a flammable liquid and also a narcotic. All flames should be extinguished during the time of this experiment.*

EXPERIMENTAL   Estimated time for the experiment: 1.0 hour.

Physical Properties of Reactants and Products

Compound	MW	Wt/Vol	mmol	bp(°C)	Density	n_D
2-Bromopentane	151.05	125 µL	1.0	117	1.21	1.4413
Magnesium	24.31	36 mg	1.5			
Iodine	253.81	1 crystal				
Diethyl ether	74.12	500 µL		34.5		
Pentane	72.15			36	0.63	1.3575

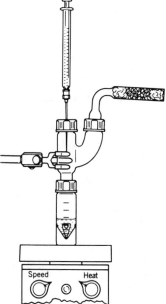

Reagents and Equipment

Note. *All the glassware used in the experiment should be cleaned, dried in an oven at 110°C for at least 30 minutes, and then cooled in a desiccator before use.*

The reagent is prepared exactly as described in Experiment 18. Reagents, amounts of reagents, order of addition, workup manipulations, and precautions are the same. The equipment is also identical.(■)

The gray-colored Grignard reagent mixture is cooled to room temperature and then analyzed by the titration method outlined in Part B.

If not used for analysis, the solution of Grignard reagent may be treated with various reagents to prepare a wide variety of compounds. For example, see Experiments 17, 18, 36, and 48B.

Mg, 36 mg + I$_2$, 1 crystal
+ (CH$_3$CH$_2$)$_2$O, 400 µL
+ CH$_3$CHBr(CH$_2$)$_2$CH$_3$, 125 µL

PART B: Analysis of the Grignard Reagent

EXPERIMENTAL   Estimated time for the experiment: 0.5 hour.

Reagents and Equipment

In a 50-mL Erlenmeyer flask is placed 10 mL of freshly boiled distilled water and one drop of phenolphthalein indicator.

Using a syringe, the cool Grignard reagent solution is transferred to the Erlenmeyer flask. The reaction vial is rinsed with 0.5 mL of diethyl ether and the rinse also added to the Erlenmeyer flask.

Note. *The addition of water to the Grignard reagent results in the hydrolysis of this reagent to form the corresponding hydrocarbon and a basic magnesium* *halide, Mg(OH)X. The water is initially boiled to remove any carbon dioxide present, which would react with the Grignard reagent.*

If any unreacted magnesium remains, it may be destroyed by addition of **HOOD** *dilute hydrochloric acid* **[hood]***. Hydrogen gas is evolved.*

This resulting mixture is now analyzed by the titration method given below.

Analysis by Titration

From a 10-mL buret is now added 5 or 6 mL of standard 0.2 N sulfuric acid solution. *The resulting solution should be acidic and colorless. If not, add another portion of the acid.*

A boiling stone is added to the flask and the mixture heated at a sand bath temperature of 90–95°C for a period of 5 minutes.

While the solution is still warm, a drop of phenolphthalein is added and the excess acid neutralized by back titration with 0.1 N sodium hydroxide solution. Back titration produces a very light pink endpoint. *It may be necessary to add an additional drop of acid and then more base to obtain the best possible endpoint.*

Data and Calculations

The difference between the initial and final buret readings is the volumes of standard acid and base used in the titration of the Grignard reagent.

From the data, calculate the equivalents of Grignard reagent present. In addition, as a percentage, determine the amount of Grignard reagent analyzed compared with the theoretical amount.

ENVIRONMENTAL DATA   The Grignard preparation emissions are listed in Experiment 18.

TLV: Iodine—0.1 ppm

Pentane—1800 mg/m^3

Sodium hydroxide—2 mg/m^3

TXDS: 2-Bromopentane—ipr-mus LD50: 150 mg/kg

Magnesium—orl-dog LDLo: 230 mg/kg

Sulfuric acid—unk-man LDLo: 135 mg/kg

QUESTIONS   **6-68.** Technical-grade ether often contains ethanol. Would you recommend this material as a suitable solvent for the preparation of Grignard reagents? If not, why not?

6-69. In the synthesis of alcohols by the Grignard reaction, the final stage of the sequence to obtain the alcohol product involves hydrolysis of the magnesium alkoxide salt with dilute acid. For the preparation of 2,3,4-trimethyl-3-pentanol by the reaction of methylmagnesium iodide with diisopropyl ketone in ether solvent, why is a saturated solution of aqueous ammonium chloride preferred in the final hydrolysis step instead of the usual dilute mineral acid solution?

6-70. In the titration procedure outlined earlier for analysis of the Grignard reagent, freshly boiled distilled water is used. As stated, this is done to remove dissolved carbon dioxide. What reaction could occur during the titration if the CO_2 were not removed?

6-71. One equivalent of o-acetylbenzoic acid was treated with two equivalents of methylmagnesium iodide in ether solvent. The product was isolated by careful neutralization of the product formed in the reaction.

    a. What was the structure of the product formed in the reaction?

    b. Why were two equivalents of Grignard reagent required?

6-72. If the Grignard reagent prepared in this experiment were hydrolyzed in D_2O solvent, what product would be formed?

REFERENCES   **1.** For references relating to the preparation of Grignard reagents, see Experiment 17.
2. For the many references related to the preparation and use of the Grignard reagent cited in *Organic Syntheses*, see the Reaction Index in Collect. Vols. I–V under Grignard Reactions.

Experiment 24

Preparation of an Enol Acetate: Cholesta-3,5-dien-3-ol Acetate

(cholesta-3,5-dien-3-ol acetate)

Preparation of the enol acetate of an α,β-unsaturated ketone is demonstrated in this experiment.

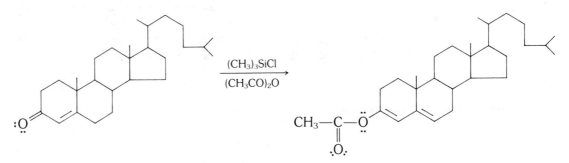

Cholesta-4-en-3-one

$(CH_3)_3SiCl$

$(CH_3CO)_2O$

Cholesta-3,5-dien-3-ol acetate

DISCUSSION   Enolate anions play a major role in synthetic organic chemistry since they are good nucleophilic reagents. They undergo nucleophilic substitution reactions with relatively electrophilic carbon atom centers, leading to the formation of C—C bonds.

   This experiment illustrates one method of stabilizing such enolates as acetate esters. An enolate is generally formed by removal of a proton from the α-carbon atom of a carbonyl compound. In the present case, the reactant is an α,β-unsaturated carbonyl species. The proton is thus removed from the γ-carbon atom to form the anion, which is stabilized by delocalization of the charge.

The enol acetate is then formed in the presence of the acylium ion generated from the acetic anhydride. The mechanism for the sequence is outlined below.

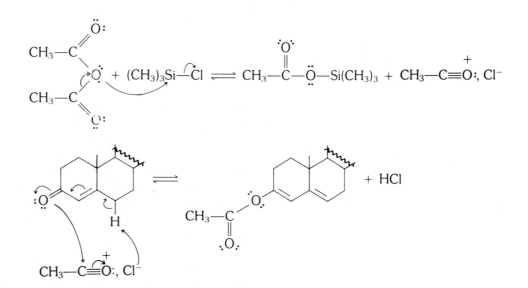

EXPERIMENTAL   Estimated time to complete the experiment: 4.0 hours.

Physical Properties of Reactants and Products

Compound	MW	Wt/Vol	mmol	mp(°C)	bp(°C)	Density	n_D
Cholesta-4-en-3-one	384.65	96 mg	0.25	81–82			
Acetic anhydride	102.09	1.0 mL		10.6	140	1.08	1.3901
Chlorotrimethylsilane	108.64	200 µL	1.58		58	0.86	
Cholesta-3,5-dien-3-ol acetate	425.68			79–80			

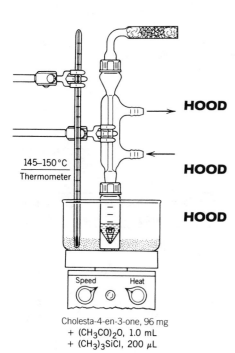

145–150 °C
Thermometer

Speed   Heat

Cholesta-4-en-3-one, 96 mg
+ (CH$_3$CO)$_2$O, 1.0 mL
+ (CH$_3$)$_3$SiCl, 200 µL

Reagents and Equipment

Note. *The glass equipment should be dried in an oven (110°C) before use.*

In a dry 5.0-mL conical vial containing a magnetic spin vane and equipped with a reflux condenser protected by a calcium chloride drying tube are placed 96 mg (0.25 mmol) of cholesta-4-en-3-one, 1.0 mL of acetic anhydride, and 200 µL of chlorotrimethylsilane **[hood]**.(■)   **HOOD**

Note. *The quality of the acetic anhydride has a significant influence on the reaction. For best results the anhydride should be distilled and stored over molecular sieves before use. It is convenient to dispense **[hood]** the anhydride and the chlorotrimethylsilane (hydrolyzes rapidly in moist air) using automatic delivery pipets. The chlorotrimethylsilane (bp 57°C) may also be purified by distillation **[hood]**.*   **HOOD**

HOOD

Reaction Conditions

The reaction mixture is heated with stirring at reflux for a period of 1–2 hours by aid of a sand bath (145–150°C). The course of the reaction is followed using TLC.

Directions. *Activated silica gel plates are used with 1:1 methylene chloride–hexane as the developing solvent. Visualization of the separated components is*

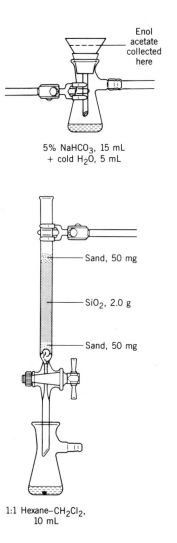

5% NaHCO₃, 15 mL
+ cold H₂O, 5 mL

Sand, 50 mg

SiO₂, 2.0 g

Sand, 50 mg

1:1 Hexane–CH₂Cl₂,
10 mL

achieved by placing the plate in a closed jar containing a few crystals of iodine. See Chapter 5, Technique 8, for further details.

Isolation of Product

The reaction mixture is allowed to cool to room temperature and then placed in an ice bath for 15–30 minutes. A solid product forms during this time. This solid is collected by vacuum filtration and the filter cake washed with 15 mL of 5% aqueous sodium bicarbonate followed by 5 mL of cold water.(■)

Purification and Characterization

The crude product is purified by column chromatography. In a 1 × 10-cm buret is placed 2.0 g of activated silica gel (100 mesh) packed wet (slurry) with methylene chloride.(■) The product is dissolved in about 1.0 mL of hexane and transferred by Pasteur pipet to the column. The material is eluted from the column using approximately 10 mL of 1 : 1 methylene chloride–hexane solvent. The eluate is collected in a tared 25-mL filter flask.

The solvent is removed by warming the solution in a sand bath under reduced pressure to give the solid product, cholesta-3,5-dien-3-ol acetate.(■) This material is recrystallized from methanol using a Craig tube, and the resulting crystals are dried on a clay plate.

Weigh the product and calculate the percentage yield. Determine the melting point and compare it with the literature value.

Obtain an IR spectrum of the material and compare it with that of an authentic sample as well as with the spectrum of the starting material.

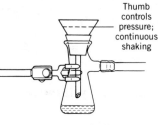

Thumb controls pressure; continuous shaking

1:1 Hexane–CH₂Cl₂, 10 mL
+ enol acetate

ENVIRONMENTAL DATA

Substance	Amount	TLV (mg/m³)	Emissions (mg)	Volume (m³)
Chlorotrimethylsilane	200 μL			
Acetic anhydride (Max., vac. filtr.)	1.0 mL	20	1080	54
Methylene chloride (Max., col. chromatog., evap.)	5.0 mL	350	6675	19
Hexane (Max., col. chromatog., evap.)	6.0 mL	180	3954	22
Methanol (SKIN) (Max., Craig)	1.0 mL	260	800	3.1

TXDS: Chlorotrimethylsilane—ipr-mus LDLo: 750 mg/kg
Silica gel—inv-mus LDLo: 234 mg/kg

QUESTIONS   **6-73.** Suggest a suitable mechanism to account for the formation of the products in the reactions given below.

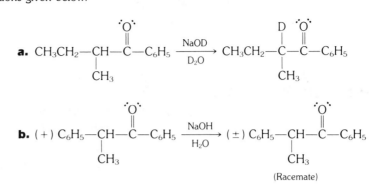

6-74. Write the structure for all possible enolate anions that could form by reaction of the ketones listed below in alkaline solution.

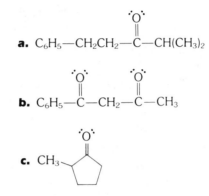

6-75. Reaction of $(CH_3)_3SiCl$ with alcohols produces a trimethylsilyl ether. For example,

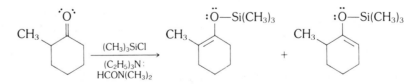

The trimethylsilyl ether is more volatile than the corresponding alcohol. Explain.

6-76. Trimethylchlorosilane reacts with enolate anions to form stable silyl enol ethers. For example,

Realizing that the reaction is run under conditions that allow equilibrium to be established in the system, predict which of the above silyl enol ethers is formed in the largest amount and why.

6-77. There are two reasons why the enols of β-dicarbonyl compounds such as 2,4-pentanedione are relatively stable. One reason is that these enols are conjugated and the π-electron overlap due to the conjugation provides additional stability. What is the other reason that these species are relatively stable? Illustrate your answer with a suitable structure.

REFERENCES   **1.** The present experiment is adapted from the work reported by Chowdhury, P. K.; Sharma, R. P.; Barua, J. N. *Tetrahedron Lett.* **1983,** *24,* 3383.

2. For information on the formation and reactivity of enolate anions, see House, H. O. *Modern Synthetic Reactions*, 2nd ed.; Benjamin: Reading, MA, 1972; Chap. 9, p 492.

Experiments 25A, 25B

Williamson Synthesis of Ethers: Propyl *p*-Tolyl Ether; Methyl *p*-Ethylphenyl Ether

(ether, propyl *p*-tolyl; anisole, *p*-ethyl)

| *p*-Cresol | Propyl iodide | Propyl *p*-tolyl ether |

The condensation of alkyl halides with sodium phenoxides to produce ethers is an example of the Williamson synthesis. The use of phase transfer catalysis is demonstrated.

DISCUSSION The two compounds whose preparations are outlined below belong to the class of ethers referred to as alkyl aryl ethers. The general method of preparation is the Williamson reaction. This reaction is often used for the synthesis of symmetrical and unsymmetrical ethers where the alkyl halide is primary. Elimination is generally observed if secondary or tertiary halides are used.

The conditions under which these reactions are conducted lend themselves to the use of phase transfer catalysis. The system involves two phases, the aqueous phase and the organic phase. In the present case, the alkyl halide reactant acts as the solvent, as does the product as it is formed. The phase transfer catalyst plays a very important role. In effect, it carries the phenolic ion (in the aqueous phase) as an ion pair across the boundary layer into the organic phase, where the reaction occurs. The ether product and the corresponding halide salt of the catalyst are produced in these conversions. The halide salt then migrates back into the aqueous phase and the process repeats itself. The catalyst can play this role since the large organic groups (the four butyl groups) increase the solubility of the ion pair in the organic phase, while the charged ionic center of the salt renders this species soluble in the aqueous phase.

In the reactions described below, the mechanism involves a nucleophilic backside attack of the phenolic ion on the alkyl halide.

It is of interest to contrast the acidity of phenols with that of simple alcohols. A phenol is more acidic than an alcohol. In a typical aliphatic alcohol, for example ethanol, loss of the proton forms a strong anionic base (ethoxide ion), the corresponding alkoxide ion.

$$\mathbf{R}-CH_2-OH \rightarrow H^+ + \mathbf{R}-CH_2-O^-$$

Alkoxide ion

The strong basic characteristic of the alkoxide species is due to the fact that the negative charge is localized on the oxygen atom. Ethanol has a $pK_a = 16$.

In contrast, the anion of a phenol has a delocalization of negative charge.

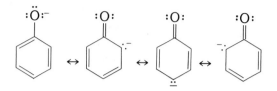

Thus, the anion of a phenol is stabilized by these resonance contributors. It is therefore a weaker base than the alkoxide ion. Conversely, the phenol is a stronger acid than a typical aliphatic alcohol. Phenol has $pK_a = 10$ and thus is one million times more acidic than ethanol.

PART A: Propyl *p*-Tolyl Ether

The reaction for Part A is shown above.

EXPERIMENTAL   Estimated time of the experiment: 2.5 hours.

Physical Properties of Reactants and Products

Compound	MW	Wt/Vol	mmol	mp(°C)	bp(°C)	Density	n_D
p-Cresol	108.15	80 μL	0.78		202	1.02	1.5312
25% NaOH solution		130 μL					
Tetrabutylammonium bromide	322.38	9 mg	0.03	103–104			
Propyl iodide	169.99	75 μL	0.77	102		1.75	1.5058
Propyl *p*-tolyl ether	150.22				210.4	0.9497	1.4958 (29°C)

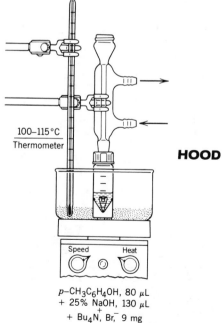

100–115°C
Thermometer

HOOD

p–CH$_3$C$_6$H$_4$OH, 80 μL
+ 25% NaOH, 130 μL
+ Bu$_4$N, Br$^-$ 9 mg
+ PrI, 75 μL

Speed   Heat

Reagents and Equipment
In a 1.0-mL conical vial containing a magnetic spin vane and equipped with a reflux condenser are placed 80 μL (84 mg, 0.78 mmol) of *p*-cresol and 130 μL of 25% aqueous sodium hydroxide. The resulting solution is thoroughly mixed. (■) Tetrabutylammonium bromide catalyst (9 mg) is added, followed by 75 μL (131 mg, 0.77 mmol) of propyl iodide.

Note. *The cresol reagent should be warmed in a hot water bath to obtain the liquid phase. This reagent and the propyl iodide are dispensed in the* **hood** *using an automatic delivery pipet.*

> **CAUTION: Propyl iodide is a suspected carcinogenic agent.**

Reaction Conditions
The reaction vessel is placed in a sand bath at 110–115°C with vigorous stirring for a period of 45–60 minutes.

Isolation of Product
The resulting two-phase mixture is cooled to room temperature, and the spin vane removed with forceps. The aqueous layer is separated using a Pasteur filter

pipet, and transferred to a 3-mL conical vial. The spin vane is rinsed with 0.5-mL diethyl ether and the rinse is added to the aqueous fraction. The vial is capped, agitated, and vented and the water layer removed. The diethyl ether extract is transferred to the 1-mL conical vial containing the product fraction. The resulting solution is extracted with a 200-μL portion of 5% aqueous sodium hydroxide solution and, after separation of the aqueous phase, washed with 100 μL of water. Removal of the aqueous phase yields the crude, wet ether product.

Purification and Characterization

The crude product is purified by chromatography on silica gel. A microchromatographic column is prepared by placing 500 mg of activated silica gel in a Pasteur filter pipet followed by 50 mg of anhydrous sodium sulfate.(■) The crude product is dissolved in 250 μL of methylene chloride and transferred to the dry column by use of a Pasteur pipet. The material is eluted with 2.0 mL of methylene chloride and the eluate collected in a tared 3.0-mL conical vial containing a boiling stone. The vial is then fitted with an air condenser and the solvent evaporated by placing the vial in a sand bath maintained at a temperature of 60–65°C.

Weigh the pure propyl *p*-tolyl ether and calculate the percentage yield. Determine the boiling point and density (optional) and compare the experimental values with those in the literature.

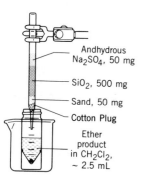

Andhydrous Na₂SO₄, 50 mg

SiO₂, 500 mg

Sand, 50 mg

Cotton Plug

Ether product in CH₂Cl₂, ~ 2.5 mL

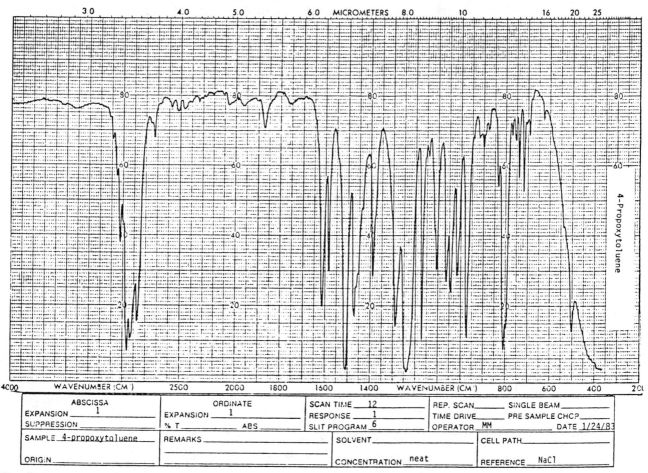

Fig. 6.12 *IR spectrum: 4-propoxytoluene.*

Obtain an IR spectrum of the compound and compare it with that shown in Figure 6.12 for 4-propoxytoluene (propyl *p*-tolyl ether).

Nuclear Magnetic Resonance Analysis

If time and facilities permit, you can obtain both ^{1}H and ^{13}C NMR spectra of your propyl *p*-tolyl ether in CDCl$_3$ and compare your spectra with those in Figures 6.13 and 6.14. There are two extraneous peaks in the ^{1}H spectrum: the small singlet at 7.24 ppm is due to residual CHCl$_3$ in the CDCl$_3$ and the small singlet at 1.55 ppm is probably due to a trace amount of water in either the sample or the NMR solvent. The 1:1:1 triplet at 77 ppm in the ^{13}C spectrum is from the CDCl$_3$ solvent.

Since the ^{1}H spectrum is essentially first order, it can readily be interpreted. You should be able to use the splitting patterns to assign peaks to the different groups of protons in the molecule. The integration can assist you.

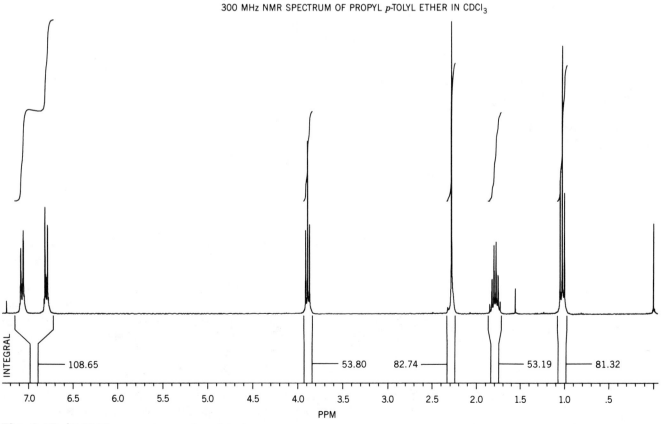

300 MHz NMR SPECTRUM OF PROPYL *p*-TOLYL ETHER IN CDCl$_3$

Fig. 6.13 ^{1}H-NMR spectrum: propyl p-tolyl ether.

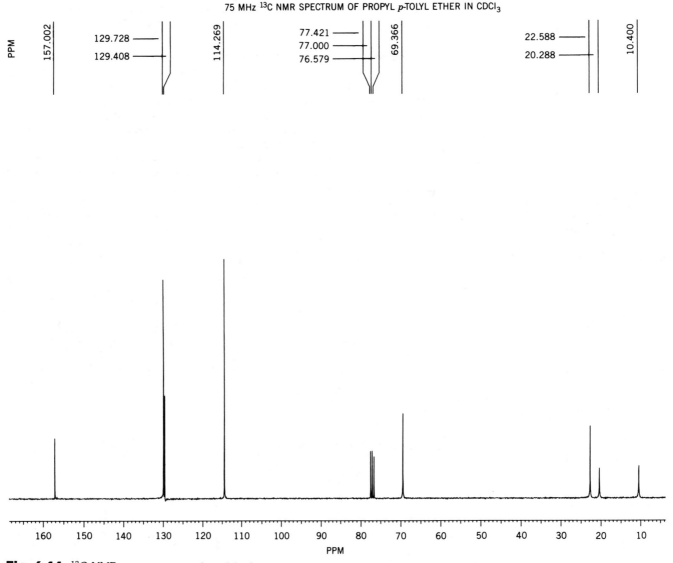

75 MHz ^{13}C NMR SPECTRUM OF PROPYL p-TOLYL ETHER IN CDCl$_3$

Fig. 6.14 ^{13}C-NMR spectrum: propyl p-tolyl ether.

Chemical Tests

Qualitative chemical tests can also be used to assist in characterizing this compound as an ether. Perform the ignition test (Chapter 7, Table 7.1) to determine whether the material contains an aromatic ring.

A key factor to investigate is the solubility characteristics of this material (see Chapter 7). Determine its solubility in water, 5% sodium hydroxide, 5% hydrochloric acid, concentrated sulfuric acid, and 85% phosphoric acid. Do the results place this compound in the solubility class of an ether containing $<C_8$ or $>C_8$ carbon atoms?

Does the Ferrox test (Chapter 7) confirm the presence of oxygen in the compound?

OPTIONAL SCALEUP   This ether may be prepared on a larger scale ($\sim$ a 200-fold increase) using a procedure similar to that just outlined, with the following exceptions:

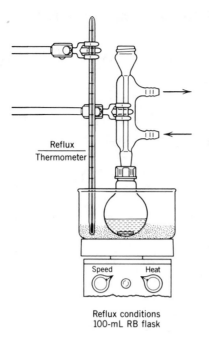

Reflux
Thermometer

Speed Heat

Reflux conditions
100-mL RB flask

a. A 100-mL round-bottom flask containing a magnetic stirrer and equipped with a reflux condenser is used.(■)

b. The reagents and solvent amounts are summarized in the table. Note that propyl bromide is used in place of propyl iodide at this scale.

Compound	MW	Wt/Vol	mol	mp(°C)	bp(°C)	Density	n_D
p-Cresol	108.15	16.3 g	0.15		202	1.02	1.5312
NaOH	40.09	6.05 g	0.15				
Tetrabutylammonium bromide	322.38	0.41 g	0.0012	103–104			
Propyl bromide	122.99	12.71 g	0.1		70.82	1.35	1.4341
Water		25 mL					

c. The reaction mixture is stirred at reflux temperature for 90 minutes.

d. The solution is allowed to cool to room temperature and transferred to a 125-mL separatory funnel, and the aqueous layer is removed. The organic layer is washed successively with 5% NaOH solution (20 mL) and distilled water (20 mL). After each washing, the aqueous phase is removed. The remaining deep red organic layer is then collected in a 125-mL Erlenmeyer flask and dried over anhydrous sodium sulfate. The crude, liquid ether product is transferred to a tared screw-capped vial using a Pasteur filter pipet. It is then weighed and the percentage yield calculated. The compound may be characterized as described in the microscale procedure. It may be further purified by distillation under reduced pressure (see Experiment 52).

PART B: Methyl *p*-Ethylphenyl Ether

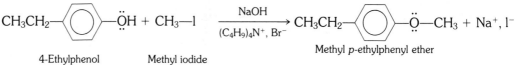

4-Ethylphenol   Methyl iodide   Methyl *p*-ethylphenyl ether

This product is the starting material for the synthesis of 4-methoxybenzoic acid (see Sequence D).

EXPERIMENTAL   Estimated time to complete the experiment: 2.5 hours.

Physical Properties of Reactants and Products

Compound	MW	Wt/Vol	mmol	mp(°C)	bp(°C)	Density	n_D
4-Ethylphenol	122.17	150 mg	1.2	42–45			
25% NaOH solution		250 μL					
Tetrabutylammonium bromide	322.3	15 mg	0.05	103–104			
Methyl iodide	141.94	205 mg	1.45		41–43	2.28	1.5304
Methyl *p*-ethylphenyl ether	136.20				195–196	0.96	1.5120

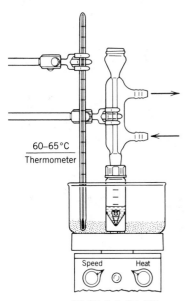

60–65°C
Thermometer

Speed   Heat

p–CH₃CH₂C₆H₄OH, 150 mg
+ 25% NaOH, 250 μL
+ Bu₄N, Br⁻ 15 mg
+ CH₃I, 205 mg

Reagents and Equipment

The apparatus in Part A is also used in this conversion. The reaction vial is charged with 150 mg (1.2 mmol) of 4-ethylphenol and 250 μL of 25% aqueous sodium hydroxide solution.(■) The mixture is stirred at room temperature until dissolution occurs. The phase transfer catalyst, tetrabutylammonium bromide (15 mg, 0.05 mmol), is now added, followed by 205 mg (1.45 mmol) of methyl iodide.

The alkaline solution is dispensed using an automatic delivery pipet, and the methyl iodide reactant by Pasteur pipet. CH₃I is used in slight excess due to its volatility.

Reaction Conditions

The reaction vial is placed in a sand bath maintained at 60–65°C and the mixture stirred for a period of 1.0 hour.

Isolation of Product

The resulting product mixture is worked up as described in Part A.

Purification and Characterization

The crude product is purified by chromatography on silica gel as described in Part A, Purification and Characterization.(■)

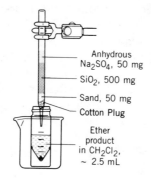

Anhydrous
Na₂SO₄, 50 mg

SiO₂, 500 mg

Sand, 50 mg

Cotton Plug

Ether
product
in CH₂Cl₂,
~ 2.5 mL

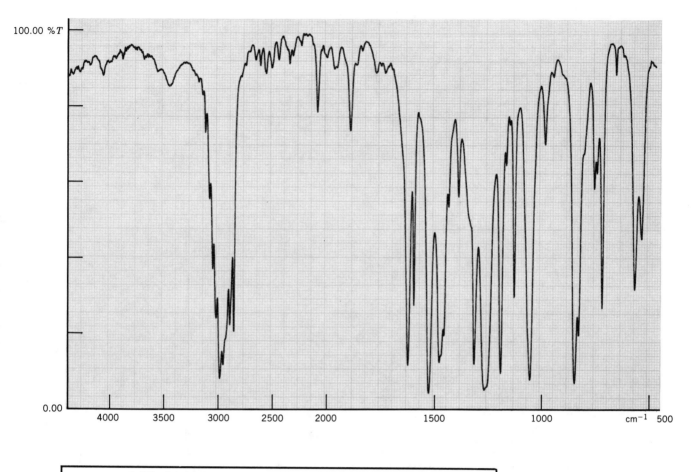

Sample ___4-Ethylanisole___

%T _X_ ABS _ Background Scans _4___   Scans _____4_____

Acquisition & Calculation Time _42 sec_   Resolution _4.0 cm⁻¹_

Sample Condition _liquid, neat_   Cell Window ___KBr___

Cell Path Length ___capillary film___   Matrix Material _____

Fig. 6.15 *IR spectrum: methyl* p-*ethylphenyl ether.*

Weigh the pure *p*-ethylanisole and calculate the percentage yield. Determine the boiling point and density (optional). Compare your results with the literature values.

Obtain the IR spectrum of the compound and compare it with that in Figure 6.15.

Nuclear Magnetic Resonance Analysis

If time and facilities permit, you can obtain both ^{1}H and ^{13}C NMR spectra of your methyl *p*-ethylphenyl ether in CDCl$_3$ and compare your spectra with those in Figures 6.16 and 6.17. There are two extraneous peaks in the ^{1}H spectrum: the small singlet at 7.24 ppm is due to residual CHCl$_3$ in the CDCl$_3$ and the small singlet at 1.55 ppm is probably due to a trace amount of water in either the sample or the NMR solvent. The 1:1:1 triplet at 77 ppm in the ^{13}C spectrum is from the CDCl$_3$ solvent.

Since the ^{1}H spectrum is essentially first order, it can readily be interpreted. You

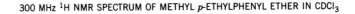

300 MHz ^{1}H NMR SPECTRUM OF METHYL p-ETHYLPHENYL ETHER IN CDCl$_3$

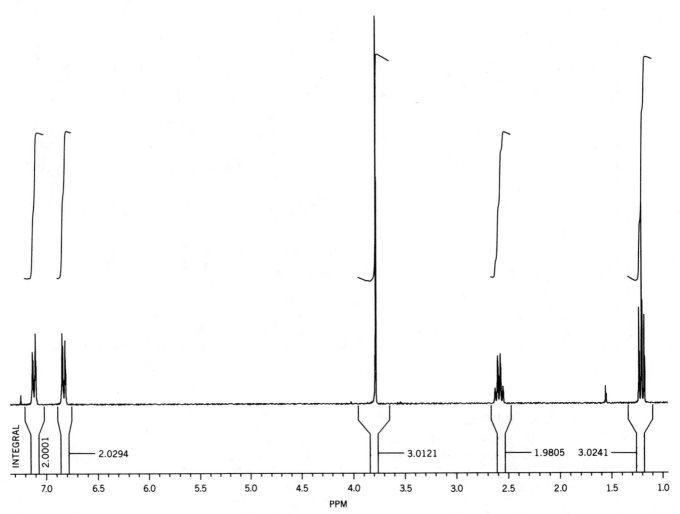

Fig. 6.16 ^{1}H-NMR spectrum: methyl p-ethylphenyl ether.

should be able to use the splitting patterns to assign peaks to the different groups of protons in the molecule. The integration can assist you.

Chemical Tests

Qualitative chemical tests can also be used to assist in characterizing this compound as an ether. Perform the ignition test (Chapter 7, Table 7.1) to determine whether the material contains an aromatic ring.

A key factor to investigate is the solubility characteristics of this material (see Chapter 7). Determine its solubility in water, 5% sodium hydroxide, 5% hydrochloric acid, concentrated sulfuric acid, and 85% phosphoric acid. Do the results place this compound in the solubility class of an ether containing $<C_8$ or $>C_8$ carbon atoms?

Does the Ferrox test (Chapter 7) confirm the presence of oxygen in the compound?

OPTIONAL SCALEUP   This experiment can be scaled up by a factor of 10 or larger so that more material is available for use in the preparation of 4-methoxyacetophenone (Experiment 52).

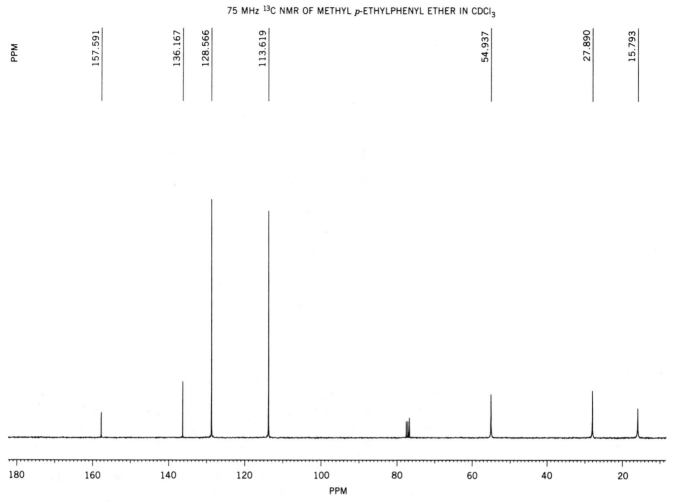

75 MHz ^{13}C NMR OF METHYL *p*-ETHYLPHENYL ETHER IN CDCl$_3$

157.591 136.167 128.566 113.619 54.937 27.890 15.793

PPM

Fig. 6.17 ^{13}C-NMR spectrum: methyl p-ethylphenyl ether.

Tenfold Scaleup

The reagent and solvent amounts are given in the table.

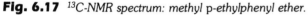

Compound	MW	Wt/Vol	mmol	mp(°C)	bp(°C)	Density	n_D
4-Ethylphenol	122.17	1.5 g	12	42–45			
Sodium hydroxide		625 mg					
Tetrabutylammonium bromide	322.3	150 mg	0.5	103–104			
Methyl iodide	141.94	2.05 g	14	41–43		2.28	1.5304

Reagents and Equipment

In a 25-mL round-bottom flask equipped with a stirring bar and a reflux condenser are placed 2.5 mL of water and 1.5 g (12 mmol) of *p*-ethylphenol (the phenol will oil out). The mixture is cooled by immersing the flask in a beaker of cold water; then, with stirring, 625 mg of sodium hydroxide is added with

CAUTION   caution.

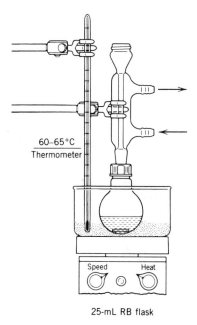

60–65°C
Thermometer

Speed   Heat

25-mL RB flask

After dissolution of the sodium hydroxide, 150 mg of the phase transfer catalyst, tetrabutylammonium bromide, is added. To the resulting solution is now added 2.05 g (900 μL) of methyl iodide using an automatic delivery pipet. (■)

Reaction Conditions The reaction vessel is placed in a sand bath and heated with stirring at 60–65°C for a period of one hour.

Isolation of Products The resulting two-phase mixture is cooled to room temperature and then transferred by Pasteur pipet to a 12-mL centrifuge tube. The reaction flask is rinsed with three 1-mL portions of diethyl ether and the rinses are also transferred to the centrifuge tube. The tube is capped, shaken, and vented (this operation may be done using a Vortex mixer if available). After separation of the layers, the lower aqueous layer is removed using a Pasteur filter pipet.

Do not remove any precipitated material that settles between the two layers.

The ether layer is extracted with one 2-mL portion of 5% sodium hydroxide solution and then with 1 mL of water.

The crude product is purified by chromatography on silica gel. A column is prepared by placing 5 g of activated silica gel in a 25-mL buret followed by 0.5 g of anhydrous sodium sulfate. The wet ether extract obtained earlier is dissolved in 2.5 mL of methylene chloride and the resulting solution transferred by Pasteur pipet to the dry column. The sample is eluted with 20 mL of additional methylene chloride. The eluate is collected in a tared 50-mL Erlenmeyer flask containing a boiling stone. The solvent is evaporated by placing the flask in a sand bath

HOOD maintained at 60–65°C **[hood]** and using a stream of nitrogen or air to hasten the process.

The resulting product is weighed and the percentage yield calculated. The product is characterized as outlined above in the microscale procedure.

Eightyfold Scaleup
The reagent and solvent amounts are given in the table.

Compound	MW	Wt/Vol	mol	mp(°C)	bp(°C)	Density	n_D
4-Ethylphenol	122.17	12.5 g	0.102	42–45			
Sodium hydroxide	40	4.0 g	0.1				
Tetrabutylammonium bromide	322.3	0.4 g	0.0012	103–104			
Methyl iodide	141.94	14.5 g	0.102		41–43	2.28	1.5304
Water		25 mL					

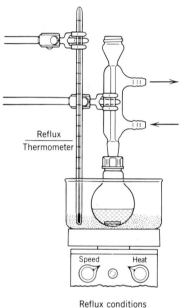

Reflux
Thermometer

Speed   Heat

Reflux conditions
100-mL RB flask

Reagents and Equipment The preparation is run using a 100-mL round-bottom flask containing a magnetic stirring bar and equipped with a reflux condenser.(■)

Reaction Conditions The resulting mixture is heated at reflux with stirring for a period of 90 minutes.

Isolation of Product The two-phase mixture is allowed to cool to room temperature, transferred to a 125-mL separatory funnel, and the aqueous layer is removed and discarded. The organic layer is washed successively with 5% sodium hydroxide solution (20 mL) and distilled water (20 mL). After each washing the aqueous layer is removed and discarded as before. Finally, the reddish-brown organic layer is collected in a 125-mL Erlenmeyer flask and dried over anhydrous sodium sulfate.

The drying agent is removed by filtration through a glass-wool plug and the product ether collected in a tared container. It is then weighed, the percentage yield is calculated, and the product is characterized as described in the microscale procedure. It may be further purified by distillation under reduced pressure (see Experiment 52).

ENVIRONMENTAL DATA

Substance	Amount	TLV (mg/m³)	Emissions (mg)	Volume (m³)
Experiment 25A				
Propyl iodide (Max. from yield)	75 µL			
p-Cresol (Max. from yield, transfer)	84 mg	22	34	1.5
Methylene chloride (Max., ɔl. chromatog., evap.)	2.25 mL	350	3000	8.6
Experiment 25B				
p-Ethylphenol (Max. from yield)	150 mg	—	78	
Methyl iodide (SKIN) (Max. from yield, transfer)	205 mg	10	127	13
Methylene chloride (Max., col. chromatog., evap.)	2.25 mL	350	3000	8.6

TLV: Sodium hydroxide—2 µg/m³
TXDS: p-Methyl anisole—orl-rat LD50: 1920 mg/kg
  Propyl iodide—ihl-rat LC50: 73,000 mg/m³/30 minutes
  Silica gel—ivn-mus LDLo: 234 mg/kg
  Sodium sulfate—orl-mus LD50: 5989 mg/kg
MTDS: Tetrabutylammonium bromide—dnd-mam:lyn: 50 mmol/L

QUESTIONS   **6-78.** Thiol ethers are often prepared using a Williamson synthesis. For example,

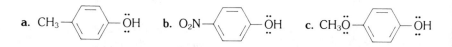

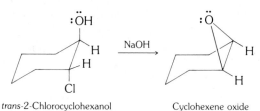

The reaction with isopropyl bromide is 16 times faster than the reaction with 2-bromo-1-nitropropane. Explain.

6-79. If 3-bromo-1-propanol is treated with NaOH, a compound of molecular formula C_3H_6O is formed. Suggest a structure for this product.

6-80. Arrange the substituted phenols given below in order of increasing reactivity toward ethyl iodide in the Williamson reaction. Explain your order.

6-81. *trans*-2-Chlorocyclohexanol reacts readily with NaOH to form cyclohexene oxide, but the cis isomer does not undergo this reaction. Explain.

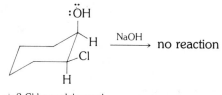

6-82. *tert*-Butyl ethyl ether might be prepared two ways using different starting materials.

Which route would you choose to prepare this ether and why?

6-83. What product(s) would you expect to form when tetrahydrofuran is treated with HI acid?

6-84. Write a suitable mechanism for the cleavage of butyl isopropyl ether with HI at 100°C to form exclusively isopropyl alcohol and 1-iodobutane. Explain why butyl alcohol and isopropyl iodide are not formed in the reaction.

6-85. There are only four lines in the aromatic region of the ^{13}C NMR spectrum of propyl *p*-tolyl ether (110–160 ppm), yet there are six aromatic carbons. Explain.

REFERENCES

1. References selected from the large number of examples of the Williamson reaction in *Organic Syntheses*:

   a. Wheeler, T. S.; Willson, F. G. *Organic Syntheses;* Wiley: New York, 1941; Collect. Vol. I, p 296.

   b. Marvel, C. S.; Tanenbaum, A. L. *Ibid.,* p 435.

   c. Fuson, R. C.; Wojcik, B. H. *Organic Syntheses;* Wiley: New York, 1943; Collect. Vol. II, p 260.

   d. Allen, C. F. H.; Gates, J. W., Jr. *Organic Syntheses;* Wiley: New York, 1955; Collect. Vol. III, p 140.

   e. *Ibid.,* p 418.

   f. Boehme, W. R. *Organic Syntheses;* Wiley: New York, 1963; Collect. Vol. IV, p 590.

   g. Vyas, G. N.; Shah, N. M. *Ibid.,* p 836.

   h. Gassman, P. G.; Marshall, J. L. *Organic Syntheses;* Wiley: New York, 1973; Collect. Vol. V, p 424.

   i. Kuryla, W. C.; Hyve, J. E. *Ibid.,* p 684.

   j. Pedersen, C. J. *Organic Syntheses;* Wiley: New York, 1972; Vol. 52, p 67.

2. Review articles on phase transfer catalysis:

   a. Gokel, G. W.; Weber, W. P. *J. Chem. Educ.* **1978,** *55,* 350.

   b. *Ibid.,* 429.

   c. Varughese, P. J. *J. Chem. Educ.* **1977,** *54,* 666.

   d. Jones, R. A. *Aldrichim. Acta* **1976,** *9,* 35.

3. The procedures used in these experiments for the preparation of the ethers were adapted from the work of

   a. Rowe, J. E. *J. Chem. Educ.* **1980,** *57,* 162.

   b. McKillop, A.; Fiaud, J. C.; Hug, R. P. *Tetrahedron* **1974,** *30,* 1379.

Experiments 26A, 26B — Amide Synthesis: Acetanilide; N,N'-Diacetyl-1,4-phenylenediamine

(acetamide, *N*-phenyl-; acetamide, *N,N'*-1,4-phenylenebis-)

One of the major routes used for the preparation of amides is the reaction of ammonia, a primary amine, or a secondary amine with an active acylating reagent. The reactions presented in Parts A and B demonstrate the use of acetic anhydride as the acylating agent.

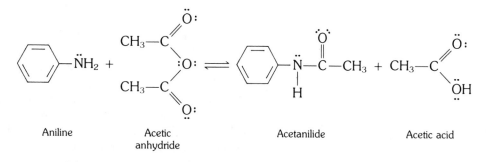

Aniline   Acetic anhydride   Acetanilide   Acetic acid

DISCUSSION   A number of important chemical and biochemical synthetic sequences are initiated by the addition of a nitrogen nucleophile to a carbonyl carbon atom. The experiments outlined in Parts A and B illustrate the attack of a primary amine on the acetyl group of acetic anhydride. Ammonia or secondary amines also react readily with active acylating reagents.

The mechanism given below is an example of the attack of a nucleophilic reagent on the trigonal carbon atom of the carbonyl unit.

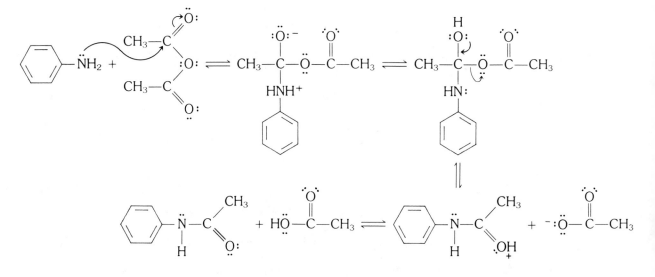

In the preparations given below the amine reagents (aniline and *p*-phenylenediamine) are purified through their hydrochloride salts. Arylamines are relatively weak bases (K_b values on the order of 10^{-10}) but, when treated with a strong acid such as HCl, they are completely protonated, yielding the corresponding water-soluble hydrochloride salt.

$$C_6H_5—NH_2 + HCl \rightarrow C_6H_5—NH_3^+, Cl^-$$

Aniline   Anilinium hydrochloride salt

(water soluble)

As directed in the experiment, decolorizing charcoal is added to the aqueous solution of the arylamine salt. The charcoal absorbs impurities, and subsequent filtration of the mixture, which removes the charcoal, yields an aqueous solution of the purified arylamine salt.

The second stage of the reaction sequence requires that a solution of sodium acetate be added to the reaction mixture after initial addition of acetic anhydride to the purified anilinium hydrochloride salt solution.

$$C_6H_5—NH_3^+, Cl^- + CH_3COO^-, Na^+ \rightleftharpoons C_6H_5—NH_2 + CH_3COOH + Na^+, Cl^-$$

Addition of the sodium acetate solution serves to liberate the arylamine so that the desired nucleophilic substitution reaction may occur.

Sodium acetate is the salt of a strong base–weak acid. Furthermore, the anilinium ion ($pK_a = 4.6$) is a slightly stronger acid than acetic acid ($pK_a = 4.8$). Thus, the equilibrium reaction is shifted to the right, liberating the arylamine.

In Part B, the *p*-phenylenediamine forms the corresponding dihydrochloride salt, Cl^-, $^+H_3N—C_6H_4—NH_3^+, Cl^-$. In a similar sequence, the aqueous solution of the salt is purified with the charcoal, and, on addition of the sodium acetate solution, the free diamine is liberated.

This overall process illustrates an important transformation for most amines. They can be converted to water-soluble ionic salts by reaction with acids and can be recovered from these acid salts by treatment with base. This technique was used in Experiment 4C to extract 4-aminoethyl benzoate from a mixture as its water-soluble salt.

PART A: Acetanilide

The reaction is shown above.

EXPERIMENTAL   Estimated time to complete the experiment: 1.5 hours.

Physical Properties of Reactants and Products

Compound	MW	Wt/Vol	mmol	mp(°C)	bp(°C)	Density	n_D
Aniline	93.13	100 μL	1.09		184	1.02	1.5863
Conc. HCl		3 drops					
Sodium acetate trihydrate	136.08	150 mg	1.10	58			
Acetic anhydride	102.09	150 μL	1.10		140	1.08	1.3901
Acetanilide	135.17			114			

Reagents and Equipment

HOOD   In a tared 10 × 75-mm test tube (standing in a small beaker or Erlenmeyer flask) fitted with a cork stopper is placed 100 μL of aniline **[hood]**.

> **WARNING: *Aniline is a toxic material and is a suspected carcinogenic agent.***

The aniline is dispensed using an automatic delivery pipet. The test tube and container are again weighed to determine the exact amount of aniline delivered.

HOOD   A 1.0-mL graduated pipet is used to add, with swirling, 0.5 mL of water followed by 3 drops of concentrated hydrochloric acid **[hood]** (Pasteur pipet). To the resulting solution is added 10 mg of powdered decolorizing charcoal or the pelletized form (Norit).

The well-mixed suspension is transferred (Pasteur pipet) to a 25-mm funnel fitted with fast-grade filter paper to remove the charcoal by gravity filtration.

Wet the filter paper in advance with distilled water and blot the excess water from the stem of the funnel.

The filtrate is collected in a 3.0-mL conical vial containing a magnetic spin vane and equipped with an air condenser.(■) An additional 0.5-mL portion of water is used to rinse the test tube and the collected charcoal. The rinse is combined with the original filtrate.

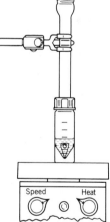

C₆H₅NH₂, 100 μL
+ H₂O, 1.0 mL
+ conc. HCl, 3 drops
+ NaOAc · 3H₂O, 150 mg
+ (CH₃CO)₂O, 150 μL

Note. *If the pelletized form of charcoal is used, transfer through a Pasteur filter pipet directly to the 3.0-mL conical vial should be sufficient.*

Tap all of the filtrate from the funnel stem into the collecting vial. As a result of this purification step a clear, colorless solution of aniline hydrochloride should be obtained.

A solution of sodium acetate trihydrate is prepared by placing 150 mg (1.10 mmol) of this material in a 10 × 75-mm test tube followed by 0.5 mL of distilled water. The tube is capped and set aside for use in the next step.

HOOD   Using an automatic delivery pipet, add **[hood]** 150 μL of acetic anhydride to the aniline hydrochloride solution, with stirring, followed quickly by addition (Pasteur pipet) of the previously prepared sodium acetate solution.

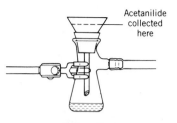

Acetanilide collected here

H_2O, ~ 2 mL
+ water-soluble reaction products

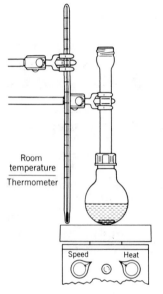

Room temperature
Thermometer

Speed   Heat

10-mL RB flask

Reaction Conditions

The reaction is very rapid and the acetanilide product begins to precipitate immediately on mixing of the reagents. After allowing the reaction mixture to stand at room temperature for approximately 5 minutes, it is placed in an ice bath for an additional 5–10 minute period to complete the crystallization process.

Isolation of Product

The acetanilide product is collected by filtration under reduced pressure using a Hirsch funnel. (■) The conical vial is rinsed with two 0.5-mL portions of water (calibrated Pasteur pipet) and the rinse used to wash the collected filter cake. The snow-white crystals are dried on a porous clay plate or on filter paper in a desiccator.

Purification and Characterization

Further purification of the product is generally not required. However, the acetanilide may be recrystallized from hot water or from ethanol–water using the Craig tube.

Acetanilide (150 mg) can be recrystallized from approximately 3 mL of water or 2 mL of ethanol–water (1 : 10 v/v) with better than 80% recovery.

Weigh the dried crystals and calculate the percentage yield. Determine the melting point of the material and compare your result with the literature value.

Obtain IR and/or NMR spectra of the product and compare them with those of an authentic sample or those recorded in the literature.

Chemical Tests

Characterization of the product may be enhanced by performing several chemical tests given in Chapter 7.

Check the solubility of acetanilide in water. Is the aqueous solution acidic or basic or does it remain neutral as indicated by pH paper? Does the ignition test indicate that an aromatic group is present? Does the soda-lime or sodium fusion test indicate the presence of nitrogen? Does the hydroxamate test for amides give a positive result?

Optional Scaleup

This experiment can be scaled up to be run with five times the amounts used in the microscale preparation. The procedure is identical to that for the microscale preparation with the following exceptions.

a. Two 13 × 100-mm test tubes are used.
b. The reaction is carried out in a 10-mL round-bottom flask containing a magnetic spin bar and fitted with an air condenser. (■)
c. The reagents and solvent amounts are increased fivefold.

Compound	MW	Wt/Vol	mmol	mp(°C)	bp(°C)	Density	n_D
Aniline	93.13	500 μL	5.45		184	1.02	1.5863
Conc. HCl		0.75 mL					
Sodium acetate trihydrate	136.08	750 mg	5.50	58			
Acetic anhydride	102.09	750 μL	7.39		140	1.08	1.3901

PART B: *N,N'*-Diacetyl-1,4-phenylenediamine

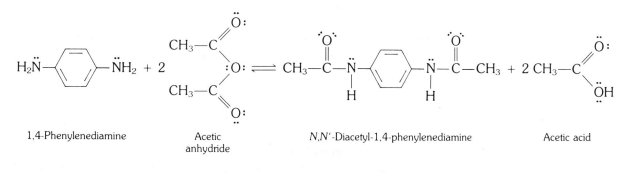

1,4-Phenylenediamine   Acetic   *N,N'*-Diacetyl-1,4-phenylenediamine   Acetic acid
anhydride

EXPERIMENTAL   Estimated time to complete the experiment: 1.5 hours.

Physical Properties of Reactants and Products

Compound	MW	Wt/Vol	mmol	mp(°C)	bp(°C)	Density	n_D
1,4-Phenylenediamine	108.14	117 mg	1.08	138			
Conc. HCl		6 drops					
Sodium acetate trihydrate	136.08	300 mg	2.20	58			
Acetic anhydride	102.09	350 μL	3.71		140	1.08	1.3901
N,N'-Diacetyl-1,4-phenylenediamine	192.24			312–315			

Reagents and Equipment
In a 10 × 75-mm test tube (standing in a small beaker or Erlenmeyer flask) is placed 117 mg (1.08 mmol) of 1,4-phenylenediamine.

> **CAUTION:** *This reagent is toxic and is a suspected carcinogenic agent.*

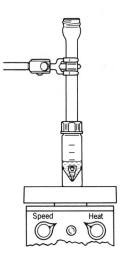

p–NH$_2$C$_6$H$_4$NH$_2$, 117 mg
+ H$_2$O, 2.5 mL
+ conc. HCl, 6 drops
+ NaOAc, 300 mg
+ (CH$_3$CO)$_2$O, 350 μL

With gentle swirling, add 1.0 mL of distilled water and 6 drops (Pasteur pipet) of concentrated hydrochloric acid. After dissolution, add 30 mg of powdered decolorizing charcoal or the pelletized form (Norit). The well-mixed suspension is transferred, by use of a Pasteur pipet, to a 25-mm funnel fitted with fast-grade filter paper previously wet with water. The charcoal is removed by gravity filtration. The filtrate, which is clear to slightly yellow, is collected in a 5-mL conical vial containing a magnetic spin vane and equipped with an air condenser.(■) Three 0.5-mL portions of water (calibrated Pasteur pipet) are used to rinse the test tube, and, in turn, are used to wash the collected charcoal. The rinse is combined with the original filtrate. *Blot the excess water from the stem of the funnel.*

Note. *If the pelletized form of charcoal is used, transfer through a Pasteur filter pipet directly to the 5.0-mL conical vial should be sufficient.*

A solution of sodium acetate trihydrate is prepared by placing 300 mg (2.20 mmol) of this material in a 10 × 75-mm test tube followed by 0.5 mL of distilled water. The mixture is stirred with a spatula to aid the dissolution process. The tube is then capped and set aside for use in the next step.

Reaction Conditions

To the 1,4-phenylenediamine dihydrochloride solution is added 350 μL of acetic anhydride and the mixture stirred briefly using a magnetic stirrer.

HOOD **Note.** *The acetic anhydride is dispensed in the **hood** using an automatic delivery pipet. A slight amount of white precipitate may be observed at this stage.*

The previously prepared sodium acetate solution is added by Pasteur pipet to the reaction mixture with stirring.

Isolation of Product

The reaction is very rapid and the desired product begins to precipitate almost immediately. After stirring briefly, the mixture is allowed to stand at room temperature for a few minutes and then is placed in an ice bath for an additional 5–10 minutes.

Purification and Characterization

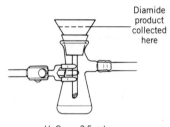

Diamide product collected here

H₂O, ~ 3.5 mL + water-soluble reaction products

The crude *N,N'*-diacetyl-1,4-phenylenediamine is collected by vacuum filtration using a Hirsch funnel.(■) The reaction vial is rinsed with two 0.5-mL portions of water and the rinse used to wash the filter cake. The collected material is placed on a clay plate or on filter paper to dry.

If this material is utilized as the substrate in Experiment 34B, recrystallization from methanol is suggested.

Weigh the dried crystals and calculate the percentage yield. Determine the melting point and compare it with the literature value. Obtain the IR spectrum and/or the NMR spectrum of the product. The infrared spectrum is shown in Figure 6.18.

Chemical characterization tests might also be run as outlined in Part A.

ENVIRONMENTAL DATA

Substance	Amount	TLV (mg/m³)	Emissions (mg)	Volume (m³)
1,4-Phenylenediamine (Emissions should be small.)	117 mg			
Aniline (SKIN) (Max. from transfer and react.)	100 μL	19	10	0.5
Acetic anhydride (transfer, open container)	150 μL	20	6	0.3

TLV: Hydrochloric acid—5 ppm
TXDS: Acetanilide—orl-rat LD50: 800 mg/kg
      Sodium acetate—orl-rat LD50: 3530 mg/kg

QUESTIONS **6-86.** What is the function of the sodium acetate in the reactions outlined in this experiment?

6-87. Which is the stronger base, aniline or cyclohexylamine? Explain.

6-88. Arrange the following substituted aniline compounds in increasing order of reactivity toward acetic anhydride. Explain the order.

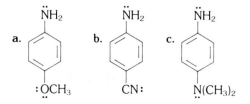

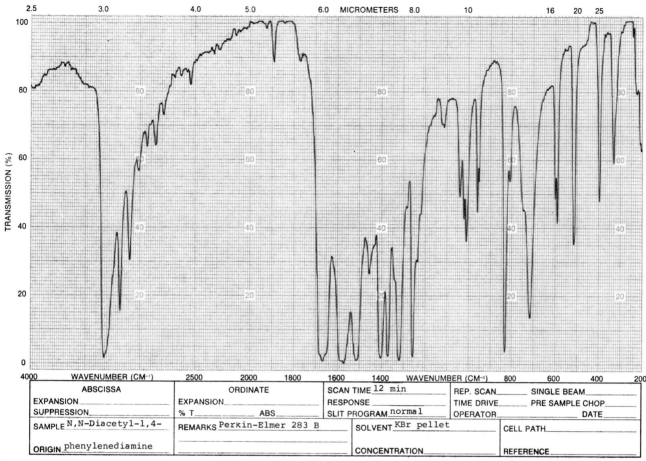

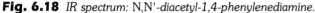

Fig. 6.18 *IR spectrum: N,N'-diacetyl-1,4-phenylenediamine.*

6-89. Suggest a mechanism for the preparation of ethanoic anhydride from ethanoic acid and ethanoyl chloride in the presence of pyridine.

6-90. Anhydrides generally react more slowly than acid chlorides with an amine to prepare amides. Explain this observation.

6-91.

$$CH_3CH_2-C\overset{\overset{\displaystyle \ddot{O}:}{\|}}{\underset{\ddot{N}H_2}{\diagdown}} \quad \text{is less basic than} \quad CH_3CH_2-\ddot{N}H_2$$

Explain.

REFERENCES   **1.** Review articles:

    a. Satchell, D. P. N. *Quart Rev.* **1963,** *17,* 160.

    b. Beckwith, A. L. J. In *The Chemistry of the Amides*; Zabicky, J., Ed.; Wiley: New York, 1966; p 86.

2. Selected acylation reactions in *Organic Syntheses* between anhydrides and amines:

    a. Noyes, W. A.; Porter, P. K. *Organic Syntheses;* Wiley: New York, 1941; Collect. Vol. I, p 457.

    b. Herbst, R. M.; Shemin, D. *Organic Syntheses;* Wiley: New York, 1943; Collect. Vol. II, p 11.

c. Jacobs, T. L.; Winstein, S.; Linden, G. B.; Robson, J. H.; Levy, E. F.; Seymour, D. *Organic Syntheses;* Wiley: New York, 1955; Collect. Vol. III, p 456.

d. Fanta, P. E.; Tarbell, D. S. *Ibid.,* p 661.

e. Wiley, R. H.; Borum, O. H. *Organic Syntheses;* Wiley: New York, 1963; Collect. Vol. IV, p 5.

f. Cava, M. P.; Deana, A. A.; Muth, K.; Mitchell, M. J. *Organic Syntheses;* Wiley: New York, 1973; Collect. Vol. V, p 944.

Experiments 27A, 27B   Imide Synthesis: *N*-Phenylmaleimide

(maleimide, *N*-phenyl)

The condensation of an anhydride with aniline to form an imide is described in this experiment. The initial reaction to give the acid amide is followed by an intramolecular condensation to produce the desired imide derivative.

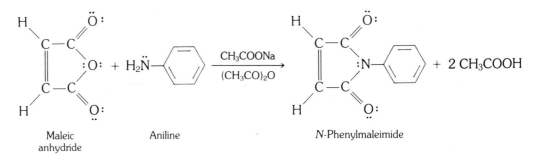

Maleic anhydride · · · · · · · · · · · · Aniline · · · · · · · · · · · · *N*-Phenylmaleimide

DISCUSSION   Imides are diacyl derivatives of ammonia or primary amines. The reaction is similar in its scope and mechanism to the acetylation of aniline or 1,4-phenylenediamine presented in Experiments 26A and 26B. As illustrated in the present experiment, cyclic anhydrides produce cyclic imides. Derivatives of imides have been suggested for use in the treatment of arthritis, tuberculosis, and epilepsy. Several also have been found to be growth stimulants. The *N*-phenylmaleimide prepared in this experiment is also a good dienophile in the Diels–Alder reaction (see Experiments 14 and 15) and in fact has been used as a reagent to characterize 1,3-dienes.

The second step in the reaction is much slower than the first. That is, attack of an amide nitrogen on the carbonyl carbon of the acid is slower than the attack of the amine nucleophile on the anhydride carbonyl carbon. The mechanistic sequence is given below.

(1)

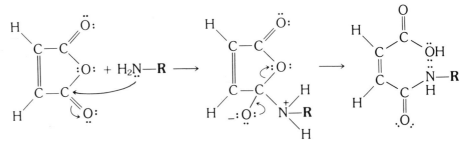

(2)

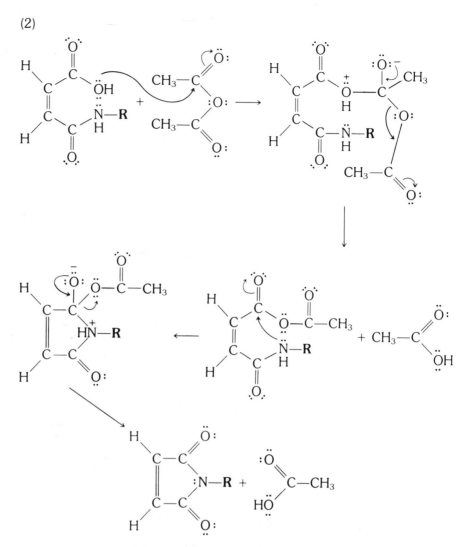

EXPERIMENTAL   Estimated time of the experiment: 2.0 hours.

PART A: Preparation of Maleanilic Acid

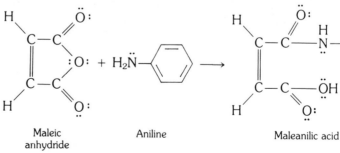

| | Maleic anhydride | Aniline | Maleanilic acid |

Physical Properties of Reactants and Products

Compound	MW	Wt/Vol	mmol	mp(°C)	bp(°C)	Density	n_D
Maleic anhydride	98.06	60 mg	0.61	60			
Diethyl ether	74.12	1.2 mL			34.5		
Aniline	93.13	56 μL	0.62		184	1.02	1.5863
Maleanilic acid	191.18			201–202			

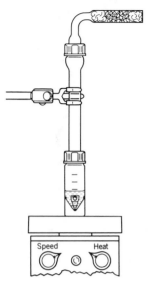

Maleic anhydride, 60 mg
+ C$_6$H$_5$NH$_2$, 56 μL
+ (CH$_3$CH$_2$)$_2$O, 1.2 mL

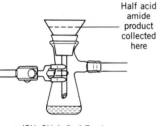

Half acid
amide
product
collected
here

(CH$_3$CH$_2$)$_2$O, 1.7 mL

Reagents and Equipment

In a 3.0-mL conical vial containing a magnetic spin vane and equipped with an air condenser protected with a drying tube place 60 mg (0.61 mmol) of maleic anhydride and 1.2 mL of anhydrous diethyl ether.(■) Stir the mixture at room temperature until all the maleic anhydride has dissolved.

HOOD

> **WARNING:** *Diethyl ether is highly flammable. All flames in the laboratory should be extinguished. Dispense this reagent in the hood using a calibrated Pasteur pipet.*

In a separate dry vial, prepare a solution of 56 μL (57 mg, 0.62 mmol) of aniline in 100 μL of anhydrous diethyl ether.

HOOD

> **WARNING:** *Aniline is highly toxic and is a suspected carcinogenic agent. It should be dispensed in the hood using an automatic delivery pipet.*

Using a Pasteur pipet, add the aniline–ether solution in one portion to the stirred maleic anhydride–ether solution. The vial is rinsed with 100 μL of anhydrous ether and this rinse is also transferred to the reaction solution.

Reaction Conditions

The reaction mixture is stirred at room temperature for a period of 15 minutes and then cooled in an ice bath for 5–10 minutes.

Isolation of Product

The deposit of fine, cream-colored powder is collected by vacuum filtration using a Hirsch funnel.(■) The maleanilic acid crystals are then washed with 0.5 mL of cold diethyl ether (calibrated Pasteur pipet) and air-dried in the funnel for 5 minutes.

Purification and Characterization

Weigh the maleanilic acid and calculate the percentage yield.

Determine the melting point and compare your value with that in the literature. Obtain an IR spectrum using the KBr pellet technique and compare it with that of an authentic sample. The air-dried product is suitable for use in the next step without further purification.

PART B: Preparation of *N*-Phenylmaleimide

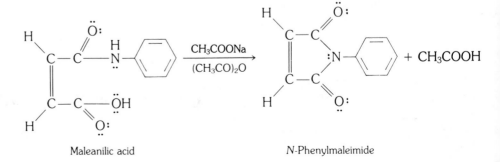

Maleanilic acid   *N*-Phenylmaleimide

Physical Properties of Reactants and Products

Compound	MW	Wt/Vol	mmol	mp(°C)	bp(°C)	Density	n_D
Maleanilic acid	191.18	100 mg	0.52	201–202			
Sodium acetate	82.03	25 mg	0.30		324		
Acetic anhydride	102.09	200 μL	2.12		140	1.08	1.3901
N-Phenylmaleimide	173.17			90–91			

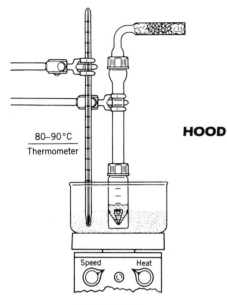

80–90°C
Thermometer

Speed   Heat

CH_3CO_2Na, 25 mg
+ $(CH_3CO)_2O$, 200 μL
+ maleanilic acid, 100 mg

HOOD

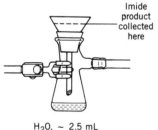

Imide
product
collected
here

H_2O, ~ 2.5 mL
+ water-soluble reaction products

Reagents and Equipment

In a 3.0-mL conical vial containing a magnetic spin vane and equipped with an air condenser protected by a drying tube are placed 25 mg (0.3 mmol) of anhydrous sodium acetate and 200 μL (216 mg, 2.12 mmol) of acetic anhydride.(■)

> **CAUTION: Acetic anhydride is corrosive and a lachrymator. It should be dispensed in the hood by use of an automatic delivery pipet.**

Maleanilic acid, prepared in Part A (100 mg, 0.52 mmol), is now added to the reaction vial.

Reaction Conditions

The reaction mixture is heated, with stirring, at a sand bath temperature of 80–90°C for a period of 30 minutes. It is then cooled to room temperature and 1.0 mL of cold water added (calibrated Pasteur pipet). The resulting mixture is stirred for a few minutes and then placed in an ice bath for 5–10 minutes.

Isolation of Product

The solid product is collected by vacuum filtration using a Hirsch funnel and the filter cake washed with three 0.5-mL portions of cold water (calibrated Pasteur pipet.)(■) It is air-dried in the Hirsch funnel for a period of 5–10 minutes.

Purification and Characterization:

The crude N-phenylmaleimide is recrystallized from cyclohexane to yield canary-yellow needles.

After drying the product on a porous clay plate or on filter paper, weigh the crystals and calculate the percentage yield. Determine the melting point and compare your result with the literature value. Obtain an IR spectrum and compare it with that of an authentic sample.

ENVIRONMENTAL DATA

Substance	Amount	TLV (mg/m³)	Emissions (mg)	Volume (m³)
Aniline (SKIN) (Max. from yield, transfer)	57 mg	19	5	0.3
Acetic anhydride (Max. from transfer)	216 mg	20	216	10.8
Ether (Max., solv., vac., filtr.)	1.2 mL	1200	857	0.7
Ether (Max., evap.)	0.5 mL	1200	357	0.3
Cyclohexane	0.5 mL	1050	780	.75

TLV: Maleic anhydride—0.25 ppm
TXDS: N-Phenylmaleimide—orl-rat LDLo: 100 mg/kg
　　　Sodium acetate—orl-rat LD50: 3530 mg/kg

QUESTIONS

6-92. As stated in the discussion section of this experiment, the second step in the reaction to form the imide is much slower than that of the first stage (formation of the acid amide). Explain.

6-93. Phthalimide has a K_a of 5×10^{-9}.

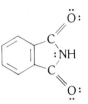

Write an equation for the reaction of phthalimide with potassium amide (a strong base) in dimethyl formamide solvent. Name the product.

6-94. Predict which of the species below is the most acidic. Explain.

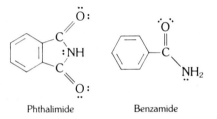

Phthalimide   Benzamide

6-95. The phthalimide anion is a strong nucleophile. It can react easily with primary alkyl halides to form substituted phthalimides.

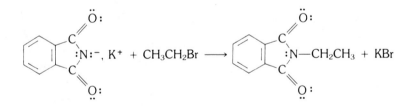

Suggest a suitable mechanism for this reaction.

6-96. *N*-Phenylmaleimide, the product prepared in Part B, can act as a dienophile in the Diels–Alder reaction (see Experiments 14 and 15). Draw the structure of the product that would be formed by the treatment of *N*-phenylmaleimide with (a) 3-sulfolene under the conditions given in Experiment 14 and (b) furan.

REFERENCES

1. Review article on cyclic imides: Hargreaves, M. K.; Pritchard, J. G.; Dave, H. R. *Chem. Rev.* **1970,** *70*, 439.
2. Selected imide preparations in *Organic Syntheses:*
   a. Noyes, W. A.; Porter, P. K. *Organic Syntheses;* Wiley: New York, 1941; Collect. Vol. I, p 457.
   b. Smith, L. I.; Emerson, O. H. *Organic Syntheses;* Wiley: New York, 1955; Collect. Vol. III, p 151.
   c. Soine, T. O.; Buchdahl, M. R. *Organic Syntheses;* Wiley: New York, 1963; Collect. Vol. IV, p 106.
   d. Cava, M. P.; Deana, A. A.; Muth, K.; Mitchell, M. J. *Organic Syntheses;* Wiley: New York, 1973; Collect. Vol. V, p 944.

Experiments 28A, 28B

Synthesis of Cyclic Carboxylic Acid Anhydrides: Succinic Anhydride; Phthalic Anhydride

(2,5-furandione,dihydro-; 1,3-isobenzofurandione)

One of the important methods for preparing cyclic carboxylic acid anhydrides is presented in this experiment. The reaction demonstrates the use of acetic anhydride, an important industrial material, as a dehydrating agent.

DISCUSSION   Five- and six-membered cyclic anhydrides can be easily formed when the corresponding dicarboxylic acid is heated in the presence of a dehydrating agent. Among the dehydrating agents most often used is acetic anhydride; anhydrides that have boiling points higher than acetic acid can be prepared by this method.

The formation of the anhydride from its corresponding acid in the presence of acetic anhydride is referred to as *anhydride exchange.*

It is possible to prepare five- and six-membered cyclic anhydrides in the absence of acetic anhydride by direct dehydration at elevated temperatures. Maleic anhydride, for example, is easily obtained by this method.

The similarity of this reaction for the preparation of anhydrides to that for the synthesis of imides (Experiment 27) should be noted. The ring closure to form the imide is mechanistically related to that of anhydride formation in that in one case an amide nitrogen makes a nucleophilic attack on a carbonyl and in the other an acid oxygen acts as the nucleophile.

The mechanistic sequence for anhydride exchange is given below.

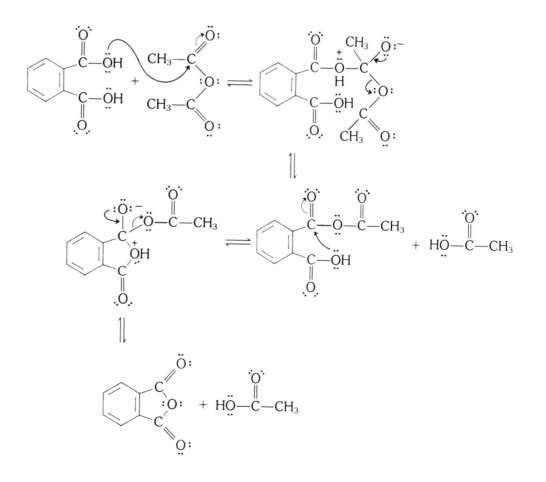

PART A: Succinic Anhydride

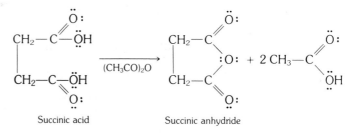

Succinic acid   Succinic anhydride

EXPERIMENTAL   Estimated time to complete the experiment: 1.5 hours.

Physical Properties of Reactants and Products

Compound	MW	Wt/Vol	mmol	mp(°C)	bp(°C)	Density	n_D
Succinic acid	118.09	150 mg	1.27	188			
Acetic anhydride	102.09	200 μL	2.12		140	1.08	1.3901
Succinic anhydride	100.08			120			

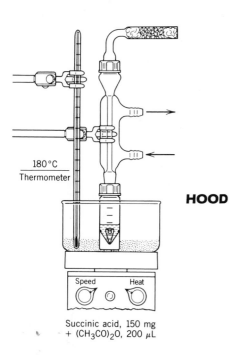

180°C
Thermometer

HOOD

Succinic acid, 150 mg
+ (CH₃CO)₂O, 200 μL

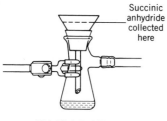

Succinic
anhydride
collected
here

(CH₃CH₂)₂O, 1.5 mL
+ CH₃CO₂H, ~100 μL

Reagents and Equipment

Note. *All equipment must be dried in an oven (110°C) half an hour before use.*

In a 1.0-mL conical vial containing a magnetic spin vane and equipped with a reflux condenser protected by a calcium chloride drying tube, place 150 mg (1.3 mmol) of succinic acid and 200 μL of acetic anhydride.(■)

> **CAUTION: Acetic anhydride is a moisture-sensitive irritant. It should be dispensed in the hood using an automatic delivery pipet.**

Reaction Conditions

The reaction mixture is heated with stirring in a sand bath at a temperature of 180°C for a period of 45 minutes. The time is measured from the point at which the succinic acid has completely dissolved.

Isolation of Product

The mixture is cooled to room temperature, whereupon a voluminous precipitate of succinic anhydride deposits. The vial is cooled further in an ice bath for 5 minutes and the solid collected by vacuum filtration using a Hirsch funnel.(■) The white needles are washed with three 0.5-mL portions of diethyl ether (calibrated Pasteur filter pipet) and then placed on a porous clay plate or on filter paper to dry.

Purification and Characterization

The succinic anhydride crystals are sufficiently pure for characterization. They may, however, be recrystallized from acetic anhydride or absolute ethanol using the Craig tube.

Weigh the product and calculate the percentage yield. Determine the melting point and obtain an infrared spectrum using the KBr pellet technique. Compare your results with those reported in the literature. What characteristic absorption pattern do you observe for the anhydride group in the carbonyl region of the IR spectrum?

PART B: Phthalic Anhydride

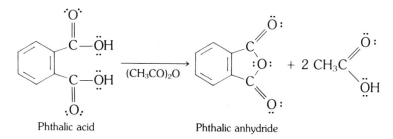

Phthalic acid   Phthalic anhydride

EXPERIMENTAL   Estimated time to complete the experiment: 1.5 hours.

Physical Properties of Reactants and Products

Compound	MW	Wt/Vol	mmol	mp(°C)	bp(°C)	Density	n_D
Phthalic acid	166.14	100 mg		0.60	210		
Acetic anhydride	102.09	200 μL	2.12		140	1.08	1.3901
Phthalic anhydride	148.12				131		

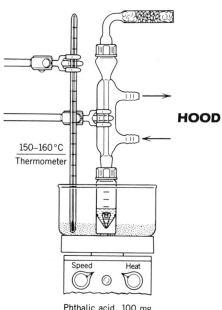

150–160°C
Thermometer

HOOD

Speed   Heat

Phthalic acid, 100 mg
(CH₃CO)₂O, 200 μL

Reagents and Equipment

To a 1.0-mL conical vial containing a magnetic spin vane and equipped with a reflux condenser protected by a calcium drying tube, add 100 mg (0.60 mmol) of phthalic acid and 200 μL of acetic anhydride.(■)

> **CAUTION: Acetic anhydride is a moisture-sensitive irritant. It should be dispensed in the hood using an automatic delivery pipet.**

Reaction Conditions

The reaction solution is heated with stirring at a sand bath temperature of 150–160°C for a period of 30 minutes.

Note. *Position the vial firmly on the bottom of the sand bath container to maintain this reaction temperature.*

Isolation of Product

The mixture is cooled to room temperature, whereupon the product crystallizes from solution. The vial and contents are cooled in an ice bath for a period of

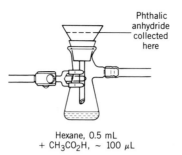

Phthalic
anhydride
collected
here

Hexane, 0.5 mL
+ CH$_3$CO$_2$H, ~ 100 μL

10 minutes, and the solid is collected by vacuum filtration using a Hirsch funnel. (■) The filter cake is rinsed carefully by dropwise addition of 0.5 mL of cold hexane (Pasteur pipet) and then air-dried in the Hirsch funnel. Final drying is achieved by placing the crystals on a porous clay plate.

Purification and Characterization

The phthalic anhydride is sufficiently pure for characterization. It may be purified further by sublimation or recrystallization from absolute ethanol using the Craig tube.

Weigh the product and calculate the percentage yield. Determine the melting point and obtain an infrared spectrum using the KBr pellet technique. Compare your results with those reported in the literature. What characteristic absorption pattern do you observe for the anhydride group in the carbonyl region of the IR spectrum?

ENVIRONMENTAL DATA

Substance	Amount	TLV (mg/m^3)	Emissions (mg)	Volume (m^3)
Acetic anhydride (transfer in hood, reflux)	216 mg	20	10	0.5
Diethyl ether (Max., evap.)	1.5 mL	1200	1071	0.9
Hexane (Max., evap.)	0.5 ml	180	330	1.8

TXDS: Phthalic anhydride—orl-rat LD50: 7900 mg/kg
Succinic anhydride—scu-rat TD: 2600 mg/kg/65W-1

QUESTIONS **6-97.** As stated in the Discussion, direct dehydration can be used as a method for the preparation of five- and six-membered cyclic anhydrides. Propose a suitable mechanism for the reaction below.

$$\text{maleic acid} \xrightarrow{\Delta} \text{maleic anhydride} + \text{H}_2\text{O}$$

6-98. Propose a suitable mechanism for the formation of the mixed anhydride obtained in the following reaction.

$$C_6H_5CH_2CO_2H + (CF_3CO)_2O \rightarrow C_6H_5CH_2\overset{\overset{\ddot{O}}{\|}}{C}-\overset{..}{\underset{..}{O}}-\overset{\overset{\ddot{O}}{\|}}{C}-CF_3 + CF_3CO_2H$$

6-99. There are two 1,3-cyclobutane dicarboxylic acids. One can form an anhydride, the other cannot. Draw the structures of these compounds and indicate which one can be converted to an anhydride. Explain.

6-100. When maleic acid is heated to about 100°C, it forms maleic anhydride. However, fumaric acid requires a much higher temperature (250–300°C) before it dehydrates. In addition it forms only maleic anhydride. Explain.

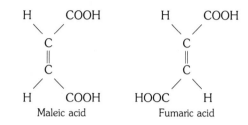

Maleic acid   Fumaric acid

6-101. What product would you expect to obtain from reaction of one equivalent of propanol with phthalic anhydride?

REFERENCES   **1.** Selected references from *Organic Syntheses* in which anhydrides are prepared, using acetic anhydride as the dehydrating agent:
   a. Clarke, H. T.; Rahrs, E. J. *Organic Syntheses;* Wiley: New York, 1944; Collect. Vol. I, p 91.
   b. Nicolet, B. H.; Bender, J. A. *Ibid.,* p 410.
   c. Grummitt, O.; Egan, R.; Buck, A. *Organic Syntheses;* Wiley: New York, 1955; Collect. Vol. III, p 449.
   d. Shriner, R. L.; Furrow, C. L., Jr. *Organic Syntheses;* Wiley: New York, 1963; Collect. Vol., IV, p 242.
   e. Cason, J. *Ibid.,* p 630.
   f. Horning, E. C.; Finelli, A. F. *Ibid.,* p 790.
   2. The synthesis of succinic anhydride is given in Fieser, L. F.; Martin, E. L. *Organic Syntheses;* Wiley: New York, 1943; Collect. Vol. II, p 560.

Experiment 29   # Heterocyclic Ring Synthesis: 1-*H*-Benzimidazole

(1,3-benzodiazole)

The condensation of a 1,2-diaminobenzene with formic acid is one of the methods used to synthesize the highly aromatic benzimidazole ring system. Benzimidazole itself is prepared in this experiment.

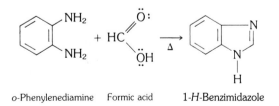

o-Phenylenediamine   Formic acid   1-*H*-Benzimidazole

DISCUSSION   This experiment illustrates the classic method of forming the benzimidazole ring system. This heterocyclic series is generally prepared from 1,2-diaminobenzenes by reaction with carboxylic acids or their derivatives, such as nitriles, under acidic conditions. The ring system is highly aromatic; it is difficult to oxidize or reduce and is stable to acids and bases. It is an important heterocyclic ring system since it occurs in vitamin B_{12} and in many biologically active compounds. Benzimidazole itself inhibits the growth of certain yeasts and bacteria.

The parent compound of this class of heterocyclic compounds is imidazole. This ring system exhibits basic properties and is protonated to give a conjugate acid with $pK_a = 6.95$. Once the imidazole ring is protonated, the two nitrogen atoms are indistinguishable since the resonance forms of the protonated species are equivalent.

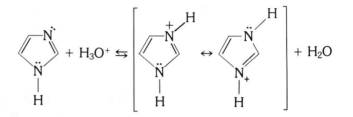

The reaction to form the ring system actually proceeds in two stages. The first involves the in situ formation of a substituted formamide; the second, an intramolecular addition and elimination reaction to form the heterocyclic ring. The sequence is outlined below.

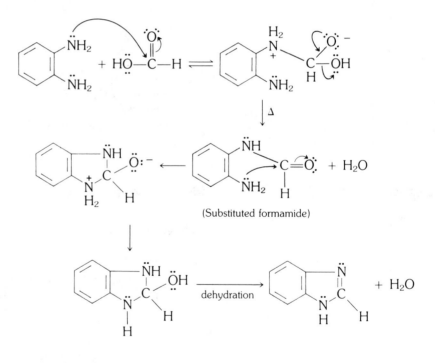

(Substituted formamide)

EXPERIMENTAL Estimated time to complete the experiment: 2.0 hours.

Physical Properties of Reactants and Products

Compound	MW	Wt/Vol	mmol	mp(°C)	bp(°C)	Density	n_D
o-Phenylenediamine	108.14	108 mg	1.0	102			
90% Formic acid	46.03	64 μL	1.7		101	1.22	1.3714
1-*H*-Benzimidazole	118.14			174			

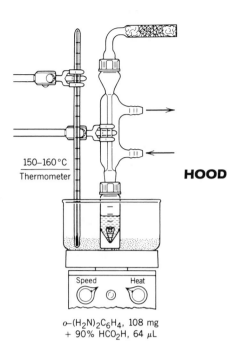

150–160°C
Thermometer

HOOD

o–(H₂N)₂C₆H₄, 108 mg
+ 90% HCO₂H, 64 μL

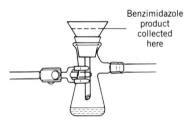

Benzimidazole
product
collected
here

H₂O, ~ 2.0 mL

Reagents and Equipment

To a 1.0-mL conical vial containing a magnetic spin vane and equipped with a reflux condenser protected by a calcium chloride drying tube, add 108 mg (1.0 mmol) of o-phenylenediamine and 64 μL (79 mg, 1.7 mmol) of 90% formic acid. (■)

WARNING: *o-Phenylenediamine is toxic and is a suspected carcinogenic agent. Formic acid is very corrosive to the skin and should be dispensed in the* hood *using an automatic delivery pipet.*

Reaction Conditions

The reaction mixture is heated, with stirring, at a sand bath temperature of 150–160°C for a period of 1 hour.

Isolation of Product

After allowing the reaction mixture to cool to room temperature, crude benzimidazole is precipitated on addition of 630 μL of 10% aqueous sodium hydroxide solution (automatic delivery pipet). The product is collected by vacuum filtration using a Hirsch funnel and the filter cake washed with three 0.5-mL portions of cold water (calibrated Pasteur pipet). (■)

Purification and Characterization

The crude material is recrystallized from water using a Craig tube and dried on a porous clay plate.

Weigh the crystals and calculate the percentage yield. Determine the melting point and compare your result with the literature value.

The UV spectrum of benzimidazole has been reported in 95% ethanol.[2]

Chemical Tests

Several tests may be run to assist in the identification of this material. Does the ignition test confirm the presence of the aromatic ring system? Does the soda-lime or sodium fusion test indicate that nitrogen is present? Is the material soluble in 10% hydrochloric acid solution?

ENVIRONMENTAL DATA

Substance	Amount	TLV (mg/m³)	Emissions (mg)	Volume (m³)
Formic acid, 90% (Max.)	64 μL	9	62	6.9
o-Phenylenediamine (relatively involatile)	108 mg		0	

TLV: Sodium hydroxide—2 μg/m³
TXDS: o-Phenylenediamine—orl-rat LD50: 1070 mg/kg
      Benzimidazole—orl-rat LDLo: 500 mg/kg

---

[2] Steck, E. A.; Nackod, F. C.; Ewing, G. W.; Gorman, N. H. *J. Am. Chem. Soc.* **1948**, *70*, 2406.

QUESTIONS

6-102. The parent compound of the imidazole series, imidazole itself **(I)**, was first prepared in 1858.

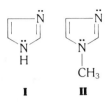

I **II**

Can you account for the fact that it has a very high boiling point (256°C) whereas 1-methyl imidazole **(II)** has a somewhat lower boiling point of 199°C?

6-103. The imidazole ring system has a great deal of aromatic character. Can you formulate two resonance hybrid structures that account for this characteristic?

6-104. Imidazole is a weak acid and reacts with strong bases to form the corresponding anion. Show this reaction and draw resonance structures to account for the stability of the generated anion.

6-105. Suggest a mechanism for the dehydration involved in the last step in the synthesis of benzimidazole.

6-106. Imidazole, acting as a nucleophile, catalyzes the hydrolysis of phenyl acetates by attack on the carbonyl carbon atom of the ester. The imidazole displaces the phenylate ion and forms acetyl imidazole. In turn, the acetyl imidazole is quite unstable in water and hydrolyzes to form acetic acid and regenerate the imidazole molecule. Write a suitable mechanism outlining these steps.

REFERENCES

The conditions of this reaction were adapted from those reported by Wagner, E. C.; Millett, W. H. *Organic Syntheses*; Wiley: New York, 1943; Collect. Vol. II, p 65.

Experiments 30A, 30B

Heterocyclic Ring Synthesis: 4-Hydroxycoumarin; Dicoumarol

(2*H*-1-benzopyran-2-one, 4-hydroxy; 3,3′-methylene bis(4-hydroxycoumarin)

This reaction illustrates the condensation of a carbon nucleophile with a carbonyl carbon to form a C—C bond. In the reaction a β-ketoester is prepared, which on cyclization forms a lactone. Further condensation with formaldehyde yields dicoumarol, a material that is the prototype for the oral anticoagulants widely used in the medical field to lower blood coagulability.

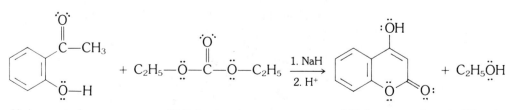

o-Hydroxyacetophenone   Diethyl carbonate   4-Hydroxycoumarin   Ethanol

DISCUSSION   This reaction sequence illustrates the well-known Claisen condensation, which is widely used to form C—C bonds. The condensation is brought about by the nucleophilic attack of a carbanion on the carbonyl carbon of an ester. The carbanion is generated by removal of a slightly acidic hydrogen from the α-carbon atom of a ketone, nitrile, or ester using a relatively strong base. In the present reaction, the methyl ketone is deprotonated by the base, NaH. The resulting *enol anion* then attacks the ester, diethyl carbonate. The β-ketoester product thus formed is esterified *intramolecularly* to form the lactone product. Note that the methylene hydrogens in this lactone **(I)** are acidic. In the alkaline medium of this reaction, the sodium enolate of the lactone **(II)** will be the species that will be formed. This enolate is water soluble and accounts for the reason that when isolating the product the aqueous solution must be acidified to precipitate the 4-hydroxycoumarin.

The condensation of 4-hydroxycoumarin with formaldehyde also involves the nucleophilic attack of an anion on a carbonyl carbon atom to produce dicoumarol. This substance is present in moldy sweet clover. It is a blood anticoagulant and causes the hemorrhagic sweet clover disease that kills cattle.

The mechanistic sequence for the reaction is given below.

1. *4-Hydroxycoumarin*

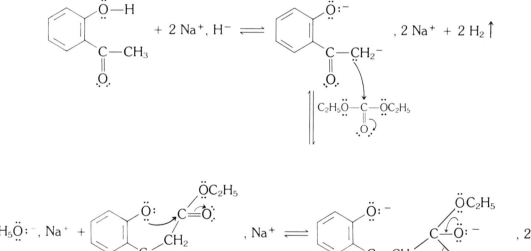

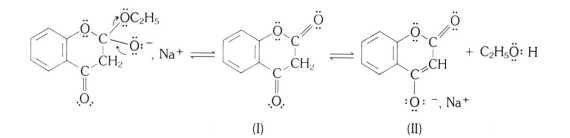

2. *Dicoumarol*

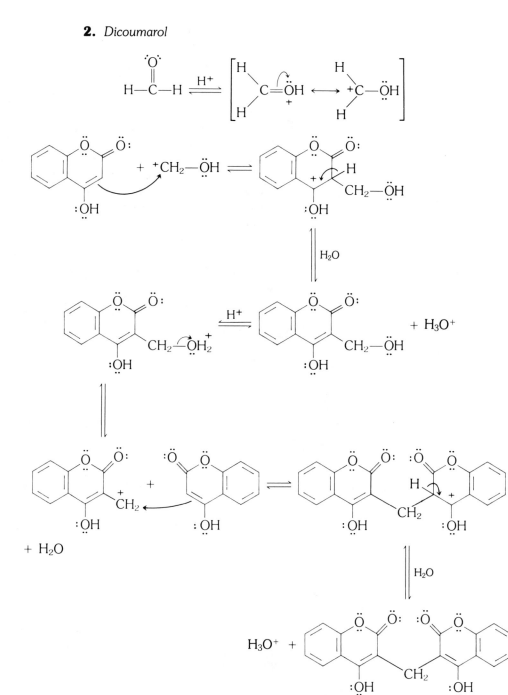

PART A: Preparation of 4-Hydroxycoumarin

The reaction is shown above.

EXPERIMENTAL   Estimated time to complete the reaction: 4.0 hours.

Physical Properties of Reactants and Product

Compound	MW	Wt/Vol	mmol	mp(°C)	bp(°C)	Density	n_D
Sodium hydride (60% in oil dispersion)		85 mg	2.13				
Toluene	92.15	6.0 mL			111		
o-Hydroxyacetophenone	136.16	133 μL	1.1		218	1.13	1.5584
Diethyl carbonate	118.13	333 μL	2.75		126	0.96	1.3845
4-Hydroxycoumarin	162.15			213–214			

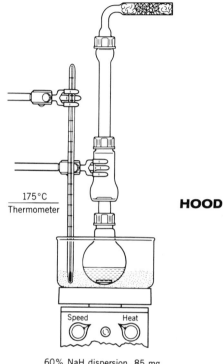

175°C
Thermometer

HOOD

Speed   Heat

60% NaH dispersion, 85 mg
+ C₆H₅CH₃, 6 mL
o–HOC₆H₄COCH₃, 150 mg
+ (CH₃CH₂O)₂CO, 324 mg

Reagents and Equipment

Important. *All equipment used in this reaction should be thoroughly dried in an oven at 110°C for half an hour just prior to use.*

In a 10-mL round-bottom flask containing a magnetic stirring bar and equipped with a Hickman still head and air condenser protected by a calcium chloride drying tube are placed 85 mg (2.13 mmol) of sodium hydride (60% in oil dispersion) and 3.0 mL of dry toluene. The Hickman still 14/10 ⊤ male joint should be wrapped with Teflon tape to prevent joint freezeup.(■)

> **CAUTION: NaH is a flammable solid. Dispense in the hood. Toluene is distilled and stored over molecular sieves. This solvent is also dispensed in the hood.**

In rapid order place 133 μL (150 mg, 1.1 mmol) of o-hydroxyacetophenone, 3.0 mL of dry toluene, and 333 μL (324 mg, 2.75 mmol) of diethyl carbonate in a stoppered 10-mL Erlenmeyer flask.

Note. *The small volumes of the liquid reagents are dispensed using an automatic delivery pipet. Be sure to dry the removable plastic tips in the oven before use. A 10-mL graduated cylinder is used to measure the toluene.*

The air condenser is removed from the distilling head and the o-hydroxyacetophenone solution added with stirring to the reaction flask as rapidly as possible using a Pasteur pipet. The resulting solution turns yellow.

Reaction Conditions
The assembled apparatus is placed in a sand bath and the temperature of the bath raised to about 175°C as rapidly as possible.

A distillate of ethanol and toluene (1.0 mL) is collected (two or three fractions) in the collar of the condenser and the apparatus is then removed from the heat source. The reaction solution is allowed to cool and 3.0 mL of water added with stirring.

Isolation of Product
The resulting two-phase solution is transferred, using a Pasteur pipet, to a 12-mL centrifuge tube. The reaction flask is rinsed with an additional 2.0 mL of water and the rinse also added to the centrifuge tube. The toluene layer is separated using a Pasteur filter pipet and transferred to a second 12-mL centrifuge tube. This organic phase is then extracted with 3.0 mL of water, and the water extract is added to the original water phase. *This aqueous solution contains the water-soluble sodium enolate* **(II).** The combined aqueous layers are cooled in an ice bath and acidified by dropwise addition of concentrated HCl delivered from a

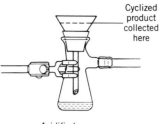

Cyclized
product
collected
here

Acidified aqueous
phase, ~ 10 mL

Pasteur pipet. The solid product precipitates from the aqueous phase. The acid is added until the yellow color of the solution disappears (about 10 drops). The product is collected by vacuum filtration using a Hirsch funnel.(■)

Purification and Characterization
The crude 4-hydroxycoumarin is recrystallized from 50% ethanol using a Craig tube and dried on a porous clay plate or on filter paper.

Weigh the product and calculate the percentage yield. Determine the melting point and obtain an infrared spectrum using the KBr pellet technique. Compare your results with those reported in the literature.

Chemical Tests
You may wish to perform the ignition test to establish that the compound is aromatic. Does the ferric chloride test (Chapter 7) for phenols give a positive result?

PART B: Dicoumarol

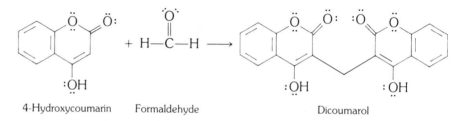

4-Hydroxycoumarin   Formaldehyde   Dicoumarol

EXPERIMENTAL   Estimated time to complete the reaction: 0.5 hour.

Physical Properties of Reactants and Products

Compound	MW	Wt/Vol	mmol	mp(°C)	bp(°C)
4-Hydroxycoumarin	162.15	50 mg	0.31	213–214	
Water		15 mL			100
37% Solution of formaldehyde		0.5 mL	6		
Dicoumarol	336.31			287–293	

Reagents and Equipment
In a 50-mL Erlenmeyer flask containing a boiling stone are placed 50 mg (0.31 mmol) of 4-hydroxycoumarin and 15 mL of water. The mixture is heated to boiling on a hot plate and to the resulting solution is added 0.5 mL (~200 mg, 6 mmol) of formaldehyde (37%).

> **WARNING: Formaldehyde is a suspected carcinogenic agent. Dispense in the hood.**

Reaction Conditions
White crystals form immediately on addition of the formaldehyde solution. The flask is cooled in an ice bath.

Isolation of Product
The solid product is collected by vacuum filtration using a Hirsch funnel.(■)

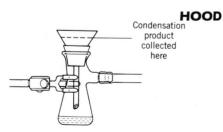

HOOD
Condensation
product
collected
here

H₂0, ~ 15 mL

Purification and Characterization

The material is recrystallized from a mixture of toluene–cyclohexanone (~2 : 1) using a Craig tube. After drying the isolated material, determine the melting point and compare your value to that recorded in the literature.

ENVIRONMENTAL DATA

Substance	Amount	TLV (mg/m³)	Emissions (mg)	Volume (m³)
Experiment 30A				
Toluene (SKIN) (Max. solv., distill.)	6 mL	375	5200	13.9
Ethanol (Max., Craig)	1.0 mL	1900	800	0.4
Diethyl carbonate	324 mg			
o-Hydroxyacetophenone	150 mg			
Experiment 30B				
Toluene (SKIN) (Max., Craig)	1.0 mL	375	870	2.3
Cyclohexanone (Max., Craig)	0.5 mL	100	500	5.0
Formaldehyde (Max.)	200 mg	1.5	200	133

TLV: Hydrochloric acid—5 ppm

TXDS: o-Hydroxyacetophenone—ipt-mus LD50: 100 mg/kg
       Diethyl carbonate—suc-rat LD50: 8500 mg/kg
       4-Hydroxycoumarin—ipr-mus LD50: 2000 mg/kg
       Dicoumarol—orl-rat LD50: 710 mg/kg

QUESTIONS

6-107. In the first step of the preceding reaction, formation of 4-hydroxycoumarin, what is the purpose of using the Hickman still to remove the ethanol generated in the reaction?

6-108. The Dieckmann condensation is an intramolecular Claisen condensation. For example,

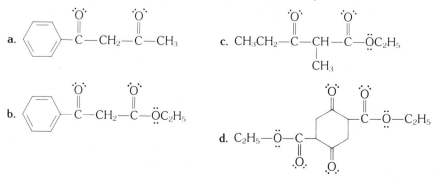

Suggest a suitable mechanism to account for the formation of the cyclic β-ketoester.

6-109. Predict the product formed in each of the following reactions.

a. Acetophenone + diethyl carbonate $\xrightarrow[\text{2. H}^+]{\text{1. NaH, toluene}}$

b. Acetophenone + ethyl formate $\xrightarrow[\text{2. H}^+]{\text{1. C}_2\text{H}_5\text{ONa, ethanol}}$

6-110. Suggest a synthesis for each of the following compounds utilizing the Claisen condensation. Any necessary organic or inorganic reagents may be used.

a. C₆H₅—C(=O)—CH₂—C(=O)—CH₃

b. C₆H₅—C(=O)—CH₂—C(=O)—OC₂H₅

c. CH₃CH₂—C(=O)—CH(CH₃)—C(=O)—ÖC₂H₅

d. C₂H₅—Ö—C(=O)—[cyclohexane ring with C(=O)—O—C₂H₅ substituent]—C(=O)—Ö

6-111. Draw structures for the possible monoenols of 1,3-cyclohexanone.

REFERENCES
1. A review of the Claisen condensation is given in Hauser, C. R.; Swamer, F. W.; Adams, J. T. *Org. React.* **1954,** *8,* 59.
2. Selected references of Claisen condensations from *Organic Syntheses:*
  a. Marvel, C. S.; Dreger, E. E. *Organic Syntheses;* Wiley: New York, 1941; Collect. Vol. I, p 238.
  b. Snyder, H. R.; Brooks, L. A.; Shapero, S. H. *Organic Syntheses;* Wiley: New York, 1943; Collect. Vol. II, p 531.
  c. Riegel, E. R.; Zwilgmeyer, F. *Ibid.,* p 126.
  d. Magnani, A.; McElvain, S. M. *Organic Syntheses;* Wiley: New York, 1955; Collect. Vol. III, p 251.
  e. Ainswoth, C. *Organic Syntheses;* Wiley: New York, 1963; Collect. Vol. IV, p 536.
  f. John, J. P.; Swaminathan, S.; Venkataramani, P. S. *Organic Syntheses;* Wiley: New York, 1973; Collect. Vol. V, p 747.

Experiment 31   Diazonium Coupling Reaction: Methyl Red

(azobenzene, 2-carboxy, 4'-*N,N*-dimethylamino-)

The generation of a diazonium salt and its use in the coupling reaction to produce an azo dye are demonstrated in this experiment.

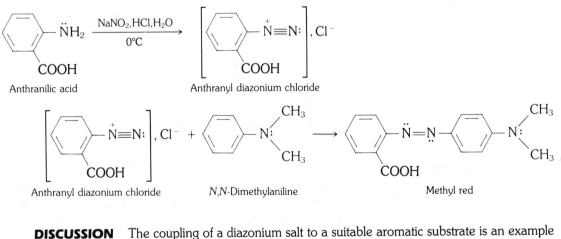

Anthranilic acid

Anthranyl diazonium chloride

Anthranyl diazonium chloride

N,N-Dimethylaniline

Methyl red

DISCUSSION The coupling of a diazonium salt to a suitable aromatic substrate is an example of an electrophilic substitution reaction on an aromatic ring. When aromatic (and also aliphatic) primary amines are treated with nitrous acid ($NaNO_2$ + HCl), they are converted into diazonium salts. In solution, nitrous acid is in equilibrium with its anhydride, dinitrogen trioxide (N_2O_3). It is this anhydride that is the actual diazotizing agent. The primary amine reacts with the dinitrogen trioxide to form a nitrosamine.

$$Ar—NH_2 + N_2O_3 \rightarrow Ar—NH—N{=}O + HONO$$
(a nitrosamine)

The nitrosamine is in equilibrium with its tautomer, a diazoic acid. The diazoic acid then undergoes dehydration to form the diazonium salt.

$$Ar—\ddot{N}H—\ddot{N}=\ddot{O}: + H_3O:^+ \rightleftharpoons Ar—\ddot{N}=\ddot{N}—\ddot{O}H + H_3O:^+$$
$$\text{(a diazoic acid)}$$

$$Ar—\ddot{N}=\ddot{N}—\ddot{O}H + H_3O:^+ \rightarrow Ar—\overset{+}{\ddot{N}}\equiv N: + 2\,H_2\ddot{O}:$$
$$\text{Diazonium ion}$$

Diazonium salts are explosive when dry and therefore are generally not isolated.

Treatment of the diazonium salt with various aromatic molecules leads to the formation of azo derivatives by what is generally called a "coupling reaction." This is an important series of reactions since it has led to the synthesis of a wide variety of commercially important organic azo dyes.

The mechanistic sequence of the coupling reaction is given below.

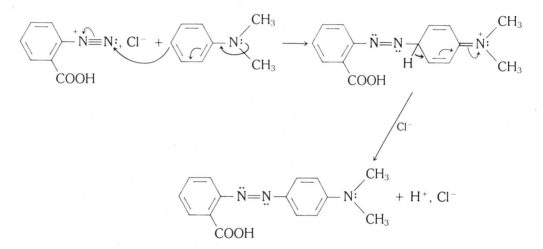

Experiment 44A describes the preparation of azobenzene by air oxidation of aniline. This is an alternate route to obtaining the azo linkage.

Anthranyl Diazonium Chloride:

The reaction is shown above.

EXPERIMENTAL   Estimated time for the completion of the experiment: 3.0 hours.

Physical Properties of Reactants and Products

Compound	MW	Wt/Vol	mmol	mp(°C)
Anthranilic acid	137.14	65 mg	0.48	146–147
Conc. HCl		150 μL		
Water		800 μL		
Sodium nitrite	69.0	36 mg	0.52	271
Anthranyl diazonium chloride		(not isolated)		

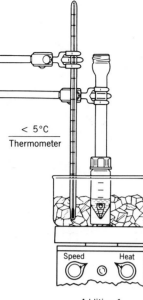

< 5°C
Thermometer

Speed   Heat

Addition 1
Anthranilic acid, 65 mg
+ conc. HCl, 150 μL
+ H₂O, 600 μL
+ NaNO₂, 36 mg
Addition 2
C₆H₅N(CH₃)₂, 85 mg
Addition 3
CH₃CO₂Na, 68 mg
+ H₂O, 200 μL

> **CAUTION:** *When dry, benzene diazonium 2-carboxylate detonates violently on being scraped or heated. It is strongly recommended that it be kept in solution at all times.*

Reagents and Equipment

Equip a 3.0-mL conical vial with a magnetic spin vane and an air condenser.(■) Add 65 mg (0.48 mmol) of anthranilic acid. A solution of 150 μL of concentrated HCl dissolved in 400 μL of water is then added to the vial, using a Pasteur pipet.

Note. *When preparing the acid solution, the acid is added to the water. These reagents are dispensed using automatic delivery pipets.*

If necessary, warm the mixture, with stirring, on a hot plate magnetic stirrer to obtain a homogeneous solution. The solution is then cooled in an ice bath, with stirring, for a period of 10 minutes.

In a 10 × 75-mm test tube or a small vial, prepare a solution of 36 mg (0.52 mmol) of sodium nitrite dissolved in 200 μL of water. This solution is also cooled in an ice bath.

Reaction Conditions

When both solutions in the ice bath are cooled to a temperature below 5°C, slowly add (dropwise) the nitrite solution to the stirred anthranilic acid solution while maintaining the temperature below 5°C. This transfer is accomplished using a Pasteur pipet. *The solution must be kept cool so that the diazonium salt will not hydrolyze to the corresponding phenol.*

After a period of 4–5 minutes, check the clear solution of anthranyl diazonium chloride for the presence of excess nitrous acid using potassium iodide–starch paper. If an excess is present, the test paper will give an immediate blue color. If no color is obtained, additional nitrite solution should be prepared and added until a positive test is observed.

The Coupling Reaction to Yield Methyl Red

Physical Properties of Reactants and Products

Compound	MW	Wt/Vol	mmol	mp(°C)	bp(°C)	Density	n_D
Anthranyl diazonium chloride solution (prepared above)							
N,N-Dimethylaniline	121.18	89 μL	0.7		194	0.96	1.5582
Sodium acetate	82.03	68 mg	0.83	324			
10% NaOH solution		100 μL					
Methyl red	269.24			183			

Reaction Conditions

To the solution of anthranyl diazonium chloride just prepared, add, fairly rapidly, 89 μL (85 mg, 0.7 mmol) of *N,N*-dimethylaniline (automatic delivery pipet).

HOOD

> **WARNING:** *This aniline compound is toxic and should be dispensed in the hood. The air condenser is removed when the addition is made; it is then reattached to the vial.*

The solution is stirred for an additional 15 minutes, keeping the temperature below 5°C.

A solution of 68 mg (0.83 mmol) of sodium acetate dissolved in 200 μL of water is prepared in a 10 × 75-mm test tube and transferred (Pasteur pipet) to the reaction mixture. This addition may be made without removing the air condenser. The resulting solution is maintained at 5°C, with stirring, for an additional 20 minutes.

The reaction vial is removed from the ice bath and allowed to warm to ambient temperature over a period of 15 minutes.

To the solution is added 100 μL of 10% aqueous NaOH solution (automatic delivery pipet). The reaction mixture is allowed to stand at room temperature for about 30 minutes. *The formation of the azo compound is a very slow reaction, but the rate is increased by raising the pH of the solution.*

Isolation of Product

The precipitate of crude methyl red dye is collected by vacuum filtration using a Hirsch funnel. (■) The reaction flask is rinsed with 0.5 mL of water and this rinse used to wash the crystals. The crystals are then washed with 0.5 mL of 3 *M* acetic acid, to remove unreacted *N,N*-dimethylaniline from the product, followed by 0.5 mL of water. This last wash is usually pale pink.

The small amounts of water and acetic acid are dispensed using a calibrated Pasteur pipet.

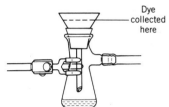

Dye collected here

Aqueous acetic acid, ~ 2.5 mL + reaction by-products

Purification and Characterization

The crude product is dissolved in 500 μL of methanol. If necessary, the mixture is warmed in a beaker of hot water to aid in the dissolution. The solution is cooled in an ice bath and the resulting crystals of methyl red are collected by vacuum filtration using a Hirsch funnel. They are then allowed to dry on a piece of filter paper.

Weigh the product and calculate the percentage yield. Determine the melting point and compare it with the literature value. If further purification is desired, the material may be recrystallized from toluene using a Craig tube.

ENVIRONMENTAL DATA

Substance	Amount	TLV (mg/m^3)	Emissions (mg)	Volume (m^3)
N,N-Dimethylaniline (SKIN) (transfer)	85 mg	25	2	<0.1
Methanol (SKIN) (Max., vac. filtr.)	0.5 mL	260	400	1.5

TLV: Hydrochloric acid—5 ppm
Sodium hydroxide—2 mg/m^3

TXDS: Anthranilic acid—orl-rat LD50: 4549 mg/kg
Methyl red—orl-rat TDLo: 12 g/kg/57W-C
Sodium nitrite—orl-hum TDLo: 85 mg/kg/56W-C
Sodium acetate—orl-rat LD50: 3530 mg/kg

QUESTIONS

6-112. In the experiment, a point is made that the formation of the azo compound is a slow reaction, but that the rate is increased by raising the pH of the solution. Why is this necessary? In other words, how does the acidity of the solution affect the reactivity of the *N,N*-dimethylaniline reagent?

6-113. In relation to Question 6-112, diazonium salts couple with phenols in slightly alkaline solution. What effect does the alkalinity of the solution have on the reactivity of the phenol?

6-114. Starting with the appropriate aromatic amine and using any other organic or inorganic reagent, outline a synthetic sequence for the preparation of the azo dyes shown below (a, b, c).

a.

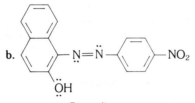

(Chrysoidine)

b.

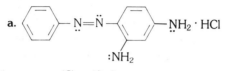

(Para red)

c.

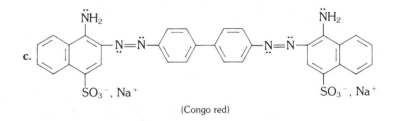

(Congo red)

6-115. What is the main feature of the azo dyes that causes them to be colored compounds?

6-116. Methyl orange is an acid–base indicator. In dilute solution at pH > 4.4, it is yellow.

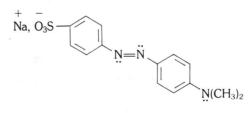

At a pH of 3.2 the solution appears red. Draw a structure of the species that is formed at the lower pH if the acid proton adds to the azo nitrogen atom adjacent to the aromatic ring containing the —SO_3^- unit. Why does the proton add to this particular N atom when two other nitrogen atoms are available in the molecule?

REFERENCES **1.** Selected coupling reactions with diazonium salts from *Organic Syntheses:*
   a. Fieser, L. F. *Organic Syntheses;* Wiley: New York, 1943; Collect. Vol. II, p 35.
   b. *Ibid.,* p 39.
   c. Hartwell, J. L.; Fieser, L. F. *Ibid.,* p 145.
   d. Conant, J. B.; Lutz, R. E.; Corson, B. B. *Organic Syntheses;* Wiley: New York, 1941; Collect. Vol. I, p 49.
   e. Santurri, P.; Robbins, F.; Stubbins, R. *Organic Syntheses;* Wiley: New York, 1973; Collect. Vol. V, p 341.

2. The synthesis of methyl red is also given in *Organic Syntheses:* Clarke, H. T.; Kirrer, W. R. *Organic Syntheses;* Wiley: New York, 1941; Collect. Vol. I, p 374.

3. The present experiment is an adaptation of that given in Vogel, A. I. *A Textbook of Practical Organic Chemistry,* 4th ed.; Longmans: London, 1978; p 716.

Experiment 32

Friedel–Crafts Acetylation: Monoacetylferrocene and Diacetylferrocene

(iron, bis[η^5-cyclopentadienyl]-1-acetyl;   iron, bis[η^5-cyclopentadienyl]-1,1'-diacetyl)

This reaction illustrates the multiple-electrophilic acyl cation attack on an aromatic nucleus. The reaction can be carried out under a variety of conditions that lead to different product ratios. The highly colored acyl derivatives are easily separated and purified by both thin-layer and dry-column chromatography.

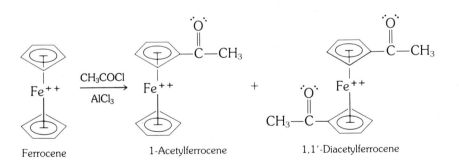

Ferrocene   1-Acetylferrocene   1,1'-Diacetylferrocene

DISCUSSION The generation of the appropriate carbocation, carbocation complex, or acylium ion in the presence of a fully delocalized ring system can lead to alkylation or acylation of an aromatic nucleus. This set of reactions, discovered by Charles Friedel and James Crafts in 1877, originally made exclusive use of aluminum chloride as the Lewis acid catalyst. The reaction now can be initiated by a range of Lewis acids other than aluminum chloride. For example, ferric chloride, zinc chloride, boron trifluoride, and strong acids such as sulfuric, phosphoric, and hydrofluoric have been used. Alkylation is accomplished by use of haloalkanes, alcohols, or alkenes; acylation, generally with acid chlorides or anhydrides.

The present experiment demonstrates the great practical value of aliquot collection followed by subsequent TLC analysis of the product.

The mechanism involves three steps: (1) ionization or partial ionization with complex formation of an alkyl halide or acyl halide; (2) attack by the formed carbocation (or complex) or acylium ion on the substrate delocalized ring; (3) loss of a proton by the arenium ion to return to a fully delocalized ring system. Ring deactivation on substitution can result in single product formation during acylation or, as in the case of ferrocene, in heteroannular substitution.

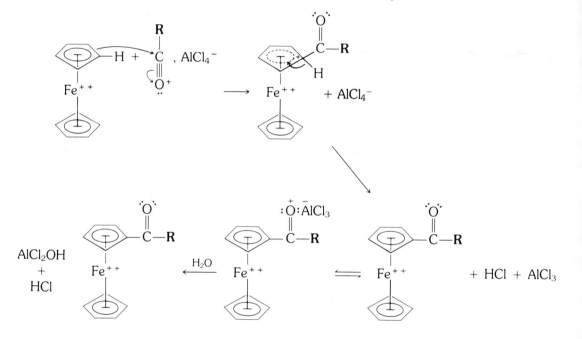

Acylium ion

EXPERIMENTAL Estimated time to complete the experiment: two 3-hour laboratory periods.

Physical Properties of Reactants and Products

Compound	MW	Wt/Vol	mmol	mp(°C)	bp(°C)	Density	n_D
Aluminum chloride	133.34	150 mg	1.12	190			
Methylene chloride		3.5 mL			40		
Acetyl chloride	78.50	80 μL	1.12		51	1.11	1.3898
Ferrocene	186.04	100 mg	0.54	173			
Monoacetyl ferrocene	228.08			85–86			
Diacetyl ferrocene	270.11			130–131			

Reagents and Equipment
Equip a tared 5.0-mL conical vial with a magnetic spin vane and a Claisen head protected by a calcium chloride drying tube and a septum cap. Add 150 mg (1.12 mmol) of fresh, anhydrous aluminum chloride.(■)

Note. *Prior to the addition of the AlCl₃, the glassware is dried in an oven at 110°C for one-half hour.*

Using a calibrated Pasteur pipet, add 2.0 mL of methylene chloride to the reaction vial. With swirling, add 80 μL (1.12 mmol) of acetyl chloride from an automatic delivery pipet **[hood].** To the resulting mixture, by use of a syringe, is added a solution of 100 mg (0.54 mmol) of ferrocene dissolved in 1.5 mL of methylene chloride.

Note. *Use a capped vial and recap it between the addition of each reagent. After addition of the acetyl chloride, the vial is attached to the Claisen head and the ferrocene solution added through the septum as shown earlier. This is done in one or two portions depending whether a 1- or 2-mL syringe is used.*

At this stage the reaction mixture turns a deep violet color.

> **WARNING:** *It is important to minimize the exposure to moist air during these transfers. Both the aluminum chloride and the acetyl chloride are highly moisture sensitive so that rapid yet accurate manipulations are necessary to minimize deactivation of these reagents and hence variable results. In addition, both chemicals are irritants. Avoid breathing the vapors or allowing them to come in contact with skin. Dispense these reagents in the hood.*

Several drops of the resulting reaction solution are now removed by Pasteur pipet and placed in a capped vial. Mark and *save*; this is to be used for TLC analysis.

Instructions
The technique for obtaining a TLC aliquot is to remove a small amount of the reaction mixture by touching the open end of a Pasteur pipet to the surface of the solution. This is done by removing the cap from the straight neck of the Claisen head and inserting the pipet down the neck so as to touch the surface of the solution. Dissolve this aliquot in 10 drops of *cold* methylene chloride in a small capped vial.

Reaction Conditions
After the addition of the ferrocene solution, note the time and initiate stirring. The reaction is allowed to proceed at room temperature for a period of 15 minutes.

Isolation of Product
The reaction is quenched by transferring the mixture by Pasteur pipet to a 12-mL capped centrifuge tube (or a 15-mL screw-capped vial) containing 5.0 mL of ice water. The tube is cooled in an ice bath. The resulting solution is neutralized by the dropwise addition (calibrated Pasteur pipet) of ~0.5 mL of 25% aqueous sodium hydroxide solution.

AlCl₃, 150 mg
+ CH₂Cl₂, 3.5 mL
+ CH₃COCl, 80 μL
+ ferrocene, 100 mg

Important. *Avoid an excess of base. Use litmus or pH paper to confirm neutrality.*

The mixture is now extracted with three 3-mL portions of methylene chloride. The capped tube is shaken and vented, and the layers are allowed to separate (a Vortex mixer may be used in this step). The bottom methylene chloride layer is then removed using a Pasteur filter pipet. The methylene chloride extracts are combined in a 25-mL Erlenmeyer flask and dried over ~200 mg of granular anhydrous sodium sulfate for 20 minutes. The dried solution is transferred to a tared 10-mL Erlenmeyer flask using a Pasteur filter pipet in aliquots of 4 mL.

HOOD After each transfer, the solution is concentrated **[hood]** under a stream of dry nitrogen gas in a warm sand bath to a volume of about 0.5 mL. The drying agent is rinsed with an additional 2.0 mL of methylene chloride and the rinse combined with the concentrate. Several drops of this solution are removed by Pasteur pipet and placed in a capped vial containing 10 drops of *cold* methylene

SAVE chloride. Mark and *save*; this is to be used for TLC analysis.

HOOD The remaining solvent is then removed by warming in a sand bath **[hood]** to yield the crude, solid product (~130 mg). Weigh the residue.

Option. *As an option, the combined extracts and wash may be left to evapo-*
HOOD *rate* **[hood]** *in a 25-mL Erlenmeyer flask with the mouth covered by filter paper until the following week.*

If the reaction is performed over a two-week period, this is a convenient point at which to stop. However, if time permits, the TLC analysis might also be performed.

Thin-Layer Chromatographic Analysis

The two aliquot samples saved earlier are analyzed by TLC. A standard mixture of the substituted ferrocenes (supplied by the instructor) is also analyzed. The developed TLC plates are used as a guide to determine the product mixture obtained in the reaction and as an aid in determining the elution solvents required for separation of the mixture by dry-column chromatography.

Information. *Thin-layer chromatography (TLC) in this experiment is carried out with Eastman Kodak silica gel–polyethylene terephthalate plates (#13179). The plates are activated at an oven temperature of 100°C for 30 minutes. They are then placed in a desiccator for cooling and stored until used. Development of the ferrocene derivatives is carried out with pure methylene chloride solvent. Visualization of unreacted ferrocene can be enhanced with iodine vapor. Each plate is approximately 1 × 4 in. Mark the plate lightly with a pencil 1.5 cm from the bottom. Using fine capillaries drawn from melting point tubes, apply the samples to the plates by spotting the marked origin one to three times. The developing chamber is a screw-capped jar or beaker (aluminum foil cover) in which 5 mL of methylene chloride gives a solvent depth of ~0.5 cm on the plate. See Chapter 5, Technique 8, for the method of TLC analysis and determination of R_f values.*

Purification and Characterization

The reaction products formed in the reaction are now purified by dry-column chromatography. *See Chapter 5, Technique 8, for a discussion of dry-column chromatography.*

The solid product residue isolated earlier is dissolved in 0.5 mL (calibrated Pasteur pipet) of methylene chloride in a small vial. This solution is mixed with 300 mg of alumina (activity III) in a tared vial and the slurry evaporated under a
HOOD stream of dry nitrogen **[hood]** to give a product–alumina mixture. A chromato-

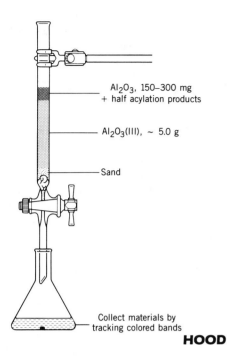

Al₂O₃, 150–300 mg
+ half acylation products

Al₂O₃(III), ~ 5.0 g

Sand

Collect materials by
tracking colored bands

HOOD

graphic column is assembled in ascending order: prewashed cotton plug, 5 mm of sand, 60–80 mm of alumina (~5.0 g, activity III), *one-half* of product–alumina mixture, and 10 mm of alumina.(■)

Information. *This procedure prevents overloading of the chromatographic column during the separation of the reaction products. If the yield of crude reaction products exceeds 75 mg (the usual case), half of the alumina admixture is introduced to the column. If the crude products, however, are obtained in quantities less than 75 mg, the entire admixture is added to the top of the column.*

If only half of the alumina–crude ferrocene acylation product mixture is on the column, it is important to reweigh the tared vial to establish a reasonably accurate estimate of overall yields obtained in the reaction.

Elution is commenced with pure hexane if TLC analysis indicated that unreacted ferrocene is present in the product mixture. The material will separate into two or three colored bands on the column during elution. Once the ferrocene band has been collected, the elution is continued with 1 : 1 and then 9 : 1 CH_2Cl_2–hexane to obtain the monosubstituted product. Further elution with 9 : 1 CH_2Cl_2–CH_3OH yields the disubstituted material. Collect and save each band separately. Remove the solvent **[hood]** under a stream of dry nitrogen gas using a warm sand bath. The volume of an eluted fraction should be in the range 2–5 mL if the band is carefully tracked down the column. Solvent that elutes without color can be discarded into the solvent waste can. During concentration of the solvent, spot each fraction on a TLC plate to verify the separation and purity.

Determine the melting point of each of the isolated products and compare your results with the literature values.

Obtain an IR spectrum of each material and compare the results with that of an authentic sample or with those spectra given in the literature. Interpretation of the spectra allows an unambiguous determination of substitution based on the presence or absence of absorption in the 1100–900 cm⁻¹ region of the spectrum.

Characterization of Fractions (total sample) Isolated from Reaction Workup

Fraction 1, Monoacetylferrocene: mp _____°C; _____ mg, _____ mmol

Fraction 2, Diacetylferrocene: mp _____ °C; _____ mg, _____ mmol

Total: _____ mmol, _____ % yield

ENVIRONMENTAL DATA

Substance	Amount	TLV (mg/m³)	Emissions (mg)	Volume (m³)
Acetyl chlorideª (Max., react.)	80 μL	10.3	88	8.6
Methylene chloride (Max., solv., evap.)	15 mL	350	20,025	57
Methylene chloride (Max., col. chromatog.)	5.5 mL	350	7,340	21
Hexane (Max., col. chromatog.)	5.0 mL	180	3,295	18
Methanol (Max., col. chromatog.)	1.0 mL	260	790	3.0

ª Treated as HCl and acetic acid.

TLV: Ferrocene—10 mg/m³
    Sodium hydroxide—2 μg/m³

TXDS: Aluminum chloride—orl-rat LD50: 3700 mg/kg
    Sodium sulfate—orl-mus LD50: 5989 mg/kg
    Acetylferrocene—inv-mus LD50: 75 mg/kg

QUESTIONS

6-117. In the formation of diacetylferrocene, the product is always the one in which each ring is monoacetylated. Why does the second acetyl group become linked to the unsubstituted ring and not the ring already containing one acetyl unit?

6-118. Ferrocene cannot be nitrated using the conventional HNO_3–H_2SO_4 mixed acid conditions even though nitration is an electrophilic aromatic substitution reaction. Explain.

6-119. In contrast to nitration, ferrocene undergoes the acetylation and sulfonation reaction. Explain.

6-120. The bonding in ferrocene involves sharing of the six electrons from each cyclopentadienyl ring with the iron atom. Based on the electronic configuration of the iron species in the compound, show that a favorable 18-electron rare gas configuration is established.

6-121. In a manner similar to that in Question 6-120, would you predict that ruthenocene and osmocene would be stable compounds containing two cyclopentadienyl rings? Explain.

6-122. Would you predict that bis(benzene)chromium(0) would be a stable compound? Give evidence to support your answer.

6-123. Predict the major product(s) in each of the following Friedel–Crafts reactions. Name each product. If the reaction does not occur, offer a reasonable explanation for that fact.

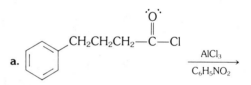

a.

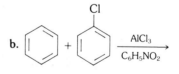

b.

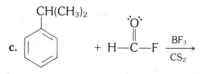

c.

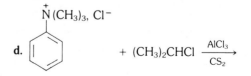

d.

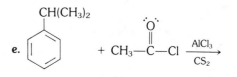

e.

REFERENCES

1. The acetylation of ferrocene has been monitored using chromatographic techniques. Several references are listed.
   a. Bozak, R. E. J. *J. Chem. Educ.* **1966,** *43,* 73.
   b. Herz, J. E. *Ibid.,* p 599.
   c. Bohen, J. M.; Joullie, M. M.; Kaplan, F. A. *J. Chem. Educ.* **1973,** *50,* 367.
2. Several reviews on the Friedel–Crafts reaction, selected from the many cited in the literature:
   a. Gore, P. H. *Chem. Rev.* **1955,** *55,* 229.
   b. Johnson, W. S. *Org. React.* **1944,** *2,* 114.
   c. Berliner, E. *Org. React.* **1949,** *5,* 229.
   d. Olah, G. A., Ed. *Friedel–Crafts and Related Reactions;* Interscience: New York, 1963–1965; Vols. I–IV.
3. Several of the large number of Friedel–Crafts acetylation reactions given in *Organic Syntheses:*
   a. Adams, R.; Noller, C. R. *Organic Syntheses;* Wiley: New York, 1941; Collect. Vol. I, p 109.
   b. Marvel, C. S.; Sperry, W. M. *Ibid.,* p 95.
   c. Fieser, L. F. *Ibid.,* p 517.
   d. Fieser, L. F. *Organic Syntheses;* Wiley: New York, 1955; Collect. Vol. III, p 6.
   e. Lutz, R. E. *Ibid.,* p 248.

Experiment 33

Halogenation: 4-Bromoacetanilide

(acetamide, *N*-(4-bromophenyl)-)

In this experiment an electrophilic aromatic substitution is described. This type of reaction constitutes a very common route by which many aromatic compounds are prepared.

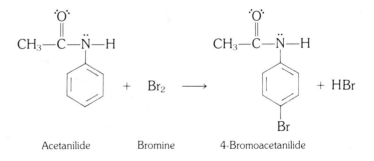

Acetanilide   Bromine   4-Bromoacetanilide

DISCUSSION   Aromatic compounds may be brominated by treatment with bromine in the presence of a Lewis acid catalyst such as ferric chloride. For very active substrates, such as amines, the reaction may proceed in the absence of a catalyst. In many cases, with amines or phenols, it is impossible to stop the bromination and all possible ortho and para positions are substituted. For this reason, primary aromatic amines are often converted to the corresponding acetanilide derivative, if the monosubstituted compound is desired. This effect is demonstrated in the present experiment. Furthermore, acetylation of the primary amine, aniline, effectively blocks the ortho positions due to steric hindrance. Electrophilic substitution by bromine is thus directed to the para position on the ring. The —NHCOCH$_3$ group is a less powerful ortho, para directing group than —NH$_2$ due to the presence of the carbonyl. Thus, substitution of the amino group renders the ring less nucleophilic. For these reasons only monosubstitution is observed.

The mechanism of the reaction is an illustration of the classic electrophilic substitution sequence on an aromatic ring. The mechanism shown below is presented as proceeding without the aid of a catalyst.

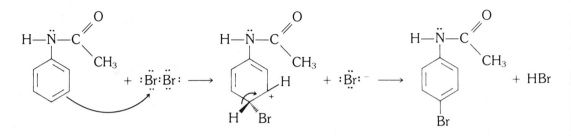

EXPERIMENTAL Estimated time for completion of the experiment: 1.5 hours.

Physical Properties of Reactants and Products

Compound	MW	Wt/Vol	mmol	mp(°C)	bp(°C)
Acetanilide	135.17	25 mg	0.19	114	
Glacial acetic acid	60.05	4 drops			118
Br_2–acetic acid reagent		3 drops			
4-Bromoacetanilide	214.08			168	

Reagents and Equipment

In a 3.0-mL conical vial fitted with a cap is placed 25 mg (0.19 mmol) of acetanilide to which are added 4 drops of glacial acetic acid using a medicine dropper. Stirring with a glass rod may be necessary to help dissolve the acetanilide. To the clear solution are added 3 drops of bromine–acetic acid reagent **[hood]**. The vial is immediately capped.

HOOD

HOOD

> **WARNING: *Bromine is a severe irritant. It is suggested that plastic gloves be worn since Br_2 burns require extended periods of time to heal. Dispense only in the hood. The reagent is prepared by mixing 2.5 mL of liquid bromine with 5.0 mL of glacial acetic acid.***

Reaction Conditions

The reddish-brown solution is allowed to stand at room temperature for 10 minutes with intermittent shaking. During this period yellow-orange crystals deposit from the solution.

Isolation of Product

Add 0.5 mL of water (calibrated Pasteur pipet) to the reaction mixture with swirling, followed by 5 drops of aqueous sodium bisulfite solution (33%). This treatment discharges the residual color, due to the presence of unreacted bromine, and results in the formation of white crystals. The reaction mixture is cooled in an ice bath for 10 minutes to maximize the product yield.

The white crystals of 4-bromoacetanilide are collected by vacuum filtration using a Hirsch funnel.(■) The filter cake is washed with three 0.25-mL portions of cold water (calibrated Pasteur pipet) and dried by drawing air through the crystals under reduced pressure for approximately 5 minutes.

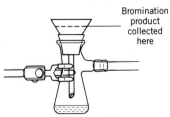

Bromination product collected here

Aqueous acetic acid, ~ 2.0 mL + other reaction by-products

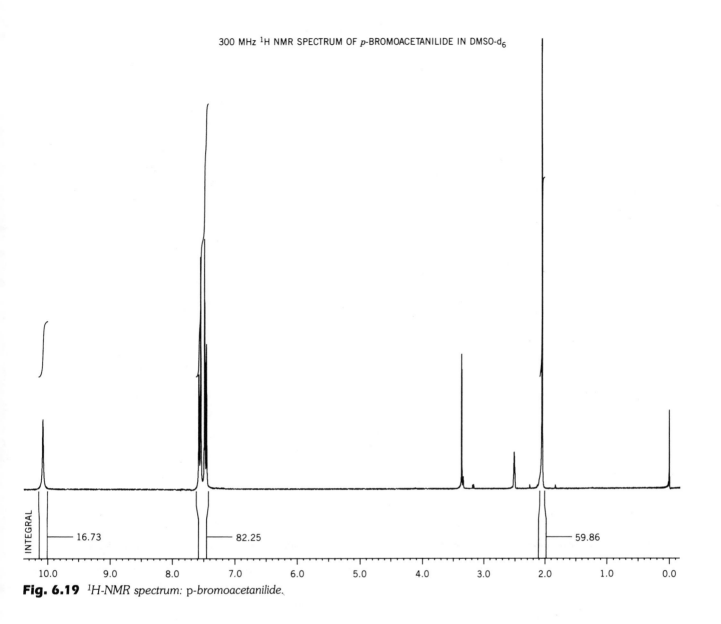

300 MHz ^{1}H NMR SPECTRUM OF *p*-BROMOACETANILIDE IN DMSO-d_6

Fig. 6.19 ^{1}H-NMR spectrum: p-bromoacetanilide.

Isolation and Characterization

The crude 4-bromoacetanilide is purified by recrystallization from 95% ethanol using the Craig tube.

Weigh the dried product and calculate the percentage yield. Determine the melting point and compare your result with the literature value.

Obtain an IR spectrum of the material and compare it with that recorded in the literature.

Nuclear Magnetic Resonance Analysis

Figures 6.19 and 6.20 are the ^{1}H and ^{13}C NMR spectra of *p*-bromoacetanilide in DMSO-d_6. These can be used to compare with NMR spectra you may obtain on your product.

In the ^{13}C spectrum, the DMSO-d_6 appears as a septet at 39.7 ppm. The resonance from the methyl group occurs at 24 ppm and the amide carbonyl carbon resonates at 169 ppm. The carbons of the benzene ring are observed between 110 and 140 ppm.

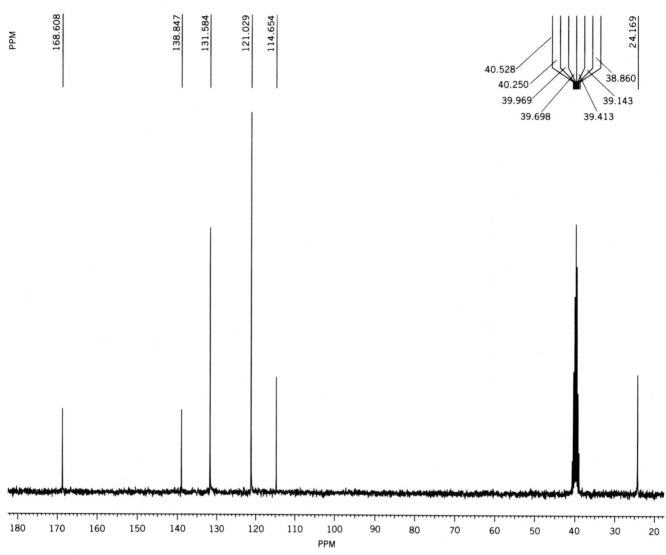

Fig. 6.20 ^{13}C-NMR spectrum: p-bromoacetanilide.

In the ^{1}H spectrum, the peak from trace amounts of DMSO-d_5 is seen at about 2.6 ppm. The peak at 3.4 ppm is probably due to water or another impurity in the sample. Note the two small peaks located equidistant to the tall singlet near 2.0 ppm. The small "satellite" peaks are the result of the 1.1% of the methyl groups that have ^{13}C instead of ^{12}C, and are thus coupled to the carbon to which the protons are attached. The amide NH proton, which is probably hydrogen bound to the basic (Lewis) sulfoxide functionality in DMSO-d_6, occurs rather downfield, near 10.1 ppm. This chemical shift may vary from your sample due to subtle differences in concentration, temperature, and moisture content of your DMSO-d_6.

Chemical Tests

Chemical classification tests may also be performed on the amide product. The ignition and the Beilstein test (Chapter 7) are used to confirm the presence of the aromatic ring and the halogen group, respectively. Does the hydroxamate test for amides (Chapter 7) give a positive result?

ENVIRONMENTAL DATA

Substance	Amount		TLV (mg/m³)	Emissions (mg)	Volume (m³)
Acetic acid (transfer)	350	mg	25	22	0.9
Bromine (Max.)	170	mg	0.7	170	243
Ethanol (Max., Craig)	0.5	mL	1900	400	0.2

TLV: Sodium bisulfite: 5 mg/m³

TXDS: Acetanilide—orl-rat LD50: 800 mg/kg
     4-Bromoacetanilide—ipr-mus LD50: 250 mg/kg

QUESTIONS

6-124. Using resonance structures show why the group shown is a less powerful ortho, para directing group than the —NH₂ unit.

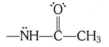

6-125. Benzene is brominated in the presence of FeBr₃ catalyst.

Suggest an appropriate mechanism for this reaction.

6-126. Draw the structure of the major monobromo products formed when each of the following compounds reacts with Br₂ in the presence of FeBr₃.

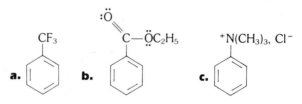

6-127. Arrange the compounds given below in order of increasing reactivity toward electrophilic substitution. Explain the reason(s) for your order.

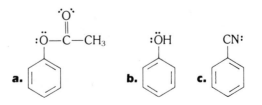

6-128. In the experiment (Isolation of Product), sodium bisulfite solution is added to discharge the unreacted bromine reagent. What reaction is occurring in this step? Is HSO_3^- acting as an oxidizing or reducing agent? Write a balanced equation as part of your answer.

6-129. Both the ¹H and ¹³C NMR spectra provide evidence that the bromination of acetanilide gave exclusively para bromination. Explain.

REFERENCES   The references listed below are selected from a large number of examples given in *Organic Syntheses* that illustrate electrophilic aromatic substitution using bromine.

a. Johnson, J. R.; Sandborn, L. T. *Organic Syntheses*; Wiley: New York, 1941; Collect. Vol. I, p 111.
b. Langley, W. D. *Ibid.*, p 127.
c. Adams, R.; Marvel, C. S. *Ibid.*, p 128.
d. Smith, L. I. *Organic Syntheses*; Wiley: New York, 1943; Collect. Vol. II, p 95.
e. Sandin, R. B.; McKee, R. A. *Ibid.*, p 100.
f. Hartman, W. W.; Dickey, J. B. *Ibid.*, p 173.
g. Coleman, C. H.; Talbot, W. F. *Ibid.*, p 592.
h. Wilson, C. V. *Ibid.*, p 575.

Experiments 34A, 34B, 34C, 34D

Nitration: 2,5-Dichloronitrobenzene; *N,N'*-Diacetyl-2,3-dinitro-1,4-phenylenediamine; 5-Nitrosalicylic Acid; *o*- and *p*-Nitrophenol

(benzene,1,4-dichloro-2-nitro-; acetamide, *N,N'*-(2,3-dinitro-*p*-phenylene)bis-; salicylic acid, 5-nitro-; phenol, 2-nitro-; phenol, 4-nitro-)

The reactions described below provide several examples of the nitration of aromatic compounds using two different nitration methods. The first uses anhydrous nitric acid and the second, a nitrated silica gel reagent.

DISCUSSION   The nitration reactions described in this experiment demonstrate one of the classic electrophilic aromatic substitution reactions. Nitration has been used extensively as a means of introducing the NO_2 substituent on an aromatic ring, which, by reduction, may be converted to the amino group. In Parts A, B, and C, anhydrous nitric acid is used in place of the usual HNO_3–H_2SO_4 mixture. In Part D, nitration is accomplished using a SiO_2–HNO_3 substrate.

The nitro group is deactivating and thus nitration usually stops at the monosubstituted product. An exception occurs when activating groups are present. This fact can be seen by comparing the results of the nitration of 1,4-dichlorobenzene (Part A) with that of *N,N'*-diacetyl-1,4-phenylenediamine (Part B). Because of the presence of the activating acetylamino groups, the dinitro derivative forms readily. In Part C (the preparation of 5-nitrosalicylic acid) the directive influence of COOH and OH substituents on the entering NO_2 group is illustrated. In fact, these two groups complement each other since they both direct the entering nitro group to the 5 position. The 5 position is favored since the COOH group is *meta* directing; the OH group is *ortho, para* directing. The nitro group prefers to attack the 5 position and not the 3 position due to steric effects.

The use of a silica gel substrate to accomplish nitration under fairly mild conditions is shown in Part D. The nitrating reagent, SiO_2–HNO_3, is prepared by treatment of silica gel with nitric acid. In the experiment, phenol is nitrated to produce a mixture of products. *Thin-layer chromatography* is used to analyze the mixture and the ortho- and para-nitrated phenols are separated by *column chromatography* using a silica gel column. If unreacted phenol is detected in the TLC analysis, an extraction technique is used to separate it from the para isomer. This separation is based on the fact that the nitrated phenol is a stronger acid than phenol itself.

It is generally accepted that the nitronium ion, NO_2^+, is the electrophilic species that attacks the aromatic ring. The overall mechanism for nitration is given below.

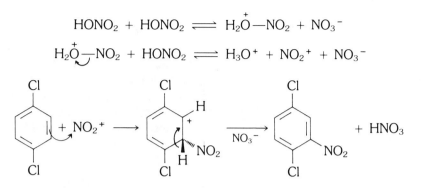

Preparation of Anhydrous HNO₃

HOOD

> **CAUTION:** *The reagents and the product of this preparation are highly corrosive. The distillation should be conducted in a hood. Prevent contact with eyes, skin, and clothing. Any spill should be neutralized using solid sodium carbonate or bicarbonate.*

EXPERIMENTAL   Estimated time of the preparation: 0.5 hour.

Important. *This acid reagent should be used immediately for the nitration experiments given below. The amount obtained is sufficient for the preparation of two or three of the nitro compounds.*

Physical Properties of Reactants and Products

Compound	MW	Wt/Vol	bp(°C)	Density
Conc. nitric acid (68%)		0.7 mL	120.5	1.41
Conc. sulfuric acid (96–98%)		1 mL	338	1.84
Anhydrous nitric acid	63.01		83	1.40

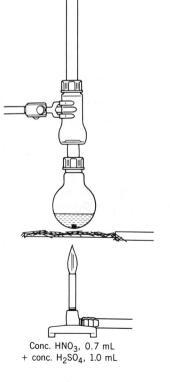

Conc. HNO₃, 0.7 mL
+ conc. H₂SO₄, 1.0 mL

Reagents and Equipment
Add 0.7 mL of concentrated nitric acid to a 10-mL round-bottom flask containing a boiling stone and equipped with a Hickman still fitted with an air condenser. Add 1.0 mL of concentrated sulfuric acid and swirl the assembly gently to mix the reagents.(■)

> **CAUTION:** *Sulfuric acid can cause severe burns. Nitric acid is a strong oxidizing agent. Prevent contact with eyes, skin, and clothing. A spill can be neutralized using sodium carbonate or bicarbonate. The acid additions are made using two clean, dry 1.0-mL graduated pipets.*

Reaction Conditions

The acid solution is heated very gently with a microburner, keeping the flame in constant motion, until approximately 0.2 mL of anhydrous nitric acid has been collected as the distillate in the collar of the still.

Purification and Characterization

The anhydrous nitric acid is used as collected. No further purification is required.

Anhydrous nitric acid (white fuming) is a colorless liquid, bp 83°C. It is estimated that the purity of the nitric acid obtained in this preparation is 99.5–100%. If it is necessary to store the distillate, the acid is removed from the collar of the still (Pasteur pipet) and placed in a 1.0-mL conical vial fitted with a glass stopper. It may be necessary to slightly bend the end of the pipet in a flame so that it can reach the collar of the still. The distillate is colorless or faintly yellow.

PART A: 2,5-Dichloronitrobenzene

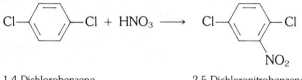

1,4-Dichlorobenzene   2,5-Dichloronitrobenzene

EXPERIMENTAL   Estimated time to complete the experiment: 0.5 hour.

Physical Properties of Reactants and Products

Compound	MW	Wt/Vol	mmol	mp(°C)	bp(°C)	Density
1,4-Dichlorobenzene	147.01	38 mg	0.26	53		
Anhydrous nitric acid	63.01	100 μL	2.4		83	1.50
2,5-Dichloronitrobenzene	192.00			56		

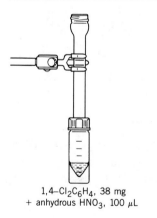

1,4–$Cl_2C_6H_4$, 38 mg
+ anhydrous HNO_3, 100 μL

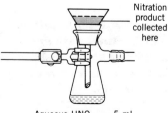

Nitration product collected here

Aqueous HNO_3, ~ 5 mL

Reagents and Equipment

Equip a 3.0-mL conical vial with an air condenser.(■) Add 38 mg (0.26 mmol) of 1,4-dichlorobenzene, followed by 100 μL of anhydrous nitric acid delivered from a calibrated Pasteur pipet (23 cm).

> **CAUTION: *The nitric acid reagent is highly corrosive. Prevent contact with eyes, skin, and clothing. A spill is neutralized using solid sodium carbonate or bicarbonate.***

Reaction Conditions

The resulting solution is allowed to stand at room temperature for a period of 15 minutes. Next add, while stirring with a thin glass rod, 1.0 mL of water (calibrated Pasteur pipet), and then place the vial in an ice bath to cool.

Isolation of Product

The crystalline precipitate is collected by vacuum filtration using a Hirsch funnel.(■) Wash the filter cake with four 1.0-mL portions of water (calibrated Pasteur pipet) and then place it on a porous clay plate or filter paper to dry.

Purification and Characterization

The product, consisting of fine, white needles, is sufficiently pure for characterization. It may be recrystallized from ethanol—water using a Craig tube if desired.

Weigh the 2,5-dichloronitrobenzene and calculate the percentage yield. Determine the melting point and compare your result with the literature value.

Note that the starting dichlorobenzene and the nitrated product have melting points 3° apart. It is recommended that a mixture melting point be carried out to establish that the desired product has been isolated (see Chapter 4).

Chemical Tests

Additional chemical tests (Chapter 7) may also be performed to further characterize the product. Does the ignition test confirm the presence of the aromatic ring? Does the Beilstein test detect the presence of chlorine? Can the sodium fusion test be used to detect the presence of nitrogen? Can the presence of the nitro group be detected by reaction with ferrous hydroxide solution?

PART B: *N,N'*-Diacetyl-2,3-dinitro-1,4-phenylenediamine

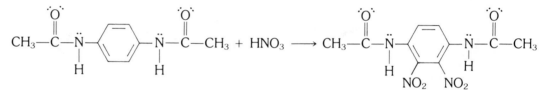

N,N'-Diacetyl-*p*-phenylenediamine   2,3-Dinitro-*N,N'*-diacetyl-*p*-phenylenediamine

EXPERIMENTAL   Estimated time to complete the experiment: 0.5 hour.

Physical Properties of Reactants and Products

Compound	MW	Wt/Vol	mmol	mp(°C)	bp(°C)	Density
N,N'-Diacetyl-1,4-phenylenediamine	192	48 mg	0.25	312–315		
Anhydrous nitric acid	63.01	100 μL	2.4		83	1.50
N,N'-Diacetyl-2,3-dinitro-1,4-phenylenediamine	282			257		

1,4–(CH₃CONH)₂C₆H₄, 48 mg
+ anhydrous HNO₃, 100 μL

CAUTION

EXOTHERMIC

Reagents and Equipment

To a 3.0-mL conical vial equipped with an air condenser add 48 mg (0.25 mmol) of *N,N*-diacetyl-1,4-phenylenediamine followed by the dropwise addition **[caution]** of 100 μL of anhydrous nitric acid from a calibrated Pasteur pipet.(■) *The N,N'-diacetyl-1,4-phenylenediamine is prepared by the procedure outlined in Experiment 26B.*

> **CAUTION: *The reaction is highly exothermic. A vigorous reaction occurs if the acid is added too rapidly. The nitric acid reagent is highly corrosive; prevent contact with eyes, skin, and clothing. A spill is neutralized using sodium carbonate or bicarbonate.***

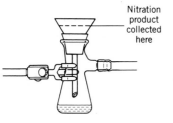

Aqueous HNO_3, ~ 5 mL

Reaction Conditions

The resulting solution is allowed to stand at room temperature for a period of 10 minutes. Water (1.0 mL) is added (calibrated Pasteur pipet) and the vial then placed in an ice bath to cool.

Isolation of Product

The resulting yellow precipitate is collected by vacuum filtration using a Hirsch funnel.(■) The filter cake is washed with four 1.0-mL portions of water (calibrated Pasteur pipet) and then dried on a porous clay plate or on filter paper.

Purification and Characterization

The product needs no further purification. Weigh the dried material and calculate the percentage yield. Determine the melting point and compare your result with the literature value. Obtain an IR spectrum of your product and compare it with that shown in Figure 6.21.

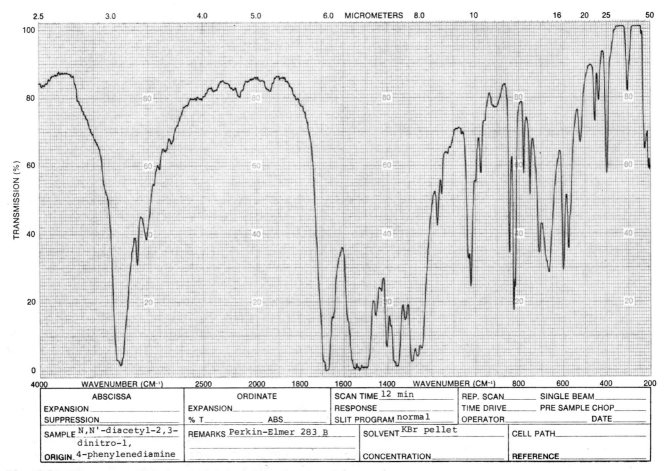

Fig. 6.21 *IR spectrum: N,N'-diacetyl-2,3-dinitro-1,4-phenylenediamine.*

PART C: 5-Nitrosalicylic Acid

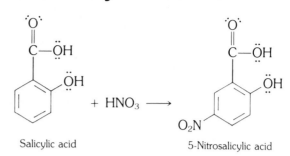

Salicylic acid   5-Nitrosalicylic acid

EXPERIMENTAL   Estimated time to complete the experiment: 0.5 hour.

Physical Properties of Reactants and Products

Compound	MW	Wt/Vol	mmol	mp(°C)	bp(°C)	Density
Salicylic acid	138.12	50 mg	0.36	159		
Anhydrous nitric acid	63.01	100 μL	2.4		83	1.5
5-Nitrosalicylic acid	183.12			229–230		

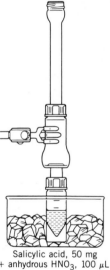

Salicylic acid, 50 mg
+ anhydrous HNO₃, 100 μL

This material may be used to prepare the caffeine 5-nitrosalicylate derivative (see Experiment 11).

HOOD

> **CAUTION: *This reaction should be conducted in a hood.***

To a 3.0-mL conical vial equipped with an air condenser, add 50 mg (0.36 mmol) of salicylic acid and place the vial in an ice bath to cool.(■) In addition, 100 μL of anhydrous nitric acid (calibrated Pasteur pipet) contained in a stoppered conical vial is also placed in an ice bath to cool.

> **CAUTION: *The nitric acid reagent is highly corrosive; prevent contact with eyes, skin, and clothing. A spill is neutralized using solid sodium carbonate or bicarbonate.***

Reaction Conditions

CAUTION   Add the cold nitric acid dropwise (Pasteur pipet) **[caution]** to the salicylic acid. The vial containing the salicylic acid is kept in the ice bath during the addition.

EXOTHERMIC

> **CAUTION: *The reaction is highly* exothermic. *A very vigorous reaction occurs if the acid is added too rapidly.***

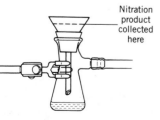

Nitration
product
collected
here

Aqueous HNO₃, ~ 5 mL

The evolution of a red-brown gas (NO₂) is observed during the addition, if the acid is not pure. Allow the vial to stand in the ice bath for an additional 20 minutes, after which 1.0 mL of distilled water is added dropwise (calibrated Pasteur pipet) to the reaction mixture.

Isolation of Product

The orange-pink solid is collected by vacuum filtration using a Hirsch funnel.(■) The filter cake is washed with four 1.0-mL portions of cold water (calibrated Pasteur pipet) and dried on a porous clay plate.

Purification and Characterization

The product is recrystallized using a Craig tube by dissolving the material in the *minimum* amount of absolute ethanol, followed by the dropwise addition of water until precipitation occurs. Cool the mixture in an ice bath and separate the light yellow crystals. They are dried on a porous clay plate or on filter paper.

Weigh the crystals and calculate the percentage yield. Determine the melting point and compare your value with that reported in the literature. Obtain an IR spectrum of the material and compare it with that recorded in the literature.

PART D: *o*- and *p*-Nitrophenol

Preparation of the SiO₂–HNO₃ Reagent

In a 250-mL Erlenmeyer flask containing a magnetic stirring bar is placed 20.0 g of silica gel (70–230 mesh; removal of fines is not necessary). Fifty milliliters of 7.5 N nitric acid is then added and the mixture stirred for 3 hours at room temperature. The nitrated silica gel is removed by gravity filtration (do not rinse), placed on a clay plate, and air-dried in a hood overnight. The product is stored in an airtight container.

The nitric acid content of the silica gel is determined by titration of a water suspension of the gel with 0.1 N NaOH solution.

The acid content of the gel should be in the range 16–20%.

Nitration of Phenol

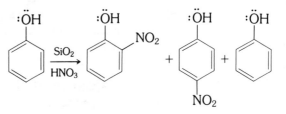

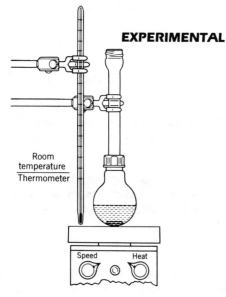

Room
temperature
Thermometer

10-mL RB flask

EXPERIMENTAL   Estimated time to complete the experiment: 2.0 hours.

Physical Properties of Reactants and Products

Compound	MW	Wt/Vol	mmol	mp(°C)	bp(°C)
Phenol	94.11	240 mg	2.55	40.5–41.5	182
Methylene chloride		5 mL			
Nitrated SiO₂ (16%)		1 g			
2-Nitrophenol	139.11			44.9	
4-Nitrophenol	139.11			114	

Reagents and Equipment

To a 10-mL round-bottom flask containing a stir bar and equipped with an air condenser are added 240 mg (2.55 mmol) of phenol and 5 mL of methylene chloride. To this solution is added 1.0 g of nitrated silica gel (~16%).(■)

HOOD

> CAUTION: *Phenol is highly toxic and corrosive. Prevent contact with eyes and skin. It is best dispensed by warming the container of phenol in a warm water bath and then using an automatic delivery pipet. This should be done in the hood.*

Reaction Conditions

The resulting mixture is stirred at room temperature for a period of 5 minutes.

Isolation of Product

The silica gel is separated from the reaction mixture by gravity filtration through a filter containing a small plug of glass wool. The filtrate is collected in a 10-mL Erlenmeyer flask containing a boiling stone. The collected silica gel is washed with two 0.5-mL portions of methylene chloride and the washings are collected in the same flask.

HOOD   The filtrate is then concentrated to a volume of ~1.0 mL using a warm sand bath under a slow stream of nitrogen **[hood]**.

Thin-layer chromatography may now be used to analyze the product mixture (see Chapter 5 Technique 8). Methylene chloride is used as the developing solvent, silica gel (fluorescent indicator) as the stationary phase, and UV light for visualization. R_f values: 4-nitrophenol, 0.04; phenol, 0.15; 2-nitrophenol, 0.58.

Note. *2,4-dinitrophenol has an R_f value of 0.33; it is not usually formed under these conditions.*

Characterization and Purification

Column chromatography is now used to separate the mixture of products. A 1.0-cm-diameter column is packed (dry) with 7.5 g of activated silica gel. The product solution is placed on the column using a Pasteur pipet and the column then eluted with 25 mL of 60/40 methylene chloride–hexane solvent. The first 20 mL of eluate is collected in a tared 25-mL Erlenmeyer flask. Concentration of this fraction to dryness in a warm sand bath under a slow stream of nitrogen yields 2-nitrophenol. Weigh the product. *Thin-layer chromatography is used to determine the purity of this material.*

The column is now eluted with ~30 mL of 1:1 ethyl acetate–methylene chloride solvent. Two or three 3-mL fractions are collected and checked using *thin-layer chromatography.* Once the 2-nitrophenol has eluted completely, the remaining elute is collected in a tared 25-mL Erlenmeyer flask containing a boiling stone. This fraction is concentrated as done earlier to a volume of ~2 mL. *Thin-layer chromatography is used to determine the purity of the final product* (see conditions outlined above). 4-Nitrophenol is the main constituent of this fraction. However, the presence of unreacted phenol or possibly quinone is often detected. The product solution is now concentrated to dryness and the weight of 4-nitrophenol recorded.

If the TLC analysis indicates the 4-nitrophenol to be impure, it may be purified as follows.

The solid residue is dissolved in 7 mL of saturated sodium bicarbonate solution and transferred to a 12-mL centrifuge tube. The resulting solution is extracted twice with 2-mL portions of methylene chloride (a Vortex mixer may be used to good advantage). The methylene chloride layer is removed using a Pasteur filter pipet.

The resulting aqueous solution is cooled in an ice bath and 6 *M* hydrochloric acid added dropwise with mixing (glass rod or Vortex mixer) until neutral or slightly acidic toward litmus or pH paper. *This step must be done carefully. Too vigorous a reaction may result in loss of product.* The resulting solution is now extracted with three 2-ml portions of methylene chloride. After each extraction, the methylene chloride layer is removed using a Pasteur filter pipet and transferred to a 10-mL Erlenmeyer flask. The wet solution is dried over anhydrous sodium sulfate and the dried solution transferred by Pasteur filter pipet to a tared 10-ml Erlenmeyer flask containing a boiling stone. The solution is concentrated using a warm sand bath under a slow stream of nitrogen gas to yield the 4-nitrophenol product. Weigh the product. The purity of the material may be checked again using TLC as outlined above. It may also be recrystallized from water using the Craig tube, if necessary.

To characterize the 2- and 4-nitrophenols, determine their melting points and obtain the IR spectra. Compare your results with those reported in the literature.

Chemical Tests

Chemical classification tests (see Chapter 7) are also of value in establishing the identity of these compounds. Perform the ferric chloride test for phenols. Is a positive result obtained for each compound? Does the sodium fusion test detect the presence of nitrogen? The test for nitro groups might also be performed.

The phenyl or α-naphthylurethane derivatives of the phenols may be prepared to further establish their identity (see Chapter 7, Preparation of Derivatives).

ENVIRONMENTAL DATA

Substance	Amount	TLV (mg/m³)	Emissions (mg)	Volume (m³)
Experiment 34				
Nitric acid (conc.) (30 minutes in open container)	0.7 mL	5	200	40
Sulfuric acid (conc.) (estimated)	1.0 mL	1	70	70
Experiment 34A				
Nitric acid (anhy.) (open container)	100 μL	5	90	18
Ethanol (Max., recryst.)	1.0 mL	1900	800	0.4
Experiment 34B				
Nitric acid (anhy.) (open container)	100 μL	5	90	18
Experiment 34C				
Nitric acid (anhy.) (open container)	100 μL	5	90	18
Ethanol (Max., recryst.)	0.5 mL	1900	400	0.2
Experiment 34D				
Methylene chloride (Max.)	16 mL	350	21,300	61
Methylene chloride (Max., col. chormatog.)	30 mL	350	40,000	114
Hexane (Max., col. chromatog.)	10 mL	180	6,300	35
Ethyl acetate (Max., col. chromatog.)	15 mL	1400	13,500	9.6

TLV: 1,4-Dichlorobenzene—75 ppm

TXDS: Nitric acid (white fuming)—inl-rat LC50: 244 ppm (NO₂)/30 *M*, 5 mg/m³
    Sulfuric acid (conc.)—unk-man LDLo: 135 mg/kg, 1 mg/m³
    2,5-Dichloronitrobenzene—orl-rat LD50: 1210 mg/kg
    Salicylic acid—orl-rat LD50: 891 mg/kg
    Phenol—orl-hmn LDLo: 140 mg/kg
    2-Nitrophenol—orl-rat LD50: 2828 mg/kg
    4-Nitrophenol—orl-rat LD50: 350 mg/kg

QUESTIONS   **6-130.** Predict the position most likely to be taken by the incoming NO_2^+ ion in the mononitration of each of the following compounds. Explain the reasons for your choice.

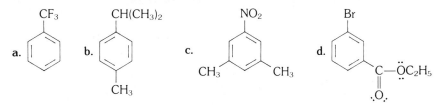

6-131. Write equations to show how nitronium ions might be formed using a mixture of nitric and sulfuric acids.

6-132. Which ring of phenyl benzoate would you expect to undergo nitration more readily? Explain.

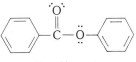

Phenyl benzoate

6-133. Arrange the following compounds in order of increasing reactivity toward nitration. Give reasons for your order.

acetanilide   acetophenone   bromobenzene   toluene

6-134. Offer a reasonable explanation of why nitration of 1,4-dichlorobenzene yields the mononitro derivative while N,N'-diacetyl-1,4-phenylenediamine forms the dinitro compound.

6-135. Explain the fact that p-nitrophenol is a stronger acid than phenol itself. Would p-methoxyphenol be a stronger or weaker acid than phenol? Explain.

REFERENCES   **1.** References related to anhydrous nitric acid:
    a. Stern, S. A.; Mullhaupt, J. T.; Kay, W. B. *Chem. Rev.* **1960**, *60*, 185.
    b. The preparation used in this experiment is an adaptation of that reported by Cheronis, N. D.; Entrikin, J. B. *Semimicro Qualitative Organic Analysis*; Crowell: New York, 1947; p 258.
  2. The preparation of the nitrated silica gel was adapted from that reported by Tapia, R.; Torres, G.; Valderrama, J. A. *Synth. Commun.* **1986**, *16*, 681.
  3. References selected from a large number of examples of nitration given in *Organic Syntheses* (none uses anhydrous nitric acid or the nitrated silica gel as the nitrating agent):
    a. Kamm, O.; Segur, J. B. *Organic Syntheses*; Wiley: New York, 1941; Collect. Vol. I, p 372.
    b. Robertson, G. R. *Ibid.*, p 376.
    c. Corson, B. B.; Hasen, R. K. *Organic Syntheses*; Wiley: New York, 1943; Collect. Vol. II, p 434.
    d. Hartman, W. W.; Smith, L. A. *Ibid.*, p 438.
    e. Huntress, E. H.; Shriner, R. L. *Ibid.*, p 459.
    f. Kobe, K. A.; Doumani, T. F. *Organic Syntheses*; Wiley: New York, 1955; Collect. Vol. III, p 653.
    g. Fitch, H. M. *Ibid.*, p 658.
    h. Fanta, P. E.; Tarbell, D. S. *Ibid.*, p 661.
    i. Howard, J. C. *Organic Syntheses*; Wiley: New York, 1963; Collect. Vol. IV, p 42.
    j. Braum, C. E.; Cook, C. D. *Ibid.*, p 711.
    k. Fetscher, C. A. *Ibid.*, p 735.
    l. Mendenhall, G. D.; Smith, P. A. S. *Organic Syntheses*; Wiley: New York, 1973; Collect. Vol. V, p 829.
    m. Newman, M. S.; Boden, H. *Ibid.*, p 1029.

Experiment 35

Quantitative Analysis of a Grignard Reagent: Phenylmagnesium Bromide

(magnesium, bromophenyl-)

This experiment outlines a technique for the analysis of Grignard reagents. It is similar in scope to Experiment 23.

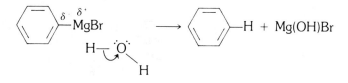

DISCUSSION For a discussion of the reaction see Experiment 23. Remember, however, that in the present experiment the hydrocarbon formed on hydrolysis of phenylmagnesium bromide is benzene.

PART A: Phenylmagnesium Bromide

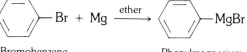

Bromobenzene   Phenylmagnesium
                                    bromide

EXPERIMENTAL   Estimated time for the experiment. 1.0 hour.

Physical Properties of Reactants and Products

Compound	MW	Wt/Vol	mmol	mp(°C)	bp(°C)	Density	n_D
Bromobenzene	157.02	76 μL	0.72		156	1.50	1.5597
Diethyl ether	74.12	800 μL			35		
Magnesium	24.31	18 mg	0.73	649			
Iodine	253.81	1 crystal					
Benzene	78.12				80	0.88	1.5011

CAUTION: *Ether is a flammable liquid and also a narcotic. All flames should be extinguished during the time of this experiment.*

Reagents and Equipment

This reagent is prepared exactly as described in Experiment 17. The reagents, amounts of reagents, order of addition, workup manipulations, and precautions are the same. The equipment is also identical.(■) The brown-gray Grignard reagent is cooled to room temperature.

PART B: Analysis of the Grignard Reagent

The titration method outlined in Experiment 23B is used for the analysis of this Grignard reagent.

ENVIRONMENTAL DATA

Grignard preparation emissions are listed in Experiment 17.

TLV: Iodine—0.1 ppm
Sodium hydroxide—2 mg/m^3

TXDS: Magnesium—orl-dog LDLo: 230 mg/kg
Sulfuric acid—unk-man LDLo: 135 mg/kg

MTDS: Bromobenzene—dnr-esc 250 mg/L

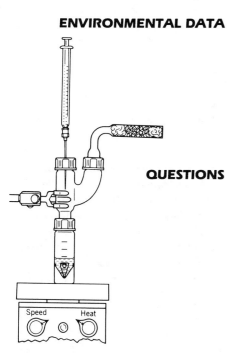

Mg, 18 mg + I$_2$, 1 crystal
+ (CH$_3$CH$_2$)$_2$O, 800 μL
+ C$_6$H$_5$Br, 76 μL

QUESTIONS

6-136. What hydrocarbon would you expect to obtain by the action of water on each of the Grignard reagents listed below?
a. Butylmagnesium bromide
b. s-Butylmagnesium bromide
c. i-Butylmagnesium bromide
d. t-Butylmagnesium bromide

6-137. What product would each of the Grignard reagents in Question 6-136 yield when treated with D$_2$O?

6-138. Consult your *Quantitative Analysis* text for the analysis of Mg^{2+} by the complexation titration method using EDTA (ethylenediaminetetraacetic acid). Do you think this technique might be adapted for the analysis of Grignard reagents? If so, outline a suitable procedure.

6-139. What product would each of the Grignard reagents listed in Question 6-136 yield when treated with ethanol? With isopropyl alcohol?

6-140. The solubility of Grignard reagents in ether plays a crucial role in their formation. The reagents are soluble because the magnesium is coordinated to the ether oxygen in a Lewis acid–base interaction. Each ether molecule donates an electron pair to the magnesium to complete an octet.

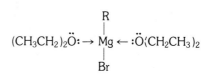

Grignard reagents are normally insoluble in hydrocarbon solvents. However, they can be rendered soluble by the addition of a tertiary amine to the hydrocarbon–Grignard reagent mixture. Explain.

REFERENCES

For references relating to the preparation of Grignard reagents, see Experiments 17 and 23.

Experiment 36

Grignard and Arene Halide Cross-Coupling Reaction: 1-Methyl-2-(methyl-d_3)-benzene

(benzene, 1-methyl-(2-methyl-d_3))

This experiment illustrates the coupling reaction of a Grignard reagent with an alkyl halide to form a C—C bond. It also illustrates the technique of "labeling" particular hydrogen atoms for identification purposes. *The use of methyl iodide in the reaction produces o-xylene. This demonstrates only the coupling reaction.*

$$CD_3I + Mg \xrightarrow{\text{ether}} D_3CMgI$$

Methyl-d_3 iodide   Methyl-d_3-
magnesium iodide

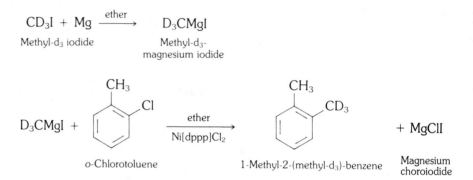

D$_3$CMgI +   *o*-Chlorotoluene $\xrightarrow[\text{Ni[dppp]Cl}_2]{\text{ether}}$ 1-Methyl-2-(methyl-d_3)-benzene   + MgClI

Magnesium
chloroiodide

DISCUSSION The selective cross-coupling of a Grignard reagent with an organic halide to produce the desired C—C bond formation is demonstrated in this reaction. The reaction is catalyzed by a phosphine–nickel catalyst and has wide application to the synthesis of unsymmetrical alkanes and alkenes. Before the discovery of nickel–phosphine activity, the cross-coupling of Grignard reagents with organic halides was seldom employed in synthetic practice. This was due to the formation of homocoupling products along with other side reactions.

The mechanism of the reaction is not fully understood at present. A discussion of the reaction is given elsewhere.[3]

EXPERIMENTAL Estimated time to complete the experiment: 5.0 hours.

Physical Properties of Reactants and Products

Compound	MW	Wt/Vol	mmol	mp(°C)	bp(°C)	Density	n_D
Magnesium	24.31	65 mg	2.7	649			
Diethyl ether	74.12	700 μL			34.5		
Methyl-d_3 iodide	144.96	250 μL	3.93		42	2.28	1.5262
Ni[dppp]Cl$_2$	540.5	10 mg	0.02				
o-Chlorotoluene	126.59	150 μL	1.28		159	1.08	1.5268
1-Methyl-2-(methyl-d_3)-benzene	109.17				144		1.5055

Reagents and Equipment

Note. *All the glassware used in the experiment should be cleaned, dried in an*

---

[3] Tamao, K.; Sumitani, K.; Kumada, M. *J. Am. Chem. Soc.* **1972,** *94,* 4374.

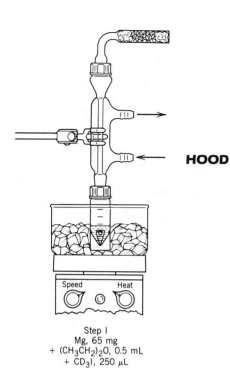

Step I
Mg, 65 mg
+ (CH_3CH_2)$_2$O, 0.5 mL
+ CD_3I, 250 μL

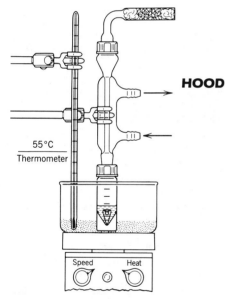

55°C
Thermometer

Step II
(C_6H_5)$_2$P(CH_2)$_3$P(C_6H_5)$_2$ · $NiCl_2$, 10 mg
+ o–$CH_3C_6H_4Cl$, 150 μL
+ (CH_3CH_2)$_2$O, 200 μL

oven at 110°C for at least 30 minutes, and then cooled in a desiccator before use.

Equip a 5.0-mL conical vial containing a magnetic spin vane with a reflux condenser protected by a calcium chloride drying tube. Add 65 mg (2.7 mmol) of magnesium and 0.3 mL of anhydrous diethyl ether.(■)

*A 4- to 5-in. piece of magnesium ribbon is scraped clean of oxide coating and cut into 1-mm-long sections. This freshly cut material is handled only with forceps. The ether is dispensed in the **hood** using an automatic delivery pipet.*

Prepare a solution of 250 μL of methyl-d_3 iodide in 200 μL of anhydrous diethyl ether in a capped vial. *These reagents may be dispensed using automatic delivery pipets.*

HOOD

CAUTION: *Methyl iodide is a suspected carcinogenic agent.*

This solution is drawn into a Pasteur pipet and transferred to the reaction flask. This is done by removing the drying tube and inserting the pipet down the length of the condenser, allowing the pipet bulb to rest on the condenser lip.

Reaction Conditions
The reaction vial is placed in an ice bath and the methyl-d_3 iodide solution is added dropwise with stirring. After the addition, the pipet is withdrawn, the drying tube reinstalled, and the mixture stirred for an additional 20 minutes.

Prepare a solution of 10 mg (0.02 mmol) of dichloro[1,3-bis(diphenylphosphino)propane]nickel(II) catalyst, 150 μL of o-chlorotoluene, and 200 μL of anhydrous diethyl ether in a capped vial. *The liquids are dispensed in the **hood** using automatic delivery pipets.*

Draw the solution into a Pasteur pipet, and in one portion add it to the reaction vial through the condenser as described above. The drying tube is reinserted as before. Place this mixture in a preheated sand bath at a temperature of 55°C and, with stirring, heat for 2 hours.(■)

On addition of the catalyst, the solution turns green. After approximately half an hour of heating, the mixture becomes dark brown.

Isolation of Product
Cool the reaction mixture in an ice bath and quench the reaction by the dropwise addition (calibrated Pasteur pipet) of 1.5 mL of 1.0 M HCL solution. *The acid is added slowly since frothing occurs.*

To the solution is added 1.0 mL of ether. Using a Pasteur filter pipet, the aqueous layer is removed. The remaining ether phase is extracted with 1.0 mL of water followed by 1.0 mL of saturated sodium bicarbonate solution and then 1.0 mL of deionized water. This is followed by extraction with two 1.0-mL portions of 1.0 N sodium thiosulfate solution and 1.0 mL of water.

The preceding amounts of extracting solutions are measured using calibrated Pasteur pipets. For each extraction operation, the vial is capped, shaken, and vented, and the layers are allowed to separate. The bottom aqueous phase is then removed.

Purification and Characterization
The reaction product is isolated and purified using column chromatography. In a modified Pasteur filter pipet place 200 mg of activated silica gel (100 mesh)

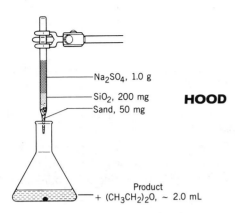

Na₂SO₄, 1.0 g

SiO₂, 200 mg
Sand, 50 mg

HOOD

Product
+ (CH₃CH₂)₂O, ~ 2.0 mL

followed by 1.0 g of anhydrous sodium sulfate. The column is wetted with 0.5 mL of ether (calibrated Pasteur pipet).(■)

The ether solution of product is transferred to the column by Pasteur pipet and the material eluted from the column with two 0.5-mL portions of ether. The eluate is collected in a tared 10-mL Erlenmeyer flask containing a boiling stone. The solvent is evaporated by warming the flask in a sand bath **[hood]** to yield the "labeled" o-xylene.

Weigh the product and calculate the percentage yield. Determine the boiling point, density, and refractive index (optional) and compare your values with those reported in the literature. Obtain an IR spectrum of the crude (dry) reaction product using the capillary film sampling technique.

In this experiment we have synthesized an isotopically labeled compound. The products of this type of synthesis are often used to obtain more detailed information about the vibrational energy levels present in a particular system, but also for many other purposes. Deuterium labeling is a particularly powerful way of investigating infrared spectra because the resultant frequency shifts (which are inversely proportional to the square root of the mass ratio; see Appendix B) are the largest obtainable with stable isotopes. In this case, o-xylene is labeled with a perdeuteromethyl group via a novel coupling reaction. The starting material is also an ortho-substituted benzene ring, so the macro group frequency trains used to characterize the reaction will be very similar.

Infrared Analysis

The spectrum of o-chlorotoluene (Fig. 6.22) nicely mimics the large macro group frequency train for an ortho-substituted alkyl benzene system, which uses the following frequency train: 3100–3000, 3000–2850, 1950, 1920, 1880, 1840, 1790, 1690, 1600, 1580, 1500, 1450, 1380, and 750 cm⁻¹.

a. 3072 & 3026 cm⁻¹: C—H stretch on aromatic ring.
b. 3000–2850 cm⁻¹: C—H stretch on alkyl substituents.
c. 1955, 1915, 1880, 1835, 1795, and 1690 cm⁻¹: combination band pattern for ortho-disubstituted rings. The fit in this case is extremely good.
d. 1596 & 1577, 1475 & 1450 cm⁻¹: two degenerate pairs of ring stretching modes. The 1450 cm⁻¹ band is overlapped by the antisymmetric methyl deformation vibrations.
e. 1382 cm⁻¹: symmetric methyl deformation mode.
f. 749 cm⁻¹: all-in-phase out-of-plane bend of four adjacent C—H groups on the aromatic ring.

The C—Cl stretch falls below the region of measurement in these aromatic systems.

The reaction replaces the C—Cl group with a —CD₃ group. Thus, the only spectral changes expected to be observed will involve the vibrations of the labeled group (Fig. 6.23). To a first approximation, the frequency shift should be by a factor of 1.41, but in practice the full shift is not observed as most vibrations are not pure modes. A factor of 1.34–1.38 is generally found. In 1-methyl-2-(methyl-d₃)-benzene, we have two sets of doublets with the doublet centers at 2220 and 2090 cm⁻¹. (Since we would expect the C—D stretching modes to give rise to only two band systems, coupling with overtones of lower lying fundamentals likely is occurring.) The corresponding C—H stretching modes also evidence some coupling. If we use the major band centers observed at 2965 and 2870 cm⁻¹, however, as the C—H stretching values, the ratios observed are 2965/2220 = 1.34 and 2870/2090 = 1.37, results that are quite reasonable to expect for this type of isotopic shift. The bending modes are moved into the cluttered fingerprint region and do not lend themselves to easy analysis.

Examine the spectrum of the reaction product you have obtained as a capillary film. Discuss the similarities and differences of the experimentally derived spectral data to the reference spectra (Figs. 6.22 and 6.23).

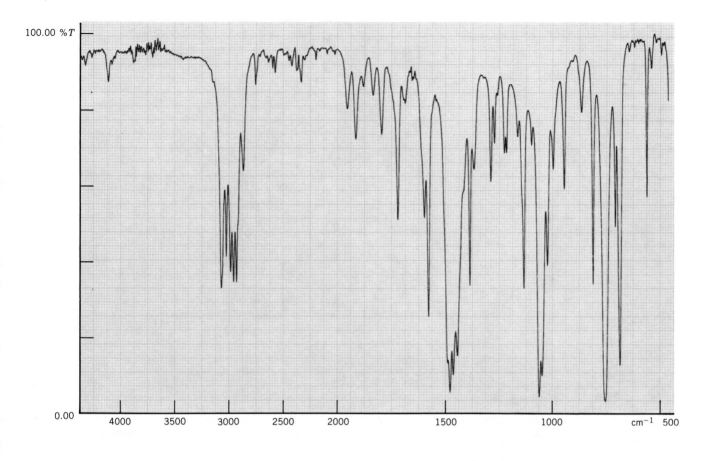

Sample	2-Chlorotoluene		
%T _X_ ABS __ Background Scans _4_		Scans __16__	
Acquisition & Calculation Time _42 sec_		Resolution _4.0 cm⁻¹_	
Sample Condition _liquid, neat_		Cell Window __KBr__	
Cell Path Length _capillary film_		Matrix Material _____	

Fig. 6.22 *IR spectrum: 2-chlorotoluene.*

ENVIRONMENTAL DATA

Substance	Amount	TLV (mg/m³)	Emissions (mg)	Volume (m³)
Methyl iodide (SKIN) (max., react., transfer, yield)	0.25 mL	10	450	47
Ether (Max., solv., col. chromatog., evap.)	3.2 mL	1200	2285	1.9
o-Chlorotoluene (Max., react., yield)	0.15 mL	250	49	0.2
Organonickel catalyst	10 mg			

TLV: o-Xylene—100 ppm
    Hydrochloric acid—5 ppm

TXDS: Magnesium—orl-dog LDLo: 230 mg/kg
    Sodium sulfate—orl-mus LD50: 5989 mg/kg
    Sodium thiosulfate—scu-rbt LDLo: 4000 mg/kg
    Sodium bicarbonate—orl-rat LD50: 4220 mg/kg
    Silica gel—inv-mus LDLo: 343 mg/kg

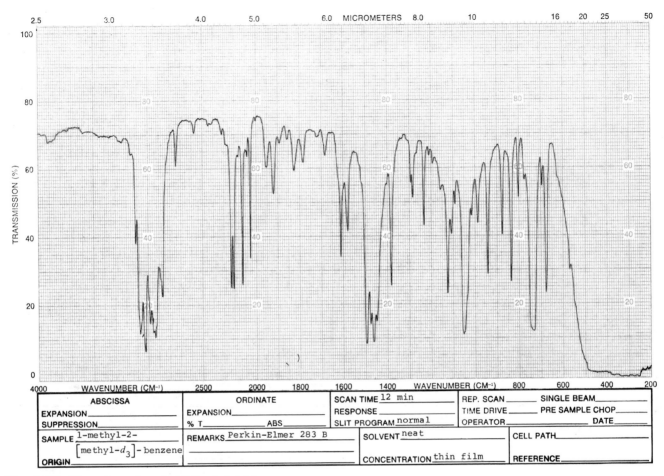

Fig. 6.23 *IR spectrum: 1-methyl-2-(methyl-d₃)-benzene.*

QUESTIONS **6-141.** Predict the reagents that could be used to prepare each of the compounds shown in the margin by the Grignard coupling reaction. Give a suitable name to each reactant.

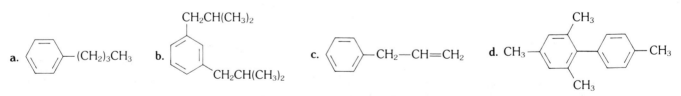

6-142. Write a structural formula for the dichloro[1,3-bis(diphenylphosphino)propane] nickel (II) catalyst used in this reaction.

6-143. The Wurtz coupling reaction involves the treatment of haloalkanes with an active metal such as sodium. What products would be formed in the following reactions? Name the products.

$$2\,\mathbf{R}\!-\!X + 2\,Na \xrightarrow[\text{solvent}]{\text{inert}} \mathbf{R}\!-\!\mathbf{R} + 2\,NaX$$

a. 2 Butyl bromide + Na →
b. 1-Bromo-3-chlorocyclobutane + Na →

6-144. In reference to Question 6-143, explain why the coupling reaction of *n*-butyl bromide and *n*-propyl bromide under conditions of the Wurtz reaction is synthetically useless for the preparation of heptane.

6-145. Deuterium has a special use in proton NMR. It is "silent." Explain.

6-146. The spectrum of *o*-chlorotoluene given in Figure 6.22 has two additional peaks that are not present in the library reference standard spectrum. Both these bands are weak. One occurs near 1714 cm^{-1} and the other at 1362 cm^{-1}. Can you offer an explanation for the presence of the extra absorption bands. How successful do you think the coupling reaction would be if carried out utilizing this sample as the starting material?

6-147. In the spectrum of *o*-chlorotoluene the lower wavenumber band of the 1596, 1577 cm^{-1} pair is the more intense member. In the labeled product, the 1610, 1580 cm^{-1} pair has the higher wavenumber peak more intense. Explain this reversal of intensities.

REFERENCES

1. For general references on Grignard reagents refer to Experiment 17.
2. The present coupling reaction is based on the work reported by Kumada, M.; Tamao, K.; Sumitani, K. *Organic Syntheses;* Wiley: New York, **1978,** *58,* p 127 and references found therein.
3. Selected Grignard coupling reactions presented in *Organic Syntheses:*
   a. Lespieau, R.; Bourguel, M. *Organic Syntheses;* Wiley: New York, 1941; Collect. Vol. I, p 186.
   b. Gilman, H.; Catlin, W. E. *Ibid.,* p 471.
   c. Gilman, H.; Robinson, J. *Organic Syntheses;* Wiley: New York, 1943; Collect. Vol. II, p 47.
   d. Smith, L. I. *Ibid.,* p 360.
   e. Turk, A.; Chanan, H. *Organic Syntheses;* Wiley: New York, 1955; Collect. Vol. III, p 121.
4. Mayo, D. W.; Bellamy, L. J.; Merklin, G. T.; Hannah, R. W. *Spectrochim. Acta* **1985,** *41A,* 355.

Experiment 37

Nucleophilic Aromatic Substitution: 2,4-Dinitrophenylthiocyanate

(thiocyanic acid, 2,4-dinitrophenyl ester)

In this experiment an example of nucleophilic substitution of an activated aromatic ring is presented. In the reaction, a bromo substituent is replaced by a thiocyanate group. The rate of the reaction is increased by the presence of a phase transfer catalyst.

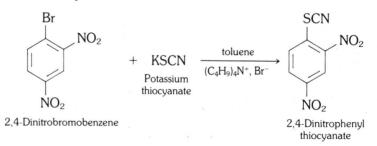

2,4-Dinitrobromobenzene   2,4-Dinitrophenyl thiocyanate

DISCUSSION   Aromatic nucleophilic substitution generally takes place only if activating groups are present on the positions ortho and/or para to the leaving group. Activating groups are those that are electron-withdrawing. Groups such as

$$-NO_2,\ -SO_2CH_3,\ -\overset{+}{N}(CH_3)_3,\ -CF_3,\ -CN,\ -SO_3^-,\ -Br,\ -Cl,\ -I,\ -CO_2^-,$$

especially if another NO_2 group is present, are strongly activating. In the reaction illustrated in this experiment, two nitro groups are present on the ring, ortho and para to the departing bromo substituent. The reaction is hindered by electron-releasing groups.

The conditions under which this reaction is conducted lend themselves to the use of phase transfer catalysis. The system involves two phases, the aqueous phase and the organic phase (toluene). The phase transfer catalyst plays a very important role. In effect, it carries the SCN^- ion (in the aqueous phase) as an ion pair, across the boundary layer into the organic phase where the reaction occurs. The product and the corresponding halide salt of the phase transfer catalyst are produced in this conversion. The halide salt then migrates back into the aqueous phase and the process repeats itself. The catalyst can play this role since the large organic groups (the four butyl groups) increase the solubility of the ion pair in the organic phase, while the charged ionic center of the salt renders this species soluble in the aqueous phase. Use of phase transfer catalysts is also seen in the preparation of ethers by the Williamson method (see Experiments 25A and 25B).

The reaction proceeds in two steps, the first step of which is generally rate determining. A tetrahedral intermediate is formed by the attack of the nucleophilic reagent on the carbon atom to which the leaving group is attached. In the subsequent step, the leaving group then departs, regenerating the aromatic nucleus. It is important to note that the reaction does not proceed by an S_N2 mechanism since an intermediate is formed. The sequence is outlined below.

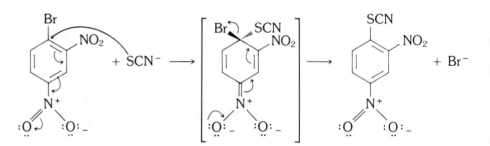

EXPERIMENTAL Estimated time to complete the reaction: 2.0 hours.

Physical Properties of Reactants and Products

Compound	MW	Wt/Vol	mmol	mp(°C)	bp(°C)
2,4-Dinitrobromobenzene	247.01	50 mg	0.20	75	
Toluene	92.15	350 μL			111
Tetrabutylammonium bromide	322.38	5 mg	0.02	103–104	
Aqueous potassium thiocyanate (50%)		150 μL			
2,4-Dinitrophenylthiocyanate	244			138–139	

Reagents and Equipment

In a 3.0-mL conical vial containing a magnetic spin vane and equipped with a reflux condenser, place 50 mg (0.20 mmol) of 2,4-dinitrobromobenzene and

350 μL of toluene.(■) To this solution add 5.0 mg (0.02 mmol) of tetrabutyl-ammonium bromide and 150 μL of a 50% aqueous potassium thiocyanate (wt/wt) solution.

> **CAUTION: *Tetrabutylammonium bromide is an irritant. HANDLE WITH CARE! It is also hygroscopic and must be protected from moisture. The toluene solvent and the thiocyanate solution are dispensed in the*** hood ***using automatic delivery pipets.***

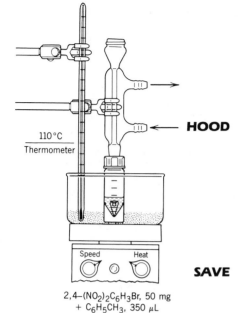

110°C
Thermometer

Speed   Heat

SAVE

2,4–$(NO_2)_2C_6H_3Br$, 50 mg
+ $C_6H_5CH_3$, 350 μL
+ $Bu_4\overset{+}{N}$, $\overset{-}{Br}$, 5 mg
+ 50% aqueous KSCN, 150 μL

HOOD

Reaction Conditions

The resulting mixture is heated at a sand bath temperature of 110°C with stirring for a period of 1.0 hour. It is then cooled to room temperature.

Isolation of Product

The toluene layer is separated from the aqueous layer (*save*) by use of a Pasteur pipet and placed in a 3.0-mL capped vial. The toluene layer is then extracted with two 1.0-mL portions of water and the water extracts are combined with the original water layer saved before. The combined aqueous layers are now extracted with two 0.5-mL portions of toluene and these toluene extracts combined with the original toluene layer.

The volumes of liquid used above are measured using calibrated Pasteur pipets. For each extraction, the mixture is shaken and vented, and the layers are allowed to separate. The transfers are made using Pasteur filter pipets.

The toluene solution is transferred by Pasteur filter pipet to a microcolumn (modified Pasteur filter pipet) containing 700 mg of anhydrous sodium sulfate previously wetted with toluene.(■) The dried toluene eluate, collected in a 10-mL Erlenmeyer flask containing a boiling stone, is concentrated by warming the solution in a sand bath under a gentle stream of nitrogen gas **[hood].** A yellow, solid product is obtained.

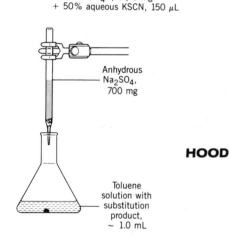

Anhydrous
Na_2SO_4,
700 mg

Toluene
solution with
substitution
product,
~ 1.0 mL

HOOD

Purification and Characterization

The product is nearly pure but may be recrystallized from chloroform, using a Craig tube, if desired.

HOOD

> **CAUTION: *Chloroform is toxic! Dispense only in the*** hood. ***Do not breath the vapors.***

Weigh the dried product and calculate the percentage yield. Determine the melting point and compare it with the literature value. Obtain an IR spectrum and compare your result with that of an authentic sample.

ENVIRONMENTAL DATA

Substance	Amount	TLV (mg/m³)	Emissions (mg)	Volume (m³)
Toluene (SKIN) (Max., evap)	1.35 mL	375	1214	3.2
Chloroform (Max., recryst.)	1.0 mL	50	1490	30

TXDS: Potassium thiocyanate—orl-hum LD50: 80 mg/kg
      Thiocyanic acid, 2,4-dinitrophenyl ester—orl-mus LD50: 2750 mg/kg

MTDS: Tetrabutylammonium bromide—dnd-mam:lyn: 50 mmol/L

QUESTIONS   **6-148.** Explain the trend in the following reactions in terms of ease of reaction.

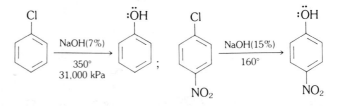

6-149. Complete the following reactions. Name the expected products.

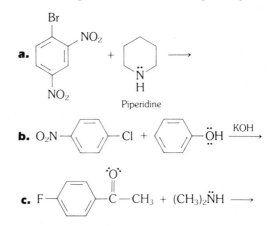

6-150. Compare the potential energy diagram of an S_N2 substitution reaction on an aliphatic halide to that of the nucleophilic substitution reaction carried out in this experiment. Discuss the main differences in the diagrams in terms of the mechanisms.

6-151. Diagram the intermediate that is formed in the aromatic nucleophilic substitution reactions given below.

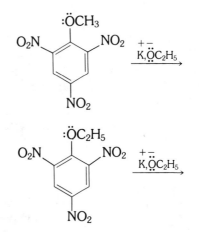

6-152. On workup of each of the reactions given in Question 6-151, what product(s) would you expect to form? If more than one, indicate their ratio.

REFERENCES   Selected examples of nucleophilic aromatic substitution reactions in *Organic Syntheses:*
a. Hartman, W. W. *Organic Syntheses;* Wiley: New York, 1941; Collect. Vol. I, p 175.
b. Reverdin, F. *Ibid.,* p 219.
c. Hartman, W. W.; Byers, J. R.; Dickey, J. B. *Organic Syntheses;* Wiley: New York, 1943; Collect. Vol. II, p. 451.
d. Brewster, R. Q.; Groening, T. *Ibid.,* p 445.

e. Skorcz, J. A.; Kuminski, F. E. *Organic Syntheses;* Wiley: New York, 1973; Collect. Vol. V, p 263.

f. Kharasch, N.; Langford, R. B. *Ibid.,* p 474.

g. Bunnett, J. F.; Connor, R. M. *Ibid.,* p 478.

h. Sahyun, M. R. V.; Cram, D. J. *Ibid.,* p 926.

Experiment 38

Halogenation Using *N*-Bromosuccinimide: 9-Bromoanthracene

(anthracene,9-bromo-)

The use of *N*-bromosuccinimide as a highly specific brominating agent is demonstrated in this experiment.

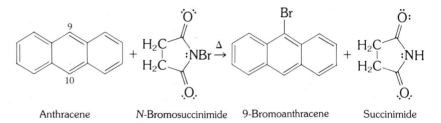

Anthracene   *N*-Bromosuccinimide   9-Bromoanthracene   Succinimide

DISCUSSION   *N*-Bromosuccinimide (NBS) is a highly specific brominating agent. Using this reagent, anthracene is brominated in the 9-position. NBS may also be used to brominate positions *alpha* to (1) a carbonyl group, (2) a triple bond, (3) an olefin (allylic position), (4) boron atoms, and (5) aromatic rings (benzylic position).

In the preparation of 9-bromoanthracene, the reaction is easily followed, since *N*-bromosuccinimide (a reactant) and succinimide (a product) are both insoluble in carbon tetrachloride. This is critical for bromination to occur when using NBS. Reactions carried out wherein the NBS is soluble in the solvent system employed yield entirely different types of products. The *N*-bromosuccinimide reactant is denser than the carbon tetrachloride solvent, and as the reaction proceeds this solid disappears from the bottom of the reaction flask and the less dense succinimide forms and floats to the surface of the reaction solution. Other polynuclear hydrocarbons that have been brominated using NBS include naphthalene, phenanthrene, and acenaphthene.

Free radicals have been implicated in the mechanism of bromination using *N*-bromosuccinimide. In fact, the reaction proceeds only under photochemical conditions, by heating or in the presence of a free-radical initiator. The NBS reagent provides a source of Br_2 in the reaction mixture. This is a result of the rapid reaction of NBS with the hydrogen bromide that is formed in one of the propagation steps (see below) of the substitution reaction.

The initiation step in bromination with NBS is the formation of a bromine radical by the homolytic dissociation of the N—Br bond in the NBS molecule itself. The bromine radical then abstracts a hydrogen from the 9-position of the anthracene species.

Propagation step

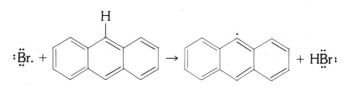

The HBr so formed then reacts with NBS to produce a bromine molecule.

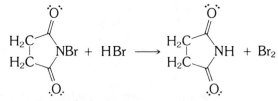

The bromine molecule then reacts with the anthracenyl radical formed above to yield the product and a bromine radical.

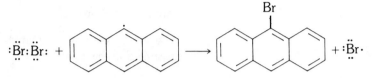

The generation of this bromine radical starts the sequence over again. That is, a chain reaction is initiated. A trace of an iodine–carbon tetrachloride solution is added to the reaction mixture. The iodine acts as a moderator or a retarder in the reaction. Thus, only the monobrominated product is formed; the 9,10-dibromoanthracene is not generated under these conditions.

Anthracene has a resonance energy of 84 kcal/mol as compared with benzene (36 kcal/mol). As can be seen, the resonance energy of anthracene is not much more than that of two benzene rings; the third ring contributes very little to the resonance stability of the system. This fact relates to the reactivity of the anthracene molecule at the 9,10-positions since aromaticity is disrupted to the least extent. For example, the Diels–Alder addition reaction (see Experiment 15) shows the ability of the molecule to act as a diene, addition occurring at the 9,10-positions. Oxidation of anthracene produces 9,10-anthraquinone. Furthermore, electrophilic aromatic substitution occurs readily at the 9-position, and if disubstitution takes place the 9,10-disubstituted product is usually obtained.

EXPERIMENTAL

Estimated time to complete the experiment: 2 hours.

Physical Properties of Reactants and Products

Compound	MW	Wt/Vol	mmol	mp(°C)	bp(°C)
Anthracene	178.24	50 mg	0.28	216	
N-Bromosuccinimide	177.99	50 mg	0.28	180–183	
Carbon tetrachloride	153.82	0.4 mL			77
Iodine–CCl₄ solution		1 drop			
9-Bromoanthracene	257.14			100–101	
Succinimide	99.09			126–127	

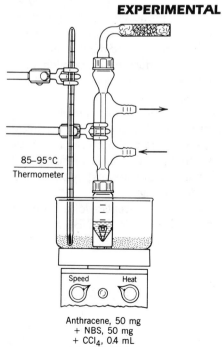

85–95°C
Thermometer

Speed   Heat

Anthracene, 50 mg
+ NBS, 50 mg
+ CCl₄, 0.4 mL
+ I₂·CCl₄, 1 drop

HOOD

Reagents and Equipment

To a 1.0-mL conical vial containing a magnetic spin vane and equipped with a reflux condenser fitted with a calcium chloride drying tube, add 50.0 mg (0.28 mmol) of anthracene and 50 mg (0.28 mmol) of N-bromosuccinimide.(■)

To this mixture add 0.4 mL of carbon tetrachloride followed by one drop of I₂–CCl₄ solution delivered from a Pasteur pipet.

> **WARNING: CCl₄ is a suspected carcinogenic agent. It is dispensed in the hood using an automatic delivery pipet.**

Instructor Preparation of I$_2$–CCl$_4$ Solution

HOOD

Iodine (0.2 g, 0.01 mol) is dissolved in 10 mL of carbon tetrachloride. The solution is placed in a **hood** for student use.

Reaction Conditions

The reaction mixture is heated, with stirring, to reflux in a sand bath (85–95°C) for a period of 1 hour. During this time the solution turns brown and crystals of succinimide appear at the surface of the reaction solution.

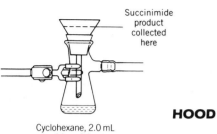

Succinimide product collected here

Cyclohexane, 2.0 mL
+ CCl$_4$, 0.4 mL
+ 9-bromoanthracene product

Isolation of Product

HOOD

The succinimide product is collected by vacuum filtration of the warm solution using a Hirsch funnel.(■) The filter cake of succinimide is washed with three or four 0.5-mL portions of cyclohexane (calibrated Pasteur pipet). The washings are combined with the original filtrate.

The filtrate is concentrated to dryness **[hood]** under reduced pressure to give yellow-green crystals of 9-bromoanthracene. *Evaporation of the solvent can be accelerated by immersing the collection flask in warm water.*

Purification and Characterization

Weigh the air-dried succinimide and calculate the percentage yield. Determine the melting point and compare your value with that in the literature. Obtain an IR spectrum and compare your spectrum with that of an authentic sample or the literature.

The crude 9-bromoanthracene is purified by recrystallization from 95% ethanol using the Craig tube. Weigh the dried product and calculate the percentage yield. Determine the melting point and compare your result with the literature value.

Chemical Tests

The ignition test should establish the presence of the aromatic nature of the substituted anthracene compound. Does the Beilstein test show the presence of bromine?

ENVIRONMENTAL DATA

Substance	Amount	TLV (mg/m^3)	Emission (mg)	Volume (m^3)
Hexane (Max., vac., evap.)	2.0 mL	180	1320	7.3
Ethanol (Max., Craig)	1.0 mL	1900	800	0.4
Carbon tetrachloride (SKIN) (reflux, open transfer up to Max.)	0.4 mL	30	150–640	5–21

TXDS: Anthracene—orl-rat TDLo: 20 g/kg
   N-bromosuccinimide—ipr-mus LDLo: 256 mg/kg
   Succinimide—orl-rat LD50: 14 g/kg

QUESTIONS **6-153.** Predict and give suitable names for the product(s) formed in the following reactions with NBS.

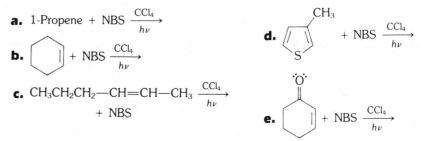

a. 1-Propene + NBS $\xrightarrow[h\nu]{CCl_4}$

b. ⬡ + NBS $\xrightarrow[h\nu]{CCl_4}$

c. CH$_3$CH$_2$CH$_2$—CH=CH—CH$_3$ $\xrightarrow[h\nu]{CCl_4}$
   + NBS

d. + NBS $\xrightarrow[h\nu]{CCl_4}$

e. + NBS $\xrightarrow[h\nu]{CCl_4}$

6-154. When 1-octene is treated with NBS three monobromo straight-chain alkenes of molecular formula $C_9H_{17}Br$ are isolated from the reaction mixture. Identify these compounds and give each a suitable name.

6-155. Benzyl bromide, $C_6H_5CH_2Br$, can be prepared by treating toluene with NBS in the presence of a peroxide initiator. Suggest a suitable mechanism to account for this reaction.

6-156. The benzyl radical

has unusual stability. Account for this fact by drawing appropriate resonance structures.

6-157. Suggest a suitable mechanism for the following reaction.

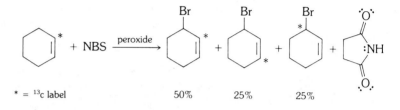

* = ^{13}c label   50%   25%   25%

6-158. The discussion indicates that other polynuclear hydrocarbons have been brominated using NBS. For the compounds shown below, give each a suitable name and draw the structure of the monosubstituted product that would form on reaction with NBS.

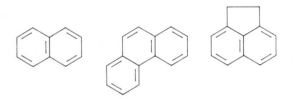

REFERENCES

1. Review articles:
   a. Djerassi, C. *Chem. Rev.* **1948,** *43,* 271.
   b. Horner, L.; Winkelmann, E. H. *Angew. Chem.* **1959,** *71,* 349.
   c. Horner, L.; Winkelmann, E. H. *Newer Methods Prep. Org. Chem.* **1964,** *3,* 151.
   d. Dauben, H. J., Jr.; McCoy, L. L. *J. Am. Chem. Soc.* **1959,** *81,* 4863.
2. Brominations in *Organic Syntheses* using NBS:
   a. Greenwood, F. L.; Kellert, M. D.; Sedlak, J. *Organic Syntheses;* Wiley: New York, 1963; Collect. Vol. IV, p 108.
   b. Campaigne, E.; Tullar, B. F. *Ibid.,* p 921.
   c. Kalir, A. *Organic Syntheses;* Wiley: New York, 1973; Collect. Vol. V, p 825.
   d. Corbin, T. F.; Hahn, R. C.; Schechter, H. *Organic Syntheses;* Wiley: New York, **1964,** *44,* 30.

Experiment 39

Hypochlorite Oxidation of an Alcohol: Cyclohexanone

(cyclohexanone)

This experiment illustrates the oxidation of a secondary alcohol to a ketone.

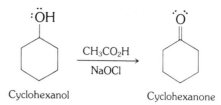

Cyclohexanol   Cyclohexanone

DISCUSSION   Sodium hypochlorite, or liquid bleach, contains 11.5–12.5% (1.8–2.0 M) available chlorine. It is used extensively in swimming pool sanitation and as a bleach in the pulp and textile industries. A less concentrated product (5%) is used in laundries and around the household. The reaction described in this experiment illustrates the use of liquid bleach (11.5–12.5%) as an oxidizing agent in the organic laboratory.

    Sodium hypochlorite is prepared commercially by passing chlorine gas through a solution of aqueous sodium hydroxide.

$$Cl_2 + Na\ddot{O}H \rightleftharpoons Na\ddot{O}Cl + NaCl$$

The actual oxidizing agent in the present experiment is the chloronium ion, Cl^+, which is reduced in the reaction to chloride ion, Cl^-. The cyclohexanol acts as a reducing agent and thus becomes oxidized to cyclohexanone.

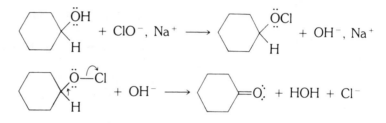

Complete details of the mechanism are not available to date.

    Distillation employing a Hickman still is used to isolate the crude cyclohexanone product from the reaction mixture. This operation is an example of the steam distillation technique (see Chapter 5, Technique 2). The crude mixture collected in the collar of the still consists of cyclohexanone, water and acetic acid. If any unreacted cyclohexanol is present in this mixture, it is removed in the subsequent chromatographic purification sequence using alumina. Gas chromatographic analysis may be used to determine the purity of the cyclohexanol product.

EXPERIMENTAL   Estimated time to complete the experiment: 3.0 hours.

Physical Properties of Reactants and Products

Compound	MW	Wt/Vol	mmol	bp(°C)	Density	n_D
Cyclohexanol	100.16	100 mg	1.0	161	0.96	1.4641
Glacial acetic acid	60.05	250 μL		118		
Sodium hypochlorite solution (~12.5% aqueous)		2 mL				
Cyclohexanone	98.15			156	0.95	1.4507

Reagents and Equipment

In a 5.0-mL conical vial containing a magnetic spin vane and equipped with a Hickman still, place 100 mg (1.0 mmol) of cyclohexanol and 250 μL of glacial

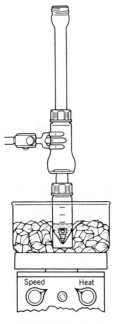

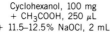

Cyclohexanol, 100 mg
+ CH$_3$COOH, 250 μL
+ 11.5–12.5% NaOCl, 2 mL

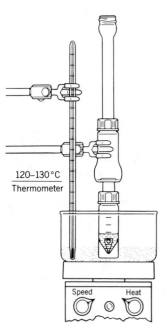

120–130 °C
Thermometer

Cyclohexanol, 100 mg
+ CH$_3$COOH, 250 μL
+ 11.5–12.5% NaOCl, 2 mL
+ NaHSO$_3$(sat. sol.) ~ 8–10 drops

HOOD acetic acid.(■) *The glacial acetic acid is dispensed in the* **hood** *by use of an automatic delivery pipet.*

The resulting solution is cooled in an ice bath. By use of a graduated pipet, there is now added dropwise, with stirring, 2.0 mL of aqueous sodium hypochlorite solution (~12.5%, 1.8–2.0 *M*). The ice bath is removed following the addition. *The NaOCl solution is added by inserting the pipet down the neck of the still just into the throat of the vial.*

Note. *Solid calcium hypochlorite (65% available chlorine) may be used as a source of chlorine in place of the sodium hypochlorite solution.*

Reaction Conditions

The resulting solution is stirred at room temperature for a period of 1 hour. *An excess of hypochlorite oxidizing agent should be maintained throughout the reaction period. The aqueous layer is monitored periodically using KI–starch paper. If a positive test is not obtained (the paper should turn blue), additional sodium hypochlorite solution (1–3 drops) is added to ensure that an excess of the oxidizing agent is present. A Pasteur pipet inserted down the neck of the still is used to add the reagent and also to remove a few drops of solution for testing.*

Isolation of Product

Using a Pasteur pipet, add saturated, aqueous sodium bisulfite solution dropwise to the reaction mixture until the solution gives a negative KI–starch test.

The crude product is now separated from the reaction mixture by steam distillation. With stirring, the mixture is heated at a sand bath temperature of 120–130°C.(■) The first 0.5–1.0 mL of distillate is collected in the ring collar of the condenser. This material is then transferred to a 3.0-mL conical vial containing a spin vane using a Pasteur pipet. The condenser collar is rinsed with 0.5 mL of diethyl ether and the rinsing is also transferred to the conical vial. *The distillate collected in the condenser collar consists of cyclohexanone, water, and acetic acid.*

To neutralize the acetic acid present in the separated product mixture, anhydrous sodium carbonate (~100 mg) is added in small portions, with stirring, to the solution until evolution of CO$_2$ gas ceases. Add 50 mg of NaCl to the mixture. The resulting two-phase system is stirred until all the solid material dissolves.

The ether layer containing the cyclohexanone product is separated from the aqueous phase using a Pasteur filter pipet and transferred to a micro drying column. The column is assembled with a Pasteur filter pipet packed first with 300 mg of alumina (activity 1), followed by 200 mg of anhydrous sodium sulfate.(■) The column is wetted with diethyl ether before the transfer.

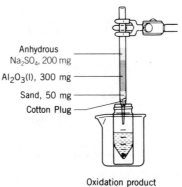

Anhydrous
Na$_2$SO$_4$, 200 mg

Al$_2$O$_3$(l), 300 mg

Sand, 50 mg

Cotton Plug

Oxidation product
in (CH$_3$CH$_2$)$_2$O, ~ 1.0 mL

The eluate is collected in a tared 3.0-mL conical vial containing a boiling stone. The aqueous layer remaining in the conical vial is extracted with three 0.5-mL portions of ether and each extract is also passed through the column and combined with the original eluate.

The 3.0-mL conical vial is fitted with an air condenser and the ether removed by gentle warming on a sand bath in the **hood.**

HOOD

Purification and Characterization

The liquid residue of cyclohexanone isolated on evaporation of the ether solvent is sufficiently pure for characterization. It may be further purified by distillation using a Hickman still or by GC (see below). Weigh the product and calculate the percentage yield.

The purity of the cyclohexanol product may be determined using gas chromatographic analysis. See Chapter 5, Technique 1 and Experiment 5A to review this technique and the experimental conditions to use. Determine the boiling point and density (optional) of your product; compare your results with the literature value.

Obtain an IR spectrum and compare it with that recorded in the literature.

Chemical Tests

Chemical classification tests may also be run to assist in the characterization of this material. The 2,4-dinitrophenylhydrazine test should give a positive result. Isolation of this derivative and the determination of its melting point would further establish the identity of the product as cyclohexanone.

ENVIRONMENTAL DATA

Substance	Amount	TLV (mg/m³)	Emissions (mg)	Volume (m³)
Cyclohexanol (Max., reactant)	100 mg	200	100	0.5
Cyclohexanone (SKIN) (Max., product)	77 mg	100	77	0.8
Acetic acid (Max., solv., distilled)	0.25 mL	25	262	10.5
Sodium hypochlorite solution (Max., treated as chlorine)	2.0 mL	3	250	83
Diethyl ether (Max., extract., evap.)	2.0 mL	1200	1171	1.0

TLV: Sodium hydrogen sulfite—5 mg/m³
    Alumina—10 mg/m³

TXDS: Sodium carbonate—orl-rat LDLo: 4000 mg/kg
    Sodium sulfate—orl-mus LD50: 5989 mg/kg

MTDS: Sodium hypochlorite—dnr-esc: 20 nL/disk

QUESTIONS

6-159. In the experiment, why is a solution of sodium bisulfite added to the reaction product mixture (Isolation of Product)? Write a reaction to account for what is happening. Is the bisulfite ion acting as an oxidizing or a reducing agent?

6-160. In the isolation of the cyclohexanone product, 50 mg of sodium chloride is added to the water–cyclohexanone–diethyl ether mixture. Explain how the addition of sodium chloride aids in isolation of the cyclohexanone product.

6-161. Predict the product(s) for each of the following oxidation reactions. Give a suitable name for each reactant and product.

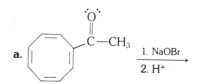

a.

b. $CH_3CH_2\ddot{O}H$ $\xrightarrow[\text{2. H}^+]{\text{1. NaOI}}$

c. $CH_3-\underset{\underset{\ddot{:}\ddot{O}H}{|}}{CH}-CH_3$ $\xrightarrow[\text{2. H}^+]{\text{1. NaOBr}}$

6-162. 2,3-Dimethyl-2,3-butanediol (*pinacol*), on heating in the presence of aqueous acid, rearranges to form 3,3-dimethyl-2-butanone (*pinacolone*). Suggest a mechanism for this reaction.

6-163. What chemical tests might be used to distinguish between pentanal and 2-pentanone? Between benzyl alcohol and diphenylmethanol?

REFERENCES   **1.** As a general reference for the use of sodium hypochlorite as an oxidizing agent see Fieser, L. F.; Fieser, M. *Reagents for Organic Synthesis;* Wiley: New York, 1967; Vol. I, p 1084.

2. The preceding experiment is based on work reported by Zuczek, N. M.; Furth, P. S. *J. Chem. Educ.* **1981,** *58,* 824. See also Stevens, R. V.; Chapman, K. T.; Weller, H. N. *J. Org. Chem.* **1980,** *45,* 2030.

Experiments 40A, 40B   **Chromium Trioxide Resin Oxidation of an Alcohol: 9-Fluorenone; Conversion to the 2,4-Dinitrophenylhydrazone**

(fluoren-9-one)

In this experiment the oxidation of a secondary (2°) alcohol to a ketone using a polymer-bound chromium trioxide oxidizing agent is demonstrated. The progress of the reaction is followed by thin-layer chromatography (TLC). The product ketone is characterized by formation of its 2,4-dinitrophenylhydrazone derivative.

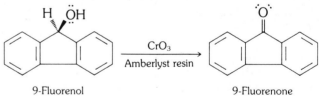

9-Fluorenol   9-Fluorenone

DISCUSSION   This experiment illustrates the oxidation of a 2° alcohol to a ketone. The oxidizing agent commonly employed for this purpose is sodium dichromate or chromic oxide in sulfuric acid. In the present case, a convenient and advantageous polymer-bound chromium trioxide reagent is used. It is not only easy to prepare, but is also easy to separate from the product mixture and can be recycled. Today, the use of polymeric reagents in organic synthesis is developing at a rapid pace. The mechanism of the oxidation is outlined below.

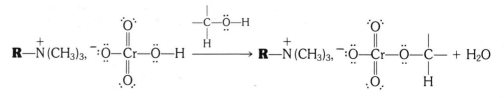

(Chromate ester)

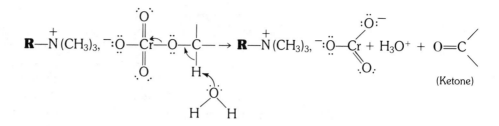

(Ketone)

It should be noted that a solution of chromic oxide in aqueous sulfuric acid is used as a test reagent for 1° and 2° alcohols. A positive test is observed when the clear orange test reagent gives a greenish opaque solution on addition of the alcohol. See Chapter 7.

PART A: 9-Fluorenone

EXPERIMENTAL   Estimated time to complete the experiment: 3–4 hours.

Physical Properties of Reactants and Products

Compound	MW	Wt/Vol	mmol	mp(°C)	bp(°C)
9-Fluorenol	182.23	100 mg	0.55	154	
Chromic oxide–resin		500 mg			
Toluene	92.15	3.5 mL			111
9-Fluorenone	180.22			84	

Reagents and Equipment
To a 10-mL round-bottom flask containing a magnetic stirring bar and equipped with a reflux condenser, add 100 mg (0.55 mmol) of 9-fluorenol and 3.5 mL of toluene.(■) The resulting solution is sampled for TLC analysis (see Chapter 5, Technique 8). Then 500 mg of the oxidizing resin is added to the flask.

Instructor Preparation
The CrO_3 resin is prepared by adding 35 g of Amberlyst A-26 resin to a solution of 15 g of CrO_3 in 100 mL of water. The mixture is stirred for 30 minutes at room temperature. The resin is then collected by vacuum filtration and successively rinsed with water and acetone. It is partially dried on the Buchner funnel by drawing air through the resin under vacuum for 1 hour and is then allowed to air-dry overnight.

Reaction Conditions
The contents of the flask are heated to reflux, with stirring, using a sand bath temperature of approximately 130°C.

Sample the solution for TLC analysis after a period of 5 minutes and every additional 15–20 minutes. In this manner, the progress of the reaction may be monitored until the conversion is complete (~35–40 minutes). This point is reached when the TLC analysis shows that the 9-fluorenol has been completely consumed.

TLC Conditions. *Eastman Kodak Fluorescent silica gel sheets (1.0 cm × 7 cm) are used. They are developed with 30% acetone in hexane and visualized by UV light. Reference R_f values for 9-fluorenol and 9-fluorenone are determined using known compounds under the conditions cited above.*

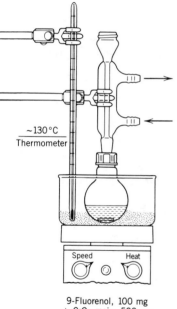

~130°C
Thermometer

Speed   Heat

9-Fluorenol, 100 mg
+ CrO_3 resin, 500 mg
+ $CH_3C_6H_5$, 3.5 mL
10-mL RB flask

Isolation of Product

After cooling the reaction mixture, the resin is removed by gravity filtration through a cotton plug placed in a small funnel. The solution is transferred to the filter funnel using a Pasteur pipet, and the reaction flask and resin are rinsed with two 1.0-mL portions of methylene chloride (calibrated Pasteur pipet). The filtrate is collected in a tared 10-mL Erlenmeyer flask containing a boiling stone. The rinse is combined with the original filtrate.

HOOD   The solvent is now removed **[hood]** from the filtrate under a stream of nitrogen by warming in a sand bath to yield a yellow residue of the crude 9-fluorenone product.

Purification and Characterization

Obtain the weight of the crude product and calculate the percentage yield. Recrystallize a 50-mg portion of the 9-fluorenone from hexane (about 1.0 mL of hexane/50 mg of ketone) using a Craig tube. Determine the melting point of this material and also calculate the percentage recovery on recrystallization.

Chemical Tests

The 2,4-dinitrophenylhydrazine test for aldehydes and ketones (see Chapter 7) may be done to confirm the presence of the carbonyl group. This hydrazone derivative may be prepared on a 38-mg scale in Part B of this experiment. Does the ignition test indicate that this compound is aromatic?

This material has a characteristic UV spectrum. The following data were obtained at a concentration of $8.66 \times 10^{-6} M$ (see Chapter 5, Technique 6).

λ_{max} 248 nm (ε_{max} = 52,656, methanol)
λ_{max} 255 nm (ε_{max} = 78,291, methanol)
λ_{max} 290 nm (ε_{max} = 3926, methanol)

PART B: Fluorenone 2,4-Dinitrophenylhydrazone

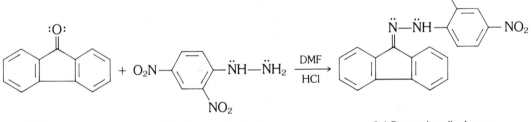

9-Fluorenone   2,4-Dinitrophenylhydrazine   2,4-Dinitrophenylhydrazone
derivative

Physical Properties of Reactants and Products

Compound	MW	Wt/Vol	mmol	mp(°C)	bp(°C)
9-Fluorenone	180.22	38 mg	0.21	84	
2,4-Dinitrophenylhydrazine	198.14	50 mg	0.25	194	
Dimethylformamide	73.09	1.0 mL			149–150
2,4-Dinitrophenylhydrazone derivative	360.56			283	

The 9-fluorenone product may be further characterized by the preparation of its 2,4-dinitrophenylhydrazone derivative.

A solution of 50 mg (0.25 mmol) of 2,4-dinitrophenylhydrazine in 1.0 mL of

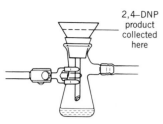

2,4–DNP
product
collected
here

DMF, 1 mL + 2 *M* HCl, 1 mL
+ H₂O, 2 mL
+ reaction by-products

dimethylformamide (DMF) is prepared in a 10×75-mm test tube. To this solution add 38 mg (0.21 mmol) of 9-fluorenone. After shaking the test tube so that all solids are in solution, 2 drops (Pasteur pipet) of concentrated HCl are added. The resulting mixture is allowed to stand at room temperature for a period of 5 minutes and then placed in an ice bath for an additional 5 minutes to complete precipitation of the derivative. The deposited crystals are collected by vacuum filtration using a Hirsch funnel.(■) They are washed with 1.0 mL of 2 *M* HCl (dropwise) to remove unreacted 2,4-dinitrophenylhydrazine and DMF, and then with two 1.0-mL portions of water to remove any residual acid. The product is dried in a desiccator.

Weigh the derivative and calculate the percentage yield. Determine the melting point and compare it with the literature value.

ENVIRONMENTAL DATA

Substance	Amount	TLV (mg/m³)	Emissions (mg)	Volume (m³)
Toluene (SKIN) (Max., solv., evap.)	3.5 mL	375	3035	8.1
Methylene chloride (Max., wash, evap.)	2.0 mL	350	2670	7.6
Hexane (Max., Craig)	2.0 mL	180	1320	7.3
Hexane (Max., TLC solv.)	4.2 mL	180	2770	15
Acetone (Max., TLC solv.)	1.8 mL	1780	1425	0.8
Dimethylformamide (Max., deriv., solv., vac. filtr.)	1.0 mL	30	945	31.5

TLV: Concentrated HCl—5 ppm

TXDS: 9-Fluorenone—scu-rat TDLo: 300 mg/kg/26W-1
CrO₃—ihl-hmn TCLo: 100 μg/m³ TFX:CAR

TXDS: 2,4-Dinitrophenylhydrazine—orl-rat LD50: 654 mg/kg

QUESTIONS

6-164. Suggest a suitable mechanism for the reaction of 9-fluorenone with 2,4-dinitrophenylhydrazine to form the corresponding 2,4-dinitrophenylhydrazone.

6-165. It is also possible to characterize 9-fluorenone by preparation of an oxime or semicarbazone. Formulate equations showing clearly the formation of these two derivatives and name each reagent used in the preparation.

6-166. As indicated in the Discussion, a solution of chromic oxide in aqueous sulfuric acid is used as a test reagent for 1° and 2° alcohols.
  a. What is this test (consult Chapter 7)?
  b. Predict which of the alcohols listed below will give a positive test with the chromic oxide reagent. Give the structure for each of the alcohols.

       1-heptanol   2,2,3-trimethyl-3-pentanol   cholesterol

       3-methyl-2-butanol   4-t-butylcyclohexanol

6-167. The compound 2-pentanone actually forms *two* isomeric 2,4-dinitrophenylhydrazones. Draw the structures of these isomers and explain why they are formed.

6-168. 2-Pentanone, in reference to Question 6-167, also forms a derivative on treatment with semicarbazide.

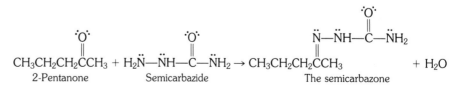

Note that semicarbazide has *two* —NH₂ groups that might react with the carbonyl of the ketone to form the semicarbazone. Explain why it reacts as depicted.

REFERENCES
1. The present experiment is an adaptation of that reported by Wade, L. G., Jr.; Stell, L. M. *J. Chem. Educ.* **1980,** *57,* 438.
2. Introduction to the use of polymer-bound reagents:
   a. Leznoff, C. C. *Acc. Chem. Res.* **1978,** *1,* 327.
   b. Hodge, P. *Chem. Br.* **1978,** *14,* 237.
3. Selected chromate oxidations reported in *Organic Syntheses:*
   a. Conant, J. B.; Quayle, O. R. *Organic Syntheses;* Wiley: New York, 1941; Collect. Vol. I, p 211.
   b. Sandborn, L. T. Ibid., p 340.
   c. Bruce, W. F. *Organic Syntheses;* Wiley: New York, 1943; Collect. Vol. II, p 139.
   d. Fieser, L. F. *Organic Syntheses;* Wiley: New York, 1963; Collect. Vol. IV, p 189.
   e. Ibid., p 195.
   f. Eisenbraun, E. J. *Organic Syntheses;* Wiley: New York, 1973; Collect. Vol. V, p 310.

Experiment 41 — Copper(II) Ion Oxidation of Benzoin: Benzil

(ethanedione, diphenyl-)

This experiment illustrates the oxidation of a secondary alcohol to a ketone using a dissolved metal ion catalyst.

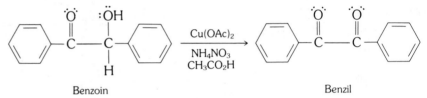

Benzoin   Benzil

DISCUSSION Benzil, a diketone, is obtained by the catalytic oxidation of benzoin using Cu^{2+} ion as the catalytic oxidant. The reaction is general for α-hydroxyketones and is the basis of Fehling's test for reducing sugars. The mechanism of the oxidation shows the catalytic effect of the Cu^{2+} ion as it is continuously reduced and oxidized in the sequence outlined below. A key ingredient is the nitrate ion, which oxidizes the Cu^+ ion to the Cu^{2+} state and is, in turn, reduced to nitrite ion. The nitrite ion decomposes to yield nitrogen gas and water.

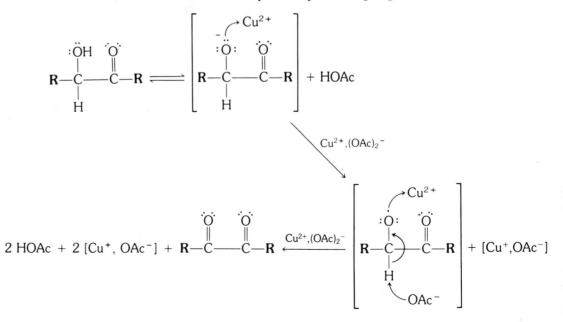

$$2\ Cu^+ + 2\ H^+ + NO_3^- \rightarrow 2\ Cu^{2+} + NO_2^- + H_2O$$

$$NH_4NO_2 \xrightarrow{H^+} N_2 + 2\ H_2O$$

Benzil is the second compound in the Sequence C series (see page 381). Benzoin, the starting material, is prepared in Experiment 20. Benzil, the product formed in the present experiment, is used as a reagent in the preparation of tetraphenylcyclopentadienone, Experiment 49. These reactions constitute the sequence of preparations that ultimately lead to the synthesis of hexaphenylbenzene.

If it is planned to continue the sequence, the Optional Scaleup procedure should be considered.

EXPERIMENTAL   Estimated time of the experiment: 2.5 hours.

Physical Properties of Reactants and Products

Compound	MW	Wt/Vol	mmol	mp(°C)
Benzoin	212.25	100 mg	0.47	137
Cupric acetate solution		350 μL		
Benzil	210.23			95

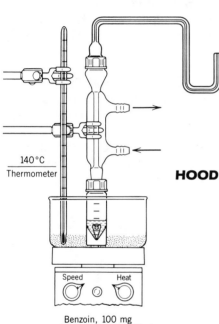

140°C
Thermometer

Speed   Heat

Benzoin, 100 mg
+ oxidation catalyst, 350 μL
1-mL conical vial

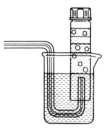

Instructor Preparation
The catalyst solution is prepared by dissolving 0.1 g of cupric acetate and 5 g of ammonium nitrate in 7.0 mL of deionized water (may require warming), followed by addition of 28 mL of glacial acetic acid. The container is placed in the **HOOD**   **hood** and the solution dispensed by use of an automatic delivery pipet.

Reagents and Equipment
A 1.0-mL conical vial containing a magnetic spin vane is equipped with a reflux condenser to which is attached a gas exit delivery tube. The system is charged with 100 mg (0.47 mmol) of benzoin and 350 μL of cupric acetate catalyst solution.(■)

The gas delivery tube is led into the neck of an inverted calibrated collection tube that is immersed in a beaker of water.(■) This facilitates the measurement of the nitrogen gas evolved during the course of the reaction.

Note. *A graduated cylinder or small buret may be used in place of the collection tube. It is absolutely necessary that all connections be tight to prevent leakage of the gas evolved. The ground-glass joint on the gas delivery tube is lightly greased.*

Reaction Conditions
The reaction mixture is heated with stirring at a sand bath temperature of ~140°C for a period of 1 hour or until the collected gas volume remains constant. *As the benzoin dissolves, the reaction mixture turns green and evolution of nitrogen gas commences. The theoretical volume of gas from 100 mg of benzoin is 10.6 mL at STP.*

Isolation of Product
If the gas delivery tube is used, disconnect it from the top of the condenser *before* removing the reaction vial from the heat source.

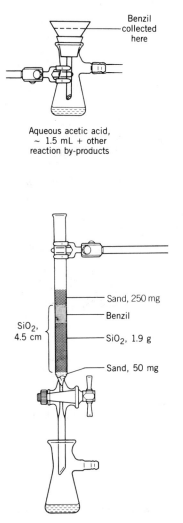

Benzil collected here

Aqueous acetic acid, ~ 1.5 mL + other reaction by-products

Sand, 250 mg

Benzil

SiO₂, 4.5 cm

SiO₂, 1.9 g

Sand, 50 mg

Benzil in ~ 8.0 mL CH₂Cl₂

The reaction mixture is cooled to room temperature, 0.5 mL of cold water (calibrated Pasteur pipet) added, and the reaction vial placed in an ice bath for 10 minutes. The yellow crystals of benzil are collected by vacuum filtration using a Hirsch funnel.(■) The reaction vial and crystals are rinsed with an additional two 0.5-mL portions of cold water.

Purification and Characterization

The crude product is purified by recrystallization from methanol or 95% ethanol using the Craig tube. The yellow crystals of benzil are dried on a clay plate or on filter paper. *The benzil obtained after recrystallization often contains a small amount of benzoin impurity. It may be purified by chromatography on silica gel using the following procedure.*

A slurry of activated silica gel in methylene chloride is packed in a 1.0-cm-diameter column to a height of 4.5 cm of the gel.(■) The sample of benzil is then introduced to the column followed by 250 mg of sand. Approximately 8.0 mL of methylene chloride are used to elute the benzil which is easily seen because of the yellow color. Concentration of the eluate gives pure benzil.

Before carrying out the column chromatography purification of benzil, the purity of the initial product may be assessed using *thin-layer chromatography* (see Chapter 5, Technique 8). Methylene chloride is used as the developing solvent, silica gel as the stationary phase, and UV light for visualization. R_f values: benzil, 0.62; benzoin, 0.33.

Weigh the dried product and calculate the percentage yield. Determine the melting point and compare your value with that reported in the literature.

Obtain an IR spectrum of the product and compare it with that of the starting material and with that reported in the literature.

Benzil also has a characteristic UV spectrum (see Fig. 6.24). It exhibits a wavelength maximum at 259 nm (ε_{max} = 16,329, methanol). It is of interest to compare this absorption spectrum of benzil with that of benzoin (Experiment 20).

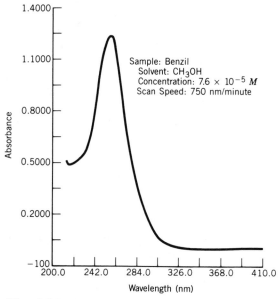

Sample: Benzil
Solvent: CH₃OH
Concentration: 7.6 × 10⁻⁵ *M*
Scan Speed: 750 nm/minute

Fig. 6.24 *UV–visible spectrum: benzil.*

Chemical Tests

Ketones and aldehydes are often characterized by the preparation of a solid derivative. To assist in the characterization of benzil, prepare its 2,4-dinitrophenylhydrazone or semicarbazone (Chapter 7). The melting points of these derivatives are listed in Appendix D, Table D.5.

Optional Scaleup

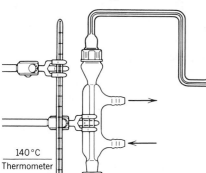

140°C
Thermometer

Speed   Heat

Benzoin, 400 mg
+ cupric acetate catalyst 1.5 mL
5-mL conical vial

The scaleup procedure is similar to that outlined for the microscale preparation with the following exceptions.

a. A 5.0-mL conical vial is substituted for the 1.0 mL vial.(■)

b. The collection of gas (if carried out) is done with a 100-mL graduated cylinder.(■)

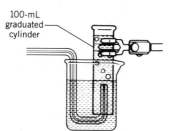

100-mL
graduated
cylinder

c. Amounts of reagents and solvents are increased fourfold.

Compound	MW	Wt/Vol	mmol	mp(°C)
Benzoin	212.25	400 mg	1.88	137
Cupric acetate solution		1.5 mL		

d. The reaction mixture is heated at 140°C until the evolution of gas ceases.

e. Two milliliters of cold water are added to the cooled reaction product, and after filtration, the material is washed with two 2-mL portions of cold water.

f. The benzil product is purified and characterized as described in the microscale procedure.

ENVIRONMENTAL DATA

Substance	Amount	TLV (mg/m^3)	Emissions (mg)	Volume (m^3)
Acetic acid, 80% (v/v)	0.35 mL	25	112	4.5
Methanol (SKIN; Max., Craig)	1.0 mL	260	800	3.1
Ethanol (Max., Craig)	1.0 mL	1900	800	0.4

TXDS: Benzoin—orl-rat TDLo: 5460 mg/kg
       Copper(II) acetate—orl-rat LD50: 595 mg/kg
       Benzil—orl-rat LD50: 2710 mg/kg

AQTX: Ammonium nitrate—TLm 96: over 1000–100

QUESTIONS

6-169. In the directions given for the experiment, it is emphasized that the gas delivery tube must be disconnected from the top of the condenser before the reaction vial is removed from the heat source. Why is this necessary?

6-170. What qualitative chemical tests would you perform to distinguish between benzoin and benzil? (See Chapter 7.)

6-171. 1,2-Dicarbonyl compounds, such as benzil, can be characterized by reaction with 1,2-phenylenediamine to form a substituted quinoxaline.

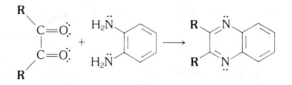

a. Write the structure for the derivative obtained when benzil is the reactant. Do you think this compound would be colored? If so, why?

b. Suggest a suitable mechanism for formation of the quinoxaline compounds based on the reaction scheme shown above.

c. What reagent would you react with 1,2-phenylenediamine to prepare un-substituted quinoxaline? Show a reaction scheme giving the structures of reactants and products.

6-172. Make a sketch of the ^1H NMR spectrum that would be observed for benzil.

6-173. Suggest a method for the preparation of $C_6H_5CH(OH)-(HO)HCC_6H_5$ from benzil. Discuss the stereochemical aspects of this 1,2-diol material.

6-174. Based on the ultraviolet data given for benzil in the Purification and Characterization section of this experiment, what concentration of benzil must have been used if a 1-cm cell was employed?

REFERENCES
1. The synthesis of benzil is reported in *Organic Syntheses:*
    a. Adams, R.; Marvel, C. S. *Org. Synth.* **1921,** *1,* 25.
    b. Clarke, H. T.; Dreger, E. E. *Organic Syntheses;* Wiley: New York, 1941; Collect. Vol. I, p 87.
2. For further information on the oxidation, see Weiss, M.; Appel, M. *J. Am. Chem. Soc.* **1948,** *70,* 3666; Depreux, P.; Bethegnies, G.; Marcincal-Lefebure, *J. Chem. Educ.* **1988,** *65,* 553.

Experiment 42

Ferric Chloride Oxidative Coupling of 2-Naphthol: 2,2'-Dihydroxydinaphthyl-1,1'

([1,1'-binaphthalene]-2,2'-diol)

The coupling reaction that aromatic phenols undergo in the presence of transition metal oxidants is demonstrated in this reaction. It mimics the biogenetic process that occurs in nature.

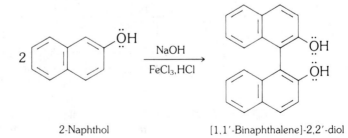

2-Naphthol   [1,1'-Binaphthalene]-2,2'-diol

DISCUSSION This reaction is important since it illustrates oxidative coupling of phenolic-type species, which is an important biogenetic pathway in nature leading to the formation of natural products.

The coupling reaction involves oxidation of 2-naphthol by electron transfer to give an aryloxy radical, which then dimerizes to yield the product. The mechanism is shown below.

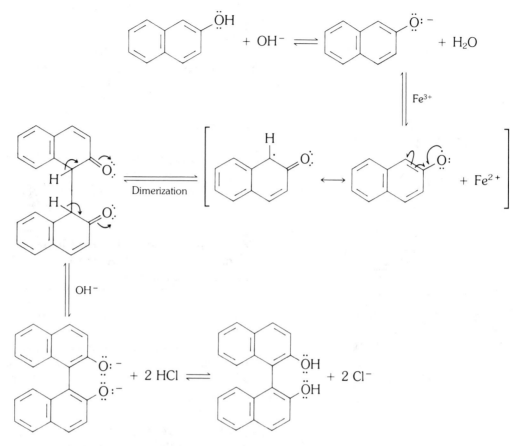

This binaphthol compound, by virtue of restricted rotation about the single bond joining the two naphthalene units, exists as a chiral molecule. That is, the molecule cannot readily exist in planar form because of the steric interference of the bulky —OH substituents and in fact it has been resolved to yield the R and S isomers of the compound. Such compounds belong to a class of materials having *atropisomerism*. The substituted 2,2′, 6,6′ biphenyls also have rotational barriers high enough that the enantiomers can be resolved, assuming that the substitution pattern confers chirality.

D. J. Cram and co-workers have incorporated this binaphthyl unit into cyclic crown ethers.

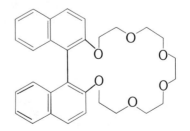

An investigation of the complexation of this class of molecules with various ionic species may lead to important insights into the catalytic nature of enzymes. For example, the crown ether similar to that above, containing two 2,2′-substituted 1,1′-binaphthyl units as chiral barriers, complexes preferentially with one enantiomer of primary amine salt racemic mixtures.

EXPERIMENTAL Estimated time to complete the experiment: 2.0 hours.

Physical Properties of Reactants and Products

Compound	MW	Wt/Vol	mmol	mp(°C)
2-Naphthol	144.19	100 mg	0.69	123–124
Sodium hydroxide	40.00	30 mg	0.75	318.4
Ferric chloride · 6H₂O	270.30	297 mg	1.1	37
Conc. HCl	36.46	100 μL		
Water	18.06	3 mL		
Dinaphthol	286.33			208–210

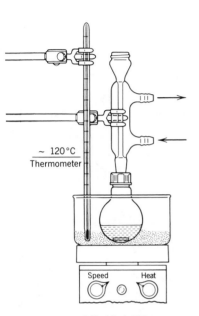

~ 120°C
Thermometer

Speed   Heat

2-Naphthol, 100 mg
+ NaOH, 30 mg
+ H₂O, 4.2 mL
+ FeCl₃·6H₂O, 297 mg
+ conc. HCl, 100 μL

Reagents and Equipment

To a 10-mL round-bottom flask containing a stirring bar and equipped with a reflux condenser, add 100 mg (0.69 mmol) of 2-naphthol, 30 mg (0.75 mmol) of sodium hydroxide, and 3.0 mL of water.(■)

> **CAUTION:** *Sodium hydroxide is a corrosive and toxic chemical. Do not allow it to touch the skin or to come in contact with the eyes.*

The reaction mixture is heated to reflux, with stirring, using a sand bath temperature of 120°C.

In a 10-mL Erlenmeyer flask prepare a solution of 180 mg (1.1 mmol) of *anhydrous* ferric chloride (MW 162) or 297 mg of ferric chloride hexahydrate (MW 270), 1.0 mL of water (calibrated Pasteur pipet), and 100 μL of concentrated hydrochloric acid.

> **WARNING:** *BE CAREFUL when mixing acid with water, especially if the acid is sulfuric. Add the acid TO the water. Avoid contact with the skin. Dispense the acid with an automatic delivery pipet.*

Using a Pasteur pipet, transfer the ferric chloride solution, through the top of the condenser, to the reaction flask. The Erlenmeyer flask is rinsed with 200 μL of water, and the rinse is also added to the reaction flask as before.

Reaction Conditions

The resulting mixture is heated at reflux, with stirring, for a period of 45–60 minutes using a sand bath temperature of 120°C. It is allowed to cool to room temperature and then placed in an ice bath to complete the crystallization of the product.

Isolation of Product

The solid product is collected by vacuum filtration using a Hirsch funnel and the filter cake washed with two 1.0-mL portions of cold water (calibrated Pasteur pipet).(■)

Purification and Characterization

The crude product is recrystallized from 95% ethanol using a Craig tube, and the resulting crystals are dried on a porous clay plate or on filter paper.

Weigh the crystals and calculate the percentage yield. Determine the melting point and compare it with the literature value.

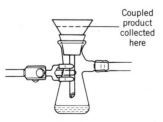

Coupled
product
collected
here

H₂O, ~ 6 mL
+ other reaction by-products

Chemical Tests

Chemical classification tests (Chapter 7) may be used to assist in the characterization of this material. Perform the ignition test and the ferric chloride test.

ENVIRONMENTAL DATA

Substance	Amount	TLV (mg/m³)	Emissions (mg)	Volume (m³)
Conc. hydrochloric acid (Max.)	100 µL	5	44	8.8
Ethanol (Max., Craig)	1.0 mL	1900	800	0.4

TLV: Ferric chloride—1 mg/m³ as (Fe)
      Sodium hydroxide—2 µg/m³

TXDS: 2-Naphthol—orl-rat LD50: 2420 mg/kg

QUESTIONS   **6-175.** In the oxidative coupling reaction of 1-naphthol, it is possible to obtain three products.

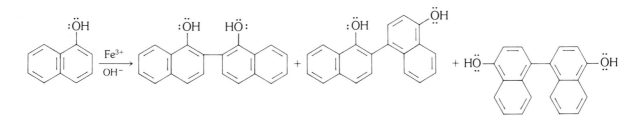

Account for the formation of this mixture by suggesting a mechanistic sequence similar to that presented in the Discussion section of this experiment.

6-176. Predict the structure of the diphenylquinone product, $C_{28}H_{40}O_2$, formed by the oxidative coupling of 2,6-di-t-butylphenol with oxygen in the presence of alkali.

6-177. Substituted phenols, such as BHT, are used as antioxidants in processed foods.

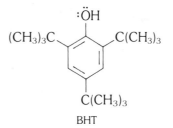

BHT

The role of the antioxidant is to stop spoilage caused by free-radical reactions brought about by reaction of oxygen with compounds containing C=C bonds.
    a. Give a suitable chemical name for BHT.
    b. Can you suggest why this compound is an effective antioxidant?

6-178. Naphthalene is the simplest of the polycyclic aromatic hydrocarbons. It can be represented by three resonance structures. Draw them. Indicate which structures are equivalent.

6-179. How many monochloro derivatives are possible for naphthalene? Give a structure for each compound.

REFERENCES   **1.** For references related to the oxidative coupling of phenolic derivatives, see
        a. Dewar, M. J. S.; Nakaya, T. *J. Am. Chem. Soc.* **1968,** *90,* 7134.
        b. Scott, A. I. *Quart. Rev.* **1965,** *19,* 1.
    2. For an overview of the work of Cram and co-workers on the binaphthyl cyclic crown ethers, see Cram, D. J.; Cram, J. M. *Science* **1974,** *183,* 803.

Experiments 43A, 43B

Hypochlorite Oxidation of Methyl Ketones by the Haloform Reaction: Benzoic Acid; p-Methoxybenzoic Acid

(benzoic acid; benzoic acid, p-methoxy-)

The well-known haloform reaction is often used to determine the presence of a methyl ketone. The reaction is also useful synthetically as a route to the preparation of organic acids.

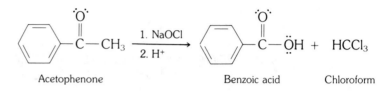

Acetophenone   Benzoic acid   Chloroform

DISCUSSION   The reaction of methyl ketones with a halogen in alkaline medium is known as the haloform reaction. In this experiment, halogen and base are present because of the equilibrium reaction

$$H_2O + NaOCl + NaCl \leftrightarrows 2 NaOH + Cl_2$$

In the haloform reaction, two products are formed: (1) a haloform ($CHCl_3$, $CHBr_3$, or CHI_3), depending on the halogen employed; (2) a carboxylic acid with one less carbon atom than the starting methyl ketone. It is the formation of the carboxylic acid that gives the reaction synthetic utility.

If I_2 in a basic solution is used as the reactant, iodoform (CHI_3) is generated in the reaction. This compound is a solid and precipitates from the reaction medium. This observation has been used extensively as a chemical test for methyl ketones (see Chapter 7).

The reaction takes place in two stages. In the first stage, the methyl group is trihalogenated. In the second, base attacks the trihaloketone to generate the haloform and the alkali metal salt of the acid. Acidification yields the carboxylic acid. The sequence is shown below.

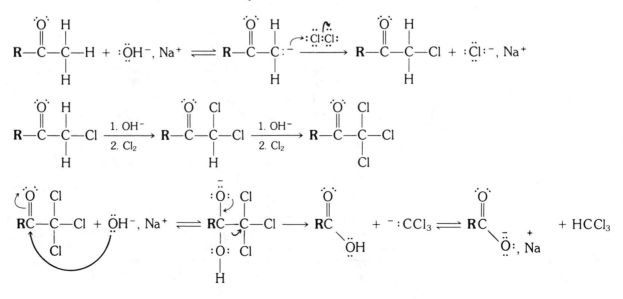

PART A: Benzoic Acid

The reaction is shown above.

EXPERIMENTAL   Estimated time to complete the experiment: 1.5 hours.

Physical Properties of Reactants and Products

Compound	MW	Wt/Vol	mmol	mp(°C)	bp(°C)	Density	n_D
Acetophenone	120.16	20 μL	0.17	20.7	202.6	1.03	1.5372
Bleach (5%)		700 μL					
Benzoic acid	122.12			122–23			

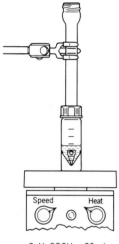

Speed   Heat

$C_6H_5COCH_3$, 20 μL
+ 5% NaOCl, 700 μL

Reagents and Equipment
To a 3.0-mL conical vial containing a magnetic spin vane and equipped with an air condenser, add 20 μL (21 mg, 0.17 mmol) of acetophenone and 700 μL of household bleach (5%).(■)

CAUTION: *Both reagents are irritants and should be dispensed in the* hood *using automatic delivery pipets.*

HOOD

Reaction Conditions
The mixture is stirred at room temperature for a period of 0.5 hour.

Isolation of Product
Sodium sulfite (5 mg) is added to destroy any unreacted bleach and the reaction mixture stirred briefly. The resulting mixture is extracted with two 0.5-mL portions of ether (calibrated Pasteur filter pipet). Each ether extract portion is separated using a Pasteur filter pipet.

Be Careful. *The ether layer is the top layer; the lower alkaline layer contains the acid product. The ether extraction removes the chloroform generated in the reaction and any unreacted acetophenone. Dispose of the ether extracts in a suitable receptacle at the end of the experiment.*

The aqueous layer is made acidic (check with pH paper) by the dropwise addition of 3 M HCl from a Pasteur pipet. A thick, white precipitate of benzoic acid appears. The solid is collected by vacuum filtration using a Hirsch funnel and the filter cake washed with three 0.5-mL portions of water (calibrated Pasteur pipet).(■) The product is dried on a porous clay plate or on filter paper.

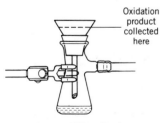

Oxidation product collected here

Aqueous HCl, ~ 2 mL
+ reaction by-products

Purification and Characterization
The product is of sufficient purity for characterization.
Weigh the material and calculate the percentage yield. Determine the melting point and compare your result with the literature value. If desired, obtain an IR spectrum using the KBr pellet technique and compare it with that of an authentic sample of benzoic acid.

Chemical Tests
Chemical classification tests may also be used to assist in product analysis. You might wish to perform the following.
a. The ignition test should indicate the presence of the aromatic ring (Chapter 7, Table 7.1).

b. The solubility of the compound in water, 5% NaOH, and in 5% NaHCO$_3$ should be checked (see the Solubility Chart, Chapter 7). Does the water solution turn blue litmus paper red? Is there evidence of CO$_2$ evolution when the benzoic acid is added to the bicarbonate solution? If positive results are obtained in these tests, how do they establish the fact that a carboxylic acid is present?

c. The preparation of an amide derivative (see Chapter 7) would also aid in establishing the identity of the acid product.

PART B: *p*-Methoxybenzoic Acid

CH$_3$Ö—⟨benzene ring⟩—C(=O)—CH$_3$ $\xrightarrow[\text{2. H}^+]{\text{1. NaOCl}}$ CH$_3$Ö—⟨benzene ring⟩—C(=O)—ÖH + HCCl$_3$

p-Methoxyacetophenone   *p*-Methoxybenzoic acid   Chloroform

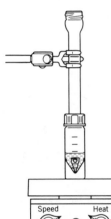

p–CH$_3$OC$_6$H$_4$COCH$_3$, 29 mg
+ 5% NaOCl, 700 μL

EXPERIMENTAL Estimated time to complete the experiment: 1.5 hours.

Physical Properties of Reactants and Products

Compound	MW	Wt/Vol	mmol	mp(°C)	bp(°C)
p-Methoxyacetophenone	150.8	29 mg	0.19	38–39	258
Bleach (5%)		700 μL			
p-Methoxybenzoic acid	152.16			185	

Note. *The starting p-methoxyacetophenone reagent may be prepared in Experiment 52.*

Reagents and Equipment

To a 3.0-mL conical vial containing a magnetic spin vane and equipped with an air condenser, add 29 mg (0.19 mmol) of *p*-methoxyacetophenone and 700 μL of household bleach (5%).(■)

HOOD

> **CAUTION:** *The bleach is an irritant to skin and eyes. It is dispensed using an automatic delivery pipet in the* **hood.**

Reaction Conditions

The mixture is stirred for 0.5 hour with very gentle heating. *The lowest possible setting on a hot plate–magnetic stirrer is used.*

Isolation of Product

Sodium sulfite (5 mg) is now added; the reaction medium is stirred briefly and then cooled to room temperature. The resulting mixture is extracted with two 0.5-mL portions of ether (calibrated Pasteur pipet). The ether extract portions are separated using a Pasteur filter pipet.

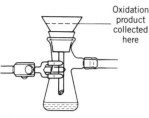

Oxidation
product
collected
here

Aqueous HCl, ~2 mL
+ reaction by-products

Be Careful. *The ether layer is the top layer; the bottom aqueous layer contains the product. The ether extraction removes the chloroform generated in the reaction and any unreacted p-methoxyacetophenone. Dispose of the ether extracts in a suitable receptacle at the end of the experiment.*

The aqueous layer is made acidic by the dropwise addition (check with pH paper) of 3 *M* HCl from a Pasteur pipet. A thick, white precipitate of *p*-methoxybenzoic acid is formed. The mixture is cooled in an ice bath for 5 minutes and the solid collected by vacuum filtration using a Hirsch funnel. (■) The filter cake is washed with three 0.5-mL portions of cold water (calibrated Pasteur pipet) and the product dried on a porous clay plate or on filter paper.

Purification and Characterization

The product is reasonably pure but may be recrystallized from ethanol–water, if desired, using a Craig tube.

Weigh the product and calculate the percentage yield. Determine the melting point and obtain an IR spectrum using the KBr pellet technique. Compare your results with those reported in the literature.

Chemical Tests

Chemical characterization tests can be used to assist in the classification of this product as outlined in Part A of this experiment.

Optional Scaleup

This preparation may be scaled up by a fifteenfold increase in reagent amounts. The procedure is similar to that given earlier for the microscale preparation with the following exceptions. A major difference is that a separatory funnel is used in the extraction step.

a. A 25-mL round-bottom flask containing a magnetic stirring bar and equipped with an air condenser is used. The reaction conditions are as given in the microscale section. (■)

b. The reagent and solvent amounts are increased fifteenfold.

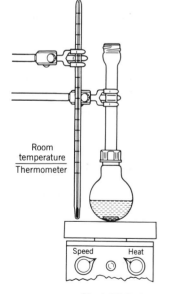

Room
temperature
Thermometer

Speed   Heat

25-mL RB flask

Compound	MW	Wt/Vol	mmol	mp(°C)	bp(°C)
p-Methoxyacetophenone	150.8	430 mg	2.85	38–39	258
Bleach (5%)		10.5 mL			
Sodium sulfite	120.6	75 mg			

c. After addition of the sodium sulfite, stirring, and cooling, the product mixture is transferred to a 125-mL separatory funnel. This mixture is extracted with two 7.5-mL portions of diethyl ether. The ether extracts are set aside for later disposal in an appropriate container. The product is in the aqueous layer.

The extraction is carried out by first returning the lower aqueous layer to the reaction flask and then the ether phase is transferred to an Erlenmyer flask for temporary storage. The aqueous layer is then returned to the separatory funnel and the reaction flask rinsed with a second portion of ether. This ether rinse is then added to the separatory funnel to complete the extraction.

d. After the product is collected by vacuum filtration, the filter cake is washed with three 7-mL portions of cold water.

e. The pale yellow product is characterized as described in the microscale procedure.

ENVIRONMENTAL DATA

Substance	Amount	TLV (mg/m^3)	Emissions (mg)	Volume (m^3)
Household bleach (as Cl$_2$)^a	18 mg	3	18	6
Diethyl ether (Max.)	1.0 mL	1200	714	0.6
Chloroform (Max.)	—	50	21	0.4

^{a} Assuming all elemental chlorine escapes as chlorine gas.

TXDS: Acetophenone—orl-rat LD50: 900 mg/kg
Benzoic acid—orl-rat LD50: 2530 mg/kg
p-Methoxyacetophenone—orl-rat LD50: 1720 mg/kg
Sodium sulfite—inv-rat LD50: 115 mg/kg

QUESTIONS

6-180. In the haloform reaction, once the first α-hydrogen atom is replaced by a halogen atom, each successive hydrogen is more easily substituted until the trihalo species is obtained. Explain.

6-181. The haloform reaction using I$_2$ and NaOH is referred to as the "iodoform" test for methyl ketones (see Chapter 7). The test also gives positive results for compounds containing the —CH(OH)CH$_3$ group. This is due to the oxidation of the alcohol to the methyl ketone. Write a balanced equation for the conversion of this 2° alcohol unit to a methyl ketone in the presence of I$_2$ and NaOH. Identify which species is being oxidized and reduced.

6-182. If you were carrying out an industrial-scale synthesis in which one step involved a haloform reaction to convert a methyl ketone into the corresponding acid having one less carbon atom, would you use NaOH and Cl$_2$, NaOH and Br$_2$, or NaOH and I$_2$ as the reagent? Give reasons for your choice.

6-183. Can you explain the fact that even though dibenzoylmethane is not a methyl ketone, it gives a positive iodoform test when treated with the NaOH and I$_2$ reagent?

6-184. Do you think that acetaldehyde would give a positive iodoform test? Explain your reasoning.

6-185. A water-soluble phenol is also acidic toward litmus paper as is a water-soluble carboxylic acid. How would you distinguish between an aromatic acid and a phenol using a chemical test?

REFERENCES

Selected examples of the haloform reaction from *Organic Syntheses*:

a. Sanborn, L. T.; Bousquet, E. W. *Organic Syntheses*; Wiley: New York, 1941; Collect. Vol. I, p 526.
b. Newman, M. S.; Holmes, H. L. *Organic Syntheses*; Wiley: New York, 1943; Collect. Vol. II, p 428.
c. Smith, L. I.; Prichard, W. W.; Spillane, L. J. *Organic Syntheses*; Wiley: New York, 1955; Collect. Vol. III, p 302.
d. Smith, W. T.; McLeod, G. L. *Organic Syntheses*; Wiley: New York, 1963; Collect. Vol. IV, p 345.
e. Staunton, J.; Eisenbraun, E. J. *Organic Syntheses*; Wiley: New York, 1973; Collect. Vol. V, p 8.

Experiment 44

Air Oxidation of Aniline: Azobenzene

(diazene, diphenyl)

This reaction illustrates the base-promoted autoxidation of aniline to azobenzene.

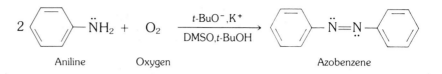

| | Aniline | Oxygen | | Azobenzene |

DISCUSSION Primary aromatic amines have been oxidized to azo compounds using a variety of oxidizing agents. This experiment illustrates room-temperature, base-promoted autoxidation of aniline in dimethyl sulfoxide solution. It is well known that solutions of alkoxides in dimethyl sulfoxide react with oxygen to form dimethyl sulfone and methane sulfonic acids. However, a mixture of dimethyl sulfoxide and *tert*-butyl alcohol is stable to oxygen in the presence of potassium *tert*-butoxide. Thus, this alternate reaction does not interfere with the oxidation of aniline to form azobenzene.

A variety of *o*-, *m*-, and *p*-substituted anilines undergo this reaction. Based on experimental evidence, it appears that the mechanism involves two stages. In the first stage, *nitrosobenzene* is formed by the reaction of the aniline molecule with oxygen. The second stage consists of the generation of a *nitranion* by reaction of the alkoxide base with aniline. Once formed, this nitranion reacts with the nitrosobenzene molecule to generate a species that then undergoes an elimination reaction to form the azobenzene product. The sequence is depicted below.

Stage I

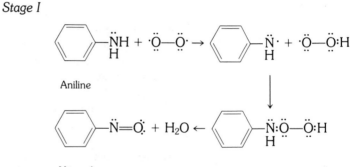

Aniline

Nitrosobenzene

Stage 2

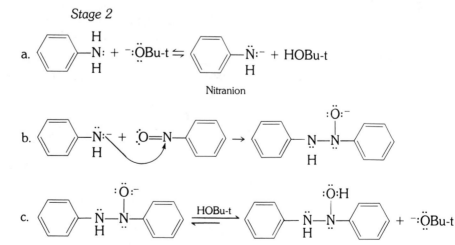

Nitranion

Azobenzene

EXPERIMENTAL   Estimated time to complete the reaction: 2.0 hours.

Physical Properties of Reactants and Products

Compound	MW	Wt/Vol	mmol	mp(°C)	bp(°C)	Density	n_D
Potassium *t*-butoxide	122.22	480 mg	3.9	256–258			
DMSO–*t*-butyl alcohol (80%)		8 mL					
Aniline	93.13	100 mg	1.08		184	1.02	1.5863
trans-Azobenzene	182.23			68.5			

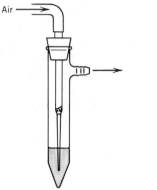

80% DMSO–(CH₃)₃COH, 8 mL
+ (CH₃)₃CO⁻ K⁺, 480 mg
+ C₆H₅NH₂, 100 mg

HOOD

GENTLY

Reagents and Equipment
To a 16 × 150-mm side-armed test tube fitted with an air inlet, add 480 mg (3.9 mmol) of potassium *tert*-butoxide and 8.0 mL of 80% dimethyl sulfoxide–*tert*-butyl alcohol solution.(■) Using an automatic delivery pipet, add 100 μL (100 mg, 1.08 mmol) of aniline to the reaction mixture.

> **CAUTION: Potassium tert-butoxide is a moisture-sensitive, flammable solid; dimethyl sulfoxide is hygroscopic; aniline is highly toxic and a suspected carcinogenic agent. These reagents are dispensed in the hood.**

Reaction Conditions
Air is now *gently* bubbled through the solution at room temperature for a period of 1 hour. During this time the reaction mixture turns red and then quickly changes to brown.

Note. *The compressed air inlet should have a filter on the line to prevent contamination of the reaction mixture.*

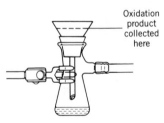

Oxidation product collected here

80% DMSO–(CH₃)₃COH, 8 mL
+ aqueous KOH, 4.5 mL
+ (CH₃)₃COH, 0.5 mL
+ other reaction by-products

Isolation of Product
Addition of 5–10 mL of water, followed by cooling of the mixture in an ice bath, yields crude, red azobenzene. The product is collected by filtration under reduced pressure using a Hirsch funnel and the filter cake washed with three 0.5-mL portions of cold water (calibrated Pasteur pipet).(■)

> **WARNING: Azobenzene is a suspected carcinogenic agent.**

Purification and Characterization
The crude material is recrystallized, using a Craig tube, from 12 : 1 (95%) ethanol–water solution and dried on a porous clay plate.

Weigh the azobenzene and calculate the percentage yield. Determine the melting point and compare your value with that reported in the literature. The purity of the compound may be determined using *thin-layer chromatography* (see Experiment 45 for conditions).

Azobenzene has a characteristic UV spectrum. The following data were obtained using a concentration of 1.14×10^{-5} M in a 1-cm cell over the range 200–375 nm (the literature also records a very weak band at 433 nm).

λ_{max} 225 nm (ε_{max} = 9035, methanol)

λ_{max} 314 nm (ε_{max} = 11,491, methanol)

ENVIRONMENTAL DATA

Substance	Amount	TLV (mg/m³)	Emissions (mg)	Volume (m³)
Dimethylsulfoxide, 80% (relatively involatile)	8.0 mL			
Aniline (SKIN) (Max., from yield.)	100 mg	19	65	3.4
Ethanol (Max., Craig)	1.0 mL	1900	800	0.4

TLV: *t*-Butyl alcohol—100 ppm

TXDS: Dimethylsulfoxide—orl-rat LD50: 19,700 mg/kg
      Azobenzene—orl-rat LD50: 1000 mg/kg

QUESTIONS

6-186. Azobenzene can also be prepared by the reduction of nitrobenzene using hot methanolic–sodium hydroxide solution with zinc dust. The stages of the reaction are outlined below.

nitrobenzene → nitrosobenzene → phenylhydroxylamine → azoxybenzene → azobenzene

Write a structural formula for each of the species involved in the transformation.

6-187. Azo compounds constitute a very useful group of initiators for free-radical polymerization reactions. The formation of a molecule of nitrogen on dissociation of the azo compound provides a strong driving force to generate an initial radical.

$$\mathbf{R}-\ddot{\mathbf{N}}=\ddot{\mathbf{N}}-\mathbf{R} \rightarrow 2\mathbf{R}\cdot + N_2$$

Careful studies of the decomposition of azoisobutyronitrile (AIBN) have shown that the compound is less than 100% efficient in the production of radicals. That is, instead of producing two radicals per mole of AIBN, fewer are formed due to recombination of the radicals.

    a. Give the structure for AIBN.

    b. One of the compounds produced on dissociation of AIBN is a tetramethyl-succinonitrile. Give a structure for this compound and account for its formation.

6-188. Based on the discussion in Question 6-187, what product would be formed when

$$D_3C-\ddot{N}=\ddot{N}-CD_3$$

is allowed to decompose in isooctane solvent?

6-189. If the two azo compounds shown below are decomposed in the same reaction mixture, three products are formed in addition to the nitrogen gas. Give the structures of the expected compounds.

6-190. Azo compounds of the type Ar—N=N—Ar′ are brightly colored compounds. Why?

6-191. As indicated earlier, the literature records a weak absorption band for azobenzene at 433 nm. To measure this absorption band a concentration of 0.1 g/L was used. The solution had an observed absorbance of 0.8. What size cell was used in this measurement if the band was known to have $\varepsilon_{max} = 729$?

REFERENCES **1.** The preparation of azobenzene is reported in Bigelow, H. E.; Robinson, D. B. *Organic Syntheses*; Wiley: New York, 1955; Collect. Vol. III, p 103.

2. Directions for the oxidation of a substituted aniline to prepare the corresponding azobenzene derivative are also given in Santuri, P.; Robbins, F.; Stubbings, R. *Organic Syntheses*; Wiley: New York, 1973; Collect. Vol. V, p 341.

3. The present experiment is based on the work of Russell, G. A.; Janzen, E. G.; Becker, H.; Smentowski, F. J. *J. Am. Chem. Soc.* **1962**, *84*, 2652.

Experiment 45 # Photochemical Isomerization: *cis*-Azobenzene

(diazene, diphenyl-)

The isomerization of *trans*-azobenzene to *cis*-azobenzene under photochemical conditions is demonstrated in this experiment. The course of the reaction is followed by thin-layer chromatography.

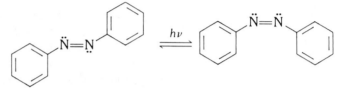

    trans-Azobenzene   *cis*-Azobenzene

DISCUSSION The photochemical conversion of *trans*- to *cis*-azobenzene proceeds by way of an excited electronic state of the trans isomer. Since both compounds are colored and have different R_f values, the conversion is conveniently followed using TLC.

Azo compounds, because of their intense color, are used commercially as synthetic dyes. Methyl red, prepared in Experiment 31, is an example of this class of materials.

EXPERIMENTAL Estimated time to complete the experiment: 1.5 hours.

Physical Properties of Reactants and Products

Compound	MW	Wt/Vol	mmol	mp(°C)	bp(°C)
trans-Azobenzene	182.23	25 mg	0.14	68.5	
Methylene chloride	84.94	1.0 mL			40
cis-Azobenzene	182.23			71	

Reagents and Equipment

In a 3-mL conical vial place 25 mg (0.14 mmol) of azobenzene followed by 1.0 mL of methylene chloride.

> **CAUTION: *Azobenzene is a suspected carcinogenic agent.***

HOOD *The azobenzene used may be prepared in Experiment 44. The methylene chloride is dispensed in the **hood** using a graduated or automatic delivery pipet.*

Determination of R_f Values

Using a capillary prepared from a melting point tube, spot an activated silica gel thin-layer chromatographic plate (1.5 × 7 cm) with one drop of the above solution.

Refer to Chapter 5, Technique 8, for directions on setting up and carrying out a TLC experiment.

The chromatogram is developed in a 4 × 2 in. screw-capped bottle using 5 mL of hexane. *Two spots are observed on the plate:*

trans-azobenzene ($R_f = 0.6$)

cis-azobenzene ($R_f = 0.1$), usual impruity in trans samples.

The Isomerization Reaction

A second thin-layer plate is spotted with the same solution used above. The plate is placed under a sunlamp (one foot away) for a period of 1 hour. At the end of this time the plate is developed, as above, using hexane solvent.

Characterization

After development of the chromatogram, only one spot near the origin should be observed.

Determine the R_f value of this material and compare it with that of the control sample developed previously. The results should show that all the trans isomer present in the starting sample has been converted to the cis isomer.

ENVIRONMENTAL DATA

TLV: Hexane—50 ppm

TXDS: Azobenzene—orl-rat LD50: 1000 mg/kg
    Silica gel—inv-mus LDLo: 234 mg/kg

QUESTIONS

6-192. Which of the two isomers of azobenzene would you expect to be the most thermodynamically stable? Explain.

6-193. Over the years the two isomers of azobenzene have been designated by various terms: cis–trans, syn–anti, and *E–Z*. Using each set of terms, assign them to the isomers of azobenzene.

6-194. Listed below are the activation energies for the decomposition of a series of azo compounds.

Compound	Activation Energy (kJ/mol)
$CH_3—\ddot{N}{=}\ddot{N}—CH_3$	210
$(CH_3)_2CH—\ddot{N}{=}\ddot{N}—CH(CH_3)_2$	170
$(C_6H_5)_2CH—\ddot{N}{=}\ddot{N}—CH(C_6H_5)_2$	110

Explain the trend in the table.

6-195. You have found that two isomers of a given compound can be separated by TLC. When the solvent front moved 60 mm above the original line where samples of each of the isomeric species had been spotted, the spot of one species (X) was at 40 mm and that of the other (Y), at 25 mm. Calculate the R_f values for the isomers X and Y.

6-196. State what you would observe after development and visualization of a TLC plate if you had made the following mistakes in carrying out the analysis.
  a. The solvent level in the developing jar was higher than in the original line where the samples were spotted.
  b. Too much sample was applied to the TLC plate.

REFERENCES

This experiment is adapted from the work of Helmkamp, G. K.; Johnson, H. W., Jr. *Selected Experiments in Organic Chemistry*, 2nd ed.; Freeman: San Francisco, 1968; p 32.

Experiment 46

Beckmann Rearrangement: Benzanilide

(benzamide, *N*-phenyl-)

This experiment illustrates the rearrangement of a ketoxime (prepared in situ) to the corresponding amide.

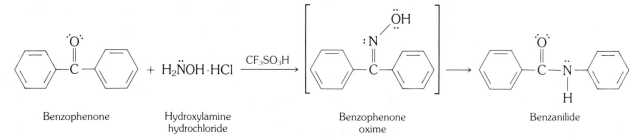

Benzophenone   Hydroxylamine hydrochloride   Benzophenone oxime   Benzanilide

DISCUSSION

The Beckmann rearrangement was discovered in 1886 by E. Beckmann and is the reaction of a ketoxime in the presence of acid to yield the corresponding amide or amides. In the present case, only one amide is formed since benzophenone is a symmetrical ketone. Because the ketoximes are prepared from ketones, the reaction is often used to determine the structure of the starting ketone. This is done by identification of the acid and amine obtained on hydrolysis of the amide formed by the rearrangement.

As depicted in the mechanism below, the acid catalyst converts the oxime hydroxy group to a good leaving group (water, a small neutral molecule).

There are two significant points concerning the stereochemistry of the reaction: (1) the group that migrates is *anti* to the hydroxyl group and (2) the stereochemical features of the migrating group are retained.

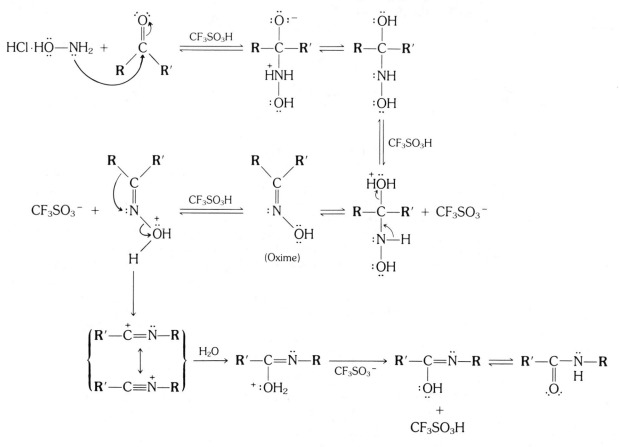

EXPERIMENTAL   Estimated time to complete the experiment: 2.5 hours.

Physical Properties of Reactants and Products

Compound	MW	Wt/Vol	mmol	mp(°C)
Benzophenone	182.21	100 mg	0.55	48.1
Hydroxylamine hydrochloride	69.49	51 mg	0.73	155–57
Triflic acid–formic acid solution		500 μL		
Benzanilide	197.24			162–64

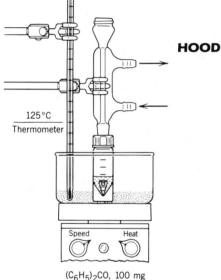

125°C
Thermometer

Speed   Heat

$(C_6H_5)_2CO$, 100 mg
+ $HCl \cdot H_2NOH$, 51 mg
+ CF_3SO_3H–HCO_2H, 0.5 mL

Reagents and Equipment
In a 5.0-mL conical vial containing a magnetic spin vane and fitted with a reflux condenser, place 100 mg (0.55 mmol) of benzophenone and 51 mg (0.73 mmol) of hydroxylamine hydrochloride.(■) Using an automatic delivery **HOOD** pipet, add 500 μL of triflic acid–formic acid solution **[hood]** to this mixture.

> **CAUTION: Triflic acid (trifluoromethanesulfonic acid) is one of the strongest acids known. It is very corrosive and toxic!**

Instructor Preparation
The acid solution is prepared by adding 2 drops of triflic acid to 5.0 mL of 90% formic acid.

Reaction Conditions
With stirring, heat the reaction mixture at reflux for 1.0 hour in a sand bath (125°C). The resulting solution is then cooled to room temperature.

Isolation of Product
To the cooled reaction solution, add 1.0 mL of water (calibrated Pasteur pipet), and extract the resulting mixture with three 1.0-mL portions of methylene chloride. The organic phase is separated using a Pasteur filter pipet and transferred to a 10-mL Erlenmeyer flask. For each extraction, after addition of the methylene chloride, the vial is capped, shaken, and vented, and the layers are allowed to separate.

The combined methylene chloride extracts are dried over granular, anhydrous sodium sulfate. Using a Pasteur filter pipet, the dried solution is transferred to a clean, dry 10-mL Erlenmeyer flask containing a boiling stone.

HOOD   The solvent is evaporated **[hood]** under a gentle stream of nitrogen in a warm sand bath.

Characterization and Purification
The crude benzanilide is removed from the Erlenmeyer flask and recrystallized from 95% ethanol using a Craig tube.

Weigh the product and calculate the percentage yield. Determine the melting point and compare your result with the literature value. Obtain an infrared spectrum using the KBr pellet technique.

In this experiment an aromatic ketone has been rearranged to a secondary amide. By examining the infrared spectra of starting material and product, we can follow this molecular transformation.

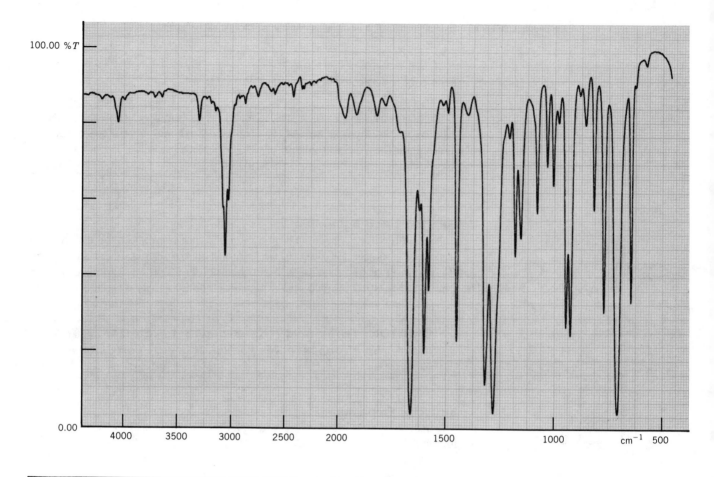

Fig. 6.25 *IR spectrum: benzophenone.*

Sample	Benzophenone		
%T _X_ ABS __ Background Scans _4_		Scans _16_	
Acquisition & Calculation Time _42 sec_		Resolution _4.0 cm⁻¹_	
Sample Condition _solid – melt_		Cell Window _KBr_	
Cell Path Length _capillary film_		Matrix Material _____	

Infrared Analysis

The spectrum of benzophenone (Fig. 6.25) possesses two macro group frequency trains:

1. *Conjugated aromatic ketone* defined by the bands at 3325 (overtone of ketone carbonyl stretch), 1663 (C=O stretch, doubly conjugated), 1601 & 1580 (ν_{8a} and ν_{8b} degenerate ring stretch), 1492 & 1450 (ν_{19a} and ν_{19b} degenerate ring stretch) cm⁻¹. The intensification of the 1580 cm⁻¹ peak confirms the conjugation of the carbonyl to the ring. The 1500 cm⁻¹ ring stretch, which is a bit variable in intensity, is quite weak in this case.

2. *Monosubstituted phenyl group* defined by weak bands located at 1969, 1913, 1823, and 1724 cm⁻¹ and strong bands recorded at 701 and 640 cm⁻¹. For discussions of these modes see Chapter 7 and Experiment 22. In the case of phenyl rings conjugated to carbonyl groups the 750 and 690 cm⁻¹ bands often appear on the low side of their range and can be down as much as 40–50 cm⁻¹ as occurs here.

The rearrangement product has incorporated a heteroatom into its structure, but the carbonyl group and the ring systems have survived. The infrared spectrum (Fig. 6.26) must be consistent with the formation of a secondary amide. We expect to observe a macro group frequency for the presence of monosubstituted phenyl systems plus a second macro group frequency for an aromatic secondary amide.

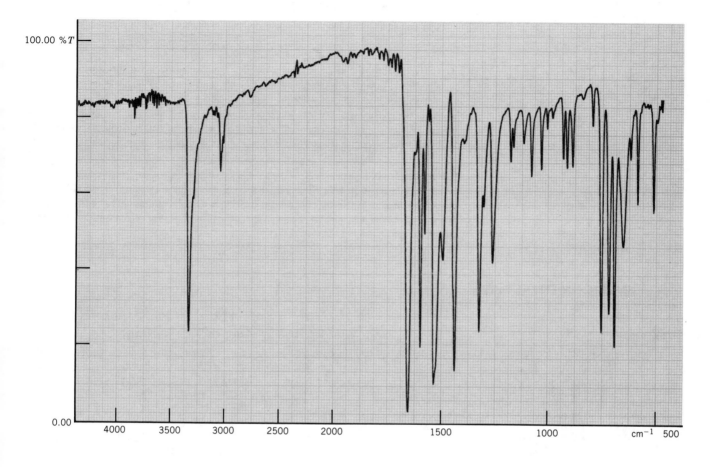

Sample ___Benzanilide___

%T _X_ ABS __ Background Scans _4_   Scans _____16_____

Acquisition & Calculation Time _42 sec_   Resolution _4.0 cm^{-1}_

Sample Condition _____solid_____   Cell Window _____

Cell Path Length _____   Matrix Material _KBr_

Fig. 6.26 *IR spectrum: benzanilide.*

The first macro group frequency is similar to that of the starting material with bands occurring at

a. 1950, 1912, 1840, 1725 cm^{-1}: All four combination bands are doubled in this case. The values given are for the centers of the doublets.

b. 752 & 718 cm^{-1}: C—H out-of-plane bend. The lower value likely corresponds to the ring conjugated directly with the carbonyl group.

c. 693 cm^{-1}: ring out-of-plane bend (puckering) vibration.

The second macro group frequency for the aromatic secondary amide utilizes the following bands: 3350, 3060, 1659, 1602, 1587, 1539, 1322, and 680 (broad) cm^{-1}.

a. 3350 cm^{-1}: This strong band corresponds to the highly H-bonded amide N—H stretch.

b. 3060 cm^{-1}: C—H stretch on sp^2 carbon—aromatic C—H.

c. 1659 cm^{-1}: C=O stretch of a conjugated and H-bonded amide. It is the most intense band in the spectrum.

d. 1602 & 1587 cm^{-1}: ring stretch degenerate pair, ν_{8a} and ν_{8b}.

e. 1539 & 1322 cm^{-1}: These two bands involve both N—H bending (in-plane) and C—N stretch. The two fundamentals fall somewhere between 1450 and 1400 cm^{-1}. Thus, they can mechanically interact and split apart with one component falling at 1539 cm^{-1} and the other near 1322 cm^{-1}. As the fundamental frequencies will vary somewhat from molecule to molecule, the interaction term that is sensitive to the frequency match (see Appendix B) will also vary in magnitude, and thus, ultimately the wavenumber separation will be affected.

f. 680 cm^{-1}: broad and weak. This ill-defined band arises from the out-of-plane bend of the N—H group. It is similar to the O—H bend found in alcohols in this spectral region (see infrared discussion in Experiments 5A, 5B, 8B, and 8C).

Examine the spectrum of the reaction product you have obtained in a potassium bromide matrix. Discuss the similarities and differences of the experimentally derived spectral data to the reference spectra (Figs. 6.25 and 6.26).

Chemical Tests

Chemical classification tests (Chapter 7) such as the ignition test and the soda-lime test (or sodium fusion test) may also be run to further establish the identity of the product. The hydroxamate test may be used to establish the presence of the amide functional group.

ENVIRONMENTAL DATA

Substance	Amount	TLV (mg/m^3)	Emissions (mg)	Volume (m^3)
Triflic acid–formic acid (for formic acid only, transfer, reflux)	0.5 mL	9	13	1.4
Methylene chloride (Max., less in hood)	3 mL	350	4000	11.4
Ethanol (Max., Craig)	1.0 mL	1900	800	0.4

TXDS: Benzophenone—orl-mus LD50: 2895 mg/kg
    Hydroxylamine hydrochloride—orl-mus LD50: 400 mg/kg

QUESTIONS   **6-197.** Draw the structure of the product expected in each of the following Beckmann rearrangements. Give a suitable name for each product.

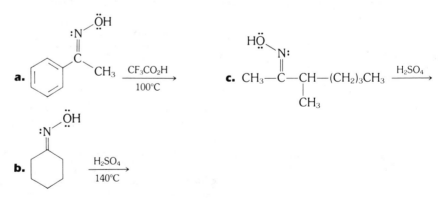

6-198. The picryl esters of oximes undergo the Beckmann rearrangement without an acid catalyst.

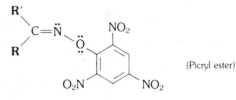

(Picryl ester)

Explain why a catalyst is not required in this rearrangement.

6-199. Compounds A and B undergo the Beckmann rearrangement with gentle heating of the solid compounds.

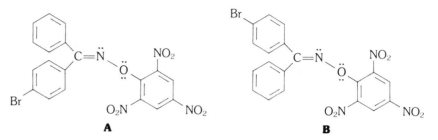

A   **B**

  a. Write the structure of the products expected.
  b. If a mixture of the two esters, A and B, is heated, what products would be formed? Explain.

6-200. When acetophenone oxime was allowed to rearrange in ^{18}O-enriched solvent, the amide product contained the same percentage of ^{18}O as the solvent. Explain this observation.

6-201. Oximes are usually crystalline materials and have been prepared as a means of identifying liquid ketones or aldehydes. It has been found in the preparation of these derivatives that if the acid concentration is too high (low pH), the oxime does not form. Explain.

6-202. The mechanical coupling between the N—H bending vibration and the C—N stretching vibration depends on a close frequency match. Normally the C—C, C—N, and C—O stretching vibrations are found in the 1200–800 cm^{-1} region. Why in the case of the amides do we find the C—N stretch approaching 1400 cm^{-1}?

6-203. In the case of benzanilide we find that the combination band region 2000–1700 cm^{-1} has a pattern corresponding to the normal set of four bands, but in this case all the bands are doubled. Explain the multiplicity of peaks between 2000 and 1700 cm^{-1}.

REFERENCES

  1. Reviews on the Beckmann rearrangement:
    a. Blatt, A. H. *Chem. Rev.* **1933**, *12*, 215.
    b. Jones, B. *Ibid.* **1944**, *35*, 335.
    c. Donaruma, L. G.; Heldt, W. Z. *Org. React.* **1960**, *11*, 1.
  2. An example of the Beckmann rearrangement using different reagents is given in Ohno, M.; Naruse, N.; Terasawa, I. *Organic Syntheses*; Wiley: New York, 1973; Collect. Vol. V, p 266.
  3. The present reaction was adapted from the work of Ganboa, I.; Palomo, C. *Synth. Commun.* **1983**, *13*, 941.

SEQUENTIAL EXPERIMENTS

The synthesis of organic molecules in academic and industrial laboratories has been one of the great achievements of the organic chemist. A vast array of compounds, which have had a profound effect on our way of life, have been devised, synthesized, and applied to many of the areas that we all too often take for granted. Just over the last 30 years, new advances in pharmaceuticals, textiles, surfactants, plastics, synthetic oils, and the like have made a great impact on our society. All this is due to the ingenuity, imagination, and knowledge of the chemist!

In practice, the synthesis of a new material usually requires a number of sequential steps. For example, the synthesis of the biologically active material cortisone requires 33 steps. Sequences of this length are rare, but syntheses requiring three to six steps are common. The objective of the organic chemist is to devise a multistep outline whereby the desired compound can be prepared efficiently, using inexpensive, readily available starting materials. For each individual step, the yield of product should be high with the minimum number of side reactions. In addition, the overall cost of the proposed project must be considered, including the time involved and the type of equipment required; safety and environmental factors must also be taken into account.

The yield of product from each individual step has a great effect on the overall yield of the final product obtained from a multistep sequence. This is so because most organic chemical transformations take place with some loss of material. *In a multistep synthesis, the overall yield is the mathematical product of the individual steps.* For example, if we assume that, in a five-step sequence for the preparation of a new organosilicone dye, each step takes place in 85% yield, the overall yield would be $(0.85)^5 \times 100$ or 44%. This result points out why a synthetic route devised for a particular compound must be well planned in order to minimize losses at each stage of the chosen route.

Experiments 47A–C; 48A–D; 41, 49, 50A–B, 51; and 25C, 52, 43B are designed as examples of multistep syntheses. When performing one or all of these multistep sequences, you will begin to understand the reason why laboratory technique is so important. The product of one reaction is used as the starting material to prepare the next compound in the sequence. When you finally isolate the *target* compound, you will feel the excitement that has come to those who have made meaningful advances in the world of chemistry.

SEQUENCE A[4]: Piperonal Reactions: The Synthesis of Piperonylonitrile from Piperonyl Alcohol (Experiments 47A–C)

Experiment 47A   Piperonal

(benzaldehyde, 3,4-(methylenedioxy-)

This experiment illustrates the selective oxidation of a primary (1°) alcohol to an aldehyde using polymer-bound chromium trioxide as the oxidizing agent.

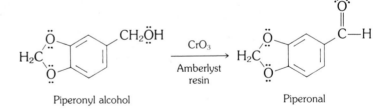

Piperonyl alcohol   Piperonal

DISCUSSION   The selective oxidation of a 1° alcohol to an aldehyde using a chromium trioxide polymer-bound oxidizing agent is demonstrated in this experiment. Normally, treatment of 1° alcohols with aqueous sodium dichromate–sulfuric acid mixtures leads to the corresponding carboxylic acids. Another selective oxidizing agent that is frequently used is pyridinium chlorochromate. With both of these selective oxidizing agents, the reaction ceases at the aldehyde stage since the oxidations are conducted in nonaqueous media.

The mechanism for the oxidation is similar to that shown in Experiment 40A.

[4] Portions of Sequence A were previously published: Mayo, D. W.; Butcher, S. S.; Pike, R. M.; Foote, C. M.; Hotham, J. R.; Page, D. S. *J. Chem. Educ.* **1985,** *62,* 149.

EXPERIMENTAL Estimated time to complete the experiment: 3.5 hours.

Physical Properties of Reactants and Products

Compound	MW	Wt/Vol	mmol	mp(°C)	bp(°C)
Piperonyl alcohol	152.16	100 mg	0.65	58	
Chromic oxide resin		750 mg			
Dioxane	88.12	1.0 mL			101
Piperonal	150.14			37	263

Reagents and Equipment
In a 5-mL round-bottom flask containing a magnetic stirring bar and equipped with a reflux condenser, place 100 mg (0.65 mmol) of piperonyl alcohol, 750 mg of the oxidizing resin, and 1.0 mL of dioxane (calibrated Pasteur pipet).

Instructor Preparation
The CrO_3 resin is prepared by adding 35 g of Amberlyst A-26 to a solution of 15 g of CrO_3 in 100 mL of water. The mixture is stirred for 30 minutes at room temperature. The resin is then collected by filtration under reduced pressure and successively rinsed with water and acetone. It is partially dried on the Buchner funnel by drawing air through the resin for 1 hour and then allowed to air-dry overnight.

Reaction Conditions
With stirring, heat the reaction mixture at reflux for a period of 1.0 hour using a sand bath temperature of 125°C.

Isolation of Product
The reaction product is cooled to room temperature and the solution transferred by Pasteur filter pipet to a funnel containing a loose cotton plug covered with 500 mg of anhydrous sodium sulfate. The filtrate is collected in a 10-mL Erlenmeyer flask. The reaction flask and resin (including the walls of the condenser) are rinsed with three 0.5-mL portions of methylene chloride solvent (calibrated Pasteur pipet). Each rinse is transferred by Pasteur filter pipet to the funnel, and the sodium sulfate is finally rinsed with an additional 0.1 mL of methylene chloride. Each rinse filtrate is combined with the original filtrate.

HOOD   A boiling stone is placed in the flask and the solvent removed **[hood]** using a sand bath maintained at a temperature of 125°C. A stream of nitrogen aids this process. The crude piperonal is obtained as an oil and is purified by column chromatography.

Purification and Characterization
A short buret column is packed with approximately 1.75 g of silica gel and the crude product transferred by Pasteur pipet to the column. The flask is rinsed with 1.0 mL of methylene chloride (calibrated Pasteur pipet) and the rinse also transferred to the column.

Methylene chloride is now added to the column, 2.0 mL at a time (calibrated Pasteur pipet), and the eluate collected in three tared 10-mL Erlenmeyer flasks. The first 4.0 mL of eluate is set aside. The next 6.0 mL is collected in one flask; this fraction contains the bulk of the product. One additional fraction of 3.0 mL is also collected in the third flask.

HOOD   The second fraction (6.0 mL) is concentrated **[hood]** using a warm sand bath and a slow stream of nitrogen to assist solvent removal. *Do not forget to add a boiling stone to the flask.*

Weigh the flask and contents. Reheat for 1 minute, cool, and weigh again. If the two weights are within 2.0 mg, the product is quite pure. If not, repeat the evaporation process until a constant weight is obtained.

Cool the product. If it does not solidify, place it in an ice bath and scratch the sides and bottom with a glass rod to induce crystallization.

Determine the melting point and compare your value with that found in the literature. IR and NMR spectra may be used to establish the identity of the material when compared with published spectra.

Chemical Tests

Chemical classification data may also be useful. The ignition test, the Tollens test, and the 2,4-dinitrophenylhydrazine test should all give a positive result for the piperonal compound.

ENVIRONMENTAL DATA

Substance	Amount	TLV (mg/m^3)	Emissions (mg)	Volume (m^3)
Dioxane (SKIN) (Max.)	1.0 mL	90	1,030	11.4
Methylene chloride (Max., evap.)	1.6 mL	350	2,140	6.1
Methylene chloride (Max., col. chromatog., evap.)	14 mL	350	18,690	53

TXDS: Piperonal—orl-rat LD50: 1700 mg/kg
      Silica gel—inv-mus LDLo: 234 mg/kg
      Chromium trioxide—ihl-hmn TCLo: 100 μg/m^3 TFX:CAR
      Sodium sulfate—orl-mus LD50: 5989 mg/kg

QUESTIONS

6-204. Primary alcohols can be oxidized to aldehydes and carboxylic acids. Often it is difficult to stop at the aldehyde stage. One technique frequently used to accomplish this (for those that boil below 100°C) is to remove the product from the reaction mixture as it is formed. This method is based on the fact that aldehydes have a lower boiling point than their corresponding alcohols. Explain this difference in boiling point.

6-205. A specific oxidizing agent for the conversion of primary alcohols to aldehydes is the pyridine–chlorochromite complex. Generally, the oxidation is run in methylene chloride solvent. For example,

$$CH_3-\underset{\underset{CH_3}{|}}{C}=CH-CH_2CH_2-CH_2\ddot{\underset{..}{O}}H \xrightarrow[\substack{CH_2Cl_2 \\ RT}]{py\cdot CrO_3{}^+Cl^-} CH_3-\underset{\underset{CH_3}{|}}{C}=CH-CH_2CH_2CHO$$

Oxidation of this alcohol with the conventional $Na_2Cr_2O_7$–H_2SO_4–water system produces the carboxylic acid. Offer an explanation for the difference in these results.

6-206. In reference to Question 6-205, can you see another advantage of the pyridine–chlorochromite reagent over the conventional conditions?

6-207. An excellent way to identify the aldehyde unit is by proton NMR. The chemical shift of the aldehydic proton occurs in the region 9–10 ppm, where few other proton signals arise. Account for this chemical shift being so far downfield.

6-208. Below are listed a series of compounds with increasing boiling points. Offer an explanation for this order.

butane (8°C)   propanal (56°C)   propanol (97°C)

REFERENCES
1. The present procedure is based on the work reported by Cainelli, G.; Cardillo, G.; Orena, M.; Sandri, S. *J. Am. Chem. Soc.* **1978,** *98*, 6737.
2. The oxidation of a primary alcohol to an aldehyde is given in Hurd, C. D.; Meinert, R. N. *Organic Syntheses*; Wiley: New York, 1943; Collect. Vol. II, p 541.

Experiments 47B, 47C Piperonal *O*-(2,4-Dinitrophenyl)oxime; Piperonylonitrile

(piperonylonitrile)

Experiments 47B and 47C illustrate the conversion of an aromatic aldehyde to an aromatic nitrile. It is an alternative to the conversion of an aromatic amine to an aromatic nitrile via the Sandmeyer reaction.

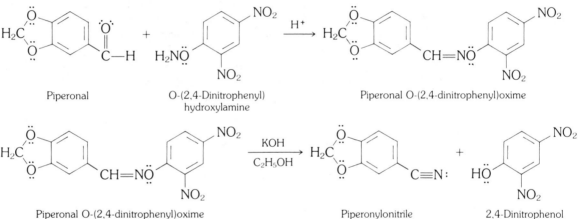

Piperonal   O-(2,4-Dinitrophenyl) hydroxylamine   Piperonal O-(2,4-dinitrophenyl)oxime

Piperonal O-(2,4-dinitrophenyl)oxime   Piperonylonitrile   2,4-Dinitrophenol

DISCUSSION An alternate method for the preparation of aromatic nitriles is demonstrated in this experiment. These compounds are usually prepared by the diazotization of the corresponding aromatic amine, followed by treatment with copper(I) cyanide. This latter sequence is known as the Sandmeyer reaction. The present reaction converts an aromatic aldehyde to an *O*-phenylated oxime, which on treatment with alcoholic base yields the nitrile by an elimination reaction. The mechanistic sequence is outlined below. (The sequence for the formation of an oxime is outlined in Experiment 46.)

EXPERIMENTAL   Estimated time to complete the experiment: two laboratory periods.

EXPERIMENT 47B: PIPERONAL
O-(2,4-DINITROPHENYL)OXIME

Physical Properties of Reactants and Products

Compound	MW	Wt/Vol	mmol	mp(°C)	bp(°C)
Piperonal	150.14	30 mg	0.20	37	263
O-(2,4-Dinitrophenyl)hydroxylamine	199.12	40 mg	0.20	111–112	
Ethanol	46.07	3.0 mL			
HCl (12 M)		2 drops			
Piperonal O-(2,4-dinitrophenyl)oxime				194–195	

Reagents and Equipment

To a 5.0-mL conical vial containing a spin vane and equipped with a reflux condenser protected by a calcium chloride drying tube, add 40 mg (0.20 mmol) of O-(2,4-dinitrophenyl)hydroxylamine and 3.0 mL of absolute ethanol. *The preparation of O-(2,4-dinitrophenyl)hydroxylamine is given in the Instructor's Manual.*

The vial is warmed gently with stirring in a sand bath to facilitate dissolution of the solid. The vial is detached from the condenser, 30 mg (0.20 mmol) of piperonal added, and the vial reattached to the condenser.

After dissolution of the aldehyde, 2 drops of 12 M HCl are added through the top of the reflux condenser by removing the drying tube. A Pasteur pipet is used for this addition. *As the HCl is delivered, the tip of the pipet should be held just above the surface of the solution. The drying tube is then reattached, and the contents are mixed by swirling.*

Reaction Conditions

The oxime forms immediately. Complete precipitation is accomplished by cooling the reaction vial in an ice bath.

Isolation of Product

The solid precipitate is collected by filtration under reduced pressure using a Hirsch funnel. The crystals are washed with three 0.5-mL portions of cold absolute ethanol (calibrated Pasteur pipet) and then air-dried on a porous clay plate or on filter paper.

Note. *Refrigeration of the filtrate for at least 24 hours produces another crop of oxime crystals. This second crop, collected by the same technique, may be combined with the initial product if its melting point is above 180 °C. Collection of the second crop is generally not required to proceed with the synthesis. This portion, however, may be used for characterization of the oxime.*

Purification and Characterization

The oxime is of sufficient purity if the melting point of the material is 187°C or greater. The oxime may be recrystallized from acetic acid or boiling acetone, if necessary. A minimum of 50 mg of the oxime should be available to proceed to Experiment 47C.

Obtain an IR spectrum of the oxime using the KBr pellet technique and compare it with that of the reference standard shown in Figure 6.27. If there is a close spectral match, proceed to Experiment 47C.

Questions and references are given at the end of Experiment 47C.

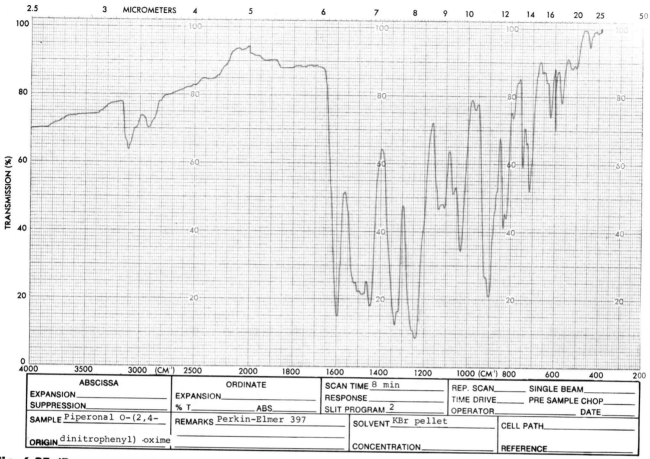

ABSCISSA		ORDINATE			SCAN TIME _8 min_		REP. SCAN_____	SINGLE BEAM_____
EXPANSION_____		EXPANSION_____			RESPONSE_____		TIME DRIVE_____	PRE SAMPLE CHOP_____
SUPPRESSION_____		% T_____	ABS_____		SLIT PROGRAM _2_		OPERATOR_____	DATE_____
SAMPLE _Piperonal O-(2,4-_		REMARKS _Perkin-Elmer 397_			SOLVENT _KBr pellet_		CELL PATH_____	
ORIGIN _dinitrophenyl) ·oxime_					CONCENTRATION_____		REFERENCE_____	

Fig. 6.27 *IR spectrum: piperonal O-(2,4-dinitrophenyl)oxime.*

EXPERIMENT 47C:
PIPERONYLONITRILE

Physical Properties of Reactants and Products

Compound	MW	Wt/Vol	mmol	mp(°C)	bp(°C)
Piperonal O-(2,4-dinitrophenyl)oxime	330.24	50 mg	0.15	194–195	
Ethanol	46.07	5 mL			78.5
Potassium hydroxide (0.2 N)		2.0 mL			
Piperonylonitrile	147.13			92–93	

Reagents and Equipment

To a 10-mL round-bottom flask containing a magnetic stirring bar and equipped with a reflux condenser, add 50 mg (0.15 mmol) of piperonal O-(2,4-dinitrophenyl)oxime, 5.0 mL of 95% ethanol, and 2.0 mL of 0.2 N ethanolic KOH.

Note. *The oxime is prepared in Experiment 47B. The 0.2 N ethanolic KOH solution is prepared using 95% ethanol.*

Reaction Conditions

The reaction mixture is slowly heated to reflux by use of a sand bath (100–110°C) and maintained at this temperature (mild reflux) for a period of 1 hour. During the initial warming period the solution turns a deep yellow.

Note. *After the heating is terminated, the solution may be cooled, and the reaction vial removed, capped, and stored until the next laboratory period.*

Isolation of Product

HOOD

The reaction flask is removed and the product mixture concentrated to a volume of 0.5 mL or less with a gentle stream of nitrogen gas while warming in a sand bath **[hood]**. *The concentration process takes a considerable length of time.*

A rinse solution is prepared by diluting 5% aqueous NaOH (1.0 mL) with distilled water (5.0 mL). This alkaline solution is used in three 2-mL portions to rinse the reaction residue. The rinses are transferred to a 12-mL capped centrifuge tube by use of a Pasteur pipet.

The resulting suspension is extracted with four 2-mL portions of methylene chloride (calibrated Pasteur pipet). The methylene chloride extract (bottom layer) is removed using a Pasteur filter pipet and the combined fractions are placed in 10-mL Erlenmeyer flask and dried over granular anhydrous sodium sulfate (0.5 g).

HOOD

By use of a Pasteur filter pipet, the dried solution is transferred to a 25-mL Erlenmeyer flask containing a boiling stone. The drying agent is rinsed with two 1-mL portions of methylene chloride, and the rinse solutions are combined with the original extract. The solvent is removed **[hood]** under a gentle stream of nitrogen while warming in a sand bath to obtain the crude piperonylonitrile.

Purification and Characterization

The crude product is purified by column chromatography using a Pasteur filter pipet filled with 300 mg of alumina (neutral, activity 1). The column is wetted with 1.0 mL of 1:1 methylene chloride–hexane solution.

The residue of crude nitrile isolated earlier is dissolved in the minimum amount of 1:1 methylene chloride–hexane solvent and the resulting solution transferred by Pasteur pipet to the column. The nitrile is eluted from the column with 2.0 mL of the methylene chloride–hexane solvent and the eluate collected in a 10-mL Erlenmeyer flask containing a boiling stone.

HOOD

The solvent is evaporated under a gentle stream of nitrogen while warming in a sand bath **[hood]**. The white needles of piperonylonitrile may be dried on a porous clay plate or on filter paper.

Weigh the product and calculate the percentage yield. Determine the melting point and compare it with the literature value. Obtain an IR spectrum and compare it with that of an authentic sample.

ENVIRONMENTAL DATA

Substance	Amount	TLV (mg/m^3)	Emissions (mg)	Volume (m^3)
Experiment 47B				
Ethanol (Max., solv., vac. filtr., cryst., wash)	4.5 mL	1900	3,550	1.9
Experiment 47C				
Ethanol, 95% (Max., solv., evap.)	7.0 mL	1900	5,250	2.8
Hexane (Max., col. chromatog.)	1.5 mL	180	990	5.5
Methylene chloride (Max., extract., evap.)	10 mL	350	13,350	38
Methylene chloride (Max., col. chromatog.)	1.5 mL	350	2,000	5.7

TLV: Potassium hydroxide—2 mg/m^3

TXDS: Piperonal—orl-rat LD50: 2700 mg/kg

     Piperonylonitrile—inv-mus LD50: 18 mg/kg

     1-Chloro-2,4-dinitrobenzene—orl-rat LD50: 1070 mg/kg

QUESTIONS

6-209. Suggest a suitable mechanism for the formation of piperonal O-(2,4-dinitrophenyl)oxime.

6-210. The Sandmeyer reaction is based on replacement of the diazonium unit in aryldiazonium salts by a chloro-, bromo-, or cyano-group. Copper salt reagents are used.

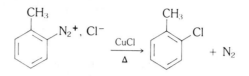

Carry out the following transformations using the Sandmeyer reaction.

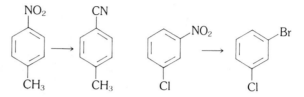

6-211. When CuCN is used in the Sandmeyer reaction, the preparation is generally carried out in a neutral medium. Can you offer an explanation of why this is done?

6-212. Outline a synthetic route for the preparation of nitriles using a carboxylic acid as the starting material.

6-213. A yellow azo dye once used to color margarine has been outlawed because it is carcinogenic. Outline a synthesis of this dye, butter yellow, starting from benzene and N,N-dimethylaniline.

$$C_6H_5-\ddot{N}=\underset{\cdot\cdot}{N}-C_6H_4-N(CH_3)_2$$

Butter yellow

REFERENCES

The procedure outlined for the preparation of piperonylonitrile is based on the work of Miller, M. J.; Loudon, G. M. *J. Org. Chem.* **1975**, *40*, 126.

SEQUENCE B: **Use of the Acetal Protecting Group to Synthesize 1,1-Diphenyl-1-buten-3-one from Ethyl Acetoacetate (Experiments 48A–D)**

Because acetals hydrolyze under very mild acidic conditions to regenerate the starting carbonyl unit, these groups are used to block or protect the carbonyl when performing a series of synthetic sequences. The series of reactions outlined in Experiments 48A–D demonstrates the use of a acetal protective group.

Experiment 48A

Ethyl Acetoacetate Ethylene Acetal

(1,3-dioxolane-2-acetic acid, 2-methyl-, ethyl ester)

The reaction of an alcohol (ethylene glycol) with a ketone in the presence of an acid catalyst forms the corresponding *acetal*, as presented in this experiment. (Previous nomenclature used the term *ketal* for these substances.)

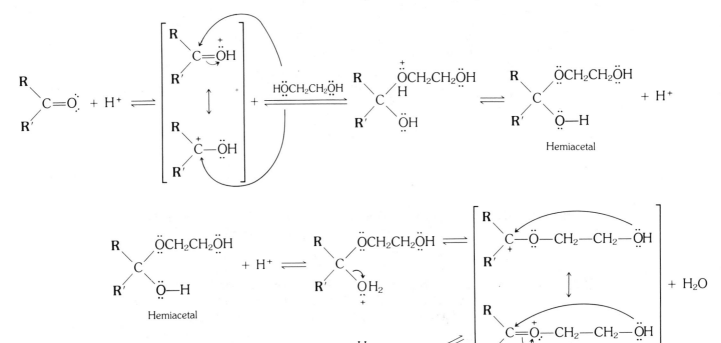

CH₃—C—CH₂—C—Ö—CH₂CH₃ + CH₂—CH₂ ⟶ CH₃—C—CH₂—C—Ö—CH₂CH₃

Ethyl acetoacetate   Ethylene glycol   Ethyl acetoacetate ethylene acetal

DISCUSSION This reaction is a classic example of the placement of a protective group in a molecule. Formation of a cyclic acetal using ethylene glycol (a 1,2-diol) is often a way to protect an aldehyde or ketone moiety.

A good protecting group must meet certain qualifications: (1) the reaction to form the group must proceed readily and in high yield, (2) it must possess stability under the conditions of further modification of the molecule, and (3) it must be readily removed when its protective function has been served.

As a protective unit, the acetal group is resistant to basic and nucleophilic reagents but is readily hydrolyzed on treatment with aqueous acid.

The acid-catalyzed mechanistic sequence is outlined below.

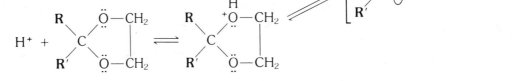

EXPERIMENTAL Estimated time to complete the reaction: 3.0 hours.

Physical Properties of Reactants and Products

Compound	MW	Wt/Vol	mmol	mp(°C)	bp(°C)	Density	n_D
p-Toluenesulfonic acid monohydrate	190.22	15 mg	0.09	106			
Ethyl acetoacetate	130.15	200 µL	1.6		180	1.03	1.4194
Ethylene glycol	62.07	100 µL	1.8		198	1.11	1.4318
Toluene	92.15	2.0 mL			111		
Ethyl acetoacetate ethylene acetal	174				109(17 torr)		1.4326

Reagents and Equipment

A 5-mL round-bottom flask containing a magnetic stirring bar and equipped with a Hickman still is charged with 15 mg (0.09 mmol) of p-toluenesulfonic acid monohydrate catalyst, 200 μL (206 mg, 1.6 mmol) of ethyl acetoacetate, 100 μL (111 mg, 1.8 mmol) of ethylene glycol, and 2.0 mL of anhydrous toluene. *It is convenient to dispense the ester and glycol using automatic delivery pipets. A graduated pipet or cylinder is sufficient to measure the toluene.*

Reaction Conditions

Using a sand bath (140–150°C), the solution is heated under vigorous reflux for a period of 1 hour. As the reaction progresses, the water formed is separated from the reaction solution by azeotropic distillation with the toluene. Then, on cooling, the water separates from the toluene in the upper portion of the Hickman still and collects on the walls and in the collar while the toluene returns to the distillation flask.

Isolation of Product

The flask is cooled to room temperature and the toluene solution extracted with 1.0 mL of 1 *M* aqueous NaOH solution followed by two 1-mL portions of water. During each extraction, the aqueous solution is added and the vial capped and then gently shaken. After separation of the layers, the cap is removed, and the lower aqueous layer is separated using a Pasteur filter pipet.

The remaining wet toluene solution is now dried over 500 mg of granular anhydrous sodium sulfate.

The dried solution is transferred by Pasteur filter pipet to a clean, dry 5.0-mL conical vial equipped with a Hickman still. The drying agent and vial are rinsed with an additional 1 mL of toluene and the rinse combined with the original toluene solution using the Pasteur filter pipet.

Using a sand bath as the heat source, the toluene solvent is removed by distillation. As the toluene distillate collects in the collar of the still, it is removed with a Pasteur pipet. The high-boiling liquid residue that remains in the pot is the ethyl acetoacetate ethylene acetal product. The toluene should distill slowly at a sand bath temperature of ~115–125°C. It is advisable not to heat the liquid residue above a sand bath temperature of 160°C.

Purification and Characterization

The acetal is sufficiently pure for characterization. It may be further purified by distillation under reduced pressure (see Experiment 52).

Determine the density and refractive index (optional) of the product and compare your values with those in the literature. Obtain an IR spectrum using the film capillary technique and compare it with that shown in Figure 6.28.

Optional Scaleup

The scaleup procedure is similar to that outlined for the microscale preparation with the following exceptions.

a. A 250-mL round-bottom flask containing a magnetic stir bar and equipped with a reflux condenser and a Dean–Stark water separator is used.

b. The amounts of reagents and solvents are increased approximately 140-fold.

Compound	MW	Wt/Vol	mol	mp(°C)	bp(°C)	Density	n_D
p-Toluenesulfonic acid monohydrate	190.22	0.13 g	0.0007	106			
Ethyl acetoacetate	130.15	30 g	0.23		180	1.03	1.4194
Ethylene glycol	62.07	15 g	0.24		198	1.11	1.4318
Toluene	92.15	100 mL			111		

c. The solution is heated at reflux until the theoretical amount of water is removed (4.2 mL).

d. The reaction mixture is cooled to room temperature, transferred to a separatory funnel, washed with 35 mL of 10% NaOH solution and two 50-mL portions of water, and then dried over anhydrous sodium sulfate.

e. The sodium sulfate is removed by filtration and the filtrate distilled under reduced pressure to yield the desired acetal product.

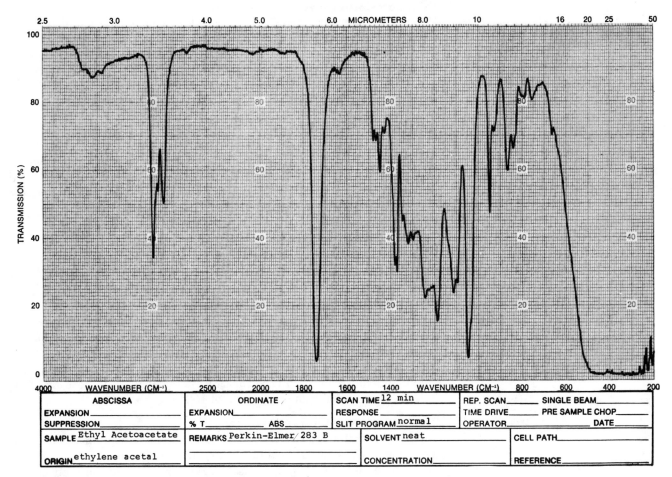

Fig. 6.28 *IR spectrum: ethyl acetoacetate ethylene acetal.*

ENVIRONMENTAL DATA

Substance	Amount	TLV (mg/m³)	Emission (mg)	Volume (m³)
Ethylene glycol (Max., low vapor pressure)	111 mg	125	111	0.9
Toluene (Max., distilled)	3.0 mL	375	2600	6.9

TLV: Sodium hydroxide—2 μg/m³
TXDS: Ethyl acetoacetate—orl-rat LD50: 3980 mg/kg
        p-Toluenesulfonic acid—orl-rat LD50: 2480 mg/kg
        Sodium sulfate—orl-mus LD50: 5989 mg/kg

QUESTIONS **6-214.** What is meant by the term *azeotropic distillation*? How is this technique used to advantage in this experiment?

6-215. Cyclic acetals are important protecting groups because they are stable in aqueous base. That is, a carbonyl unit can be protected in alkaline solutions and regenerated in aqueous acid. Based on this fact, carry out the following transformation.

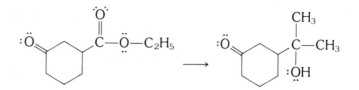

6-216. Aldehydes and ketones can also form cyclic thioacetals when $HSCH_2CH_2SH$ is used in place of $HOCH_2CH_2OH$. These thioacetals are useful because they undergo reduction with hydrogen in the presence of Raney nickel [Ni(Al)] catalyst to yield hydrocarbons.

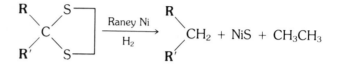

Knowing this fact, carry out the following transformations.

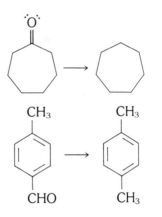

6-217. Heptanal, when treated with 1,2,3-propanetriol (glycerol), forms four different isomers. What are the structures of these species?

6-218. Outline simple chemical tests that would distinguish between the following pairs of compounds.
   a. Pentanal and 2-pentanone
   b. 2-Pentanol and 2-pentanone
   c. Benzaldehyde and benzyl alcohol

REFERENCES

1. This experiment is adapted from Paulson, D. R.; Hartwig, A. L.; Morgan, G. F. *J. Chem. Educ.* **1973**, *50*, 216, and Rivett, D. E. A. Ibid., **1980**, *57*, 751.
2. References for the preparation of acetals in *Organic Syntheses*:
   a. Renoll, M.; Newman, M. S. *Organic Syntheses*; Wiley: New York, 1955; Collect. Vol. III, p 502.
   b. Issidorides, C. H.; Gulen, R. *Organic Syntheses*; Wiley: New York, 1963; Collect. Vol. IV, p 679.
   c. Daignault, R. A.; Eliel, E. L. *Organic Syntheses*; Wiley: New York, 1973; Collect. Vol. V, p 303.

Experiment 48B

1,1-Diphenyl-1-hydroxy-3-butanone Ethylene Acetal

(1,3-dioxolane, 4-hydroxy-4,4-diphenyl-2-methyl-)

The addition of a Grignard reagent to an ester is a classic method for the synthesis of *tert*-alcohols. This experiment demonstrates the reaction.

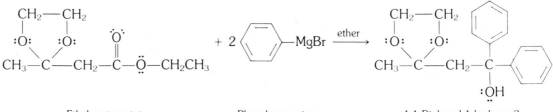

| Ethyl acetoacetate ethylene acetal | Phenylmagnesium bromide | 1,1-Diphenyl-1-hydroxy-3-butanone ethylene acetal |

DISCUSSION This reaction is the second stage in this sequence. Note that the ketone group is protected by the cyclic acetal unit and thus is not vulnerable to attack by the Grignard reagent. For discussion of the reaction of Grignard reagents with carbonyl compounds, see Experiments 17 and 18. The mechanism is outlined below.

The intermediate ketone reacts with another equivalent of Grignard reagent to form the *tert*-alcohol. The mechanism of this stage is outlined in Experiment 17.

EXPERIMENTAL Estimated time to complete the experiment: 3.5 hours.

Physical Properties of Reactants and Products

Compound	MW	Wt/Vol	mmol	mp(°C)	bp(°C)	Density	n_D
Bromobenzene	157.02	155 µL	1.4		156	1.50	1.5597
Diethyl ether	74.12	1.1 mL			34.5		
Iodine	253.81	1 crystal					
Magnesium	24.31	35 mg	1.4	649			
Ethyl acetoacetate ethylene acetal	174	100 mg	0.58		109 (17 torr)		1.4326
1,1-Diphenyl-1-hydroxy-3-butanone ethylene acetal	284			90–91			

Reagents and Equipment

Preparation of Phenylmagnesium Bromide
This reagent is prepared exactly as described in Experiment 17. The order of addition, manipulations, and precautions is the same. The equipment is also identical.

The following reagents and amounts are used.

1. Magnesium—35 mg (1.4 mmol)
2. Iodine—one small crystal
3. Anhydrous diethyl ether—800 μL
4. Bromobenzene—225 mg (1.4 mmol, 155 μL)

The Acetal-Ester Reagent
A solution of 100 mg (0.58 mmol) of the acetal-ester, prepared in Experiment 48A, dissolved in 300 μL of diethyl ether is prepared in a dry 1-mL conical vial. The solution is immediately drawn into a 1.0-mL syringe, and the syringe is then inserted into the rubber septum on the Claisen head. An additional 100 μL of ether is placed in the vial as a rinse to be added later.

HOOD **Note.** *The ether is measured in the* **hood** *using an automatic delivery pipet. The acetal-ester reagent must be free of any acidic impurities. An ether solution of the material may be passed through a column of alumina (500 mg in a Pasteur filter pipet) if necessary.*

Reaction Conditions
With stirring, add the ether solution of the acetal-ester dropwise to the warm Grignard solution so as to keep the ether solvent under steady reflux. The rinse saved earlier is then added in like manner in one portion. The resulting mixture is then stirred for a period of 15 minutes while the reflux conditions are maintained by warming the reaction mixture with a sand bath.

The resulting mixture is allowed to cool to room temperature. The conical vial is finally removed and capped.

Isolation of Product
CAREFUL The alkoxide magnesium salt is hydrolyzed by the *careful* dropwise addition of 1 mL of saturated ammonium chloride solution using a Pasteur pipet.

Caution: *Addition of the dilute acid solution is accompanied by the evolution of heat. An ice bath should be kept handy to cool the solution if necessary. A two-layer reaction solution (ether–water) forms as the solid dissolves.*

The spin vane is now removed with forceps and set aside to be rinsed with an ether wash. The vial is capped tightly, shaken, and vented, and the layers are allowed to separate.

Using a Pasteur filter pipet, transfer the bottom aqueous layer to a clean 5.0-mL conical vial.

SAVE The ether layer is **SAVED**; *it contains the crude reaction product.*

The aqueous layer transferred to the 5.0-mL vial is now extracted with 0.5 mL of diethyl ether, rinsing the spin vane with the ether as it is added to the vial by means of a Pasteur filter pipet. The vial is capped, shaken, and vented, and the layers are allowed to separate. Using a Pasteur filter pipet, the bottom aqueous layer is removed. The ether layer is then combined with the ether extracts previously saved.

The combined ether solution is partitioned (using the technique just described) with several 1.0-mL portions of saturated ammonium chloride solution until the aqueous extracts are no longer alkaline to litmus. The wet ether solution is dried by passing the solution through a Pasteur filter pipet packed with 500 mg of anhydrous sodium sulfate. The vial is rinsed with an additional 0.5-mL portion of ether and this also is transferred to the drying column. The eluate is collected in a tared 10-mL Erlenmeyer flask containing a boiling stone.

HOOD   The ether is evaporated by gently warming the Erlenmeyer flask in a sand bath in the **hood** to yield a light yellow oil. Cooling this oil in an ice bath, while scratching the walls of the flask with a small glass rod, produces a yellow solid. The flask may be weighed to obtain the crude weight of the product.

Purification and Characterization

The crude yellow product is recrystallized from methanol using the Craig tube. The light yellow, cubic crystals are dried on a clay plate or on filter paper.

Weigh the crystals and calculate the percentage yield. Determine the melting point and compare it with the literature value. Obtain an IR spectrum using the KBr pellet technique and compare it with that in Figure 6.29.

Optional Scaleup   **a.** A 500-mL round-bottom flask equipped with a reflux condenser, addition funnel, mechanical stirrer, and protected by calcium chloride drying tubes is used.

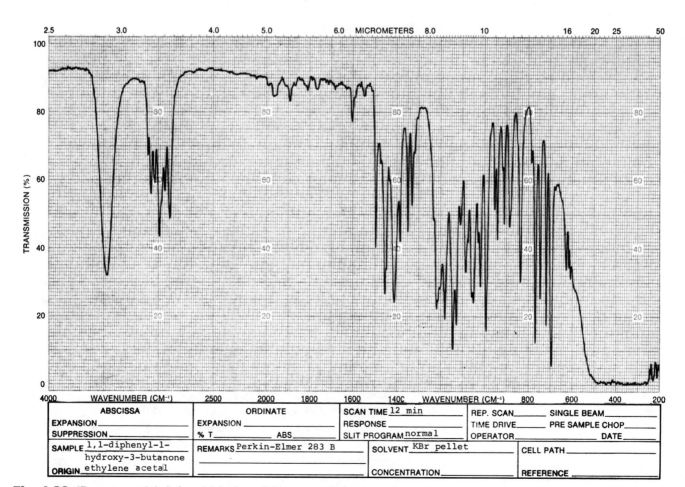

Fig. 6.29 *IR spectrum: 1,1-diphenyl-1-hydroxy-3-butanone ethylene acetal.*

b. The amounts of reagents and solvents are increased approximately 140-fold.

Compound	MW	Wt/Vol	mol	mp(°C)	bp(°C)	Density	n_D
Bromobenzene	157.02	31.4 g	0.2		156	1.50	1.5597
Diethyl ether	74.12	130 mL			34.5		
Iodine	253.81	2 crystals					
Magnesium	24.31	5.35 g	0.22	649			
Ethyl acetoacetate ethylene acetal	174	17.4 g	0.1		109 (17 torr)		1.4326

c. The Grignard reagent is prepared first by adding a solution of bromobenzene dissolved in 50 mL of ether to a mixture of 30 mL of ether, 5.35 g magnesium, and the iodine crystals in the reaction flask. All the normal precautions for running the Grignard reaction on a large scale are taken.

d. The Grignard solution is cooled in an ice bath and, with stirring, the acetal-ester dissolved in 50 mL of ether is added dropwise. A thick white gum forms at this stage. After the addition is complete the reaction mixture is stirred at room temperature for a 30-minute period.

e. A mixture of 50 mL of water and 50 g of ice is now added. When the ice has melted, 50 mL of ether is added and the mixture stirred until the gummy residue dissolves. The layers are separated using a separatory funnel and the water layer is washed with 50 mL of ether. The ether wash is added to the original ether solution. The combined ether solution is now washed with 50 mL of water and the ether extract dried over anhydrous sodium sulfate.

f. The dried ether solution is filtered to remove the sodium sulfate and then concentrated on a rotary evaporator to yield the crystalline product. It may be recrystallized from petroleum ether.

ENVIRONMENTAL DATA

Substance	Amount	TLV (mg/m³)	Emissions (mg)	Volume (m³)
Diethyl ether (Max., evap.)	2.1 mL	1200	1500	1.25
Methanol (SKIN) (Max., recryst., Craig)	2.0 mL	260	1580	6.1

TLV: Iodine—0.1 ppm
Ammonium chloride—10 mg/m³
TXDS: Magnesium—orl-dog LDLo: 230 mg/kg
Sodium sulfate—orl-mus LD50: 5989 mg/kg
TMDX: Bromobenzene—dnr-esc 250 mg/L

QUESTIONS

6-219. In the reaction of Grignard reagents with esters, a ketone is formed in the first stage of the reaction. Since ketones are more reactive toward Grignard reagents than esters, the ketone reacts immediately with additional Grignard reagent to yield the tertiary alcohol after hydrolysis. Offer an explanation of why ketones are more reactive than esters toward Grignard reagents.

6-220. Can you suggest a reason why Grignard reagents do not add to the carbonyl group of amides?

6-221. Give the structure and name of the product formed by reaction of propylmagnesium bromide with ethyl formate in ether solvent followed by hydrolysis using ammonium chloride solution.

6-222. Addition of the Grignard reagent to an ester is another useful preparation of alcohols. Can you indicate at least three other methods by which alcohols may be prepared?

6-223. In this experiment, the synthesis of a *tert*-alcohol is illustrated. Indicate the reagents one would use to synthesize *primary* and *secondary* alcohols by use of the Grignard reaction.

6-224. What chemical classification tests would you use to distinguish between the following pairs of alcohols (see Chapter 7).
- **a.** 1-Pentanol and 2-pentanol
- **b.** *tert*-Butyl alcohol and 2-butanol
- **c.** 1-Butanol and 2-methyl-2-propanol

REFERENCES
1. This experiment was adapted from Paulson, D. R.; Hartwig, A. L.; Morgan, G. F. *J. Chem. Educ.* **1973,** *50,* 216, and Rivett, D. E. A. Ibid., **1980, 57,** 751.
2. For a general reference on the reaction of Grignard reagents with esters see Kharasch, M. S.; Reinmuth, O. *Grignard Reactions of Non-metallic Substances;* Prentice–Hall: New York, 1954; p 557.

Experiment 48C · 1,1-Diphenyl-1-hydroxy-3-butanone

(3-butanone, 1-hydroxy-1,1-diphenyl-)

Acid hydrolysis of a acetal protecting group to regenerate the original ketone carbonyl unit is demonstrated in this experiment.

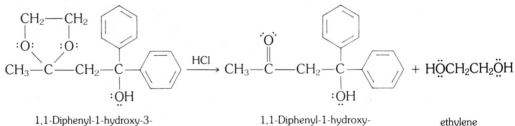

| 1,1-Diphenyl-1-hydroxy-3-butanone ethylene acetal | 1,1-Diphenyl-1-hydroxy-3-butanone | ethylene glycol |

DISCUSSION This reaction illustrates the removal of the acetal protecting group from a carbonyl unit. It is the third stage in this sequence of reactions. If the Grignard reaction conducted in Experiment 48B were used without first protecting the keto function, it too would have reacted with the Grignard reagent since a ketone carbonyl is more reactive than an ester to nucleophilic attack. The formation of the acetal and its hydrolysis constitute an equilibrium reaction. The mechanism for the hydrolysis reaction is shown in Experiment 48A.

EXPERIMENTAL Estimated time to complete the experiment: 2.0 hours.

Physical Properties of Reactants and Products

Compound	MW	Wt/Vol	mmol	mp(°C)	bp(°C)
1,1-Diphenyl-1-hydroxy-3-butanone ethylene acetal	284	50 mg	0.19	90–91	
Acetone	58.08	2.0 mL			56
1 *M* HCl solution		100 µL			
1,1-Diphenyl-1-hydroxy-3-butanone	238			85–86	

Reagents and Equipment

HOOD

To a 5.0-mL conical vial containing a magnetic spin vane and equipped with an air condenser are added 50 mg (0.19 mmol) of the recrystallized hydroxyacetal prepared in Experiment 48B, 2.0 mL of acetone, and 100 μL of 1 *M* HCl solution. *The HCl is dispensed using an automatic delivery pipet. A graduated cylinder or pipet may be used to deliver the acetone* **[hood]**.

Reaction Conditions

The mixture is now heated at vigorous reflux for a period of 15 minutes using a sand bath maintained at a temperature of 80–90°C.

Isolation of Product

Saturated, aqueous sodium bicarbonate solution (1.0 mL) is added to the cooled reaction solution (calibrated Pasteur pipet) and the aqueous mixture extracted with three 1.0-mL portions of diethyl ether. For each extraction the ether is added, and the vial is capped, shaken, and vented. The ether layer is separated, using a Pasteur filter pipet, and transferred to a 5.0-mL conical vial.

The combined ether extracts are partitioned with a 1-mL portion of saturated, aqueous sodium bicarbonate solution followed by a 1-mL portion of water. For each of these two extractions, the vial is capped, shaken, and vented, and the lower aqueous layer separated (Pasteur filter pipet). The wet ether layer is dried over ~500 mg of anhydrous sodium sulfate.

The dried ether solution is now transferred (Pasteur filter pipet) to a 10-mL Erlenmeyer flask. The drying agent is rinsed with an additional 1 mL of diethyl ether and the rinse solution also transferred to the Erlenmeyer flask.

HOOD

The ether is evaporated in the **hood** using a filtered stream of air while warming in a sand bath to yield the crude hydroxy ketone.

Purification and Characterization

Recrystallization of the crude material from hexane using the Craig tube yields pure 1,1-diphenyl-1-hydroxy-3-butanone. The crystals are dried on a porous clay plate.

Weigh the product and calculate the percentage yield. Determine the melting point and compare your result with the literature value.

Obtain the IR spectrum using the KBr pellet technique and compare it with that in Figure 6.30.

Optional Scaleup **a.** The reaction is run using a 100-mL round-bottom flask containing a magnetic stirrer and equipped with a reflux condenser.

b. The amounts of reagents and solvents are increased approximately 20-fold.

Compound	MW	Wt/Vol	mol	mp(°C)	bp(°C)
1,1-Diphenyl-1-hydroxy-3-butanone ethylene acetal	284	1.0 g	0.0035	90–91	
Acetone	58.08	50.0 mL			56
0.02 *M* HCl solution		1.0 mL			

c. The reaction mixture is heated at reflux with stirring for 15 minutes, cooled in an ice bath, and poured into a 125-mL Erlenmeyer flask containing 10 mL of saturated sodium bicarbonate solution and 10 mL of water.

d. Using a separatory funnel, the aqueous layer is separated and washed with two 50-mL portions of ether. The original ether layer is combined with the ether washings and dried over anhydrous sodium sulfate.

e. The dried ether solution is filtered to remove the sodium sulfate and the filtrate concentrated to dryness on a rotary evaporator to yield the crystalline product. It may be recrystallized from hexane.

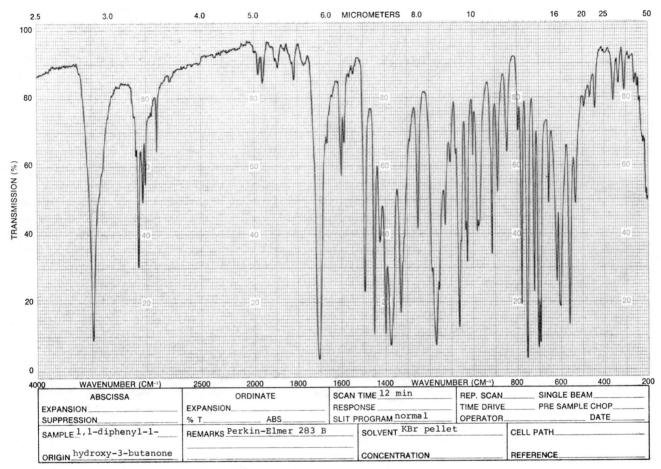

Fig. 6.30 *IR spectrum: 1,1-diphenyl-1-hydroxy-3-butanone.*

ENVIRONMENTAL DATA

Substance	Amount	TLV (mg/m^3)	Emissions (mg)	Volume (m^3)
Hydrochloric acid (1 M)	0.1 mL	7	3.7	0.5
Acetone (Max., solvent, water soluble, reflux)	2 mL	1780	1580	0.9
Diethyl ether (Max., extract, evap.)	4 mL	1200	2860	2.4
Hexane (recryst., Craig)	2 mL	180	1320	7.3

TLV: Sodium bicarbonate—5 mg/m^3
TXDS: Sodium sulfate—orl-mus LD50: 5989 mg/kg

QUESTIONS **6-225.** How could you distinguish between the starting reagent of this experiment and the isolated product by using chemical tests (see Chapter 7)?

6-226. Aldehydes and ketones form hemiacetals when treated with alcohols in the presence of acid or base. Draw the structures of the cyclic hemiacetals generated in the following reactions.

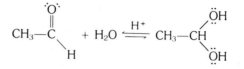

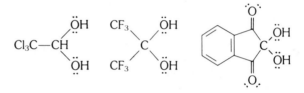

6-227. Aldehydes and ketones form hydrates (gem-diols) on addition of water under acid or base conditions.

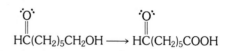

For most aldehydes the equilibrium lies far to the left so the reaction is not synthetically important. Can you explain why each of these hydrates is stable and can be isolated?

6-228. Acetals are stable in alkaline solution. Why are they not hydrolyzed as in acid solution?

6-229. Write the structures of the cyclic acetals formed by the following reagents.
    a. Benzaldehyde and ethylene glycol
    b. Cyclohexanone and 1,3-propanediol
    c. Isobutyl methyl ketone and 1,2-ethanediol

6-230. Primary alcohols are readily oxidized to carboxylic acids by potassium permanganate in neutral solution. Aldehydes are also converted to carboxylic acids under these conditions. Keeping these facts in mind, suggest a way to complete the following conversion.

$$\overset{\overset{\cdot\cdot}{\underset{\cdot\cdot}{O}}}{\overset{\|}{\text{HC}}}(CH_2)_5CH_2OH \longrightarrow \overset{\overset{\cdot\cdot}{\underset{\cdot\cdot}{O}}}{\overset{\|}{\text{HC}}}(CH_2)_5COOH$$

REFERENCES This experiment is adapted from the work reported by Rivett, E. A. *J. Chem. Educ.* **1980,** *57,* 751.

Experiment 48D   1,1-Diphenyl-1-buten-3-one

(1,1-diphenyl-1-buten-3-one)

The dehydration of a β-keto alcohol to give an α,β-unsaturated ketone is demonstrated in this experiment.

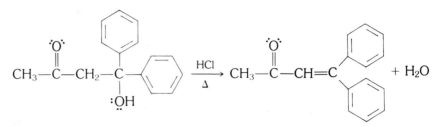

1,1-Diphenyl-1-hydroxy-3-butanone   1,1-Diphenyl-1-buten-3-one

DISCUSSION   This experiment illustrates a very important reaction in organic chemistry: the dehydration of a 3° alcohol to yield an alkene. The reaction is classified as an acid-catalyzed elimination, and in the present case elimination may proceed in only one direction. Thus, only one olefin is produced. This completes the total sequence of reactions begun in Experiment 48A. The mechanism for the dehydration of an alcohol to an alkene is outlined in Experiment 9.

EXPERIMENTAL   Estimated time to complete the reaction: 3.0 hours.

Physical Properties of Reactants and Products

Compound	MW	Wt/Vol	mmol	mp(°C)	bp(°C)
1,1-Diphenyl-1-hydroxy-3-butanone	238	35 mg	0.15	85–86	
Conc. HCl		100 μL			
Acetone	58.08	0.5 mL			56
1,1-Diphenyl-1-buten-3-one	220			34–36	192–194(15 torr)

Reagents and Equipment

In a 3.0-mL conical vial containing a magnetic spin vane and equipped with a reflux condenser are placed 35 mg (0.15 mmol) of 1,1-diphenyl-1-hydroxy-3-butanone (prepared in Experiment 48C), 100 μL of concentrated HCl, and **HOOD** 0.5 mL of acetone. *The reagents are dispensed in the **hood**. An automatic delivery pipet is used for the acid and a graduated pipet for the acetone.*

Reaction Conditions

With stirring, the reaction mixture is heated at gentle reflux, using a sand bath temperature of ~70°C, for a period of 30 minutes. *The solution develops a deep red color about 5–10 minutes after the solution reaches the reflux temperature.*

Isolation of Product

The reaction solution is cooled to room temperature and diluted with 1 mL of water and then extracted with two 0.5-mL portions of diethyl ether. With each extraction after the addition of the ether, the vial is capped, shaken, and vented. The ether extracts are transferred to a 3.0-mL conical vial with a Pasteur filter pipet.

The ether solution is then extracted with two 0.5-mL portions of saturated sodium bicarbonate solution followed by one portion (0.5 mL) of water. The extraction procedure is identical to that outlined earlier. The resulting wet ether fraction is dried over 500 mg of anhydrous sodium sulfate.

The dried ether solution is transferred, with a Pasteur filter pipet, to a 10-mL Erlenmeyer flask containing a boiling stone. The drying agent is washed with an additional 0.5 mL of ether which is also transferred to the Erlenmeyer flask.

HOOD   The ether solvent is evaporated in the **hood** using a warm sand bath to yield the unsaturated ketone product as a yellow oil.

Purification and Characterization

The crude material is purified by column chromatography. In a Pasteur filter pipet is placed 0.5 g of activated silica gel followed by 0.5 g of anhydrous sodium sulfate. The crude material isolated earlier is dissolved in 100 μL of diethyl ether and the solution then transferred by Pasteur pipet to the column. The material, which appears as a yellow band on the column, is eluted with toluene ($\sim$5.0 mL) and the eluate collected in a tared 25-mL filter flask.

The toluene solvent is removed under reduced pressure while the flask is
HOOD   warmed on a sand bath in the **hood** (see Chapter 5, Technique 8).

Weigh the pure 1,1-diphenyl-1-buten-3-one and calculate the percentage yield. Obtain an IR spectrum using the hot-melt technique (IR lamp) and compare it with that shown in Figure 6.31.

ENVIRONMENTAL DATA

Substance	Amount	TLV (mg/m^3)	Emissions (mg)	Volume (m^3)
Hydrochloric acid (conc.)	0.1 mL	7	46	6.5
Acetone (Max., pipet, heat, dilute with water)	1.0 mL	1780	790	0.4
Diethyl ether (Max., extract, evap.)	1.6 mL	1200	1140	1.0
Toluene (Max., col. chromatog., evap.)	5.0 mL	375	4335	11.6

TLV: Sodium bicarbonate—5 mg/m^3
TXDS: Sodium sulfate—orl-mus LD50: 5989 mg/kg
          Silica—inv-mus LDLo: 234 mg/kg

QUESTIONS   **6-231.** Why is 1,1-diphenyl-1-buten-3-one colored?

6-232. The relative ease with which alcohols undergo dehydration is $3° > 2° > 1°$. Offer a reasonable explanation for this order.

6-233. In relation to Question 6-232, draw potential energy diagrams that reflect the relative energy of activation for the dehydration of the alcohols.

6-234. Citric acid is converted to isocitric acid in the Krebs cycle through a dehydration–hydration sequence catalyzed by aconitase enzyme.

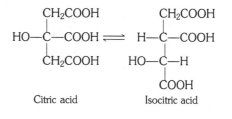

Citric acid   Isocitric acid

The reaction proceeds by way of formation of an intermediate olefin, Z-aconitic acid. Give the structure of this acid intermediate.

6-235. Identify the alkene(s) formed on dehydraton of the following compounds.

    a. 2-Butanol   **b.** 3-Ethyl-3-pentanol   **c.** 2,3-Dimethyl-2-butanol

REFERENCES   **1.** The reaction given is adapted from Paulson, D. R.; Hartwig, A. L.; Moran, G. F. *J. Chem. Educ.* **1973,** *50,* 216, and Rivett, D. E. A. Ibid., **1980,** *57,* 751.

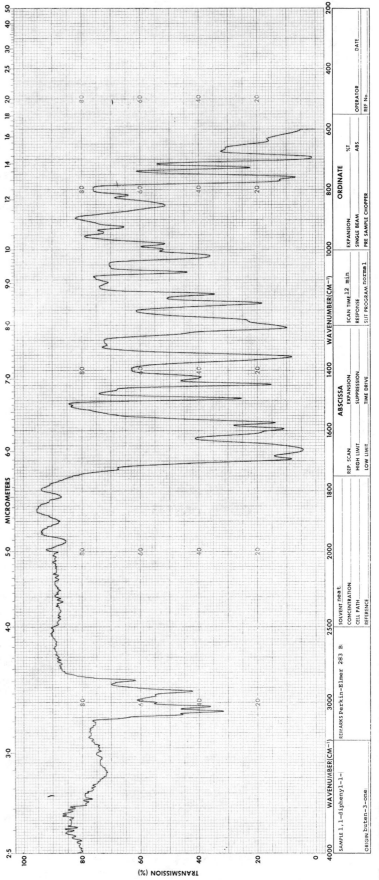

Fig. 6.31 IR spectrum: 1,1-diphenyl-1-buten-3-one.

2. Selected references from *Organic Syntheses* that illustrate the dehydration of alcohols to alkenes:
   a. Allen, C. F. H.; Converse, S. *Organic Syntheses;* Wiley: New York, 1941; Collect. Vol. I, p 226.
   b. Conant, J. B.; Tuttle, N. Ibid., p 345.
   c. Norris, J. F. Ibid., p 431.
   d. Adkins, H.; Zartman, W. *Organic Syntheses;* Wiley: New York, 1943; Collect, Vol. II, p 606.
   e. Wiley, R. H.; Waddey, W. E. *Organic Syntheses;* Wiley: New York, 1955; Collect. Vol. III, p 560.
   f. Grummitt, O.; Becker, E. I. *Organic Syntheses;* Wiley: New York, 1963; Collect. Vol. IV, p 771.

SEQUENCE C: Synthesis of Hexaphenylbenzene from Benzoin and *Z*-Stilbene (Experiments 49, 50A, 50B, 51)

The preparation of hexaphenylbenzene demonstrates the manner in which a variety of basic organic reactions can be integrated to prepare a desired end product. An outline of the overall scheme is presented below (Experiment numbers are given in brackets.)

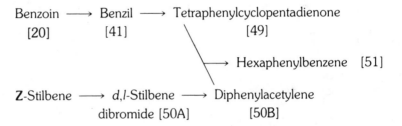

Benzoin $\longrightarrow$ Benzil $\longrightarrow$ Tetraphenylcyclopentadienone
[20]   [41]   [49]
                                                    $\searrow$
                                                        $\longrightarrow$ Hexaphenylbenzene [51]
                                                    $\nearrow$
Z-Stilbene $\longrightarrow$ *d,l*-Stilbene $\longrightarrow$ Diphenylacetylene
                dibromide [50A]   [50B]

As the flow chart illustrates, the total synthesis of hexaphenylbenzene involves an oxidation followed by an aldol condensation to obtain a diene and a bromination reaction followed by dehydrohalogenation to obtain a dienophile. A pericyclic Diels–Alder reaction using these end products completes the sequence.

Experiment 49   Tetraphenylcyclopentadienone

(2,4-cyclopentadien-1-one, 2,3,4,5-tetraphenyl-)

Use of the aldol condensation for the synthesis of a five-membered carbocyclic ring system is demonstrated in this experiment.

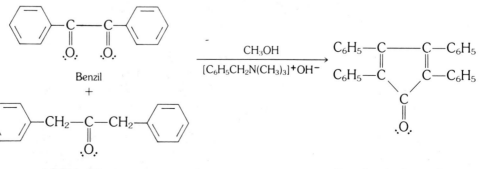

Benzil
+
1,3-Diphenylacetone

$\xrightarrow[\text{[C}_6\text{H}_5\text{CH}_2\text{N(CH}_3)_3]^+\text{OH}^-]{\text{CH}_3\text{OH}}$

Tetraphenylcyclopentadienone

DISCUSSION   This experiment is a further example of the aldol condensation (see Experiment 22 for discussion). The reaction carried out in this sequence differs in that two ketones, one of which has no α-hydrogen atoms, are the reactants. It is unique because the selected reagents lead to the formation of a carbocyclic ring system. The aldol product initially formed undergoes an elimination reaction to yield a material that has a highly conjugated system of double bonds. In general, the more conjugation in a molecule, the less energy required to promote the π electrons to a higher energy level. In this case, energy in the visible region of the spectrum is absorbed, resulting in a product with a deep purple color. The UV–visible spectrum exhibits two fairly intense absorption maxima at 510 and 345 nm.

The mechanism is a sequence of two aldol condensations. The first is intermolecular; the second, intramolecular. The mechanism is similar to that outlined in Experiment 22.

The product is the diene used for the preparation of hexaphenylbenzene (see Experiment 51).

EXPERIMENTAL   Estimated time to complete the experiment: 1.5 hours.

Physical Properties of Reactants and Products

Compound	MW	Wt/Vol	mmol	mp(°C)	bp(°C)
1,3-Diphenylacetone	210.28	50 mg	0.24	35	
Benzil	210.23	50 mg	0.24	95	
Triethylene glycol	150.18	0.25 mL			278
Benzyltrimethylammonium hydroxide (40% solution in methanol)		50 μL			
Tetraphenylcyclopentadienone	384.48			220–221	

Reagents and Equipment

In a 1.0-mL conical vial containing a magnetic spin vane and equipped with an air condenser, place 50 mg (0.24 mmol) of 1,3-diphenylacetone and 50 mg (0.24 mmol) of benzil followed by 0.25 mL of triethylene glycol.

Note. a. *The benzil used in this reaction must be free of benzoin impurity. If benzil is prepared according to Experiment 41, it should be purified by the chromatographic procedure cited therein.*

b. *1,3-Diphenylacetone is a low-melting solid. If the reactant is in the half liquid–half solid form when opened, use only the solid material for this experiment. The liquid material is quite impure and noticeably lowers the yield of tetraphenylcyclopentadienone.*

HOOD   **c.** *1,3-Diphenylacetone has an unpleasant odor. It should be kept in the* **hood**.

Reaction Conditions

The mixture is heated with stirring in a sand bath maintained at a temperature of 155–165°C for a period of 10 minutes. The benzil dissolves during this time.

The reaction solution is removed from the bath and immediately 50 μL of a 40% benzyltrimethylammonium hydroxide–methanol solution is added (automatic delivery pipet) to the hot reactants with gentle shaking. The vial and contents are again heated for 2–3 minutes at 150–160°C and then allowed to cool. As the cooling occurs, the dark purple crystals of tetraphenylcyclopentadienone are evident. *Cooling may be accelerated by placing the vial under a stream of cold water.*

Isolation of Product

With stirring, 0.7 mL of cold methanol is added and the mixture cooled in an ice bath for 5–10 minutes. The dark crystals are collected by filtration under reduced pressure by use of a Hirsch funnel. The reaction vial and crystals are rinsed with several drops of cold methanol. The dropwise addition of cold methanol to the crystals is continued until the product appears purple and not brown. The crystalline product is dried on a porous clay plate or on filter paper.

Purification and Characterization

The product is of sufficient purity for use in the preparation of hexaphenylbenzene (Experiment 51).

If a purer material is desired, it may be recrystallized from triethylene glycol.

Weigh the tetraphenylcyclopentadienone product and calculate the percentage yield. Determine the melting point and compare it with the literature value.

The comparison of the UV–visible spectra of benzil (see data given in Experiment 41) and the product may be used to demonstrate the shift of absorption bands with increased conjugation in the molecule.

The UV–visible data for tetraphenylcyclopentadienone are summarized below and the spectrum is presented in Figure 6.32.

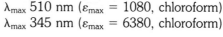

λ_{max} 510 nm (ε_{max} = 1080, chloroform)
λ_{max} 345 nm (ε_{max} = 6380, chloroform)

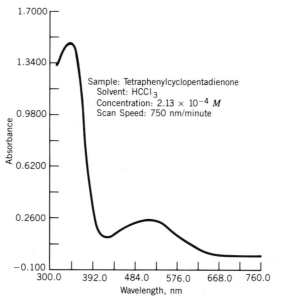

Fig. 6.32 *UV-visible spectrum: tetraphenylcyclopentadienone.*

Optional Scaleup   If additional material is desired for subsequent reactions in the sequence, this preparation may be run at a scale five times larger than that given in the microscale procedure.

The scaleup procedure is similar to that given for the microscale run, with the following exceptions.

a. A 10-mL round-bottom flask containing a magnetic spin bar and fitted with an air condenser is used.

b. The reagent and solvent amounts are increased fivefold.

Compound	MW	Wt/Vol	mmol	mp(°C)	bp(°C)
1,3-Diphenylacetone	210.28	250 mg	1.2	35	
Benzil	210.23	250 mg	1.2	95	
Triethylene glycol	150.18	1.5 mL			278
Benzyltrimethylammonium hydroxide (40% solution in methanol)		250 μL			

c. After addition of the catalyst, the mixture is heated at 155°C with stirring for 5 minutes. On addition of the catalyst, the solution turns a deep purple color.

d. After the reaction mixture is cooled, 4 mL of methanol is added.

e. The product may be characterized as outlined in the microscale procedure.

ENVIRONMENTAL DATA

Substance	Amount	TLV (mg/m^3)	Emissions (mg)	Volume (m^3)
Methanol (SKIN) (Max., solv., recryst.)	2.0 mL	260	1580	6.1

TXDS: Benzyltrimethylammonium hydroxide—suc-mus LDLo: 35 mg/kg
Triethylene glycol—orl-rat LD50: 17 g/kg
Benzil—orl-rat LD50: 2710 mg/kg

QUESTIONS

6-236. Outline a complete mechanistic sequence to account for formation of the tetraphenylcyclopentadienone compound.

6-237. Cyclopentadienone is unstable and rapidly undergoes the Diels–Alder reaction with itself. Write the structure for this Diels–Alder adduct.

6-238. The Diels–Alder adduct of Question 6-237 undergoes a fragmentation reaction on heating to produce a bicyclotrienone compound plus carbon monoxide. Suggest a structure for this product.

6-239. Using the Hückel $4n + 2$ rule for aromaticity, predict which of the following species might be expected to show aromatic properties.

6-240. Based on Questions 6-237 and 6-238, why is tetraphenylcyclopentadienone such a stable compound?

6-241. Explain why in the UV–visible absorption spectrum of benzil as compared with that for tetraphenylcyclopentadienone, the maximum absorption for the tetraphenylcyclopentadienone appears at a longer wavelength.

REFERENCES

1. For references on the aldol condensation see Experiment 22.

2. An *Organic Syntheses* preparation of tetraphenylcyclopentadienone is available: Johnson, J. R.; Grummitt, O. *Organic Syntheses;* Wiley, New York, 1955; Collect. Vol. III, p 80.

Experiment 50A   *d,l*-Stilbene Dibromide

(*d,l*-ethane, 1,2-dibromo-1,2-diphenyl-)

The bromination of an alkene is presented in this experiment. In the reaction, two chiral centers are generated yielding a mixture of optically active bromides. This demonstrates that the reaction is *stereospecific*.

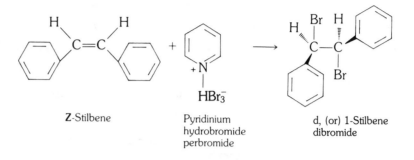

Z-Stilbene   Pyridinium   d, (or) 1-Stilbene
                     hydrobromide   dibromide
                     perbromide

DISCUSSION   Bromination of an alkene (cyclic or acyclic) is an example of an electrophilic addition reaction. The reaction proceeds in two stages. The first stage involves the formation of a cyclic *bromonium ion* intermediate. In a special case, this intermediate has actually been isolated as a tribromide salt.[5] The second stage is a nucleophilic attack by a bromide ion on the intermediate, with inversion occurring at the carbon atom attacked, to yield a *vic*-dibromo derivative. This second stage is an S_N2 sequence. Cyclic alkenes provide evidence that the reaction is an anti addition, the bromine atoms being introduced trans to one another. It is important to realize that if two different groups are present on one or both of the sp^2 carbon atoms of the alkene linkage, chiral carbon centers are generated on bromination of these carbon atoms. For Z-stilbene, bromination leads to formation of a racemic mixture of enantiomeric bromides. The absence of the *meso* isomer confirms that the reaction is stereospecific. On the other hand, bromination of E-stilbene leads to formation of the *meso* isomer.

Bromination of alkenes using a Br_2–CCl_4 solution (a red-brown color) is frequently used as a qualitative test for the presence of unsaturation in a compound. Rapid loss of color from the reagent solution is a positive test (see Chapter 7).

The mechanism of the bromination reaction is outlined below.

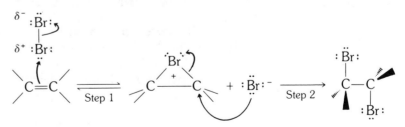

EXPERIMENTAL   Estimated time to complete the reaction: 1.0 hour.

---

[5] See Strating, J.; Wieringa, J. H.; Wynberg, H. *J. Chem. Soc. D* **1969**, 907.

Note. *As an alternate experiment, E-stilbene may be brominated to form meso-stilbene dibromide. The alternate procedure is given after the method for bromination of the Z isomer outlined below.*

Physical Properties of Reactants and Products

Compound	MW	Wt/Vol	mmol	mp(°C)	bp(°C)	Density	n_D
Z-Stilbene	180.25	100 μL	0.55	5–6	141(12 torr)	1.01	1.6130
Glacial acetic acid		2 mL			118		
Pyridinium hydrobromide perbromide	319.83	200 mg	0.63	205			
d,l-Stilbene dibromide	340.07			110			

Reagents and Equipment

In a 5.0-mL conical vial containing a magnetic spin vane and equipped with an air condenser, place 100 μL (0.55 mmol) of Z-stilbene followed by addition of 1 mL of glacial acetic acid.

HOOD

> **WARNING: *Glacial acetic acid is corrosive and toxic. It is dispensed in the* hood *with an automatic pipet.***

To the cooled solution add 200 mg (0.63 mmol) of pyridinium hydrobromide perbromide. Any perbromide that adheres to the side of the flask is washed down with an additional 1 mL of glacial acetic acid.

> **WARNING: *This solid compound is a mild lachrymator. It should be dispensed in the* hood.**

Note. *A stable, nontoxic brominating agent, tetra-N-butylammonium bromide, may be used with comparable results. If used in the preceding reaction sequence, 300 mg (0.62 mmol) is employed.*

Reaction Conditions

With stirring, heat the reaction mixture in a sand bath at a temperature of 90–100°C until the yellow color of the perbromide disappears or fades to a pale yellow (about 5–10 minutes).

Isolation of Product

The vial is removed from the heat source and allowed to cool to approximately 40–50°C (water bath). Water (2 mL) is then added and the vial placed in an ice bath for a period of 15 minutes. The resulting crystalline solid is collected by vacuum filtration using a Hirsch funnel.

Purification and Characterization

The solid material is recrystallized from methanol using a Craig tube. The final product is dried on a porous clay plate or on filter paper.

Weigh the stilbene dibromide product and calculate the percentage yield. Determine the melting point and compare your result with the literature value.

Chemical Tests

It may be of interest to perform several classification tests to assist in the characterization of the product (see Chapter 7).

a. Test the solubility characteristics of the product in water, 5% NaOH, 5% HCl, and concentrated sulfuric acid. Do the results concur with the solubility chart?

b. Carry out the Beilstein test. Do the results confirm the presence of a halogen? You may also wish to do the sodium fusion test to confirm that the halogen in the compound is bromine.

c. An alternate test for alkyl halides is the silver nitrate test. Do the results of this test on the product indicate that the material contains bromine?

Alternate Preparation: *meso*-Stilbene Dibromide

The procedure is similar to that given above with the following exceptions.

a. The initial reaction mixture of *E*-stilbene and glacial acetic acid may be warmed on a sand bath if necessary, to obtain a homogeneous solution.

b. The reagent amounts are summarized in the following table:

Compound	MW	Wt/Vol	mmol	mp(°C)	bp(°C)
E-Stilbene	180.25	100 mg	0.54	124–125	
Glacial acetic acid		2 mL			118
Pyridinium hydrobromide perbromide	319.83	200 mg	0.63	205	
meso-Stilbene dibromide	340.07			238	

c. Since the product sublimes, the melting point is taken using the evacuated technique (Chapter 4).

Optional Scaleup (*d,l*-Stilbene Dibromide)

The preparation may be scaled up by a factor of 10. The procedure is similar to that for the microscale preparation with the following exceptions.

a. A 50-mL Erlenmeyer flask containing a boiling stone is used.

b. The reagent and solvent amounts are increased tenfold.

Compound	MW	Wt/Vol	mmol	mp(°C)	bp(°C)	Density	n_D
Z-Stilbene	180.25	1.0 mL	5.5	5–6	141(12 torr)	1.01	1.6130
Glacial acetic acid		20 mL			118		
Pyridinium hydrobromide perbromide	319.83	2 g	6.3	205			

HOOD

c. With occasional swirling, the reaction mixture is heated at 140°C on a sand bath in the **hood**.

d. The mixture is cooled to room temperature and 20 mL of water added.

ENVIRONMENTAL DATA

Substance	Amount	TLV (mg/m³)	Emissions (mg)	Volume (m³)
Acetic acid (Max., heat, vac. filtr.)	2.0 mL	25	2000	80
Methanol (Max., Craig)	1.0 mL	260	790	3.1

TXDS: Stilbene—ipr-mus LD50: 1150 mg/kg

QUESTIONS

6-242. Use Newman projections to show the addition of bromine to *E*-stilbene and *Z*-stilbene.

6-243. What is the stereochemical relationship between the products formed in the two reactions (Question 6-242)?

6-244. Bromine undergoes addition to ethylene in the presence of a high concentration of chloride ion to give 1-bromo-2-chloroethane, as well as 1,2-dibromoethane. Chloride ion does not add to the C=C unless bromine is present. Suggest a suitable mechanism to explain these results. Is the rate of bromination significantly affected by the presence of the chloride ion?

6-245. The silver nitrate test was suggested as a classification test for the product of this experiment. Would the "sodium iodide in acetone" test give similar results? Why or why not?

6-246. Write the mechanism, showing the stereochemistry, for the addition of bromine–water to cyclohexene to form 2-bromocyclohexanol.

6-247. A student adds a few drops of Br_2–CCl_4 solution to an unknown organic compound. The color of the bromine solution disappears. The student reports that the unknown contains a C=C. Would you arrive at the same conclusion? If not, why not?

REFERENCES   Many examples of the bromination of alkenes appear in *Organic Syntheses*. Selected references are given here.

a. Allen, C. F. H.; Abell, R. D.; Normington, J. B. *Organic Syntheses;* Wiley: New York, 1941; Collect. Vol. I, p 205.
b. Snyder, H. R.; Brooks, L. A. *Organic Syntheses;* Wiley: New York, 1943; Collect. Vol. II, p 171.
c. Rhinesmith, H. S. Ibid., p 177.
d. Cromwell, N. H.; Benson, R. *Organic Syntheses;* Wiley: New York, 1955; Collect. Vol. III, p 105.
e. McElvain, S. M.; Kundiger, D. Ibid., p 123.
f. Fieser, L. F. *Organic Syntheses;* Wiley: New York, 1963; Collect. Vol. IV, p 195.
g. Khan, N. A. Ibid., p 969.
h. Paquette, L. A.; Barrett, J. H. *Organic Syntheses;* Wiley: New York, 1973; Collect. Vol. V, p 467.

Experiment 50B   Diphenylacetylene

(acetylene, diphenyl-)

This reaction illustrates dual dehydrohalogenation to yield an alkyne. The product is the dienophile used in the preparation of hexaphenylbenzene.

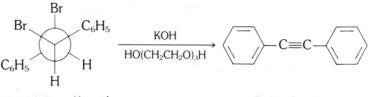

d, (or) l-Stilbene dibromide   Diphenylacetylene

DISCUSSION   This reaction illustrates the dehydrohalogenation of a *vic*-dibromo compound to form an acetylenic linkage. It is a useful reaction for generation of the C≡C bond since the starting dibromides are readily available from alkenes. The reaction is usually run in the presence of a strong base and proceeds in two stages. In the first stage an intermediate bromoalkene is formed, which may be isolated if the conditions employed are not too drastic. In fact, it is a valuable route for the preparation of vinyl halides. The reaction involves the removal of the hydrogen on the carbon atom *beta* to the carbon to which the bromine is attached. The important E_2 mechanism operates at this stage of the reaction where the hydrogen being removed is anti-periplanar to the departing bromo group. The second stage of the reaction is more difficult, but at high temperatures and in the presence of a strong base, the bromoalkene undergoes further E_2 elimination to form the alkyne linkage. The different stages of the mechanism are presented below.

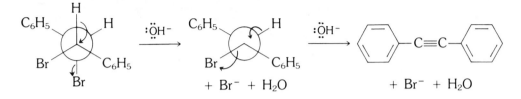

EXPERIMENTAL   Estimated time to complete the reaction: 1.0 hour.

Physical Properties of Reactants and Products

Compound	MW	Wt/Vol	mmol	mp(°C)	bp(°C)
d,l-Stilbene dibromide	340.07	80 mg	0.24	110	
Potassium hydroxide	56.11	75 mg	1.3	360	
Triethylene glycol	150.18	400 µL			278
Diphenylacetylene	178.23			61	

Reagents and Equipment
In a 3.0-mL conical vial containing a boiling stone and equipped with an air condenser, place 80 mg (0.24 mmol) of *d,l*-stilbene dibromide and 75 mg (1.3 mmol) of KOH flakes. Using an automatic delivery pipet, add 400 µL of triethylene glycol to the vial.

Reaction Conditions
The reaction mixture is heated using a sand bath maintained at a temperature of 190°C for a period of 5 minutes.

Isolation of Product
The resulting dark reaction mixture is cooled to approximately 40–50°C (water bath) and 1.0 mL of water added. The vial is then placed in an ice bath for 15 minutes. The solid product is collected by filtration under reduced pressure using a Hirsch funnel.

Purification and Characterization
The solid diphenylacetylene is recrystallized from 95% ethanol (Craig tube) and dried on a porous clay plate.

   Weigh the product and calculate the percentage yield. Determine the melting point and compare your result with the literature value. Obtain an IR spectrum of the sample using the KBr pellet technique and compare your result with that recorded in the literature.

Chemical Tests
Perform the Br_2–CH_2Cl_2 test for unsaturation. Do the results indicate the presence of unsaturation in the product? Should the Baeyer test lead to the same conclusion? If you are undecided, run this test also.

Optional Scaleup This preparation may be scaled up by a factor of 10. The procedure is similar to that of the microscale method with the following exceptions.

a. The reaction is run in a 25-mL Erlenmeyer flask containing a boiling stone.
b. The amounts of reactants and solvents used are summarized in the table below.

Compound	MW	Wt/Vol	mmol	mp(°C)	bp(°C)
d,l-Stilbene dibromide	340.07	800 mg	2.35	110	
Potassium hydroxide	56.11	750 mg	13	360	
Triethylene glycol	150.18	4 mL			278

HOOD **c.** The reaction is run in the **hood**.
d. On isolation, 10 mL of water is used. The light tan, crude product is recrystallized from 95% ethanol.

ENVIRONMENTAL DATA

Substance	Amount	TLV (mg/m³)	Emissions (mg)	Volumes (m³)
Ethanol (Max., Craig)	1.0 mL	1900	800	0.4

TLV: Potassium hydroxide—2 μg/m^3
TXDS: Triethylene glycol—orl-rat LD50: 17 g/kg

QUESTIONS **6-248.** Both *E*- and *Z*-2-chlorobutendioic acids dehydrochlorinate to give acetylene dicarboxylic acid.

$$E,Z\text{—HOOC—C(Cl)}\text{=}\text{CH—COOH} \longrightarrow \text{HOOC—C}\equiv\text{C—COOH}$$

The *Z* acid reacts about 50 times faster than the *E* acid. Explain.

6-249. Compounds containing the acetylenic bond undergo the Diels–Alder reaction. Formulate the product formed by the reaction of (1*E*,3*E*)-1,4-diphenyl-1,3-butadiene with diethyl acetylenedicarboxylate by heating at 140–150°C.

6-250. Acetylenes can be hydrated in the presence of acid and HgSO$_4$ by electrophilic addition of a molecule of water to the triple bond. The reaction proceeds by way of a carbocation intermediate. Hydration of acetylene produces acetaldehyde. Outline the steps that occur in this transformation.

6-251. In reference to Question 6-250, explain why acetaldehyde is isolated instead of the vinyl alcohol.

6-252. Use the IR tables to locate the absorption bands of the stretching frequencies of the alkyne C$\equiv$C—H bond, the alkyne —C$\equiv$C— bond, and the alkene C=C—H bond. Using these data, explain how you would distinguish between 1-butyne, 2-butyne, and 2-butene.

REFERENCES **1.** For a review on the preparation of alkynes see Jacobs, T. L. *Org. React.* **1949**, *5*, 1.
2. A large number of elimination reactions leading to the formation of acetylenes appear in *Organic Syntheses*. Selected references are given here:
   a. Allen, C. F. H.; Abell, R. D.; Normington, J. B. *Organic Syntheses;* Wiley: New York, 1941; Collect. Vol. I, p 205.

b. Hessler, J. C. Ibid., p 438.

c. Abbott, W. T. *Organic Syntheses;* Wiley: New York, 1943; Collect. Vol. II, p 515.

d. Synthesis of diphenylacetylene: Smith, L. I.; Falkof, M. M. *Organic Syntheses;* Wiley: New York, 1955; Collect. Vol. III, p 350.

e. Guha, P. C.; Sankaren, D. K. Ibid., p 623.

f. Khan, N. A., *Organic Syntheses;* Wiley: New York, 1963; Collect. Vol. IV, p 967.

g. Campbell, K. N.; Campbell, B. K. Ibid., p 763.

Experiment 51

Hexaphenylbenzene

(benzene, hexaphenyl-)

This experiment illustrates the Diels–Alder reaction, a method that is used extensively to form six-membered cyclic ring systems. The reaction is the final stage of Sequence C.

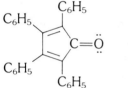
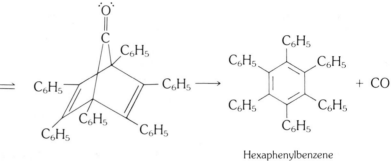

Tetraphenylcyclopentadienone   Diphenylacetylene   Hexaphenylbenzene

DISCUSSION   The Diels–Alder reaction is one of the most useful synthetic tools in organic chemistry. It is an example of a cycloaddition reaction between a conjugated diene and a dienophile (an -ene or -yne), which leads to the formation of six-membered cyclic rings.

A very large number of structures can be prepared using the Diels–Alder reaction by varying the nature of the diene and dienophile. In the majority of cases, carbocyclic rings are generated, but ring closure can also occur with reactants containing heteroatoms. This leads to the synthesis of compounds containing heterocyclic rings.

The present reaction is important in that the condensation of tetraphenylcyclopentadienone with diphenylacetylene leads to the formation of an aromatic ring with the evolution of carbon monoxide. Decarbonylation of the type of cyclopentenone intermediate involved in this experiment is easy, common, and synthetically useful. Kinetic data on systems of this type are consistent with molecular orbital correlation diagram predictions that decarbonylation occurs by a concerted mechanism with evidence of a disrotatory stereochemical course for the reaction.

For further discussion and examples of the Diels–Alder reaction see Experiments 14 and 15.

EXPERIMENTAL   Estimated time to complete the experiment: 1.0 hour.

Physical Properties of Reactants and Products

Compound	MW	Wt/Vol	mmol	mp(°C)
Tetraphenylcyclopentadienone	384.48	20 mg	0.05	220–221
Diphenylacetylene	178.23	20 mg	0.11	61
Hexaphenylbenzene	534.66			465

Reagents and Equipment

In an 8 × 80-mm Pyrex tube sealed at one end, place 20 mg (0.05 mmol) of tetaphenylcyclopentadienone and 20 mg (0.11 mmol) of diphenylacetylene.

The tube is prepared from 8-mm Pyrex glass tubing. Tetraphenylcyclopenta-dienone is prepared in Experiment 49 and diphenylacetylene is prepared in Experiment 50B.

Reaction Conditions

The tube is clamped and the mixture heated with a microburner until a molten mass is obtained (see Fig. 6.33). Heating is continued for approximately 3–5 minutes, during which time the dark purple color of the reactants gradually fades to white. *If the heating is too vigorous, the product will darken and decompose.*

Isolation of Product

The unreacted diphenylacetylene is separated from the mixture by clamping the tube at an angle and gently heating the contents with a microburner (Fig. 6.33). As the diphenylacetylene condenses on the walls of the tube, gentle heating is applied just below this liquid phase, forcing the acetylene compound to the top of the tube. The tube is tipped at an angle and the liquid diphenylacetylene condensed at the top is removed by touching the end of the tube to a piece of filter paper (Fig. 6.33). *The product remains in the bottom of the tube as an off-white solid mass.*

Purification and Characterization

The tube is allowed to cool and 100 μL of diphenyl ether added to the crude product. The mixture is heated to dissolve the crystalline mass and the resulting solution then allowed to cool. *It may be necessary to add an additional 50 μL of diphenyl ether to dissolve the mass.*

The cooled tube is now cut approximately 2–3 cm above the thick crystalline mass (Fig. 6.33).

Directions. *Hold the tube upright, score with a file, and break using a towel to protect the hands.*

While stirring with a spatula, the mixture is diluted with 500 μL of cold toluene, and the crystals are collected by filtration under reduced pressure using a Hirsch funnel. The tube and crystals are then rinsed with an additional 0.5 mL of cold toluene and the product is dried on a porous clay plate or on filter paper.

Weigh the hexaphenylbenzene product and calculate the percentage yield. Obtain an IR spectrum using the KBr pellet technique and compare it with that recorded in the literature.

ENVIRONMENTAL DATA

Substance	Amount	TLV (mg/m³)	Emissions (mg)	Volume (m³)
Diphenyl ether (Max., involatile)	100 μL	7	109	15.6
Toluene (SKIN) (Max., recryst.)	0.5 mL	375	2170	5.8
Toluene (SKIN) (Max., cryst., wash)	2.5 mL	50	3725	75

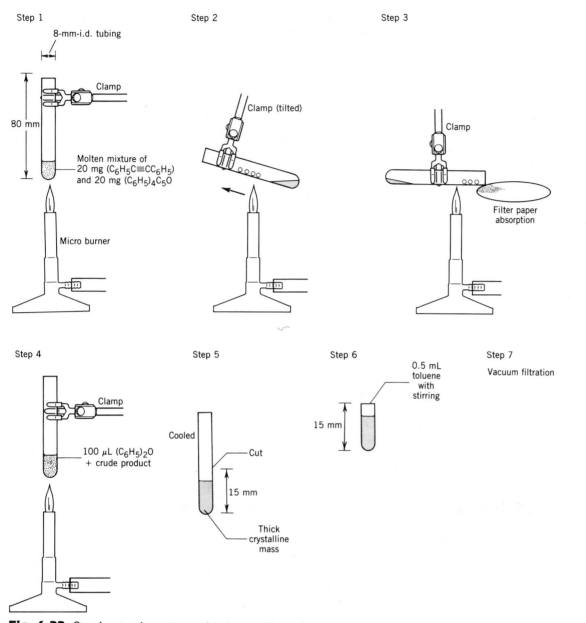

Fig. 6.33 *Step-by-step formation and isolation of hexaphenylbenzene.*

QUESTIONS   **6-253.** What starting materials would you use to prepare the following compounds by the Diels–Alder reaction?

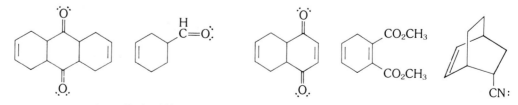

6-254. Diels–Alder reactions with benzene are rare. Two are shown below. Give the structures of the products produced in these reactions.

6-255. Below are shown two heteroatom compounds that undergo the Diels–Alder reaction. Formulate the product obtained in each reaction.

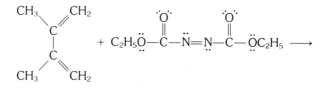

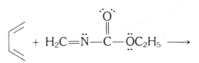

6-256. Explain why (2Z, 4Z)-hexadiene is a very poor dienophile in the Diels–Alder reaction.

6-257. One of the main features of the Diels–Alder reaction is that it is *stereospecific*. Explain.

REFERENCES

1. Review articles:
   a. Norton, J. A. *Chem. Rev.* **1942,** *31,* 319.
   b. Kloetzel, M. C. *Org. React.* **1948,** *4,* 1.
   c. Holm, H. L. Ibid., p 60.
   d. Butz, L. W.; Rytina, A. W. *Org. React.* **1949,** *5,* 136.
   e. Sauer, J. *Angew. Chem. Int. Ed. Engl.* **1966,** *5,* 211.
   f. Ibid., **1967,** *6,* 16.
2. An *Organic Syntheses* preparation using tetraphenylcyclopentadienone in a Diels–Alder reaction to obtain tetraphenylphthalic anhydride: Grummitt, O. *Organic Syntheses;* Wiley: New York, 1955; Collect. Vol. III, p 807.

SEQUENCE D: **Synthesis of 4-Methoxybenzoic Acid from 4-Ethylphenol**

This series of reactions involves the synthesis of an ether using the Williamson reaction followed by two selective oxidations to obtain the final product.

4-Ethylphenol → 4-Ethylanisole →
[25B]
4-Methoxyacetophenone → 4-Methoxybenzoic acid
[52]   [43B]

Experiment 52   4-Methoxyacetophenone

(acetophenone, 4-methoxy-)

This reaction illustrates the selective oxidation of a methylene group to a ketone function.

$$CH_3\ddot{O}\!-\!\!\langle\ \rangle\!\!-\!CH_2CH_3 + (NH_4)_2S_2O_8 \xrightarrow[H_2O]{Ag^+} CH_3\ddot{O}\!-\!\!\langle\ \rangle\!\!-\!\overset{\overset{\ddot{O}}{\|}}{C}\!-\!CH_3$$

| 4-Ethylanisole | Ammonium peroxydisulfate | 4-Methoxyacetophenone |

DISCUSSION   This reaction utilizes the silver–persulfate couple as an oxidizing agent to selectively oxidize substituted ethylbenzenes to the corresponding methyl ketones. The reaction is carried out in aqueous medium with a trace of silver ion. The mechanism is complicated, but a radical ion species, $SO_4^{\bar{\cdot}}$, apparently is involved. The silver ion assists in the electron transfer process to generate this radical ion. The complete details of the mechanism, however, are not well understood.

EXPERIMENTAL   Estimated time of the experiment: 5 hours (two laboratory periods).

Physical Properties of Reactants and Products

Compound	MW	Wt/Vol	mmol	mp(°C)	bp(°C)	Density	n_D
4-Ethylanisole	136.20	50 mg	0.37		195–196	0.96	1.5120
Ammonium peroxydisulfate	228.18	167 mg	0.73	120			
Silver nitrate (0.2% aqueous solution)		750 μL					
4-Methoxyacetophenone	150.8			38–39	258		

Note. *A reaction time of 2.5–3.5 hours is required. It is recommended that another experiment be conducted during this time. The oxidation product can be isolated during the following laboratory period.*

Reagents and Equipment

In a 3.0-mL conical vial containing a magnetic spin vane and equipped with an air condenser, place 50 mg (0.37 mmol) of 4-ethylanisole, 167 mg (0.73 mmol) of ammonium peroxydisulfate, and 750 μL of 0.2% aqueous silver nitrate solution.

Note. *4-Ethylanisole is prepared in Experiment 25B and is of sufficient purity for use in this preparation. The silver nitrate solution is dispensed using an automatic delivery pipet.*

WARNING: *Silver nitrate will blacken the skin!*

Reaction Conditions

The reaction mixture is stirred at room temperature for 2.5–3.5 hours.

Isolation of Product

The resulting aqueous solution is extracted with three 0.5-mL portions of chloroform (calibrated Pasteur pipet). The wet chloroform extracts are separated with a Pasteur filter pipet and dried by transferring the solution to a Pasteur filter pipet packed with 1.0 g of anhydrous sodium sulfate.

To separate the two phases, draw them into the Pasteur filter pipet and gently transfer the bottom layer to the column. The aqueous layer is then returned to the original vial for the next extraction with additional chloroform.

The eluate is collected in a tared 5.0-mL conical vial containing a boiling stone. An additional 1.0 mL of chloroform is added to wash the column and this eluate combined with the original material collected in the vial.

The solution is concentrated by placing the vial in a sand bath maintained at HOOD   75–85°C in the **hood**. A gentle stream of nitrogen gas impinging on the surface of the solution speeds up the evaporation process.

Purification and Characterization

The crude product is a brown liquid that may crystallize on standing. Weigh the material and calculate the crude yield. Obtain an IR spectrum, using the capillary film technique, and compare your result with that reported in the literature. The analysis of the spectrum is given below.

Infrared Analysis

The Williamson reaction of 4-ethylphenol and methyl iodide is expected to produce a mixed aryl–alkyl ether. The infrared spectrum of the reaction product supports this argument. If time is available, obtain the spectrum of the product of Experiment 25B (or new starting material if Experiment 25 has not been carried out) by the capillary film technique. Compare your data with the reference spectrum (Fig. 6.15). This spectrum contains several macro group frequencies: methoxyl aryl ether [2838, 1601, 1585 (strong), 1518, 1460 (broad), 1255, and 1041 cm^{-1}]; para-substituted benzene ring (1881, 1762, 1720, 832 cm^{-1}); and alkyl substituent [2965–2850, 1450 (broad), 1375 cm^{-1}].

Examine these peaks and assign the appropriate vibrational modes (see Chapter 7 and discussions in earlier experiments).

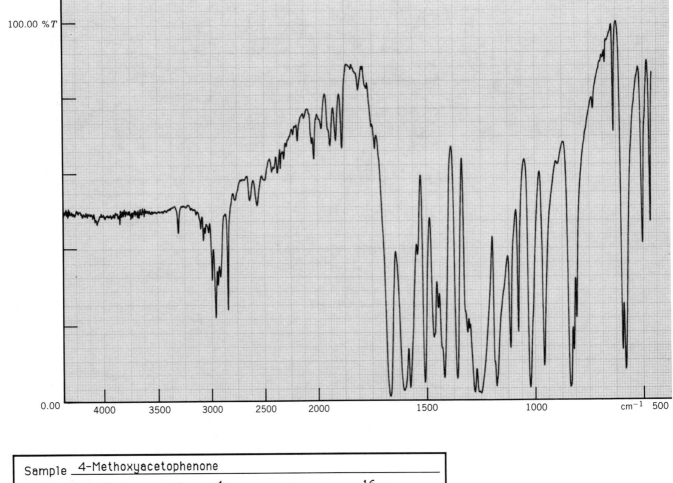

Fig. 6.34 *IR spectrum: 4-ethylanisole.*

The product of this novel oxidation reaction is thought to be the para-substituted aromatic ketone, 4-methoxyacetophenone. Examine the product spectrum (Fig. 6.34) by the macro group frequency approach. The conjugated aromatic methyl ketone train utilizes peaks at 3330, 3095–3010, 2975, 1673, 1600, and 1580 (strong) 1366 cm^{-1}. A second macro train covers the surviving aromatic methoxyl system: 2850 (strong), 1600, 1580 (strong), 1255, 1020 cm^{-1}. Finally, the original para-substituted benzene ring macro group frequency train, observed in the starting material, is reduced to a single specific frequency as the combination band region is badly distorted by the ketone C=O stretch, 837 cm^{-1}.

As with the starting material, assign the peaks to specific vibrational modes present in the product ketone.

Chemical Tests

Chemical tests (Chapter 7) may also be used to aid in the characterization of this product. The ignition test (Table 7.1) should establish the presence of the aromatic ring system. The presence of the carbonyl group can be confirmed by the 2,4-dinitrophenylhydrazine test. The isolated product of this test can serve as a derivative, if desired.

4-Methoxyacetophenone is a methyl ketone. The iodoform test can be run to confirm this fact.

Optional Scaleup   This reaction has been run with 20 times the amounts of reagents used in the microscale operation. At this scale a separatory funnel is employed during the extraction step. The preparation is outlined below and is recommended as a two-laboratory-period procedure.

The reagent and solvent amounts are summarized below.

Compound	MW	Wt/Vol	mmol	mp(°C)	bp(°C)	Density	n_D
4-Ethylanisole	136.20	1.0 g	7.4		195–196	0.96	1.5120
Ammonium peroxydisulfate	228.18	3.34 g	14.6	120			
Silver nitrate (0.2% aqueous solution)		15 mL					

Reagents and Equipment

In a 50-mL round-bottom flask equipped with a magnetic stirring bar and an air condenser, place 1.0 g (~1 mL) of p-ethylanisole, 3.34 g of ammonium peroxydisulfate, and 15 mL of 0.2% aqueous silver nitrate.

Reaction Conditions

The mixture is stirred at room temperature for ~3–3.5 hours or may be run overnight.

Isolation of Product

The contents of the reaction flask are transferred to a 125-mL separatory funnel using a Pasteur pipet. The flask is rinsed with a 10-mL portion of methylene chloride and this rinse also transferred to the separatory funnel.

Shake, vent, and allow the layers to separate in the funnel. The lower methylene chloride layer is then transferred to a 25-mL buret containing 20 g of anhydrous sodium sulfate. The dry methylene chloride elutate is collected in a tared 50-mL Erlenmeyer flask containing a boiling stone. The aqueous layer remaining in the funnel is extracted in like manner with two additional 10-mL portions of methylene chloride. These extracts are also transferred to the sodium sulfate column. Finally, the column is rinsed with an additional 10-mL portion of methylene chloride.

HOOD   The resulting solution is concentrated **[hood]** using a sand bath maintained at 60–65°C and a slow stream of nitrogen to assist solvent removal. Continue the evaporation process until a constant weight of the brown liquid product is obtained.

Calculate the yield of crude product and obtain the IR spectrum using the capillary film technique. This material is of sufficient purity for the conversion to p-methoxybenzoic acid (Experiment 43B) by the haloform oxidation reaction.

Purification and Characterization

The crude material may be further purified by distillation under reduced pressure using an aspirator or vacuum pump.

In a 5-mL round-bottom flask (containing a magnetic spin bar) attached to a Hickman still equipped with an air condenser, place approximately 150 mg of the 4-methoxyacetophenone.

Using a Water Aspirator   The distillation is carried out at ~16 mm pressure. Attach tubing from the aspirator to a water trap (see Chapter 5, Technique 5) and then to the top of the air condenser on the Hickman still. The system is hand held (to help control bumping) and the assembly placed in a sand bath at 165–175°C. The 4-methoxyacetophenone collects in the collar of the still as a pale yellow liquid. During this operation make sure the magnetic bar continues to spin rapidly.

Using a Vacuum Pump   The distillation is carried out at ~3.0 mm pressure. The Hickman still is attached to the reduced pressure apparatus as described above. Under these conditions, the 4-methoxyacetophenone distills at a sand bath temperature of 120°C; bumping is not a problem at this lower temperature.

Obtain an IR spectrum of the distallate using the capillary film technique and compare it with that of the crude material (see Infrared Analysis section).

ENVIRONMENTAL DATA

Substance	Amount	TLV (mg/m³)	Emissions (mg)	Volume (m³)
Chloroform (Max., extract, evap.)	2.5 mL	50	3725	75

TLV: Ammonium peroxydisulfate—2 mg/m³
TXDS: 4-Methoxyacetophenone—orl-rat LD50: 1720 mg/kg
2,4-Dinitrophenylhydrazine—orl-rat LD50: 654 mg/kg
Silver nitrate—unk-man LDLo: 29 mg/kg
Sodium sulfate—orl-mus LD50: 5989 mg/kg

QUESTIONS   **6-258.** Selenium dioxide is often used to effect oxidation of aldehydes and ketones to yield α-dicarbonyl compounds. Predict the product in each of the following reactions.

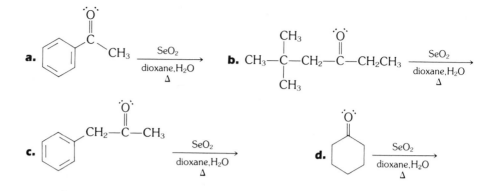

6-259. Under the oxidation conditions with SeO_2 described in Question 6-258, it was found that with unsymmetrical ketones oxidation occurs at the —CH_2— group that is most readily enolized. Based on this fact would you change any of the structures of the products you formulated in question 6-258?

6-260. The Ag^+ ion acts to generate the SO_4^- ion radical from $S_2O_8^{2-}$. Propose an oxidation–reduction reaction sequence to account for the formulation of the SO_4^-.

6-261. The second step in the haloform oxidation reaction is a nucleophilic substitution reaction at the $C=O$ group: the replacement of —CX_3 by —OH. Give a suitable mechanism to account for this transformation.

6-262. Which of the following substances would give a positive iodoform test?
acetone acetic acid 2-pentanone isopropyl alcohol 3-pentanone

REFERENCES **1.** Peroxydisulfate as an oxidizing agent:
a. House, D. A. *Chem. Rev.* **1962,** *62,* 185.
b. Daniher, F. A. *Org. Prep. Proc.* **1970,** *2,* 207, and references therein.
2. Preparation of 2,4-dinitrophenylhydrazones:
a. Howell, B. A. *J. Chem. Educ.* **1984,** *61,* 176.
b. Shriner, R. L.; Fuson, R. C.; Curtin, D. Y.; Morrill, T. C. *The Systematic Identification of Organic Compounds,* 6th ed.; Wiley: New York, 1980.

Chapter 7

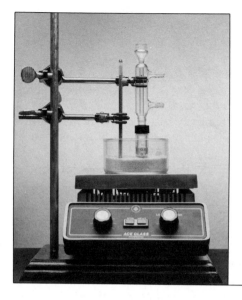

Identification of Organic Compounds

I. Organic Qualitative Analysis

Millions of organic compounds have been recorded in the literature. It may seem a bewildering task to attempt to identify one certain compound from this vast array; however, it is important to realize that the majority of these substances can be easily grouped into any one of a number of classes. One of the exciting challenges that a chemist faces on a regular basis is the identification of organic compounds or the characterization and determination of the structure of hitherto unknown species. The chemist, working as an analyst, has at his or her disposal an enormous data base of chemical and spectroscopic information that has been correlated and organized over the years. The area of forensic chemistry, the detection of environmental pollutants, progress in the industrial research and development areas, to name a few, all depend to a large extent on the ability of the chemist to isolate, purify, and identify specific chemicals.

The cost of purchasing and maintaining sophisticated analytical instruments is a major budgeting factor in many academic and industrial laboratories. Nevertheless, it is *critical* that a strong analytical capability be maintained as a support for the chemical research and development programs. If chemists trained in "characterization techniques" are not available, basic work in these vital areas is drastically curtailed. The *importance* of this area of chemical instrumentation is *increasing* as problems become more complex. In today's laboratories, an analyst needs not only to be able to identify and in many cases quantify a single component, but also to be able to determine the precise microstructure of specific materials. Chapter 7 presents an introduction to this vital area of "chemical characterization". The ability to master the techniques used in the separation and characterization of materials, especially using modern instrumental techniques, is one of the important areas of your training.

The object of organic qualitative analysis is to place a given compound, through screening tests, into one of a number of specific classes, which in turn greatly simplifies the *identification* of the compound. This is usually done using a series of preliminary observations and chemical tests in conjunction with the instrumentation that developments in the spectroscopic area of analysis have made available to the analyst.

The systematic approach taken in this text for the identification of an unknown organic compound is as follows.

1. Preliminary tests are performed to determine the physical nature of the compound.
2. The solubility characteristics of the unknown species are determined. This can often lead to valuable information related to the structural composition of the unknown organic compound.
3. Chemical tests, mainly to assist in identifying elements other than C, H, or O, may also be performed.
4. Classification tests to detect *common functional groups* present in the molecule are carried out. The majority of these tests may be done using a few drops of a liquid or a few milligrams of a solid. There is an added benefit, especially in relation to the chemical detection of functional groups, that an introductory student in the organic chemistry laboratory may obtain. That is, that an incredible amount of chemistry can be *observed* and *learned* in performing these tests. Successful application of these tests requires that you develop the ability to think in a logical manner and that, based on your observations, you learn to interpret the significance of each result. Later, as the *spectroscopic techniques* are introduced, *the number of chemical tests performed is usually curtailed.*
5. The spectroscopic method of analysis is utilized. As you develop further in your knowledge of chemistry, you will appreciate more and more the revolution that has taken place in chemical analysis over the past 25–30 years and what powerful methods are at your disposal for the identification of organic compounds. For the introductory laboratory, the techniques related to *infrared, nuclear magnetic resonance, and ultraviolet–visible spectroscopy* are generally developed.

It is important to realize that *negative* findings are often as important as *positive* results in identifying a given compound. Cultivate the habit of following a *systematic pathway* or *sequence* so that no clue or bit of information is lost or overlooked along the way. It is also important to develop the *attitude* and *habit* of planning ahead. Outline a logical plan of attack, depending on the nature of the unknown, and follow it. As you gain more and more experience in this type of investigative endeavor, the planning stage will become easier. At this initial phase of your development, the unknowns to be identified will be relatively pure materials. *Record all observations and results of the tests in your laboratory notebook.* Review these data as you execute the sequential phases of your plan. This serves to keep you on the straight and narrow path to success.

A large number of texts have been published on organic qualitative analysis. Excellent sources listing physical and spectral data are also available as well as texts related to identification by spectral means. Several are cited below.

REFERENCES **1.** Organic qualitative analysis:

 a. Shriner, R. L.; Fuson, R. C.; Curtin, D. Y.; Morrill, T. C. *The Systematic Identification of Organic Compounds,* 6th ed.; Wiley: New York, 1980.

 b. Schneider, F. L. *Monographien aus dem Gebiete der Qualitativen Mikroanalyse,* Vol. II; *Qualitative Organic Microanalysis;* Benedetti-Pichler, A. A., Ed.; Springer-Verlag: Vienna, 1964.

 c. Vogel, A. I. *Qualitative Organic Analysis,* Part 2 of *Elementary Practical Organic Analysis;* Wiley: New York, 1966.

 d. Cheronis, N. D.; Entrikin, J. B.; Hodnett, E. M. *Semimicro Qualitative Organic Analysis,* 3rd ed.; Interscience: New York, 1965.

 e. Cheronis, N. D.; Ma, T. S. *Organic Functional Group Analysis by Micro and Semimicro Methods;* Interscience: New York, 1964.

 f. Feigl, F.; Anger, V. *Spot Tests in Organic Analysis;* Elsevier: New York, 1966.

 g. Kamm, O. *Qualitative Organic Analysis,* 2nd ed.; Wiley: New York, 1932.

 2. Spectral methods:

 a. *CRC Handbook of Data in Organic Compounds,* 2nd ed.; Weast, R. C.; Grasselli, J. G., Eds.; CRC Press, Inc.: Boca Raton, FL, 1988.

 b. Lambert, J. B.; Shurvell, H. F.; Lightner, D. A.; Cooks, R. G. *Introduction to Organic Spectroscopy;* Macmillan: New York, 1987.

 c. Silverstein, R. M.; Bassler, G. C.; Morrill, T. C. *Spectrometric Identification of Organic Compounds,* 4th ed.; Wiley: New York, 1981.

 d. Parikh, V. M. *Absorption Spectroscopy of Organic Molecules;* Addison-Wesley: Reading, MA, 1973.

 e. Cooper, J. W. *Spectroscopic Techniques for Organic Compounds;* Wiley: New York, 1980.

 f. McFarlane, W. In *Elucidation of Organic Substances by Physical and Chemical Methods;* Bently, K.; Kirby, G. W., Eds.; Wiley: New York, 1972.

PRELIMINARY TESTS **Overview**

The objective of the preliminary tests is to provide a pathway or sequence to assist you in developing the route to follow in order ultimately to identify the unknown material at hand. It must be emphasized, however, that these tests frequently consume material. Given the amounts generally available at the micro or semimicro range, judicious selection of the tests to perform must be made; as, in some tests, the material may be recovered. You should always be aware of this fact. Each preliminary test that can be conducted with *little expenditure of time and material* can offer very valuable clues as to the class to which a given compound belongs.

Nonchemical Tests

Physical State

If the material is a *solid,* a few milligrams of the sample may be viewed under a magnifying glass or microscope. This may give some indication as to the homogeneity of the material. Crystalline shape often is an aid in classifying the compound.

 Determine the melting point using a few milligrams of the solid material. If a narrow melting point range (1–2°C) is observed, it is a good indication that the material is quite pure. If a broad range is observed, the compound must be recrystallized from a suitable solvent before proceeding. If the material undergoes decomposition on heating or if any evidence indicates that sublimation is occurring, an evacuated melting point should be run. Furthermore, this indicates that sublimation might be used to purify the compound, if necessary.

 If the material is a *liquid,* the boiling point is determined by the ultramicro method. If sufficient material is on hand and the boiling point reveals that the material is relatively pure, the *density* and the *refractive index* add valuable information for identification purposes.

Color

Since the majority of organic compounds are colorless, color can occasionally provide a clue to the nature of the sample. Use caution, however, since some impurities can give a substance color. Aniline is a classic example. When freshly distilled it is colorless, but on standing a small fraction oxidizes and turns the entire sample reddish brown.

Colored organic compounds contain a *chromophoric group*, usually indicating extended conjugation in the molecule. For example, trans-1,2-dibenzoylethylene (Experiments 2 and 6) is yellow, 5-nitrosalicylic acid (Experiment 34C) is light yellow, azobenzene (Experiment 44) is red, and tetraphenylcyclopentadienone (Experiment 49) is purple.

Can you identify the chromophore that causes these compounds to be colored? Note that a colorless liquid or white solid would *not contain* these units. Thus, compounds containing these groups would be excluded from consideration as possible candidates.

Odor

A compound's odor can occasionally be of assistance, since the vast majority of organic compounds have no definitive odor. You should become familiar with the odors of the common compounds or classes. For example, aliphatic amines have a fishy smell, benzaldehyde (like nitrobenzene and benzonitrile) has an almond odor (Experiment 20), and esters have fruity odors (Experiments 8A–D). Common solvents such as acetone, diethyl ether, and toluene all have distinctive odors. Butyric and caproic acids have rancid odors. In many cases, extremely small quantities of certain compounds can be detected by their odor. For example, a C_{16} unsaturated alcohol released by the female silkworm moth elicits a response from male moths of the same species at concentrations of 100 molecules/cm^3.

Odor detection involves your olfactory capabilities and thus can be considered a helpful lead, but very rarely can this property be used to strictly classify or identify a substance. As mentioned above, contamination by a small amount of an odorous substance is always a possibility.

> **CAUTION: *You should be very cautious when detecting odors. Any odor of significance can be detected several inches from the nose. Do not place the container closer than this to your eyes, nose, or mouth. Open the container of the sample and gently waft the vapors toward you.***

Ignition Test[1]

Valuable information can be obtained by carefully noting the manner in which a given compound burns. The ignition test is carried out by placing 1–2 mg of the sample on a spatula and then heating with a micro burner. Do not hold the sample directly in the flame; heat the spatula about 1 cm from the flat end and move the sample slowly into the flame (see Fig. 7.1)

Important observations to be made concerning the ignition test are summarized in Table 7.1.

As the sample is heated you should make the following observations.

1. Any melting or evidence of sublimation gives an approximate idea of the melting point by the temperature necessary to cause melting.

---

[1] For an extensive discussion on examination of ignition residues see Feigl, F.; Anger, V. *Spot Tests In Organic Analysis*, 7th ed.; Elsevier: New York, 1966; p 51.

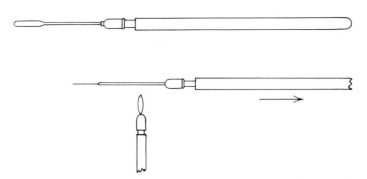

Fig. 7.1 *Heating on the microspatula. (Courtesy of Springer-Verlag, Vienna, Austria.)*

2. Color of the flame as the substance begins to burn (see Table 7.1).
3. Nature of the combustion (flash, quiet, or an explosion). Rapid, almost instantaneous combustion indicates high hydrogen content. Explosion indicates presence of nitrogen or nitrogen–oxygen-containing groups, for example, azo units (Experiment 44) or nitro groups (Experiment 34).
4. Nature of a residue, if present, after ignition.
   a. If a black residue remains and disappears on further heating at higher temperature, the residue is carbon.
   b. If the residue undergoes swelling during formation, the presence of a carbohydrate or similar compound is indicated.
   c. If the residue is black initially but still remains after heating, an oxide of a heavy metal is indicated.
   d. If the residue is white, the presence of an alkali or alkaline earth carbonate or SiO_2 from a silane or silicone is indicated.

SEPARATION OF IMPURITIES   If the preliminary tests outlined above indicate that the unknown in question contains impurities, it may be necessary to carry out one of several purification steps. These techniques are discussed in earlier chapters and are summarized below for correlation purposes.

1. For a liquid, distillation is generally used (see Experiments 2, 3A–3C)
2. For a solid, recrystallization is generally used (see Experiment 6).
3. Extraction is used especially if the impurity is insoluble in a solvent in which the compound itself is soluble (see Experiment 4).
4. Sublimation is a very efficient technique, if the compound sublimes (see Experiment 11).
5. Chromatography—gas, column, or thin-layer—is often used (see Experiments 1 and 8).

Table 7.1 Ignition Test Observations[a]

Type of Compound	Example	Observation
Aromatic compounds, unsaturated, or higher aliphatic compounds	Toluene	Yellow, sooty flame
Lower aliphatic compounds	Hexane	Yellow, almost nonsmoky flame
Compounds containing oxygen	Ethanol	Clear bluish flame
Polyhalogen compounds	Chloroform	Generally do not ignite until burner flame is applied directly to the substance
Sugars and proteins	Sucrose	Characteristic odor
Acid salts or organometallic compounds	Ferrocene	Residue

[a] Cheronis, N. D.; Entrikin, J. B. *Semimicro Qualitative Analysis*; Interscience: New York, 1947; p 85.

DETECTION OF ELEMENTS OTHER THAN CARBON, HYDROGEN, AND OXYGEN

The elements other than carbon, hydrogen, and oxygen that are most often present in organic compounds are nitrogen, sulfur, and the halogens (F, Cl, Br, I). To detect the presence of these elements, the organic compound is generally fused with metallic sodium. This reaction converts them to the water-soluble inorganic compounds NaCN, Na_2S, and NaX. Inorganic qualitative analysis techniques enable the investigator to determine the presence of the corresponding anions.

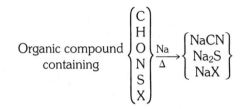

$$\text{Organic compound containing} \begin{Bmatrix} C \\ H \\ O \\ N \\ S \\ X \end{Bmatrix} \xrightarrow[\Delta]{Na} \begin{Bmatrix} NaCN \\ Na_2S \\ NaX \end{Bmatrix}$$

Sodium Fusion: Lassaigne's Test[2]

The procedure recommended uses a sodium–lead alloy[3] in place of metallic sodium. This reagent is quite stable in air, is easily stored in a screw-capped bottle, and does not give a vigorous reaction in water. Any metallic residues obtained from running the fusion reaction should be disposed of properly to avoid lead pollution.

HOOD

Important. *The fusion reaction is carried out in the* **hood.**

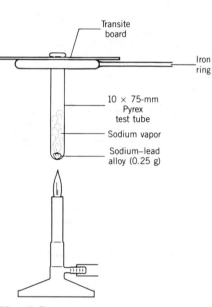

In a small (10×75 mm) test tube supported in a transite board (see Fig. 7.2) is placed ~0.25 g of the sodium–lead alloy. The tube is heated with a flame until the alloy melts and sodium vapor is observed condensing on the walls of the tube.

Note. *Do* **NOT** *heat the alloy to redness.*

A small sample of your unknown compound is carefully added to the tube. If your unknown is a *liquid*, add 2–3 drops from a Pasteur pipet; if it is a solid, add ~5 mg using a spatula. Be careful not to get any sample on the sides of the tube.

If your sample is a volatile liquid (bp < 100°C) or if it contains nitrogen or sulfur, mix the material with ~20–30 mg of powdered sucrose prior to its addition to the test tube. This aids in the reduction of various nitrogen or sulfur compounds. In addition, it absorbs volatile materials so that they may undergo the desired reaction before vaporization can occur.

The tube is heated gently to initiate the reaction with sodium. Remove the flame until the reaction subsides and then heat to redness for 1–2 minutes.

A boiling stone is now added to the cooled test tube followed by 2.0 mL of distilled water. The resulting solution is reheated with stirring and filtered while warm through filter paper. The filtrate is collected in a 10-mL Erlenmeyer flask.

Note. *The filtration step may be eliminated if the aqueous solution is clear.*

If the filtration is performed, wash the filter paper with an additional 2.0 mL of water. This wash is combined with the original filtrate.

The fusion solution collected in the Erlenmeyer flask is used to test for the

Fig. 7.2 *Apparatus for sodium fusion.*

Transite board

Iron ring

10 × 75-mm Pyrex test tube

Sodium vapor

Sodium–lead alloy (0.25 g)

[2] Lassaigne, J. L. *Ann. Chem.* **1843,** *48,* 367; also see Campbell, K. N.; Campbell, B. K. *J. Chem. Educ.* **1950,** *27,* 261.
[3] Vinson, J. A.; Grabowski, W. T. *J. Chem. Educ.* **1977,** *54,* 187.

presence of CN^- (nitrogen), S^{2-} (sulfur), and X^- (halogens except F^-) as described in the following sections.

Sulfur   Place 2–3 drops (Pasteur pipet) of the fusion solution on a white spot plate followed by 2 drops of water. Now add 1 drop of dilute (2%) aqueous sodium nitroprusside solution. Formation of a deep blue-violet color is a positive test for sulfur.

$$Na_2S + Na_2Fe(CN)_5NO \rightarrow Na_4[Fe(CN)_5NOS]$$

Sodium nitroprusside   Blue-violet complex

Nitrogen[4]   **Using the Fusion Solution**

Reagents

1. 1.5% solution of *p*-nitrobenzaldehyde in 2-methoxyethanol
2. 1.7% solution of *o*-dinitrobenzene in 2-methoxyethanol
3. 2.0% solution of NaOH in distilled water

All reagent drops are dispensed using Pasteur pipets.

On a white spot plate are placed 5 drops of reagent 1, 5 drops of reagent 2, and 2 drops of reagent 3. This mixture is gently stirred with a glass rod.

One drop of the fusion solution is now added. Formation of a deep purple color is a positive test for the presence of CN^- ion; a yellow or tan color is negative. If a positive result is obtained, nitrogen is present in the sample.

The test is valid in the presence of halogens (NaX) or sulfur (Na_2S). It is much more sensitive than the traditional Prussian blue test.[5]

The Soda-Lime Test

Mix in a 10×75-mm test tube ~50 mg of soda lime and 50 mg of MnO_2. Add 1 drop of liquid unknown or ~10 mg of a solid unknown. Place over the mouth of the tube a moist strip of brilliant yellow paper (moist red litmus paper may be used if necessary). Using a test tube holder, hold the tube at an incline, and heat the contents gently at first and then quite strongly. Nitrogen-containing compounds will usually evolve ammonia.

A positive test for nitrogen is the deep red coloration of the brilliant yellow paper (or blue color to the litmus paper).

Halogens (Except Fluorine)

Using the Fusion Solution

In a 10×75-mm test tube containing a boiling stone, place 0.5 mL (calibrated Pasteur pipet) of the fusion solution. This solution is carefully acidified by the dropwise addition of dilute HNO_3 acid, delivered from a Pasteur pipet (test acidity with litmus paper). If a positive test for nitrogen or sulfur was obtained, the resulting solution is heated to a gentle boil (stir with a microspatula to prevent

HOOD   bumping) for 1 minute over a micro burner (in the **hood**) to expel any HCN or H_2S that might be present. The tube is then cooled to room temperature.

---

[4] Adapted from Guilbault, G. G.; Kramer, D. N. *Anal. Chem.* **1966,** *39,* 834. *Idem, J. Org. Chem.* **1966,** *31,* 1103. See also Shriner, R. L.; Fuson, R. C.; Curtin, D. Y.; Morrill, T. C. *The Systematic Identification of Organic Compounds,* 6th ed.; Wiley: New York; 1980, p 80.
[5] See Vogel, A. I. *Elementary Practical Organic Chemistry,* Part 2, 2nd ed.; Wiley: New York, 1966, p 37.

To the fusion solution is now added 2 drops (Pasteur pipet) of aqueous $0.1\ M$ $AgNO_3$ solution.

A heavy curdy-type precipitate is a positive test for the presence of Cl^-, Br^-, or I^- ion. A faint turbidity is a negative test.

AgCl precipitate is white.

AgBr precipitate is pale yellow.

AgI precipitate is yellow.

AgF is not detected by this test since it is relatively soluble in water.

The silver halides have different solubilities in dilute ammonium hydroxide solution.

Centrifuge the test tube and contents and remove the supernatant liquid using a Pasteur filter pipet. Add 0.5 mL (calibrated Pasteur pipet) of dilute ammonium hydroxide solution to the precipitate and stir with a glass rod to determine whether the solid is soluble.

AgCl is soluble in ammonium hydroxide due to the formation of the complex ion $[Ag(NH_3)_2]^+$.

AgBr is slightly soluble in the reagent.

AgI is insoluble in this solution.

Further Test[6]

Once the presence of a halide ion has been established, a further test is also available to aid in distinguishing between Cl^-, Br^-, and I^-.

As described earlier, acidify 0.5 mL of the fusion solution with dilute HNO_3. To this solution add 5 drops (Pasteur pipet) of a 1.0% aqueous $KMnO_4$ solution and shake the test tube for ~ 1 minute.

Now add 10–15 mg of oxalic acid, enough to decolorize the excess purple permanganate, followed by 0.5 mL of methylene chloride solvent. The test tube is stoppered, shaken, and vented, and the layers are allowed to separate. Observe the color of the CH_2Cl_2 (lower) layer.

A clear methylene chloride layer indicates Cl^-.

A brown methylene chloride layer indicates Br^-.

A purple methylene chloride layer indicates I^-.

The colors may be faint and should be observed against a white background.

The Beilstein Test[7]

Organic compounds that contain chlorine, bromine, or iodine and hydrogen are decomposed on ignition in the presence of copper oxide to yield the corresponding hydrogen halides. These gases react to form the volatile cupric halides that impart a green or blue-green color to a nonluminous flame. It is a very sensitive test, but some nitrogenous compounds and carboxylic acids also give positive results.

---

[6] For further tests to distinguish between the three halide ions see Shriner, R. L.; Fuson, R. C.; Curtin, D. Y.; Morrill, T. C. *The Systematic Identification of Organic Compounds*, 6th ed.; Wiley: New York, 1980; p 81. Also see this reference (p. 85) for a specific test for the F^- ion.

[7] Beilstein, F. *Ber.* **1872,** *5*, 620.

Pound the end of a copper wire to form a flat surface that can act as a spatula. Stick the other end of the wire (~4 in. long) in a cork stopper to serve as a handle.

Heat the flat tip of the wire in a flame until coloration in the flame is negligible.

On the cooled flat surface is placed a drop (Pasteur pipet) of liquid unknown or a few milligrams of solid unknown. Gently heat the material in the flame. The carbon present in the compound will burn first, and thus the flame will be luminous, but then the characteristic green or blue-green color will be evident. It may be fleeting, so watch carefully.

It is recommended that a known compound containing a halogen be tested so that you become familiar with the appearance of the expected color.

F^- is not detected by this test since copper fluoride is not volatile.

SOLUBILITY CHARACTERISTICS

Determination of the solubility characteristics of an organic compound can often give valuable information as to its structural composition. It is especially useful when correlated with *spectral analysis*.

Several schemes have been proposed that place a substance in a definite group according to its solubility in various solvents. The scheme presented here is similar to that outlined by Shriner et al.[8]

There is no sharp dividing line between soluble and insoluble, and an arbitrary ratio of solute to solvent must be selected. We suggest that a compound be classified as soluble if its solubility is greater than 15 mg/500 μL of solvent. The solubility determinations are carried out at ambient temperature in 10 × 75-mm test tubes. The sample (15 mg) is placed in the test tube and a total of 0.5 mL of solvent added in three portions from a graduated or calibrated Pasteur pipet. After addition of each portion, the sample is stirred vigorously with a glass stirring rod for 1.5 to 2 minutes. If the sample is water-soluble, test the solution with litmus paper to assist in classification according to the Solubility Scheme.

To test with litmus paper, dip the end of a small glass rod into the solution and then gently touch the litmus paper with the rod. DO NOT DIP THE LITMUS PAPER INTO THE TEST SOLUTION.

In doing the solubility tests follow the Solubility Scheme in the order given. *Keep a record of your observations.*

Step 1. Test for water solubility. If soluble, test with litmus paper.

Step 2. If water soluble, determine the solubility in diethyl ether. This test further classifies water-soluble materials.

Step 3. Water-insoluble compounds are tested with 5% aqueous NaOH solution. If soluble, determine the solubility in 5% aqueous $NaHCO_3$. The use of the $NaHCO_3$ solution aids in distinguishing between strong (soluble) and weak (insoluble) acids.

Step 4. Compounds insoluble in 5% aqueous NaOH are tested with 5% HCl.

Step 5. Compounds insoluble in 5% aqueous HCl are tested with concentrated H_2SO_4. If soluble, further differentiation is made using 85% H_3PO_4 as shown in the scheme.

Step 6. Miscellaneous neutral compounds containing sulfur or nitrogen are normally soluble in strong acid solution.

[8] Shriner, R. L.; Fuson, R. C.; Curtin, D. Y.; Morrill, T. C. *The Systematic Identification of Organic Compounds*, 6th ed.; Wiley: New York, 1980.

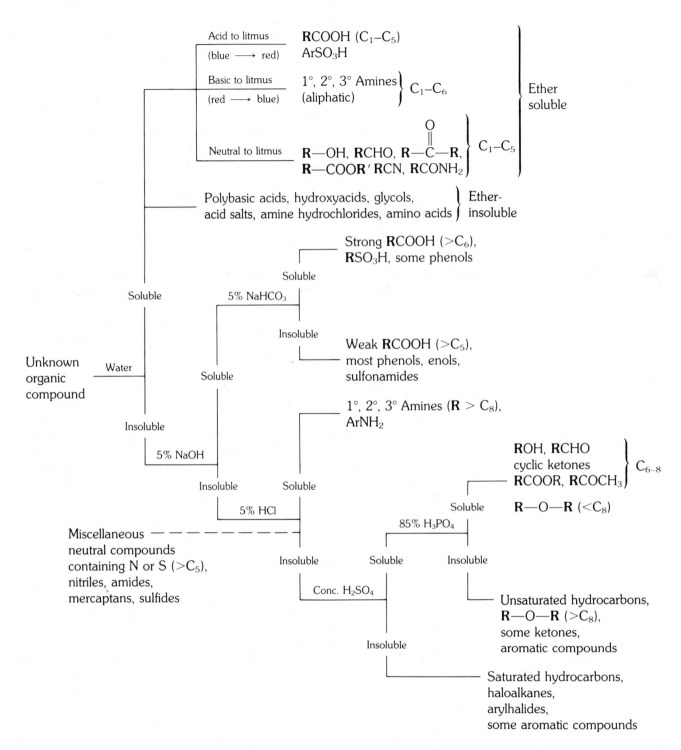

Note that to classify a given compound it may not be necessary to test its solubility in every solvent. *Do only the tests that are required to place the compound in one of the solubility groups.* Make your observations with care, and proceed in a logical sequence as you make the tests.

CLASSIFICATION TESTS[9]   **Note.** *For all tests given in this section, drops of reagents are measured and delivered using Pasteur pipets.*

---

[9] For a detailed discussion of classification tests see Shriner, R. L.; Fuson, R. C.; Curtin, D. Y.; Morrill, T. C. *Systematic Identification of Organic Compounds*, 6th ed.; Wiley: New York, 1980.

Alcohols

Ceric Nitrate Test

Primary, secondary, and tertiary alcohols having fewer than 10 carbon atoms give a positive test as indicated by a change in color from *yellow* to *red*.

$$(NH_4)_2Ce(NO_3)_6 + RCH_2OH \rightarrow \text{[alcohol + reagent]}$$

$$\text{Yellow} \qquad\qquad\qquad\qquad\qquad \text{Red complex}$$

On a white spot plate, place 5 drops of test reagent. (*The reagent is prepared by dissolving 4.0 g of ceric ammonium nitrate in 10 mL of 2 N HNO₃. Warming may be necessary.*) Add 1–2 drops of the unknown sample (5 mg if a solid). Stir with a thin glass rod to mix the components and observe any color change.

1. If the alcohol is water-insoluble, 3–5 drops of dioxane may be added, but run a blank to make sure the dioxane is pure. Efficient stirring gives positive results with most alcohols.
2. Phenols, if present, give a brown color or precipitate.

Chromic Anhydride Test: The Jones Oxidation

The Jones oxidation test is a rapid method to distinguish primary and secondary alcohols from tertiary alcohols. A positive test is indicated by a color change from *orange* (the oxidizing agent, Cr^{6+}) to *blue-green* (Cr^{3+}).

$$\left.\begin{array}{c} RCH_2OH \\ \text{or} \\ R_2CHOH \end{array}\right\} + CrO_3 \xrightarrow{H_2SO_4} Cr_2(SO_4)_3 + \begin{array}{c} RCO_2H \\ \text{or} \\ R_2C{=}O \end{array} + H_2O$$

$$\text{Orange} \qquad\qquad \text{Green}$$
$$\text{solution}$$

The test is based on oxidation of a primary alcohol to an aldehyde or acid and of a secondary alcohol to a ketone.

On a white spot plate, place 1 drop of the liquid unknown (10 mg if a solid). Add 5–8 drops of acetone and stir the mixture with a thin glass rod. To the resulting solution is added 1 drop of the test reagent. Stir and observe any color change within a 2-second time period. (*The reagent is prepared by slowly adding a suspension of 5.0 g of CrO₃ in 5.0 mL of concentrated H₂SO₄ to 15 mL of water. Allow the solution to cool to room temperature before using.*)

1. Run a blank to make sure the acetone is pure.
2. Tertiary alcohols, unsaturated hydrocarbons, amines, ethers, and ketones give a negative test within the 2-second time period for observing the color change. Aldehydes give a positive test.

HCl/ZnCl₂ Test: The Lucas Test

The Lucas test is used to distinguish between primary, secondary, and tertiary monofunctional alcohols having fewer than six carbon atoms.

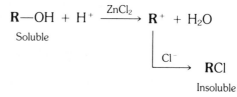

The test requires that the alcohol initially be in solution. As the reaction proceeds, the corresponding alkyl chloride is formed, which is insoluble in the reaction mixture. As a result, the solution becomes cloudy. In some cases a separate layer may be observed.

1. Tertiary alcohols react to give an immediate cloudiness to the solution. You may be able to see a separate layer of the alkyl chloride after a short time.
2. Secondary alcohols generally produce a cloudiness within 3–10 minutes. The solution may have to be heated to obtain a positive test.
3. Primary alcohols dissolve in the reagent but react very, very slowly.

In a small test tube prepared by sealing a Pasteur pipet off at the shoulder (■), place 2 drops of the unknown (10 mg if a solid) followed by 10 drops of the Lucas reagent. (*The test reagent is prepared by dissolving 13.6 g of anhydrous $ZnCl_2$ in 10.5 g of concentrated HCl with cooling in an ice bath.*) Shake or stir the mixture with a thin glass rod and allow the solution to stand. Observe the results. Based on the times given above, classify the alcohol.

1. Certain polyfunctional alcohols also give a positive test.
2. If an alcohol having three or fewer carbon atoms is expected, a 1-mL conical vial equipped with an air condenser should be used to prevent low-molecular-weight alkyl chlorides (volatile) from escaping and thus remaining undetected.

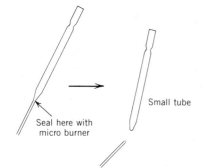

Seal here with micro burner

Small tube

Aldehydes and Ketones

The 2,4-Dinitrophenylhydrazine Test

Aldehydes and ketones react rapidly with 2,4-dinitrophenylhydrazine to form 2,4-dinitrophenylhydrazones. These derivatives range in color from *yellow* to *red* depending on the degree of conjugation in the carbonyl compound.

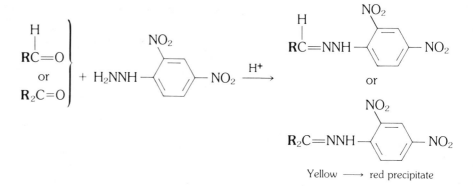

Yellow ⟶ red precipitate

On a white spot plate, place 7–8 drops of 2,4-dinitrophenylhydrazine reagent. (*The reagent is prepared by dissolving 1.0 g of 2,4-dinitrophenylhydrazine in 5.0 mL of concentrated sulfuric acid. This solution is slowly added, with stirring, to a mixture of 10 mL of water and 35 mL of 95% ethanol.*) Add 1 drop of a liquid unknown. If the unknown is a solid, add 1 drop of a solution prepared by dissolving 5 mg of the material in 5 drops of ethanol. The mixture is stirred with a thin glass rod. The formation of a red to yellow precipitate is a positive test.

Esters and amides do not interfere with the test.

Silver Mirror Test for Aldehydes: Tollen's Reagent

The reaction in the silver mirror test involves the oxidation of aldehydes to the corresponding carboxylic acid using an alcoholic solution of silver ammonium hydroxide. A positive test is the formation of a *silver* mirror or a black precipitate of finely divided silver.

$$RC\overset{\overset{\displaystyle H}{|}}{=}O + 2\,Ag(NH_3)_2OH \longrightarrow 2\,Ag\downarrow + R-C\underset{O^-,\ NH_4^+}{\overset{O}{\diagdown}} + H_2O + 3\,NH_3$$

In a small test tube prepared from a Pasteur pipet (*see* the Lucas test), place 1.0 mL of a 5% aqueous solution of AgNO$_3$ followed by 1 drop of aqueous 10% NaOH solution. Now add concentrated aqueous ammonia drop by drop (2–4 drops), with shaking, until the precipitate of silver oxide just dissolves. Add 1 drop of the unknown (10 mg if a solid), with shaking, and allow the reaction mixture to stand for 10 minutes at room temperature. If no reaction occurs, place the test tube in a sand bath at 40°C for 5 minutes. Observe the result.

1. Avoid a large excess of ammonia.
2. Reagents must be well mixed. Stirring with a thin glass rod is recommended.
3. *This reagent is freshly prepared for each test. It should not be stored since decomposition occurs with the formation of AgN$_3$, which is explosive.*
4. This oxidizing agent is very mild and thus alcohols are not oxidized under these conditions. Some sugars, acyloins, hydroxylamines, and substituted phenols do give a positive test.

Ammonium Salts, Amides, and Nitriles

Ammonium salts, amides, and nitriles undergo hydrolysis in alkaline solution to form ammonia gas or an amine.

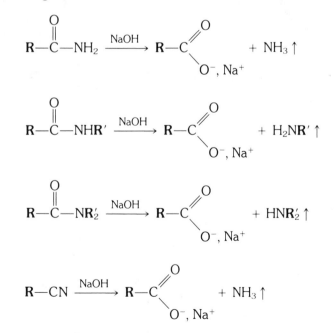

Ammonia from ammonium salts, primary amides, and nitriles can be detected by a color test using copper sulfate solution. The same test may also be used for secondary and tertiary amides that can generate low-molecular-weight (volatile) amines on hydrolysis.

In a 1.0-mL conical vial containing a boiling stone and equipped with an air condenser are placed 1–2 drops of the unknown liquid (10 mg if a solid) and

0.5 mL of 20% aqueous NaOH solution. This mixture is heated to *gentle* reflux on a sand bath. A strip of filter paper moistened with 2 drops of 10% aqueous copper sulfate solution is placed over the top of the condenser. Formation of a *blue* color (copper ammonia or amine complex) is a positive test.

The filter paper may be held in place using a small test tube holder or other suitable device.

Amines: The Hinsberg Test

The Hinsberg test is useful for distinguishing between primary, secondary, and tertiary amines. The reagent used is *p*-toluenesulfonyl chloride in alkaline solution.

Primary amines with fewer than seven carbons form a sulfonamide that is soluble in the alkaline solution. Acidification of the solution results in precipitation of the insoluble sulfonamide.

$$CH_3-\langle\text{benzene}\rangle-SO_2Cl + R-NH_2 \xrightarrow{NaOH}$$

$$CH_3-\langle\text{benzene}\rangle-SO_2NR^-, Na^+ \underset{\substack{\text{excess} \\ \text{base}}}{\overset{\substack{\text{excess} \\ \text{acid}}}{\rightleftharpoons}} CH_3-\langle\text{benzene}\rangle-SO_2NHR + NaCl + H_2O$$

(Soluble)   (Insoluble)

Secondary amines form an insoluble sulfonamide in the alkaline solution.

$$CH_3-\langle\text{benzene}\rangle-SO_2Cl + R_2-NH \xrightarrow{NaOH} CH_3-\langle\text{benzene}\rangle-SO_2NR_2 + NaCl + H_2O \xrightarrow{\substack{\text{excess} \\ \text{base}}} \text{no change}$$

(Insoluble)

Tertiary amines normally give no reaction under these conditions.

$$CH_3-\langle\text{benzene}\rangle-SO_2Cl + R_3N \xrightarrow{NaOH} CH_3-\langle\text{benzene}\rangle-SO_3^- + NR_3 + 2 Na^+ + Cl^- + H_2O$$

(Soluble)   (Oil)

In a 1.0-mL conical vial containing a boiling stone and equipped with an air condenser place 0.5 mL of 10% aqueous sodium hydroxide solution, 1 drop of the sample unknown (10 mg if a solid), followed by 30 mg of *p*-toluenesulfonyl

HOOD chloride **[hood]**. The mixture is heated to reflux for 2–3 minutes on a sand bath and then cooled in an ice bath. Test the alkalinity of the solution using litmus paper. If it is not alkaline, add additional 10% aqueous NaOH dropwise.

Using a Pasteur filter pipet, separate the solution from any solid that may be

SAVE present. Transfer the solution to a clean 1.0-mL conical vial **[save]**.

Note. *If an oily upper layer is obtained at this stage, remove the lower alkaline*

SAVE *phase* **[save]** *using a Pasteur filter pipet. To the remaining oil add 0.5 mL of cold water and stir vigorously to obtain a solid material.*

If a solid is obtained it may be (1) the sulfonamide of a secondary amine, (2) recovered tertiary amine if the original amine was a solid, or (3) the insoluble salt of a primary sulfonamide derivative (if the original amine had more than six carbon atoms).

1. If the solid is a tertiary amine, it is soluble in aqueous 10% HCl.
2. If the solid is a secondary sulfonamide, it is insoluble in aqueous 10% NaOH.
3. If no solid is present, acidify the alkaline solution by adding 10% aqueous HCl. If the unknown amine is primary, the sulfonamide will precipitate.

Unsaturated Hydrocarbons: Alkenes and Alkynes

Bromine in Methylene Chloride
Unsaturated hydrocarbons readily add bromine. An example of this reaction is given in Experiment 50A.

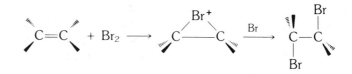

The test is based on the decolorization of a red-brown bromine–methylene chloride solution.

In a 10 × 75-mm test tube or a small tube prepared from a Pasteur pipet (see Lucas test), place 2 drops of a liquid unknown (15 mg if a solid) followed by 0.5 mL of methylene chloride, in the **hood**. Add dropwise, with shaking, a 2% solution of Br_2 in methylene chloride solvent, also in the **hood**. The presence of an unsaturated hydrocarbon will require 2–3 drops of the reagent before the reddish-brown color of bromine persists in the solution.

HOOD
HOOD

> **CAUTION: *Bromine is highly toxic and can cause burns.***

1. Methylene chloride is used in place of the usual carbon tetrachloride since it is less toxic.
2. Phenols, enols, amines, aldehydes, and ketones interfere with this test.

Permanganate Test: The Baeyer Test for Unsaturation
Unsaturation in an organic compound can be detected by the decolorization of permanganate solution. The reaction involves the cis hydroxylation of the olefin to give a 1,2-diol (glycol).

$$\text{C=C} + 2\,MnO_4^- + 4\,H_2O \longrightarrow \underset{\text{OH OH}}{\text{C—C}} + 2\,MnO_2 + 2\,OH^-$$

On a white spot plate is placed 0.5 mL of *alcohol-free* acetone followed by 2 drops of the unknown compound (15 mg if a solid). There is now added dropwise (2–3 drops), with stirring, a 1% aqueous solution of $KMnO_4$. A positive test is the discharge of purple permanganate from the reagent and the precipitation of brown manganese oxides.

Any functional group that undergoes oxidation with permanganate interferes with the test (phenols, aryl amines, most aldehydes, primary and secondary alcohols, etc.)

Phenols and Enols

Most phenols and enols form colored complexes in the presence of ferric ion, Fe^{3+}.

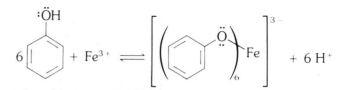

Phenols give a red, blue, purple, or green color. Sterically hindered phenols give a negative test. Enols generally give a tan, red, or red-violet color.

On a white spot plate, place 2 drops of water, or 1 drop of water plus 1 drop of ethanol, or 2 drops of ethanol, depending on the solubility characteristics of the unknown. To this solvent system add 1 drop (10 mg if a solid) of the substance to be tested. Stir the mixture with a thin glass rod to complete dissolution. Add 1 drop of 2.5% aqueous ferric chloride solution (light yellow in color). Stir and observe any color formation. If necessary, a second drop of the $FeCl_3$ solution may be added.

1. The color developed may be fleeting or it may last for many hours. A slight excess of the ferric chloride solution may or may not destroy the color.
2. An alternate procedure using $FeCl_3$–CCl_4 solution in the presence of pyridine is available.[10]

Esters: Hydroxamate Test

Esters of carboxylic acids can be identified by conversion to hydroxamic acid salts. Acidification of this salt produces the corresponding hydroxamic acid, which is identified by formation of a red to purple color in the presence of Fe^{3+} ion.

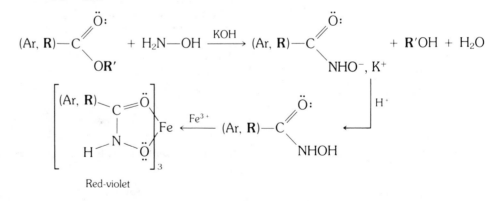

Red-violet

In a 3.0-mL conical vial containing a boiling stone and equipped with an air condenser, place 1 drop of the liquid unknown (10 mg if a solid) followed by 0.5 mL of 1.0 M ethanolic hydroxylamine hydrochloride solution. To this solution add dropwise 10% methanolic KOH until the resulting solution is pH ~10 (pH paper). This mixture is heated to reflux temperature on a sand bath for 5 minutes, cooled to room temperature, and acidified to pH 3–4 by dropwise addition of 5% aqueous HCl solution. Add 2 drops of 5% aqueous $FeCl_3$ solution. Formation of a red to purple color is a positive test.

1. It is suggested that a blank be run for comparison purposes.
2. Acid chlorides, anhydrides, lactones, and imides also give a positive test.

---

[10] Soloway, S; Wilen, S. H. *Anal. Chem.* **1952,** *4,* 979.

Amides: Hydroxamate Test

Like esters, unsubstituted amides and the majority of substituted amides will give a positive hydroxamate test, although more drastic conditions are required.

$$(Ar, \mathbf{R}) - \overset{\overset{\ddot{O}:}{\|}}{C} \diagdown_X + H_2N - OH \xrightarrow{\text{Propylene glycol}} (Ar, \mathbf{R}) - \overset{\overset{\ddot{O}:}{\|}}{C} \diagdown_{NHOH} + HX$$

$$(X = NH_2, NH\mathbf{R'}, N\mathbf{R'R''})$$

The hydroxamic acid is identified by formation of a red to purple color in the presence of Fe^{3+} ion as for the test with esters.

In a 3.0-mL conical vial containing a boiling stone and equipped with an air condenser is placed one drop of a liquid unknown (10 mg if a solid) followed by 0.5 mL of 1 M hydroxylamine hydrochloride–propylene glycol solution. The resulting mixture is heated to reflux temperature (~190°C) on a sand bath for 3–5 minutes. The solution is now cooled to room temperature and 2 drops of 5% aqueous $FeCl_3$ solution added. Formation of a red to purple color is a positive test.

Methyl Ketones, Methyl Carbinols: Iodoform Test

The iodoform test involves hydrolysis and cleavage of methyl ketones to form a yellow precipitate of iodoform.

$$R - \overset{\overset{\cdot\overset{\cdot}{O}\cdot}{\|}}{C} - CH_3 + 3\,I_2 + 3\,KOH \longrightarrow R - \overset{\overset{\cdot\overset{\cdot}{O}\cdot}{\|}}{C} - CI_3 + 3\,KI + 3\,H_2O$$

$$\downarrow KOH$$

$$R - \overset{\overset{\cdot\overset{\cdot}{O}\cdot}{\|}}{C} - O{,}K^{-+} + CHI_3 \downarrow$$

Yellow

It is also a positive test for compounds that on oxidation generate methyl ketones under these reaction conditions. For example, methyl carbinols (secondary alcohols having at least one methyl group attached to the carbon atom to which the OH unit is linked), acetaldehyde, and ethanol give positive results.

In a 3.0-mL conical vial equipped with an air condenser is placed 2 drops of the unknown liquid (10 mg if a solid) followed by 5 drops of 10% aqueous KOH solution.

Note. *If the sample is insoluble in the aqueous phase, either mix vigorously or add dioxane or bis(2-methoxyethyl) ether to obtain a homogeneous solution.*

The mixture is warmed on a sand bath to 50–60°C and the KI–I_2 reagent is added dropwise until the solution becomes dark brown (~1.0 mL). Additional 10% aqueous KOH is now added (dropwise) until the solution is again colorless. (*The KI–I_2 reagent is prepared by the instructor. Mix 3 g of KI and 1 g I_2 in 20 mL of water.*)

CAUTION: *Iodine is highly toxic and can cause burns.*

After a period of 2 minutes, cool the solution and determine whether a yellow precipitate (HCl_3) has formed. If a precipitate is not observed, reheat as before for another 2 minutes. Cool and check again for the appearance of iodoform.

1. The iodoform test is reviewed elsewhere.[11]
2. An example of the general haloform reaction using bleach to oxidize a methyl ketone is given in Experiments 43A and 43B.

Alkyl Halides

Silver Nitrate Test

Alkyl halides that undergo S_N1 substitution react with alcoholic silver nitrate to form a precipitate of the corresponding silver halide.

Secondary and primary halides react slowly or not at all at room temperature. However, they do react at elevated temperatures.

Tertiary halides react immediately at room temperature.

In a 1.0-mL conical vial place 0.5 mL of 2% ethanolic $AgNO_3$ solution and 1 drop of unknown (10 mg if a solid). A positive test is formation of a precipitate within 5 minutes. If no reaction occurs, add a boiling stone and equip the vial with an air condenser. Heat the solution at *gentle* reflux for an additional 5 minutes on a sand bath. Cool the solution.

If a precipitate forms, add 2 drops of dilute HNO_3. AgX will not dissolve in nitric acid solution.

1. The order of reactivity for **R** groups is allyl = benzyl > tertiary > secondary $>>>$ primary. For the halide leaving groups, I > Br > Cl.
2. Acid halides, α-haloethers, and 1,2-dibromo compounds also give a positive test at room temperature. The activated aryl halides only give a positive test at elevated temperatures.

Sodium Iodide in Acetone

Primary alkyl chlorides and bromides can be distinguished from aryl and vinylic halides by reaction with sodium iodide in acetone.

$$\underset{(X = Cl,\ Br)}{\textbf{R}\!-\!X} + NaI \xrightarrow{\text{acetone}} \textbf{R}\!-\!I + NaX \downarrow$$

Primary alkyl bromides undergo an S_N2 displacement reaction within 5 minutes at room temperature, primary alkyl chlorides only at 50°C.

In a 1.0-mL conical vial place 1 drop of a liquid unknown (10 mg if a solid) and 3 drops of acetone. To this solution add 0.5 mL of NaI-acetone reagent. (*The reagent is prepared by dissolving 3 g of NaI in 25 mL of acetone*). A positive test is the appearance of a precipitate of NaX within 5 minutes. If no precipitate is observed, add a boiling stone and equip the vial with an air condenser. Warm the reaction mixture in a sand bath at ~50°C for 5 minutes. Cool to room temperature and determine whether a reaction has occurred.

1. Benzylic and allylic chlorides and bromides, acid chlorides and bromides, and α-halo ketones, esters, amides, and nitriles also give a positive test at room temperature.

---

[11] Fuson, R. C.; Bull, B. A. *Chem. Rev.* **1934** *15*, 275.

2. Primary and secondary alkyl chlorides and secondary and tertiary alkyl bromides react at 50°C under these conditions.

3. If the solution turns red-brown, I_2 is being liberated.

Ethers: Ferrox Test

The ferrox test is a color test sensitive to oxygen that may be used to distinguish ethers from some hydrocarbons, which like most ethers are soluble in sulfuric acid.

In a dry 10×75-mm test tube, using a glass stirring rod, is ground a crystal of ferric ammonium sulfate and a crystal of potassium thiocyanate. The ferric hexathiocyanatoferriate that is formed adheres to the rod.

In a second clean 10×75-mm test tube, place 2–3 drops of a liquid unknown. If dealing with a solid, use ~10 mg and add toluene until a saturated solution is obtained. Now using the rod with the ferric hexathiocyanatoferriate attached, stir the unknown. *If the unknown contains oxygen, the ferriate compound dissolves and a reddish-purple color is observed.*

Some high-molecular-weight ethers do not give a positive test.

Nitro Compounds

Many nitro compounds give a positive test based on the following reaction.

$$\mathbf{R}-NO_2 + 4\,H_2O + 6\,Fe(OH)_2 \rightarrow \mathbf{R}-NH_2 + 6\,\underset{\text{Red-brown}}{Fe(OH)_3} \downarrow$$

The nitro derivative oxidizes the iron(II) hydroxide to iron(III) hydroxide, the latter being a red-brown solid.

In a 1.0-mL conical vial is placed 5–10 mg of the unknown compound followed by 0.4 mL of freshly prepared 5% aqueous ferrous ammonium sulfate. After mixing, 1 drop of 3 N sulfuric acid is added followed by 10 drops of methanolic 2 N KOH. The vial is capped, shaken vigorously, vented, and allowed to stand over a 5-minute period. Formation of a red-brown precipitate, usually within 1 minute, is a positive test.

Acids

The presence of a carboxylic acid is detected by its solubility behavior. An aqueous solution of the acid will be acidic to litmus paper. Since a sulfonic acid would also give a positive test, the test for sulfur (sodium fusion) is used to distinguish between the two types of acids. A water-soluble phenol is acidic toward litmus paper but also would give a positive ferric chloride test.

PREPARATION OF DERIVATIVES

Based on the preliminary and classification tests carried out to this point, you should have established the type of functional group present (or lack of one) in the unknown organic sample. The next step is to consult a set of tables containing a listing of known organic compounds by functional group and/or physical properties or both. Using the physical property data for your compound, you can select a few possible candidates that appear to "fit" the data you have collected. On a chemical basis, the final step in the identification sequence is to prepare one or two *crystalline derivatives* of your compound. Selection of the specific compound and thus final confirmation of its identity can then be made from the extensive derivative tables that have been accumulated. With the advent of spectral analysis, the preparation of derivatives is often not necessary, but the wealth of chemistry that can be learned by the beginning student in carrying out these procedures is extensive and important. Methods of preparing selected derivatives for the most common functional groups are given below. Condensed tables of compounds and their derivatives are summarized in Appendix D. For extensive tables see the references cited below.

Important. *In each of the procedures outlined below, drops of reagents are measured using Pasteur pipets.*

REFERENCES
1. Shriner, R. L.; Fuson, R. C.; Curtin; D. Y.; Morrill, T. C. *The Systematic Identification of Organic Compounds,* 6th ed.; Wiley: New York, 1980.
2. Rappoport, Z.; *Handbook of Tables for Organic Compound Identification,* 3rd ed.; CRC Press: Boca Raton, FL, 1967.

ACIDS (Tables D.1, D.2) **Preparation of Acid Chlorides**

$$R-\overset{\overset{\cdot\cdot{O}\cdot\cdot}{\|}}{C}-\overset{\cdot\cdot}{O}H \ + \ Cl-\overset{\overset{\cdot\cdot{O}\cdot\cdot}{\|}}{S}-Cl \ \xrightarrow{\text{DMF}} \ R-\overset{\overset{\cdot\cdot{O}\cdot\cdot}{\|}}{C}-Cl + HCl\uparrow \ + \ SO_2\uparrow$$

In a 1.0-mL conical vial containing a boiling stone and equipped with a reflux condenser are placed 20 mg of the unknown acid, 4 drops of thionyl chloride, and 1 drop of *N,N*-dimethylformamide.

HOOD

> **CAUTION: *This reaction is run in the* hood *since hydrogen chloride and sulfur dioxide are evolved. Thionyl chloride is an irritant and is harmful to breathe.***

After allowing the mixture to stand at room temperature for 10 minutes, heat it at gentle reflux in a sand bath for a period of 15 minutes and then cool to room temperature. It is then diluted with 5 drops of methylene chloride solvent.

The acid chloride is not isolated but is used directly in the preparations given below.

Amides

$$R-\overset{\overset{\cdot\cdot{O}\cdot\cdot}{\|}}{C}-Cl \ + \ 2\,NH_3 \ \longrightarrow \ R-\overset{\overset{\cdot\cdot{O}\cdot\cdot}{\|}}{C}-NH_2 \ + \ NH_4Cl$$

HOOD

In a 3.0-mL conical vial containing a magnetic spin vane and equipped with an air condenser is placed 10 drops of concentrated aqueous ammonia **[hood]**. The flask is cooled in an ice bath and the acid chloride solution prepared earlier transferred by Pasteur pipet, *dropwise* with stirring, to the ammonia solution. *It is convenient to make this addition down the neck of the condenser.* The amide may precipitate during this operation. After the addition is complete, the ice bath is removed and the mixture stirred for an additional 5 minutes. Methylene chloride (10 drops) is added and the mixture stirred to dissolve any precipitate. The methylene chloride layer is separated from the aqueous layer using a Pasteur filter pipet and transferred to a Pasteur filter pipet containing 200 mg of anhydrous sodium sulfate. The eluate is collected in a Craig tube containing a boiling stone. The aqueous phase is extracted with 0.5 mL of methylene chloride and the methylene chloride layer separated as before and also passed through the column. Both eluates are combined. The methylene chloride solu-

HOOD   tion is now evaporated using a sand bath **[hood]**. The amide product is recrystallized from an ethanol–water mixture using the Craig tube. Dissolve the material in ~0.5 mL of ethanol, add water (dropwise) to the cloud point, cool in an ice bath, and collect the crystals in the usual manner. Dry the crystalline amide on a porous clay plate or on filter paper and determine the melting point.

Anilides

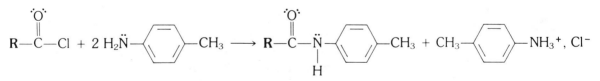

In a 3.0-mL conical vial containing a magnetic spin vane and equipped with an air condenser are placed 5 drops of aniline and 10 drops of methylene chloride. The solution is cooled in an ice bath and the acid chloride solution prepared earlier transferred by Pasteur pipet, *dropwise with stirring*, to the aniline solution

HOOD   **[hood]**. *It is convenient to make this addition down the neck of the condenser.* After the addition is complete, the ice bath is removed and the mixture stirred for an additional 10 minutes.

The methylene chloride layer is now transferred to a 10 × 75-mm test tube and washed with 0.5 mL of water, 0.5 mL of 5% aqueous HCl solution, 0.5 mL of 5% aqueous sodium hydroxide solution, and finally 0.5 mL of water. For each washing, the test tube is stoppered, shaken, and the top aqueous layer removed by Pasteur filter pipet. The wet methylene chloride layer is transferred to a Pasteur filter pipet containing 200 mg of anhydrous sodium sulfate and the eluate collected in a Craig tube containing a boiling stone. The original test tube is rinsed with an additional 10 drops of methylene chloride, and this rinse is also passed through the column. Both eluates are combined.

HOOD   The methylene chloride solvent is evaporated using a sand bath in the **hood**. The crude anilide is recrystallized from an ethanol–water mixture using the Craig tube. Dissolve the material in ~0.5 mL of ethanol, add water (dropwise) to the cloud point, cool in an ice bath, and collect the crystals in the usual manner. Dry the purified derivative product on a porous clay plate or on filter paper and determine the melting point.

Toluidides

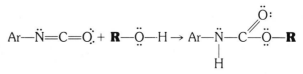

The same procedure described for the preparation of anilides is used except that *p*-toluidine replaces the aniline.

ALCOHOLS (Table D.3)   Phenyl- and α-Naphthylurethans

$$Ar\!-\!\ddot{N}\!=\!C\!=\!\ddot{O}: \ + \ \mathbf{R}\!-\!\ddot{O}\!-\!H \rightarrow Ar\!-\!\underset{\underset{H}{|}}{\ddot{N}}\!-\!C\!\overset{\overset{\ddot{O}:}{\parallel}}{}\!\ddot{O}\!-\!\mathbf{R}$$

Isocyanate   Urethan

In a 1.0-mL conical vial containing a boiling stone and equipped with an air condenser protected by a calcium chloride drying tube is placed 15 mg of an anhydrous alcohol or phenol. The air condenser is removed from the vial and 2 drops of phenyl isocyanate or α-napthyl isocyanate is added. The air condenser is replaced immediately. If the unknown is a phenol, one drop of pyridine is added in a similar manner.

HOOD

> **CAUTION:** *This addition must be done in the hood. The isocyanates are lachrymatory! Pyridine has a strong odor of the amines.*

If a spontaneous reaction does not take place, the vial is heated at ~80–90°C, on a sand bath, for a period of 5 minutes and then cooled in an ice bath. It may be necessary to scratch the sides of the vial to induce crystallization. The solid product is collected by vacuum filtration using a Hirsch funnel and purified by recrystallization from ligroin. For this procedure the solid is placed in a 10 × 75-mm test tube and dissolved in 1.0 mL of warm (60°–80°C) ligroin. If diphenyl (or dinaphthyl) urea is present, which is formed by reaction of the isocyanate with water, it is insoluble in the solvent. The warm ligroin solution is transferred to a Craig tube using a Pasteur filter pipet. The solution is cooled in an ice bath and the resulting crystals are collected in the usual manner. After drying the product on a porous clay plate, determine the melting point.

3,5-Dinitrobenzoates

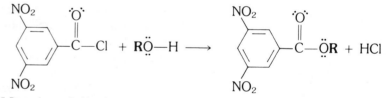

3,5-Dinitrobenzoyl chloride

In a 1.0-mL conical vial containing a boiling stone and equipped with an air condenser protected by a calcium chloride drying tube are placed 25 mg of 3,5-dinitrobenzoyl chloride and 2 drops of the unknown alcohol. The mixture is then heated to ~10°C below the boiling point of the alcohol (but not over 100°C) on a sand bath for a period of 5 minutes. Water (0.3 mL) is added and the vial placed in an ice bath to cool. The solid ester is collected by vacuum filtration using a Hirsch funnel and the filter cake washed with three 0.5-mL portions of 2% aqueous sodium carbonate solution followed by 0.5 mL of water. The solid product is recrystallized from an ethanol–water mixture using a Craig tube. Dissolve the material in ~0.5 mL of ethanol. Add water (dropwise) to the cloud point, cool in an ice bath, and collect the crystals in the usual manner. After drying the product on a porous clay plate or on filter paper determine the melting point.

ALDEHYDES AND KETONES (Tables D.4, D.5)

2,4-Dinitrophenylhydrazones

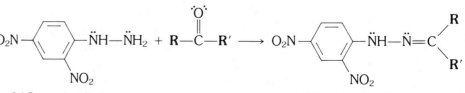

2,4-Dinitrophenylhydrazine   A 2,4-dinitrophenylhydrazone

The procedure outlined in the Classification Tests section for aldehydes and ketones is used. If necessary, the derivative can be recrystallized from 95% ethanol.

Semicarbazones

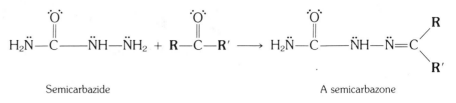

Semicarbazide   A semicarbazone

In a 1.0-mL conical vial, place 12 mg of semicarbazide hydrochloride, 20 mg of sodium acetate, 10 drops of water, and 12 mg of the unknown carbonyl compound. Cap the vial, shake vigorously, vent, and allow the vial to stand at room temperature until crystallization is complete. Cool the vial in an ice bath if necessary. Collect the crystals by vacuum filtration using a Hirsch funnel and wash the filter cake with 0.2 mL of cold water. Dry the crystals on a porous clay plate or on filter paper. Determine the melting point.

AMINES (Tables D.6, D.7)

Primary and Secondary Amines: Acetamides

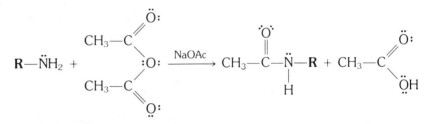

In a 1.0-mL conical vial equipped with an air condenser are placed 20 mg of the unknown amine, 5 drops of water, and 1 drop of concentrated HCl.

In a small test tube a solution of 40 mg of sodium acetate trihydrate dissolved in 5 drops of water is prepared. The stoppered solution is set aside for use in the next step.

The solution of amine hydrochloride is warmed to ~50°C in a sand bath. It is **HOOD** then cooled and 40 μL of acetic anhydride added in one portion **[hood]** through the condenser. This is followed *immediately* by addition of the sodium acetate solution using a Pasteur pipet. Swirl the contents of the vial to ensure complete mixing.

The reaction mixture is allowed to stand at room temperature for ~5 minutes and then placed in an ice bath for an additional 5–10 minutes. The white crystals are collected by vacuum filtration using a Hirsch funnel and the filter cake is washed with two 0.1-mL portions of water. The crystals are dried on a porous clay plate or on filter paper. Determine the melting point.

Primary and Secondary Amines: Benzamides

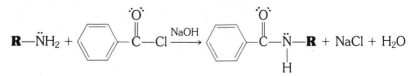

In a 1.0-mL conical vial are placed 0.4 mL of 10% aqueous NaOH solution, **HOOD** 25 mg of the amine, and 2–3 drops of benzoyl chloride **[hood]**. The vial is capped and shaken over a period of ~10 minutes. It is vented periodically to release any pressure buildup.

The crystalline precipitate is collected by vacuum filtration using a Hirsch funnel and the filter cake washed with 0.1 mL of dilute HCl followed by 0.1 mL of water. It is generally necessary to recrystallize the material from methanol or aqueous ethanol using the Craig tube. Dry the product on a porous clay plate or on filter paper and determine the melting point.

Primary, Secondary, and Tertiary Amines: Picrates

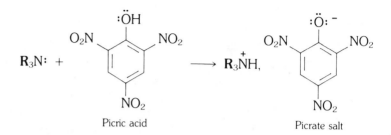

In a 1.0-mL conical vial containing a boiling stone and equipped with an air condenser are placed 15 mg of the unknown amine and 0.3 mL of 95% ethanol.

Note. *If the amine is not soluble in the ethanol, shake the mixture to obtain a saturated solution and then transfer this solution, using a Pasteur filter pipet, to another vial.*

There is now added 0.3 mL of a saturated solution of picric acid in 95% ethanol.

> **CAUTION:** *Picric acid explodes when rapidly heated or by percussion.*

The mixture is heated at reflux, on a sand bath, for ~1 minute and then allowed to cool slowly to room temperature. The yellow crystals of the picrate are collected by vacuum filtration using a Hirsch funnel. The material is dried on a porous clay plate and the melting point determined.

ACID CHLORIDES AND ANHYDRIDES (Table D.8)

Amides

$$R-\overset{\overset{\cdot\cdot}{O}\cdot\cdot}{\underset{\|}{C}}-Cl + 2\,NH_3 \longrightarrow R-\overset{\overset{\cdot\cdot}{O}\cdot\cdot}{\underset{\|}{C}}-\ddot{N}H_2 + NH_4Cl$$

HOOD

In a 10×75-mm test tube, place 0.4 mL of ice-cold, concentrated ammonium hydroxide solution. To the solution, add slowly **[hood]**, with shaking, 15 mg of the unknown acid chloride or anhydride. Stopper the test tube and allow the reaction mixture to stand at room temperature for approximately 5 minutes. Collect the crystals by vacuum filtration using a Hirsch funnel, and wash the filter cake with 0.2 mL of ice-cold water. Recrystallize the material, from water or an ethanol–water mixture, using a Craig tube. Dry the purified crystals on a porous clay plate and determine the melting point.

AROMATIC HYDROCARBONS | **Picrates**
(Table D.9)

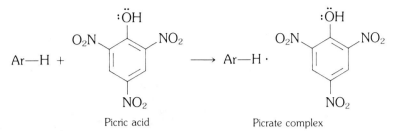

Picric acid   Picrate complex

The procedure outlined under Amines (Primary, Secondary, and Tertiary Amines: Picrates) is used to prepare these derivatives.

PHENOLS (Table D.10) | **α-Naphthylurethans**

The procedure outlined under Alcohols (Phenyl- and α-Naphthylurethans) is used to prepare these derivatives.

Bromo Derivatives

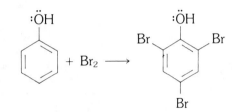

HOOD

In a 1.0-mL conical vial is placed 10 mg of the unknown phenol followed by 2 drops of methanol and 2 drops of water. To this solution is added 3 drops **[hood]** of brominating agent from a Pasteur pipet. [*The brominating reagent is prepared by adding 1.0 mL (3 g) of bromine* **(caution! hood)** *to a solution of 4.0 g of KBr in 25 mL of water*]. The addition is continued (dropwise) until the yellow color of bromine persists. Water (4 drops) is now added, and the vial is capped, shaken, vented, and then allowed to stand at room temperature for 10 minutes. The crystalline precipitate is collected by vacuum filtration using a Hirsch funnel and the filter cake washed with 0.5 mL of 5% aqueous sodium bisulfite solution. The solid is recrystallized from ethanol or an ethanol–water mixture using a Craig tube. Dissolve the material in ~0.5 mL of ethanol, add water to the cloud point, cool in an ice bath, and collect the crystals in the usual manner. Dry the purified product on a porous clay plate and determine the melting point.

HYDROCARBONS, HALOGENATED HYDROCARBONS, NITRILES, AMIDES, NITRO COMPOUNDS, ETHERS, AND ESTERS (Tables D.11, D.12, D.13, D.14, D.15, D.16, D.17)

These compounds do not give derivatives directly. They are usually converted into another material that can then be derivatized. The procedures are for the most part lengthy and frequently give mixtures of products. It is recommended that compounds belonging to these classes be identified using spectral methods.

QUESTIONS | **7.1.** The following six substances have approximately the same boiling point and all are colorless liquids. Suppose you were given six unlabeled bottles, each of which contained one of these compounds. Explain how you would use simple chemical tests to identify the contents of each bottle.

ethanoic acid	toluene
propyl butanoate	diisobutylamine
1-butanol	styrene

7-2. A colorless liquid, C_4H_6O, having a boiling point of 97–98°C was found to be soluble in water and also in ether. It gave a negative test for the presence of halogens, sulfur, and nitrogen. It did, however, give a positive test with the Baeyer reagent and also gave a positive test with the 2,4-dinitrophenylhydrazine reagent. It gave a negative result when treated with ceric nitrate solution and also with Tollen's reagent. Treatment with ozone followed by hydrolysis in the presence of zinc gave formaldehyde as one of the products. What are the structure and name of the colorless liquid?

7-3. A colorless liquid A, C_3H_6O, was soluble in water and ether and had a boiling point of 94–96°C. It decolorized a Br_2–CH_2Cl_2 solution and gave a positive ceric nitrate test. On catalytic hydrogenation it formed compound B, C_3H_8O, which did not decolorize the bromine solution but did give a positive ceric nitrate test. Treatment of compound A with ozone followed by hydrolysis in the presence of zinc gave formaldehyde as one of the products. Compound A formed an α-naphthylurethan having melting point 109°C. What are the names and structures of compounds A and B?

7-4. A compound of formula $C_{14}H_{12}$ gave a positive Baeyer test and burned with a yellow, sooty flame. Treatment with ozone followed by hydrolysis in the presence of zinc gave formaldehyde as one of the products. Also isolated from the ozonolysis reaction was a second compound, $C_{13}H_{10}O$, which burned with a yellow, sooty flame and readily formed a semicarbazone having melting point 164°C. The NMR spectrum of the $C_{13}H_{10}O$ compound showed a single absorption peak at 7.5 ppm. What are the structures and names of the two compounds?

7-5. Compound A, $C_7H_{14}O$, burned with a yellow, nonsooty flame and did not decolorize a bromine–methylene chloride solution. It did give a positive 2,4-dinitrophenylhydrazine test, but a negative Tollen's test. Treatment of the compound with lithium aluminum hydride followed by neutralization with acid produced compound B, which gave a positive Lucas test in ~5 minutes. Compound B also gave a positive ceric nitrate test. The NMR analysis for compound A gave the following data:

1.02 ppm	9H, singlet
2.11 ppm	3H, singlet
2.31 ppm	2H, singlet

Give suitable structures for compounds A and B.

7-6. A friend of yours, a graduate student attempting to establish the structure of a chemical species from field clover, isolated an alcohol that he found to have an optical rotation of +49.5°. Chemical analysis gave a molecular formula of $C_5H_{10}O$. It was also observed that this alcohol readily decolorized bromine–methylene chloride solution. On this basis, the alcohol was subjected to catalytic hydrogenation and it was found to absorb 1 mol of hydrogen gas. The product of the reduction gave a positive ceric nitrate test, indicating that it too was an alcohol. However, the reduced compound was optically inactive. Your friend has come to you for assistance in writing the structures of the two alcohols. What do you believe the structures are?

7-7. An unknown compound burned with a yellow, nonsmoky flame and was found to be insoluble in 5% sodium hydroxide solution but soluble in concentrated sulfuric acid. Measurement of its boiling point gave a range of 130–131°C. Analysis gave a molecular formula of C_5H_8O. It was found to give a semicarbazone having melting point 204–206°C. However, it gave a negative result when treated with Tollen's reagent and did not decolorize the Baeyer reagent. It also gave a negative iodoform test. Identify the unknown compound.

7-8. An unknown organic carboxylic acid of melting point 139–141°C burned with a yellow, sooty flame. The sodium fusion test showed that nitrogen was present. It did not react with p-toluenesulfonyl chloride but did give a positive test when treated with 5% aqueous ferrous ammonium sulfate solution, acidified with 3 N sulfuric acid and then followed by methanolic potassium hydroxide solution. A 200-mg sample of the acid neutralized 12.4 mL of 0.098 N sodium hydroxide solution. Identify the acid. Does your structure agree with the calculated equivalent weight?

7-9. An unknown organic liquid, A, was found to burn with a yellow, sooty flame and give a positive Lucas test (~5 minutes). On treatment with sodium dichromate–sulfuric acid solution it produced compound B, which also burned with a yellow, sooty flame. Compound B gave a positive 2,4-dinitrophenylhydrazine test but a negative result when treated with Tollen's reagent. However, compound B did give a positive iodoform test. The NMR spectrum for compound A showed the following absorption data:

1.4 ppm	3H (doublet)
1.9 ppm	1H (singlet)
4.8 ppm	1H (quartet)
7.2 ppm	5H (singlet)

Give structures and suitable names for compounds A and B.

7-10. A hydrocarbon A, C_6H_{10}, burned with a yellow, almost nonsmoky flame. On catalytic hydrogenation over platinum catalyst it absorbed 1 mol of hydrogen to form compound B. It also decolorized bromine–methylene chloride solution to yield a dibromo derivative, C. Ozonolysis of the hydrocarbon A gave only one compound, D. Compound D gave a positive iodoform test when treated with I_2–NaOH solution. On treatment of compound D with an alcoholic solution of silver ammonium hydroxide, a silver mirror formed within a few minutes. Identify the hydrocarbon A and compounds B, C, and D.

7-11. A high-boiling liquid had a boiling point of 202–204°C and burned with a yellow, sooty flame. Sodium fusion indicated that halogens, nitrogen, and sulfur are not present. It was not soluble in water, dilute sodium bicarbonate solution, or dilute hydrochloric acid. However, it proved to be soluble in 5% aqueous sodium hydroxide solution. The compound gave a purple color with ferric chloride solution and a precipitate when reacted with bromine water. Treatment with hydroxylamine reagent did not give a reaction, but a white precipitate was obtained when the compound was treated with α-naphthyl isocyanate. On drying, this white, solid derivative had a melting point of 127–129°C. Identify the original liquid and write a structure for the solid derivative. After identifying the unknown liquid, can you indicate what the structure of the precipitate obtained on reaction with bromine might be?

7-12. A colorless liquid had a boiling point of 199–201°C and burned with a yellow, sooty flame. The sodium fusion test proved negative for the presence of halogens, nitrogen, and sulfur. It was not soluble in water, 5% aqueous sodium hydroxide, or 5% hydrochloric acid. However, it dissolved in sulfuric acid with evolution of heat. It did not give a precipitate with 2,4-dinitrophenylhydrazine solution and it did not decolorize bromine–methylene chloride solution. The unknown liquid did give a positive hydroxamate test and was found to have a saponification equivalent of 136. Identify the unknown liquid.

7-13. Your friend in Question 7-6 still needs your help. A week later a low-melting solid A was isolated, and combustion analysis showed it had composition $C_9H_{10}O$. The substance gave a precipitate when treated with 2,4-dinitrophenylhydrazine solution. Furthermore, when reacted with iodoform reagent a yellow precipitate of CHI_3 was observed. Acidification of the alkaline solution from the iodoform test produced a solid material, B. Reduction of compound A with $LiAlH_4$ gave compound C, $C_9H_{12}O$. This material, C, also gave compound B when treated with iodoform reagent. Vigorous oxidation of A, B, or C with sodium dichromate–sulfuric acid solution gave an acid having melting point 121–122°C. Your friend needs your assistance in writing the structures for compounds A, B, and C. Can you identify the three compounds?

7-14. An organic compound, $C_9H_{10}O$, showed strong absorption in the IR spectrum at 1735 cm^{-1} and gave a semicarbazone having melting point 198°C. It burned with a yellow, sooty flame and also gave a positive iodoform test. The NMR spectrum of the compound provided the following information:

2.11 ppm	3H (singlet)
3.65 ppm	2H (singlet)
7.20 ppm	5H (singlet)

Identify the unknown organic compound.

7-15. An unknown compound, A, was soluble in ether but only slightly soluble in water. It burned with a clear blue flame and analysis showed it to have the molecular formula $C_5H_{12}O$. It gave a positive test with the Jones reagent producing a new compound, B, having formula $C_5H_{10}O$. Compound B gave a positive iodoform test and formed a semicarbazone. Compound A on treatment with sulfuric acid produced a hydrocarbon, C, of formula C_5H_{10}. Hydrocarbon C readily decolorized bromine–methylene chloride solution and on ozonolysis produced acetone as one of the products. Identify the structures of compounds A, B, and C.

7-16. Compound A, C_7H_{14}, decolorized bromine–methylene chloride solution. It reacted with BH$_3$–THF reagent followed by alkaline peroxide solution to produce compound B. Compound B, on treatment with chromic acid–sulfuric acid solution, gave carboxylic acid C which could be resolved into enantiomers. Compound A on treatment with ozone followed by addition of hydrogen peroxide produced compound D. Compound D was identical to that material isolated from the oxidation of 3-hexanol with chromic acid–sulfuric acid reagent. Identify the structures of compounds A, B, C, and D.

7-17. Compound A, C_8H_{16}, decolorized bromine–methylene chloride solution. Ozonolysis produced two compounds, B and C, which could be separated easily by gas chromatography. Both B and C gave a positive 2,4-dinitrophenylhydrazine test. Carbon–hydrogen analysis and molecular weight determination of B gave a molecular formula of $C_5H_{10}O$. An NMR analysis revealed the following information for B.

0.92 ppm	3H, triplet
1.6 ppm	2H, multiplet
2.17 ppm	3H, singlet
2.45 ppm	2H, triplet

Compound C was a low-boiling liquid (bp 56°C). The NMR of this material showed only a single peak. Identify compounds A, B, and C.

II. The Interpretation of Infrared Spectra

INTRODUCTION TO GROUP FREQUENCIES

Studies of the vibrational spectra of thousands of molecules have revealed that many of the normal modes associated with particular atomic arrangements may be transferred from one molecule to another. These vibrational frequencies are associated with small groups of atoms that are essentially uncoupled from the rest of the molecule. The absorption bands that result from these modes, therefore, are characteristic of the small group of atoms regardless of the composition of other parts of the molecule. These vibrations are known as the *group frequencies,* and interpretation of infrared spectra of complex molecules based on group frequency assignments is an extremely powerful aid in the elucidation of molecular structure.

Note. *An introduction to the theory of this effect is given in Appendix B.*

We will discuss group frequencies in the following sequence:

Part A. Group Frequencies of the Hydrocarbons
Alkanes
Alkenes
Alkynes
Arenes

Part B. Factors Affecting the Carbonyl Frequencies
Mass effects
Geometric effects
Electronic effects (inductive and conjugative effects)
Interaction effects

Part C. Group Frequencies of the Functional Groups

Hydrocarbons	Amines
Alcohols	Nitriles
Aldehydes	Amides, primary
Ketones	Amides, secondary
Esters	Isocyanates
Acid halides	Thiols
Carboxylic acids	Halogens
Anhydrides	Phenyl
Ethers	

Part D. Strategies for Interpreting Infrared Spectra

PART A: Group Frequencies of the Hydrocarbons

CHARACTERISTIC GROUP FREQUENCIES OF ALKANES

The saturated hydrocarbons and the alkane section of mixed structures contain only C—C and C—H bonds. The fundamental modes derived from the hydrocarbon portion of these molecules, therefore, are limited to C—C and C—H stretching and bending vibrations. These two types of oscillators are good examples of structural units that give rise to both excellent and very poor group frequencies. We have established (Appendix B) that a change in dipole moment during the vibration is essential for the absorption process to occur in the infrared. The exact relationship between the vibrational displacements and the observed band intensities is rather complex. It is related to the slope of the curve of the variation of dipole moment with the normal coordinate at the equilibrium point. The intensity of the fundamental is proportional to the square of the derivative of the dipole moment with respect to the normal coordinates. The absorptivity of infrared bands, therefore, usually can be gauged from a rough estimate of the magnitude of the oscillating dipole moment.

The C—C *group* is an oscillator that, at best, will possess a very small dipole moment because of the symmetry inherent in the bond. Because of the low bond polarization, the absorption bands associated with this system can be expected to be quite weak and difficult to identify. In addition, the C—C oscillator, in most cases, will be directly connected to other C—C oscillators with similar or identical frequencies. In such an arrangement mechanical coupling effects are to be expected. This coupling will give rise to a very complex absorption pattern unique to a particular compound. Although these highly coupled vibrations have little value as group frequencies, they are the most powerful means of identifying organic materials by modern chemical instrumentation. The region of the infrared spectrum where these frequencies predominate, 1500 to 500 cm^{-1}, is often referred to as the "fingerprint region."

The C—H *oscillator* is at the other extreme from the C—C case. It gives rise to excellent group frequencies. The light terminal H atom, which is connected to a relatively massive carbon atom by a strong bond (large force constant), possesses a high natural frequency. Because of the separation of this frequency from other frequencies, the only coupling that can influence the oscillator is that of other C—H groups connected to the same carbon. A rule of thumb to remember is "Coupling generates as many modes as there are coupled oscillators." The methyl group, CH$_3$—, has three stretching modes; the methylene group, —CH$_2$—, two stretching modes; and the methine group, —CH—, a single stretching mode. Since there is very little mechanical coupling of C—H oscillators beyond the local carbon atom, the natural frequencies of the methyl, methylene, and methine groups remain relatively constant when these groups are transferred from compound to compound or from group to group within the same system. The dipole moment of the C—H bond is not large, but its derivative is sufficient to give rise to reasonably identifiable absorption bands. Because many compounds will contain several C—H bonds of similar character, they will also possess an equal number of nearly equivalent C—H frequencies. These C—H stretching modes overlap, often to produce the most intense collection of absorption bands observed in the infrared spectrum of a material.

The three *coupled C—H stretching modes of methyl groups* can be described in terms of two antisymmetric vibrations that are degenerate, or nearly degenerate (Appendix B), depending on the symmetry of the system, plus a symmetric mode (Fig. 7.3). In most cases, the two antisymmetric methyl stretching modes, although not rigorously degenerate, will give rise to a pair of very close-lying bands that are seldom resolved. The antisymmetric methyl modes are the high-

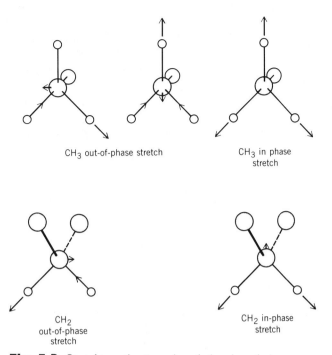

CH₃ out-of-phase stretch   CH₃ in phase stretch

CH₂ out-of-phase stretch   CH₂ in-phase stretch

Fig. 7.3 *Stretching vibration of methyl and methylene groups.*

est frequency vibrations of the purely sp^3 hybridized C—H bonds. These fundamentals occur near 2960 cm^{-1}. The methyl symmetric stretching mode is found close to 2870 cm^{-1} (Fig. 7.4).

The two *coupled stretching modes of the methylene group* are very similar in displacement pattern to the fundamental vibrations of the water molecule (Fig. B.5). The higher frequency mode, as in water, is the antisymmetric stretch. This fundamental occurs near 2925 cm^{-1} in hydrocarbons. The symmetric stretching mode (which is particularly sensitive to adjacent heteroatoms bearing lone-pair electrons) in saturated hydrocarbons has the lowest frequency of the sp^3 hybridized coupled C—H oscillators. The symmetric stretch occurs close to 2850 cm^{-1} (Fig. 7.4).

The *methine group* has a *single uncoupled mode*. As relatively few groups of this type are normally present in a structure, compared with methyl and methyl-

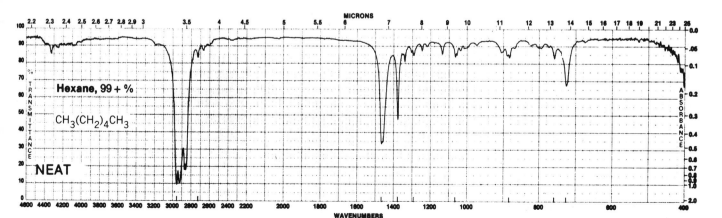

Hexane, 99 + %

CH₃(CH₂)₄CH₃

NEAT

Fig. 7.4 *IR spectrum: hexane*

ene groups, this vibration gives rise to a weak fundamental, usually masked by the absorption of the other alkane groups. The absorption of the tertiary C—H bonds occurs in the 2900 cm^{-1} region.

With very few exceptions the sp^3 hybridized C—H bonds have their fundamental modes in the 3000–2800 cm^{-1} region.

Since the hydrogen atom is much lighter than other atoms, it undergoes most of the displacement. The mass term in the expression for frequency of a diatomic molecule is actually a *reduced* mass. In the case of hydrogen and carbon this is defined by

$$\frac{1}{\mu} = \frac{1}{m_H} + \frac{1}{m_C} \quad \text{since } m_C \gg m_H, \quad \mu \cong m_H$$

It is possible, therefore, to express to a very good first approximation the vibrational frequency of this system by a simple Hooke's law relationship assuming an infinite mass for the carbon atom.

$$\tilde{\nu} = \frac{1}{2\pi c} \sqrt{\frac{k}{m_H}}$$

This expression makes it possible to predict frequency shifts on substitution by deuterium for hydrogen of $\tilde{\nu}_H/\tilde{\nu}_D = 1.41$. In practice these shifts are somewhat less than the theoretical values, usually falling in the range 1.32–1.38 (Fig. 7.5).

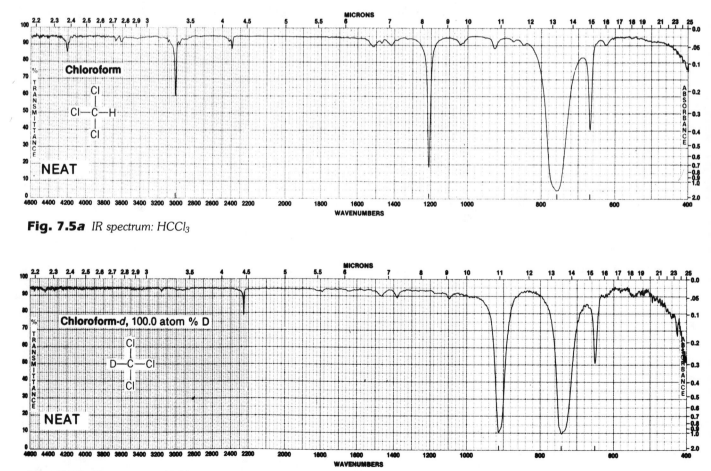

Fig. 7.5a *IR spectrum: HCCl₃*

Fig. 7.5b *IR spectrum: DCCl₃*

If the $\tilde{\nu}_H/\tilde{\nu}_D$ ratio departs significantly from this range, the result can be taken as an indication that one or possibly both of the fundamentals are not behaving in harmonic fashion and that coupling is present. (See also Experiment 36 for an example of the shift of methyl frequencies on deuterium substitution.) Chloroform is one of the rare exceptions to the 3000 cm^{-1} rule (see p. 432). The single uncoupled C—H stretching mode occurs at 3022 cm^{-1}. The rise in frequency in this case results from extensive substitution by the strongly electronegative chlorine atoms.

The *bending frequencies of the C—H oscillator* will now be discussed. As in the case of stretching modes of the alkyl groups, the bending fundamentals are coupled only with those oscillators directly bonded to the carbon. Since there are three independent H—C—H bond angles, there are *three deformation or bending vibrations* associated with the *methyl group* (Fig. 7.6). As in the stretching fundamentals, these modes can be described in terms of two antisymmetric degenerate or nearly degenerate vibrations that occur near 1460 cm^{-1}. The third vibration is the symmetric bending mode (umbrella) close to 1375 cm^{-1} (Figs. 7.4 and 7.6), which is easily identified.

The *methylene group* has a single H—C—H bond angle, and a *single deformation mode* (scissoring, Fig. 7.7) directly analogous to that of the water molecule. The symmetric bend of the —CH$_2$— group occurs near 1450 cm^{-1} in hydrocarbons (Figs. 7.4 and 7.7).

Three other bending modes are available to the *methylene group* (Fig. 7.7). When the methylene unit is fused into the molecule, three vibrational degrees of freedom develop, which are related to rotational motion of the isolated system, for example, the rotational motion of the water molecule (see Appendix B). These fundamentals (wag, twist, and rock) are subject to significant coupling to adjacent methylene groups. The transitions of these modes are also rather weak in intensity. Thus, wag, twist, and rock are not useful group frequencies with the exception of a component of the rocking vibration (Fig. 7.7) in certain structures. In those molecules having four or more methylene groups in a row, the coupled mode corresponding to the *all-in-phase* rocking vibration develops a significant

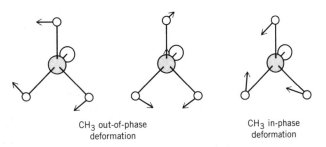

CH$_3$ out-of-phase
deformation

CH$_3$ in-phase
deformation

Fig. 7.6 *Bending vibrations of the methyl group.*

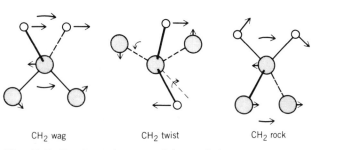

CH$_2$ wag   CH$_2$ twist   CH$_2$ rock

CH$_2$ scissoring
deformation

Fig. 7.7 *Bending vibrations of the methylene group.*

Table 7.2 Alkane Normal Modes

C—H Vibrational Modes	$\tilde{\nu} \pm 10$ (cm^{-1})
Methyl groups	
Antisymmetric (degenerate) stretch	2960
Symmetric stretch	2870
Antisymmetric (degenerate) deformation	1460
Symmetric (umbrella) deformation	1375
Methylene groups	
Antisymmetric stretch	2925
Symmetric stretch	2850
Symmetric deformation (scissor)	1450
Rocking mode (all-in-phase)	720

dipole moment change and a stable frequency. Thus, the in-phase rock that occurs near 720 cm^{-1} (Fig. 7.4) in the fingerprint region gives rise to an absorption band of sufficient intensity to allow for confident assignment.

The bending modes associated with the *methine hydrogen* are hard to identify in most hydrocarbons. (The in-plane bend of an isolated sp^2 C—H group can be particularly important, however, in certain heteroatom systems; see discussion of the aldehyde functional group.)

The C—H vibrational modes of the alkanes (or mixed compounds containing alkyl groups) that are characteristic and reliable group frequencies can be summarized as in Tables 7.2 and B.1.

CHARACTERISTIC GROUP FREQUENCIES OF ALKENES

Alkenes (olefins) possess the carbon–carbon double bond, C=C. The group frequencies of these molecules will be those of the alkanes for the saturated portion of the molecule, plus those modes contributed by the unsaturated group.

C=C Stretching

The stretching fundamental of the C=C group is a useful group frequency. The increase in the force constant in going from the single bond to the double bond moves the frequency to sufficiently high values ($\tilde{\nu}_{C=C} = 1616$ cm^{-1}, ethylene) to decouple this mode from adjacent C—C vibrations. Substitution of carbon for hydrogen on the double bond tends to raise the C=C stretching frequency, as the effective force constant has been shown to include an increased compression term (Fig. 7.8). It is possible to classify open-chain unsaturated systems into two groups as in Tables 7.3 and B.2.

The high-frequency set has quite weak bands unless conjugation occurs. Indeed, in the tetrasubstituted case, if the four groups are identical, the C=C band is formally forbidden in the infrared. When the alkyl groups are similar but not identical, the magnitude of the dipole moment is so small that it becomes difficult to detect an absorption band. In this situation the molecule is considered to have a ''pseudo-center of symmetry.'' In the low-frequency set more intense absorption bands occur, with the average intensity falling in the medium-to-strong range.

Olefinic C—H

A number of the fundamental modes associated with the olefinic C—H groups are good group frequencies.

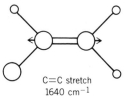

C=C stretch
1640 cm^{-1}

Fig. 7.8 *Vinyl group double-bond stretching vibration.*

Table 7.3 Substitution Classification of C=C Stretching Frequencies

C=C Normal Modes	$\tilde{\nu}$(cm^{-1})
trans-, tri-, tetrasubstituted	1680–1665
cis-, vinylidene- (terminal-1,1-), vinyl-substituted	1660–1620

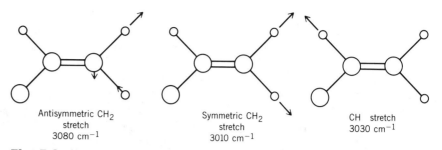

Fig. 7.9 *C—H stretching vibrations of the vinyl group.*

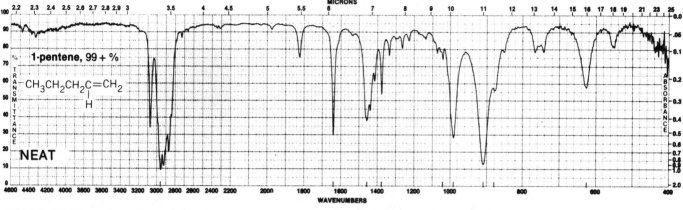

Fig. 7.10 *IR spectrum: 1-pentene.*

The C—H stretching frequencies occur above 3000 cm^{-1} and are localized in two regions as follows. (1) If two C—H groups are present on an sp^2 carbon, two coupled (antisymmetric and symmetric) vibrations occur near 3080 and 3010 cm^{-1}. (2) if a single C—H group is attached to an sp^2 hybridized carbon, it gives rise to a single mode near 3030 cm^{-1}. (3) A vinyl group will have both sets of bands, but the lower frequency modes are seldom resolved (Figs. 7.9 and 7.10).

The olefinic C—H bending frequencies fall into two categories. (1) There are bending modes that occur in the plane of the double bond. These fundamentals are not useful group frequencies. (2) There are bending modes that occur out of the plane of the double bond. These fundamentals are useful group frequencies.

Out-of-Plane Deformation Modes

Vinyl Groups

The vinyl group has three C—H bonds and therefore three out-of-plane bending modes. Two of these normal modes are good group frequencies. The fundamental with the two trans hydrogens bending in-phase occurs near 990 cm^{-1} (Fig. 7.11). A second vibration involving primarily the two hydrogens attached

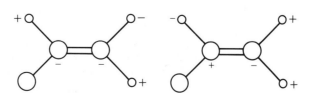

Fig. 7.11 *Out-of-plane C—H bending vibrations of the vinyl group.*

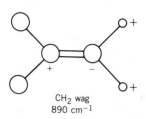

Fig. 7.12 *Out-of-plane C—H bending vibration of the vinylidene group.*

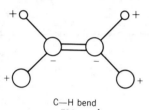

Fig. 7.13 *Out-of-plane C—H bending vibration of cis-substituted HC═CH group.*

to the terminal carbon, bending in-phase together (wag), is located close to 910 cm^{-1} (Fig. 7.11). The absorption bands resulting from both of these fundamentals are strong and easily detected in the fingerprint region.

Vinylidene Groups

The 1,1-substituted system will have two bending frequencies. One of these bending modes is a good group frequency. When both hydrogens wag out of the C═C plane together, a strong dipole moment change develops and gives rise to an intense absorption band near 890 cm^{-1}. This mode is related to the low-frequency mode found in the vinyl group at 910 cm^{-1} (Figs. 7.11 and 7.12).

Trans Olefins

A trans olefin has two out-of-plane bending modes, but again only one gives rise to a good group frequency. The mode that involves both hydrogens moving in-phase together occurs close to 965 cm^{-1}. This mode is directly related to the high-frequency mode of the vinyl group near 990 cm^{-1} (Fig. 7.11).

Cis Olefins

A cis-substituted C═C group does not possess a very good out-of-plane group frequency. The only mode with reasonable intensity involves the in-phase bend of the two hydrogens. This fundamental is derived from a rotational-type motion that couples to the rest of the system (Fig. 7.13). Therefore, the cis mode is not localized, but occurs in a broad region near 700 cm^{-1}.

Trisubstituted Olefins

In the case of **R**$_2$C═CH**R** systems we have a single out-of-plane bending vibration that occurs near 820 cm^{-1}. The mode is uncoupled and gives rise to a medium-intensity band.

Tetrasubstituted Olefins

These groups have no C—H bending modes.

Overtones

The overtones of the fundamentals that involve wagging of the terminal hydrogen atoms in the vinyl and vinylidene groups occur with unusual intensity. These bands are observed at 1825 and 1785 cm^{-1}, slightly more than double the fundamental frequency. Thus, these harmonics exhibit *negative anharmonicity* in addition to unusual intensity for forbidden vibrations.

The group frequencies of the olefinic C—H modes can be summarized as in Tables 7.4 and B.3.

CHARACTERISTIC GROUP FREQUENCIES OF ALKYNES

C≡C Stretching Vibration

Triple bond formation further increases the force constant involved in the C—C stretching vibration. Thus, acetylenes possess the highest of all observed C—C stretching frequencies. The triple bond group frequency is located near 2120 cm^{-1} in monosubstituted alkynes. Compression effects on the force constant similar to those observed in substituted olefins raise the stretching mode into the 2225 cm^{-1} region in disubstituted acetylenes. The alkyne stretching vibration involves a relatively small dipole moment change. The resulting bands, therefore, are particularly weak in the infrared. In the disubstituted case, if the two groups are identical, the vibration is infrared inactive. Here, as in the olefins, a pseudo-center of symmetry can operate to significantly suppress the intensity of the mode. Even though these bands occur in a region that is essentially devoid of other absorptions, their inherent weakness and variable intensity

Table 7.4 Alkene Normal Modes

C—H Vibrational Modes	$\tilde{\nu} \pm 10$ (cm^{-1})
Stretching modes	
Antisymmetric stretch (=CH$_2$)	3080
Symmetric stretch (=CH$_2$)	3020
Uncoupled stretch (=CH)	3030
Out-of-plane bending modes	
Vinyl group	
Trans hydrogens (in-phase)	990
Terminal hydrogens (wag)	910
Vinylidene group	
Terminal (wag)	890
Trans group	
Trans hydrogens (in-phase)	965
Cis group	
Cis hydrogens (in-phase)	~700
Trisubstituted group	
Uncoupled hydrogen	820
Tetrasubstituted group: no modes	

present significant problems in making confident band assignments. The high frequency of the mode effectively decouples the vibration from the rest of the system. Thus, triple bonds show little evidence of any first-order coupling. On the other hand, these vibrations are prone to second-order effects that complicate their interpretation. Triple bond assignments must be handled with care. These vibrations constitute a set of group frequencies relatively difficult to deal with, if they can be observed at all.

Alkyne C—H Vibrations

The stretching of acetylenic C—H bonds gives rise to the highest carbon–hydrogen vibrations observed. These are relatively intense uncoupled single sharp modes that occur near 3300 cm^{-1}. Although this normal mode occurs in the same region as O—H and N—H fundamental vibrations, acetylenic C—H stretches usually can be distinguished by the sharpness of the band. They are highly reliable group frequencies.

The bending modes of the C—H acetylenic group do not give rise to reliable group frequencies.

The group frequencies of the alkynes are summarized in Tables 7.5 and B.4.

CHARACTERISTIC GROUP FREQUENCIES OF ARENES

Aromatic ring systems represent the final class of hydrocarbons to be considered in this section. The discussion will center on the benzene ring, but many of the more complicated systems have been examined in detail.

The infrared spectra of aromatic compounds possess many needle-sharp bands. This characteristic sets these spectra apart from the spectra of aliphatic compounds. It arises from the fact that aromatic systems are tightly bound rigid

Table 7.5 Alkyne Normal Modes

C≡C, C—H Vibrational Modes	$\tilde{\nu} \pm 10$ (cm^{-1})
Triple bond stretch (monosubstituted)	2120
Triple bond stretch (disubstituted)	2225
C—H bond stretch (monosubstituted)	3300

molecules having little opportunity for rotational isomerism. With aliphatic compounds the observed spectrum, in reality, often contains the spectrum of a complex mixture of rotamers. These isomers all exhibit very similar but not identical spectra that overlap and result in band broadening.

GROUP FREQUENCIES OF THE PHENYL GROUP

The group frequencies of the phenyl group can be classified as carbon–hydrogen vibrations consisting of stretching, and out-of-plane bending modes, plus carbon–carbon ring stretching, and out-of-plane bending modes. The in-plane bending modes in both cases are not effective group frequencies.

C—H Stretching Modes

The C—H stretching vibrations occur as a series of weak bands in the region 3100 to 3000 cm^{-1}. This is consistent with sp^2 hybridization of the carbon atom. These modes directly overlap the olefinic C—H stretching fundamentals. Substitution of heteroatoms into the ring can significantly perturb these frequencies (oxygen raises the mode into the 3200 to 3100 cm^{-1} region). As the bands are generally weak, they may be masked by strong aliphatic absorption in mixed compounds if they lie close to 3000 cm^{-1}. Care must be taken in the assignment of these modes.

C=C Stretching Modes

The phenyl ring modes, which possess excellent group frequency properties, involve two pairs of closely related C=C stretching vibrations. These vibrations are related to degenerate fundamentals in unsubstituted benzene. On ring substitution the degeneracy is removed because of the lowered symmetry. The ring vibrations ν_{8a} and ν_{8b} (the numbering has been carried over from the benzene fundamental assignments by Wilson) involve displacements in three carbon units at each end of the ring, which are analogous to the symmetric and antisymmetric stretching modes of water. The two modes result from the two sets of displacements, which are in-phase (Fig. 7.14).

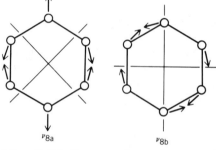

Fig. 7.14 *Ring stretching vibrations of benzene, ν_{8a}, ν_{8b}.*

These vibrations are degenerate and inactive in the infrared spectrum of the unsubstituted ring. In general, the higher-frequency vibration, ν_{8a}, is the more intense of the pair. If a substituent is conjugated to the ring system, however, the lower-frequency component gains in intensity and sometimes becomes the most intense member of the pair. These two modes are also substituent-independent with para-disubstitution; however, if the groups are identical or nearly identical, the modes are infrared forbidden or active with greatly suppressed intensity. These modes, ν_{8a} and ν_{8b}, occur near 1600 and 1580 cm^{-1}, respectively. The second pair of vibrations corresponds to the identical displacements of the first pair of modes with the sets now out-of-phase (Fig. 7.15). Substitution on the ring removes the degeneracy, giving two bands corresponding to ν_{19a} and ν_{19b} in benzene.

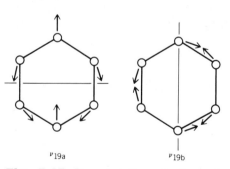

Fig. 7.15 *Ring stretching vibrations of benzene, ν_{19a}, ν_{19b}.*

These fundamentals, as with ν_{8a} and ν_{8b}, are substituent-independent on monosubstitution or parasubstitution. The dipole moment change associated with these normal modes generally gives rise to rather intense absorption bands. The high-frequency component, however, is sensitive to electron-withdrawing substituents that can significantly suppress the intensity of this mode. These fundamentals occur near 1500 and 1450 cm^{-1}, respectively.

C—H Bending Vibrations

The C—H bending normal modes of group frequency value are the out-of-plane vibrations. These fundamentals are useful guides to the substitution pattern on the ring system. There will be as many out-of-plane vibrations as there are C—H groups. The modes of interest, however, are those fundamentals in

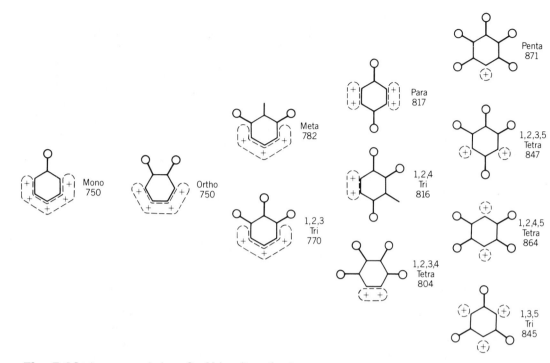

Fig. 7.16 *Arene out-of-plane C—H bending vibrations.*

which all the hydrogens move in-phase. These vibrations have substantial dipole moment changes and, thus, give rise to intense absorption bands (Fig. 7.16). The very strong intensity of the out-of-plane deformation modes plays a key role in our ability to make confident assignments for these fundamentals, as they fall in the heart of the fingerprint region. The five all-in-phase bending vibrations are as presented in Figure 7.16 and Tables 7.6 and B.5.

Although there is considerable overlap of the ranges, the uncertainty of the assignment can often be reduced by the identification of an additional strong band in the 690 cm^{-1} region. This band results from a carbon–carbon out-of-plane ring deformation, ν_4, of benzene (Fig. 7.17). This mode is substituent-insensitive to mono-, meta-, and 1,3,5-substitution. Thus, a band will seldom occur in this region in ortho-disubstituted systems.

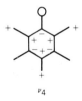

Fig. 7.17 *Phenyl out-of-plane bending vibration, ν_4.*

Sum Tone Patterns

Some of the out-of-plane C—H bending modes can be excited simultaneously with other low-frequency fundamentals. As these *sum tone* transitions are formally forbidden, the resulting absorption bands are very weak. With optically thick samples, however, weak bands are observed in the 2000–1650 cm^{-1} region. The pattern of these bands is highly characteristic of the substitution

Table 7.6 Arene Out-of-Ring Plane C—H Deformation Modes

Number of Adjacent H's	$\tilde{\nu}$ range (cm^{-1})
5	770–730
4	770–735
3	810–750
2	860–800
1	900–860

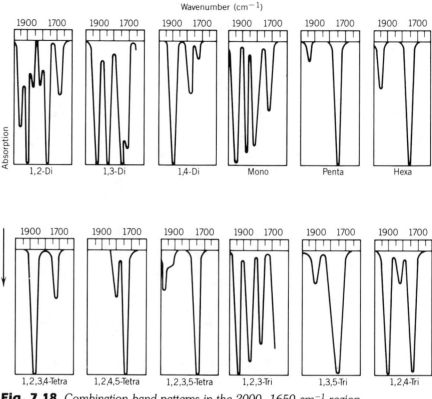

Fig. 7.18 *Combination band patterns in the 2000–1650 cm⁻¹ region.*

arrangement on the ring, since they have their origin in the out-of-plane C—H bending frequencies. Unless carbonyl groups are present in the molecule, the 2000–1650 cm⁻¹ region will be open for observation. These combination band patterns can be used to remove ambiguity about the ring substitution pattern based on assignments of the out-of-plane fundamentals (Fig. 7.18).

The group frequencies of the arenes can be summarized as in Tables 7.7 and B.6.

Table 7.7 Arene Group Frequencies

Arene Fundamentals	$\tilde{\nu}$ **(cm⁻¹)**
C—H stretch	3100–3000
C=C ring stretch (ν_{8a})	1600 ± 10
C=C ring stretch (ν_{8b})	1580 ± 10
C=C ring stretch (ν_{19a})	1500 ± 10
C=C ring stretch (ν_{19b})	1450 ± 10
C—H out-of-plane bend (1H)	900–860
C—H out-of-plane bend (2H)	860–800
C—H out-of-plane bend (3H)	810–750
C—H out-of-plane bend (4H)	770–735
C—H out-of-plane bend (5H)	770–730
C—C ring out-of-plane bend (1; 1,3; 1,3,5)	690 ± 10
C—H out-of-plane bend sum tones	2000–1650

PART B: Characteristic Group Frequencies of the Carbonyl

The carbonyl unit is perhaps the single most important functional group in organic chemistry. It is certainly the most commonly occurring functionality. Infrared spectroscopy can play a powerful role in the characterization of the carbonyl because this group possesses all of the properties that give rise to an excellent group frequency. (1) The carbonyl group has a large dipole moment derivative, which gives rise to very intense absorption bands. (2) As a result of the large force constant, it has a stretching frequency that occurs at high values outside the fingerprint region. In addition, this portion of the spectrum is devoid of most other fundamentals. (3) Its stretching fundamental occurs in a range that is reasonably narrow (little coupling), 1750 ± 150 cm^{-1}, but sensitive enough to the local environment to allow for considerable interpretation of the surrounding structure. (4) The range of frequencies is determined by a number of factors that are now well understood in terms of the effects outlined below.

MASS EFFECTS

The mode of principal interest is the stretching vibration. In this oscillator the C and O atoms undergo comparable displacements; thus, we must replace the simplified mass expression in the Hooke's law approximation that was applied to the C—H stretching fundamentals by the reduced mass μ (where $\mu = m_C m_O / (m_C + m_O)$, see above). On substitution of the isotopes ^{13}C and ^{18}O, the predicted small frequency shifts ($\sim$30–40 cm^{-1}) are observed. These results are consistent with a relatively low degree of mechanical coupling to the rest of the system. This lack of coupling is expected, because the force constant of the multiple bond is significantly different from the values of the force constants of the bonds that connect the carbonyl to the rest of the molecule.

GEOMETRIC EFFECTS

Geometric effects can play a major role in determining the location of the carbonyl frequency within the 1750 cm^{-1} region. Although the displacement of the oxygen atom involves simply stretching or compressing the C=O bond, the displacement of the carbon atom is more complex. This latter movement also contains a compression component of the force constants in the two connecting single bonds as the carbonyl carbon is being stretched. In the opposite phase of the vibration, a stretching component of the force constants in the two connecting bonds is required, as the carbonyl carbon is being compressed. The magnitude of these additional force constant components is angle dependent. As the angle between the single bonds (C—CO—C angle) decreases, the contribution of the single bond components to the effective C=O stretching force constant will be raised. Since the frequency of the vibration is directly proportional to the square root of the force constant, a decrease in the internal carbonyl bond angle will raise the frequency (Table 7.8). Alternatively, an increase in the internal carbonyl bond angle will lower the carbonyl frequency.

RESONANCE AND INDUCTIVE EFFECTS

Resonance and inductive effects can profoundly influence the vibrational frequency of the carbonyl group. In the following discussion the carbonyl stretching mode for acetone ($\tilde{\nu}_{C=O} = 1715$ cm^{-1}, liq.) will be used as a reference frequency representative of simple alkyl substitution on the carbonyl group. Effects that perturb this reference fundamental to either higher or lower values will be examined.

Table 7.8 Variation of Carbonyl Frequency (cm^{-1}) Versus Bond Angle

Ring Size:	7	6	5	4	3	(2)
Lactones	1727	1740	1775	1832		
Ketones	1699	1710	1744	1782	1906	2049
Lactams		1670	1695	1750		

Electronic Effects That Raise the Carbonyl Frequency

When an alkyl substituent is replaced by a more electronegative system, the balance of contributing resonance forms in the carbonyl is slightly shifted away from dipolar forms by strong inductive effects. This shift results in a larger effective C=O force constant and higher frequencies. For example, in hexanoyl chloride, $\tilde{\nu}_{C=O} = 1805$ cm^{-1} (Fig. 7.23).

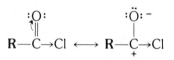

Electronic Effects That Lower the Carbonyl Frequency

Direct conjugation of the carbonyl via α,β-unsaturation will introduce new dipolar carbonyl resonance forms that lower the effective force constant values and, thus, decrease the carbonyl stretching frequency. For example, consider cyclohexene methyl ketone (**I**), $\tilde{\nu}_{C=O} = 1685$ cm^{-1}, and acetophenone (**II**), $\tilde{\nu}_{C=O} = 1687$ cm^{-1}.

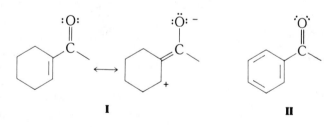

The Case of Ester Carbonyl Vibrations: Competing Inductive and Resonance Effects

The substitution of the more electronegative oxygen for carbon in going from ketones to esters will raise the carbonyl frequency in esters as a result of inductive influences (see Electronic Effects That Raise the Carbonyl Frequency, above). On the other hand, the lone-pair electrons present on the ether oxygen of the ester will be in direct conjugation with the carbonyl. This latter interaction will generate dipolar resonance forms that will tend to drop the C=O frequency. The balance between these two competing effects in esters, which might be difficult to anticipate, appears to favor the inductive effects. The ester carbonyls commonly are located 40–20 cm^{-1} higher than the simple aliphatic ketones, in the range 1755–1735 cm^{-1}.

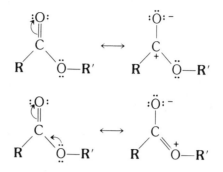

Now let us examine ester carbonyl frequencies in somewhat greater detail.

If the ester carbonyl is directly involved with α,β-unsaturation, the normal ester frequency is lowered by 30–20 cm^{-1}. Thus, unsaturated ester carbonyl frequencies occur very nearly in the same region as simple aliphatic ketone frequencies. For example, in ethyl benzoate, $\tilde{\nu}_{C=O} = 1720$ cm^{-1}.

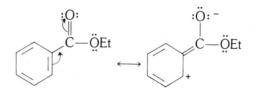

If the ester is conjugated, but the conjugation is located adjacent to the ether oxygen rather than alpha to the carbonyl group, then the carbonyl frequency is raised. The higher $\tilde{\nu}_{C=O}$ results from resonance competition for the lone-pair electrons of the ether oxygen by the carbonyl and the new conjugating group, for example, phenyl acetate, $\tilde{\nu}_{C=O} = 1769$ cm^{-1}. These frequency shifts support the arguments concerning competition between inductive and resonance effects within the ester ether group.

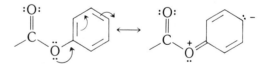

If the ester group is directly conjugated on both sides, then the resonance effects should cancel, and we would expect this type of system to exhibit near-normal carbonyl frequencies, for example, phenyl benzoate, $\tilde{\nu}_{C=O} = 1743$ cm^{-1} as compared with ethyl acetate, $\tilde{\nu}_{C=O} = 1742$ cm^{-1}.

INTERACTION EFFECTS   Interaction effects vary from those that have a dramatic impact on the spectra to those that are barely detectable. Our understanding of these terms completes the discussion of the major factors affecting the carbonyl group frequencies. We can roughly divide this area of discussion into intramolecular and intermolecular types of interactions.

Intramolecular Carbonyl Interactions

First-order coupling effects are rarely observed as this oscillator is rather effectively decoupled from the rest of the molecule by differences in frequency. The most spectacular example (see Appendix B) is the case of carbon dioxide. In CO_2 the two coupled oscillators are aligned for maximum interaction and possess identical frequencies. The splitting between the antisymmetric and symmetric levels is very large, approximately 1000 cm^{-1}. A second, much less dramatic, example is the case of anhydrides. In this instance the two oscillators are joined through a central oxygen. Delocalization across the connecting atom operates to maintain planarity of the system and thereby to increase the coupling. Even so, since the carbonyls are no longer held at the optimum angle and are vibrationally insulated by an intervening atom, the first-order coupling drops to ~70 cm^{-1}, less than 10% of the CO_2 value, for example, hexanoic anhydride, $\tilde{\nu}_{C=O} = 1817, 1750$ cm^{-1}. The uncoupled vibration would be expected to occur near 1770 cm^{-1}. This latter value is consistent with an oxygen-substituted carbonyl in which conjugation of the lone-pair electrons on the ether oxygen has been nearly canceled. The full inductive effect of the ether oxygen atom on the carbonyl stretching vibration can be inferred from these data. Equalized resonance competition by both carbonyl systems of the anhydride for the ether lone-pair electrons might be expected to bring about just such an effect.

Intramolecular H bonding of carbonyls can be enhanced by resonance interactions and, thus, significantly perturb the stretching frequency. For example, let us consider the anthraquinone series shown below. In structure **I**, H bonding is not present and a single stretching frequency for the quinone is observed at 1675 cm^{-1}. This value reflects direct conjugation with the aromatic ring and with the methoxyl groups conjugated in equivalent fashion with both carbonyls. Little coupling between the carbonyls is observed through the ring. In structure **II** a phenol group replaces one of the methoxyl groups. Resonance forms operating through a tightly hydrogen-bonded six-membered ring act to reduce the effective force constant of the carbonyl, with the result that the observed frequency, $\tilde{\nu}_{C=O} = 1636$ cm^{-1}, is lowered nearly 40 cm^{-1}. In structure **III** both methoxyl groups have been replaced by phenolic groups. This change results in both carbonyls undergoing strong H bonding, which involves resonance forms similar to those found in structure **II**. Thus, a single band is observed at $\tilde{\nu}_{C=O} = 1627$ cm^{-1}. This downward shift approaching 50 cm^{-1} can be ascribed primarily to strong internal H bonds present in the anthraquinone system.

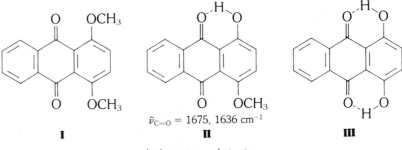

$\tilde{\nu}_{C=O} = 1675, 1636$ cm^{-1}

I   **II**   **III**

Anthraquinone derivatives

Several other examples of very strong intramolecular H bonding are known. In tropolone, where the hydrogen appears to be essentially equidistant from the two oxygen atoms, the absorption occurs at 1605 cm^{-1}.

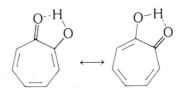

Tropolone (H is equidistant from oxygens)

Table 7.9 Field Effects in Chloroacetones

Compound	$\tilde{\nu}_{C=O}(cm^{-1})$
Acetone	1715
Chloroacetone	1752, 1726
1,1-Dichloroacetone	1743, 1724
1,1,1-Trichloroacetone	1729

Second-order coupling (Fermi resonance) often occurs with carbonyl vibrations in complex organic molecules. In the large majority of cases, the frequency match of overtone and fundamental is relatively poor so that the frequency of the fundamental is not affected. The main evidence for the interaction in these cases will be weak shoulders associated with the main carbonyl peak. An example of the coupling not being trivial is the case of cyclopentanone. The Fermi interaction involves an overtone or combination of a C—H level, as the splitting collapses to a singlet in 2,2,5,5-tetra-d₄-cyclopentanone.

Field effects will also perturb the carbonyl frequency. The classic case is that of the chloroacetones. In those rotamers, in which the chlorine atom is in the eclipsed position with respect to the oxygen, repulsive lone-pair interactions occur. Field effects result in suppression of the contribution of the dipolar carbonyl resonance form, and therefore cause a rise ($\sim$30 cm^{-1}) in the stretching frequency (Table 7.9).

These arguments are supported by the observation that two carbonyl frequencies are present in the mono- and dichloroacetones, and single frequencies in acetone and trichloroacetone. *The small frequency rise observed in the low-frequency component is attributed to inductive effects*, which, therefore, can have only a minimal influence on the high-frequency component.

Transannular interactions occur when cyclic carbonyl groups are sterically positioned so that the carbon atom of the carbonyl is oriented toward an electron-rich center lying across the ring. The interaction can greatly enhance the dipolar resonance form of the carbonyl and result in a significant drop in the stretching frequency. The effect has a major impact on the carbonyl frequency of the alkaloid protopine ($\tilde{\nu}_{C=O}$ = 1660 cm^{-1}).

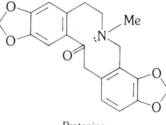

Protopine

Most interesting, however, are the results obtained from a number of model compounds, for example, the cyclooctaaminoketone (**I**) $\tilde{\nu}_{C=O}$ = 1666 cm^{-1}, and its perchlorate salt (**II**), which exhibits no carbonyl absorption band at all!

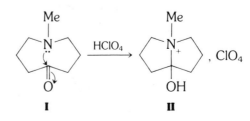

I   **II**

Intermolecular Carbonyl Interactions

Strong intermolecular H bonding can significantly perturb carbonyl frequencies. It is known in the case of aliphatic carboxylic acids that these substances form strongly H-bonded dimers when neat or in highly concentrated solutions. Association through the carbonyl groups leads to the formation of a symmetric eight-membered ring containing two H bonds. Coupling through the tightly bonded ring results in a splitting of the carbonyl levels of approximately the same magnitude as found in anhydrides ($\Delta\tilde{\nu}_{C=O} = \sim70$ cm^{-1}). As the dimer possesses a center of symmetry, the in-phase mode will not be active in the infrared ($\tilde{\nu}_{C=O} = \sim1650$ cm^{-1}). The out-of-phase stretch of the carbonyls, however, will be active. The antisymmetric C=O stretch gives rise to a strong band in the infrared ($\tilde{\nu}_{C=O} = \sim1720$ cm^{-1}). In very dilute solution, it is sometimes possible to observe these systems in the monomeric state. Under these conditions the carbonyl frequencies return to expected values ($\tilde{\nu}_{C=O} = \sim1770$ cm^{-1}). As the discussion of group frequencies expands, we will see a number of other examples of the effect of intermolecular H bonding on the carbonyl group frequency.

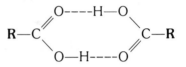

The interaction of weak H bonds is relatively hard to detect in the infrared, as the shifts are measured in terms of a few wavenumbers. One of the better examples is the effect on the carbonyl stretch of acetone ($\tilde{\nu}_{C=O} = 1722$ cm^{-1}) as measured in hexane solution. When the hydrocarbon solvent is replaced by chloroform, weak H bonds (O···H—C) develop, and the carbonyl mode drops 12 wavenumbers to 1710 cm^{-1}.

Weak dipolar interactions between carbonyls can also be observed in the infrared. The frequency shifts caused by these interactions parallel the development of polarization in the carbonyl group, as can be judged by the data in Table 7.10.

The major factors perturbing carbonyl frequencies can be summarized as follows:

Factors That Raise the C=O Frequency

1. Electronegative substitution
2. Decrease in C—CO—C internal bond angle

Factors That Lower the C=O Frequency

1. Conjugation
2. Hydrogen bonding

Table 7.10 Carbonyl Dipolar Interactions[a]

Compound	$\Delta\tilde{\nu}_{C=O}$ (cm^{-1})
Acetyl chloride	15
Phosgene	13
Acetone	21
Acetaldehyde	23
Dimethylformamide	50

[a] Shift measured between dilute nonpolar solution and neat sample.

Table 7.11 Carbonyl Group Frequencies

Functional Group	$\tilde{\nu}_{C=O}$ (cm^{-1})
Ketones, aliphatic, open chain (**R**$_2$CO)	1725–1700
Ketones, conjugated	1700–1675
Ketones, ring	(see Table 7.8)
Acid halides	>1800
Esters, aliphatic	1755–1735
Esters, conjugated	1735–1720
Esters (conjugated to oxygen)	1780–1760
Lactones	(see Table 7.8)
Anhydrides, aliphatic, open chain	1840–1810 and 1770–1740
Acids, aliphatic	1725–1710
Amides	(see Part C)
Lactams	(see Table 7.8)
Aldehydes	1735–1720

As several of these factors may be operating simultaneously, careful judgment as to the contribution of each individual effect must be exercised in predicting carbonyl frequencies. This judgment develops rapidly with practice at interpretation.

This completes the discussion of factors affecting carbonyl group frequencies. A number of additional examples will be discussed in detail in Part C, in which a survey of the infrared spectra of functional groups is considered. Carbonyl frequencies are summarized in Tables 7.11 and B.7.

PART C: Characteristic Frequencies of Functional Groups

Now that we have examined the major group frequencies associated with the common hydrocarbon platforms (platform = hydrocarbon structural unit supporting a functional group) and the principal parameters affecting the carbonyl group, let us consider the vibrations associated with the common functional groups that lead to good group frequency correlations. For the most part we will use, in these discussions, a series of infrared spectra derived from straight-chain aliphatic C$_6$ compounds.

HEXANE   The spectrum of normal hexane (Fig. 7.4) obtained with the pure liquid, as expected, contains simply the group frequencies of an aliphatic hydrocarbon. The antisymmetric and symmetric methyl stretching modes occur below 3000 cm^{-1} at $\tilde{\nu}$ = 2960 and 2876 cm^{-1}. The antisymmetric and symmetric methylene stretching fundamentals occur near $\tilde{\nu}$ = 2938 and 2860 cm^{-1}. The antisymmetric methyl deformation ($\tilde{\nu}$ = 1467 cm^{-1}) overlaps the symmetric methylene scissoring vibration, which is found as a difficult-to-identify shoulder at $\tilde{\nu}$ = 1455 cm^{-1}. The symmetric methyl bend (umbrella mode) is easily assigned to the sharp band at 1379 cm^{-1}. Finally, the all-in-phase rocking mode of a sequence of four or more methylene groups can be identified by its intensity in the fingerprint region near 725 cm^{-1}. We often will be able to identify this collection of platform group frequency bands as we progress through the infrared spectra of the following series of compounds (Fig. 7.4 and Table 7.2).

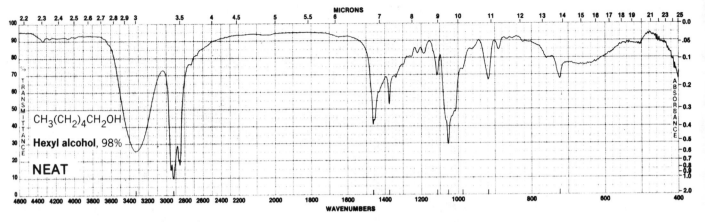

Fig. 7.19 *IR spectrum: 1-hexanol.*

HEXANOL   If an oxygen atom is inserted across one of the terminal C—H bonds, we obtain the alcohol, 1-hexanol. The change in the infrared spectrum obtained with a sample path length of less than half that used to obtain the spectrum of hexane is remarkable (Fig. 7.19). A very intense band appears at 3350 cm^{-1}, which is assigned to the stretching mode of the single O—H group (Tables 7.12 and B.8). The very broad and intense properties of this absorption are characteristic of the stretching of H-bonded hydroxyl groups. The increase in intensity of this mode also reflects an increase in the polarity of the bond involved in the vibration over that of the C—H bond in hexane. A second strong band in the spectrum is located near 1058 cm^{-1}. This absorption has been identified as the C—O stretching mode. The vibrational displacements of this fundamental are similar to the antisymmetric stretch of water. Since the vibration involves significant displacement of the adjacent C—C oscillator, the vibration will be substitution-sensitive. These latter shifts can be of value in determining the nature of the alcohol (primary, secondary, tertiary, see Table 7.13).

The only other new modes observed in going from hexane to 1-hexanol are the O—H bending vibrations. Two types of bending vibrations would be expected, the in-plane and out-of-plane displacements. The in-plane bend of the O—H oscillator is not a very good frequency because it is coupled to adjacent methylene group bending vibrations (wagging). It can be found because of its breadth (O—H bonding) as an underlying absorption running across the 1500–

Table 7.12 Normal Modes of the Hydroxyl Group

$\tilde{\nu}$ (cm^{-1})	Intensity	Mode Description
3500–3200	Very strong	O—H stretch (only strong when H—bonded)
1500–1300	Medium to strong	O—H in-plane bend (overlaps CH_2 and CH_3 bend)
1260–1000	Strong	C—C—O antisymmetric stretch
650	Medium	O—H out-of-plane bend

Table 7.13 Substitution Effects on C—O Stretch of Alcohols

Type of —OH Substitution	$\tilde{\nu}_{C-O}$ (cm^{-1})
RCH_2—OH	1075–1000
R_2CH—OH	1150–1075
R_3C—OH	1200–1100
C_6H_5—OH	1260–1180

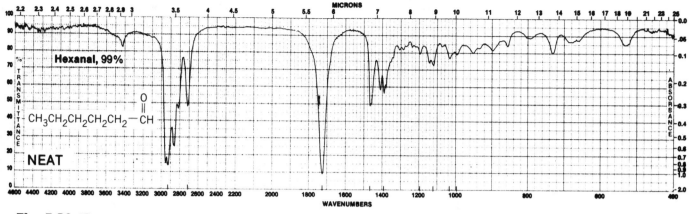

Fig. 7.20 *IR spectrum: hexanal.*

1300 cm^{-1} region (Table 7.12 or B.8). The out-of-plane O—H bending funda-mental occurs at lower frequencies as a broad band (H bonding) near 650 cm^{-1}. The group frequencies of the hydrocarbon portion of the molecule are easily identified, including the rocking fundamental (727 cm^{-1}), which is superim-posed on the broad O—H out-of-plane bending mode. The normal modes of the hydroxyl group possess many of the characteristics that lead to excellent group frequency correlations (Fig. 7.19).

HEXANAL   If two terminal hydrogen atoms of hexane are replaced by a single oxygen atom, we have hexanal. The aldehyde functional groups gives rise to several good group frequencies (Fig. 7.20 and Table 7.14 or B.9). The system has strong bonds and a large dipole moment, and it is essentially decoupled from the rest of the molecule by the low-frequency C—C connecting bond. A component of the C—H stretching mode of the aldehyde group can be assigned to a band of weak to medium intensity at $\tilde{\nu} = 2723$ cm^{-1}. The low frequency of this mode is interpreted on the basis of a Fermi resonance interaction. The aldehyde in-plane C—H bending fundamental found at 1390 cm^{-1} would be expected to generate an overtone very close to the aldehyde group C—H stretching mode, which must occur near 2775 cm^{-1}. The two levels interact and split to give two compo-nents. The more easily identified low-frequency component is very characteristic of the aldehyde group and is located near 2730 cm^{-1}. The higher-frequency component often is masked by other aliphatic C—H stretching absorptions in the 2900 to 2800 cm^{-1} region. In the case of hexanal, the upper band is observed as a distinct shoulder occurring at 2823 cm^{-1}. The carbonyl stretching frequency is the most intense band in the spectrum and is located at 1728 cm^{-1}. The structural change in going from ketone to aldehyde will produce mass and inductive effects that will lower the frequency and bond angle and hyperconju-gative effects that will raise the carbonyl frequency. The outcome of this compe-

Table 7.14 Normal Modes of the Aldehyde Group

$\tilde{\nu}$ (cm^{-1})	Intensity	Mode Description
2750–2720	Weak to medium	C—H stretch in Fermi resonance with C—H bend
1735–1720	Very strong	C=O stretch
1420–1405	Medium	CH$_2$ symmetric bend, —CH$_2$— alpha to —CHO carbonyl
1405–1385	Medium	C—H in-plane bend

tition is that aliphatic aldehyde carbonyl stretching modes generally occur at slightly higher values than do those of the saturated ketones. The only other identifiable group frequency associated with the aldehyde system is the C—H in-plane bending fundamental ($\tilde{\nu} = 1390$ cm^{-1}) responsible for the overtone that undergoes Fermi resonance with the aldehyde C—H stretching mode. It should be noted that the aldehyde group perturbs one of the aliphatic chain group frequencies. Thus, the frequency of the symmetric deformation (scissoring) of the methylene group alpha to the carbonyl is lowered ($\tilde{\nu} = 1408$ cm^{-1}), and intensified by hyperconjugation with the carbonyl. It is also evident that the vibrational modes of the aliphatic portion of the molecule now contribute a much smaller fraction of the overall absorption by the sample. Hence the spectrum is obtained with a path length even shorter than that of 1-hexanol.

3-HEPTANONE

If we insert a carbonyl group between the first two methylene groups of hexane, we have the ketone 3-heptanone. The only group frequency mode associated with this group is the stretching frequency ($\tilde{\nu}_{C=O} = 1718$ cm^{-1}), which occurs within the expected region for an aliphatic ketone (Fig. 7.21 and Table 7.15 or B.10). As in the case of aldehydes having α-methylene groups, the symmetric bending modes (scissoring) of the adjacent methylene groups at carbons 2 and 4 in 3-heptanone are perturbed to lower frequencies ($\tilde{\nu} = 1425$ cm^{-1}) by hyperconjugation with the carbonyl. The shift of the methylene bending frequency is a useful indication of the substitution surrounding the ketone. The very weak band at 3425 cm^{-1} can be confidently assigned as the overtone of the carbonyl stretching frequency ($2\tilde{\nu}_{C=O} = 3444$ cm^{-1}). The drop in intensity from that of the fundamental and the frequency contraction are typical of these forbidden transitions. Note that in this molecule the sequence of methylene groups has dropped below 4, and the rocking vibration ($\sim$720 cm^{-1}) is no longer easily detectable.

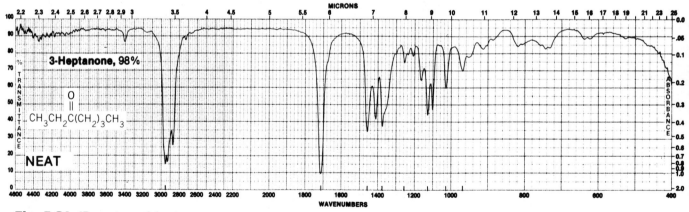

Fig. 7.21 IR spectrum: 3-heptanone.

Table 7.15 Normal Modes of the Ketone Group

$\tilde{\nu}$ (cm^{-1})	Intensity	Mode Description
3430–3410	Very weak	Not fundamental, overtone of carbonyl stretch
1725–1700	Very strong	C=O stretch
1430–1415	Medium	CH$_2$ symmetric bend, —CH$_2$— α to ketone carbonyl

1-HEXYL ACETATE   Replacing the O—H hydrogen of hexanol with CH_3CO— gives 1-hexylacetate. The very strong band found at 1743 cm^{-1} is typical of the carbonyl frequency of an aliphatic ester, particularly acetate esters (Fig. 7.22 and Table 7.16 or B.11) Two very intense bands occur in the spectra of acetate esters in the 1250–1000 cm^{-1} region. In primary acetates these bands are found near 1250 and 1050 cm^{-1} (hexylacetate, 1242, 1042 cm^{-1}). The higher frequency mode is assigned to the antisymmetric C—CO—O stretch (similar to that in water), and the lower frequency mode to the antisymmetric O—CH_2—C stretch. Although there is some coupling of these vibrations to the adjacent structure, resonance through the carbonyl group by the ether oxygen tends to localize the higher of the two vibrations. The lower mode, which would be expected to be more highly coupled, is in fact more subject to substitution effects. The upper mode is found in most saturated esters at frequencies slightly lower than those that occur in acetates (1210–1160 cm^{-1}). The only other absorption to note in acetate esters is the methyl symmetric bending mode. Here the umbrella deformation of the methyl adjacent to the carbonyl occurs at a slightly lower (hyperconjugation) frequency ($\tilde{\nu}$ = 1366 cm^{-1}), and the band is significantly intensified by the interaction with the carbonyl as compared with the deformation of the methyl at the end of the hexyl chain ($\tilde{\nu}$ = 1384 cm^{-1}).

HEXANOYL CHLORIDE   Hexanoyl chloride can be formed from hexane by the exchange of three terminal methyl hydrogens for an oxygen and a chlorine. The carbonyl stretching mode dominates the spectrum (Fig. 7.23 and Tables 7.17 and B.12). It is an extremely intense band occurring near 1802 cm^{-1}. The high frequency and intensity result from inductive effects of chlorine substitution directly on the carbonyl group. The chlorine modes, on the other hand, are not easy to identify,

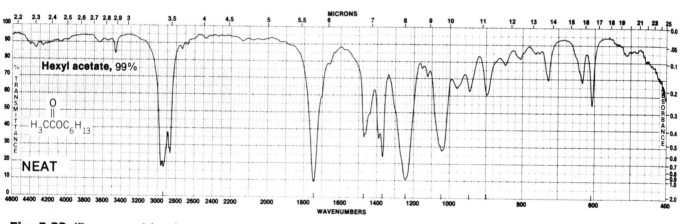

Fig. 7.22 *IR spectrum:* 1*-hexylacetate.*

Table 7.16 Normal Modes of the Ester Group

$\tilde{\nu}$ (cm^{-1})	Intensity	Mode Description
1755–1735	Very strong	C=O stretch
1370–1360	Medium	CH_3 symmetric bend alpha to ester carbonyl
1260–1230	Very strong	C—CO—O antisymmetric stretch—acetates
1220–1160	Very strong	C—CO—O antisymmetric stretch—higher esters
1060–1030	Very strong	O—CH_2—C antisymmetric stretch—primary acetates
1100–980	Very strong	O—CH_2—C antisymmetric stretch—higher esters (may overlap with upper band)

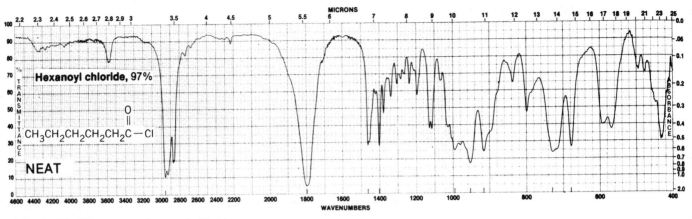

Fig. 7.23 *IR spectrum: hexanoyl chloride.*

Table 7.17 Normal Modes of the Acid Halide Group

$\tilde{\nu}$ (cm^{-1})	Intensity	Mode Description
1810–1800	Very strong	C=O stretch, acid chlorides
1415–1405	Strong	CH$_2$ symmetric bend, alpha to —COCl carbonyl

and these vibrations do not develop good group frequencies. The symmetric deformation (scissoring) of the methylene group adjacent to the carbonyl group again shows the same frequency decrease resulting from hyperconjugation as observed with the ketones and aldehydes. In this case the mode ($\tilde{\nu}$ = 1408 cm^{-1}) gains considerable intensity via interaction with the highly polarized carbonyl group.

HEXANOIC ACID Hexanoic acid is obtained from hexane by substituting an oxygen and an —OH group for the terminal hydrogens. The acid possesses a very intense band with a width at half peak height of ~1000 cm^{-1}, which covers the region 3500 to 2200 cm^{-1} (Fig. 7.24 and Table 7.18 or B.13). This absorption is characteristic of very strongly H-bonded carboxylic acid groups. The relatively weak C—H stretching absorption of the aliphatic chain is superimposed on the O—H stretch between 3000 and 2800 cm^{-1}. Also occurring along this broad absorption are a characteristic set of weak overtone and combination bands running from ~2800

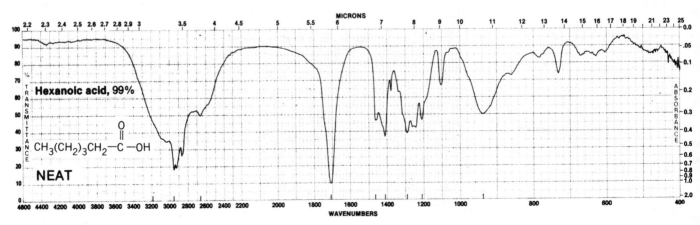

Fig. 7.24 *IR spectrum: hexanoic acid.*

Table 7.18 Normal Modes of the Carboxylic Acid Group

$\tilde{\nu}$ (cm^{-1})	Intensity	Mode Description
3500–2500	Very, very strong	O—H stretch intensified by H bonding
2800–2200	Very weak	Overtone and sum tones
1725–1710	Very strong	C=O antisymmetric H-bonded dimer stretch
1450–1400	Strong	CH$_2$—CO—O antisymmetric stretch mixed with O—H bend
1300–1200	Strong	CH$_2$—CO—O antisymmetric stretch mixed with O—H bend
950–920	Medium	Out-of-plane O—H bend, acid dimer

to 2200 cm^{-1}. The spectrum of the neat material will be that of the H-bonded dimer, which as noted earlier (see p. 446), has a center of symmetry. The carbonyl mode (out-of-phase stretch) is found at 1709 cm^{-1}. Two rather broad and intense bands located between 1450 and 1400 and between 1300 and 1200 cm^{-1} are associated with the in-plane O—H bend and the antisymmetric —CH$_2$—CO—O— stretch. These modes show evidence of considerable mixing, a situation quite different from the case of alcohols. A strong broad absorption band at 930 cm^{-1} is assigned to an out-of-plane bending mode of the H-bonded dimer ring. Thus, this latter band is present only when detectable concentrations of the dimers exist. In dilute solution the dimer band vanishes. This low wavenumber absorption is termed the *acid dimer band*. It is worth noting that even in carboxylic acids, in which the normal modes of a large, highly polarized functional group dominate the spectrum, the group frequencies of the aliphatic molecular backbone are still identifiable.

HEXANOIC ANHYDRIDE   Hexanoic anhydride can be formed from two molecules of hexanoic acid by removing the elements of water. Coupling of the carbonyls through the ether oxygen splits the carbonyls ($\tilde{\nu}_{C=O}$ = 1831, 1761 cm^{-1}) by ~70 cm^{-1} (see Fig. 7.25 and Tables 7.19 and B.14). In this instance the higher-frequency mode is the in-phase vibration. Strong bands occur in aliphatic anhydrides, near

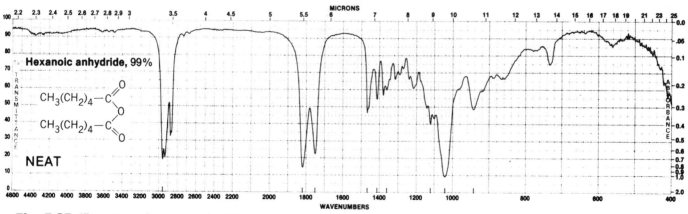

Fig. 7.25 *IR spectrum: hexanoic anhydride.*

Table 7.19 Normal Modes of the Anhydride Group (Open Chain)

$\tilde{\nu}$ (cm^{-1})	Intensity	Mode Description
1840–1810	Very strong	C=O in-phase stretch
1770–1740	Very strong	C=O out-of-phase stretch
1420–1410	Strong	CH$_2$ symmetric bend alpha to carbonyls
1100–1000	Very strong	C—O stretch, mixed modes

1050 cm^{-1}, which are directly related to C—O stretching modes. The scissoring deformation of the —CH$_2$— groups alpha to the carbonyls is assigned to the band near 1415 cm^{-1}.

DIHEXYL ETHER   Dihexyl ether is obtained by removing the elements of water from two molecules of 1-hexanol. Long-chain aliphatic ethers have many physical properties similar to those of the hydrocarbons, and the infrared spectrum of dihexyl ether (Fig. 7.26) is not very different from that of tridecane. The single major departure in the spectrum is the presence of a very strong band near 1100 cm^{-1}. In dihexyl ether this absorption is found at 1130 cm^{-1}. The vibration responsible for this absorption band must involve a considerable amount of antisymmetric C—O—C stretch. Heavy coupling, however, is also involved in this level. Because of the extensive mechanical coupling with the chain carbons, substitution adjacent to the ether linkage can rather significantly shift this frequency. Fortunately, the large intensity associated with this fundamental relative to the other bands occurring in this region makes it possible, in most cases, to assign with confidence the antisymmetric C—O—C stretching mode (see Table 7.20 or B.15).

1-HEXYLAMINE   If an NH unit is inserted across one of the terminal C—H bonds of *n*-hexane, we obtain 1-hexylamine. Compare the infrared spectrum of 1-hexylamine (Fig. 7.27) with those of 1-hexanol (Fig. 7.19) and *n*-hexane (Fig. 7.4). The difference in the region above 3310 cm^{-1} is quite remarkable. Although *n*-hexane is essentially devoid of absorption in this region and 1-hexanol exhibits a very strong band, 1-hexylamine possesses two bands ($\tilde{\nu}_{N-H} = 3380, 3290$ cm^{-1}) of medium to weak intensity. These latter bands are the antisymmetric and symmetric N—H stretching modes, respectively, of the primary amino group (Table 7.21 or B.16). H-bond intensification of the N—H group in simple primary amines does not equal that of the hydroxyl system. A band of medium intensity just above 1600 cm^{-1} is assigned to the symmetric (scissoring) deformation of the amino group. This vibration is directly related to the methylene mode near 1450 cm^{-1}. The occurrence of this fundamental requires that the amino group

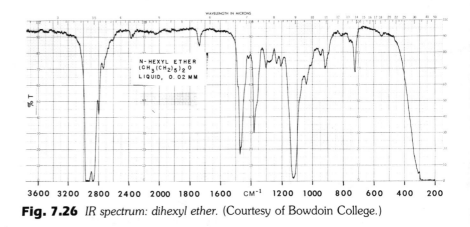

Fig. 7.26 *IR spectrum: dihexyl ether. (Courtesy of Bowdoin College.)*

Table 7.20 Normal Mode of the Ether Group

$\tilde{\nu}$ (cm^{-1})	Intensity	Mode Description
1150–1050	Strong	C—O—C antisymmetric stretch, mixed mode

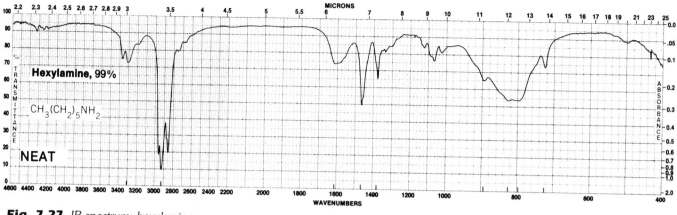

Fig. 7.27 *IR spectrum: hexylamine.*

Table 7.21 Normal Modes of the Primary Amine Group

$\tilde{\nu}$ (cm^{-1})	Intensity	Mode Description
3400–3200	Weak to medium	NH$_2$ stretch (antisymmetric and symmetric)
1630–1600	Medium	NH$_2$ symmetric bend
820–780	Medium	NH$_2$ wag

be unsubstituted, as the two hydrogen atoms undergo major displacements during the vibration. A second bending mode of primary amino groups can sometimes be observed. This vibration resembles the methylene wagging motion in which the hydrogen atoms are displaced more or less parallel to the molecular axis. As is usual for H-bonded bending fundamentals, the mode occurs as a fairly broad and quite strong band near 800 cm^{-1}. Frequencies related to the normal modes of single-bonded C—N systems are highly coupled to the surrounding C—C oscillators and cannot be assigned with any confidence.

HEXANENITRILE   Replacement of three terminal hydrogens of hexane by a nitrogen gives hexanenitrile. The nitrile group is a very simple two-atom oscillator. The very strong triple bond (as in the case of the alkynes) contributes to an unusually high stretching frequency (Fig. 7.28 and Table 7.22 or B.17), and the polar character

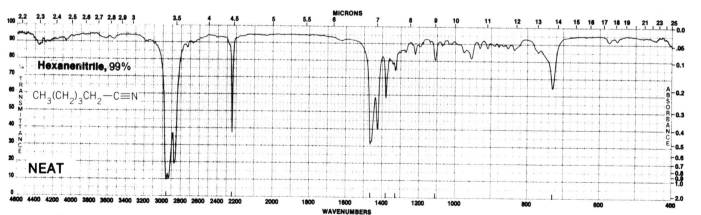

Fig. 7.28 *IR spectrum: hexanenitrile.*

Table 7.22 Normal Modes of the Nitrile Group

$\tilde{\nu}$ (cm^{-1})	Intensity	Mode Description
2260–2240	Strong	C—N stretch, aliphatic
2240–2210	Strong	C—N stretch, conjugated

of the group gives rise to very strong bands. These two factors allow for easy distinction of nitrile bands from acetylenic absorption. The stretching fundamental of saturated nitriles falls in the region 2260–2240 cm^{-1}. As expected, conjugation lowers this fundamental (2240–2210 cm^{-1}). In hexanenitrile the normal mode occurs at 2255 cm^{-1}. Primary nitrile groups can interact through hyperconjugation with the adjacent methylene groups so that the symmetric deformation (scissoring) mode is lowered into the 1425 cm^{-1} region. In hexanenitrile this bending vibration is assigned to a band at 1430 cm^{-1}. The remaining absorption bands are the aliphatic group frequencies.

HEXANAMIDE   Hexanamide is obtained by replacing the terminal hydrogens of hexane with the elements ONH$_2$. The highly polar amide group leads to very strong H bonding, which in turns leads to greatly intensified N—H antisymmetric and symmetric stretching modes ($\tilde{\nu}_{\text{N—H}}$ = 3375, 3200 cm^{-1}; Fig. 7.29 and Tables 7.23 and B.18). The carbonyl stretch occurs at low values (1675 cm^{-1}) for a saturated system substituted with an electronegative atom. Resonance between the carbonyl and the nitrogen lone pair, plus strong hydrogen bonding, appears to overcome the inductive effect. The carbonyl band is accidentally degenerate (two fundamentals occurring at the same frequency by chance rather than being required to have the same frequency by symmetry restrictions), but does not interact to any appreciable extent with the symmetric bending mode of the

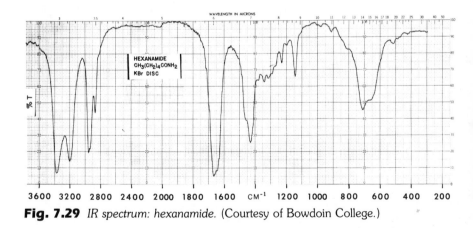

Fig. 7.29 *IR spectrum: hexanamide.* (Courtesy of Bowdoin College.)

Table 7.23 Normal Modes of the Primary Amide Group

$\tilde{\nu}$ (cm^{-1})	Intensity	Mode Description
3400–3150	Very strong	NH$_2$ antisymmetric and symmetric stretch, H bonded
1680–1650	Very strong	C=O stretch, H-bonded
1660–1620	Strong	NH$_2$ symmetric bend (overlap with C=O stretch)
1430–1410	Strong	CH$_2$ symmetric bend alpha to amide carbonyl
750–650	Medium	NH$_2$ wag

—NH$_2$ group in hexanamide (symmetry constraints restrict the interaction in this case). In some cases both bands can be resolved, but they will often occur as a single band. The wagging vibration of the amino group in which the hydrogen atoms are displaced parallel to the chain axis is found in the 700 cm^{-1} region as a strong broad band. The symmetric deformation of the methylene group adjacent to the amide carbonyl undergoes the conventional frequency drop to the 1425 cm^{-1} region. The perturbed scissoring fundamental, however, gains considerable intensity from the interaction with the highly polarized amide carbonyl. Because of the presence of very strong intermolecular H bonds, the overall spectra of solid or pure liquid amides exhibit broad, rather ill-defined, absorption bands.

N-METHYLHEXANAMIDE

If we substitute on the amide group by replacing a hydrogen with a methyl group, we obtain the secondary amide, N-methylhexanamide. The single N—H group gives rise to a very strong band at ~3300 cm^{-1}, which is very indicative of strong H bonding (Fig. 7.30). A medium-intensity band near 3100 cm^{-1} is the overtone of the N—H bending mode identified at 1570 cm^{-1} in Fermi resonance with the N—H stretching fundamental. The overtone does not match the fundamental particularly well, but it is close enough to acquire substantial intensity enhancement. The carbonyl stretching mode at 1650 cm^{-1} exhibits the very intense and broad characteristics of the amide C=O system. The N—H in-plane bend of the single oscillator occurs near 1570 cm^{-1}. The drop in frequency from that of the primary scissoring mode near 1600 cm^{-1} allows for confident assignment of the substitution on secondary amide groups (Table 7.24 or B.19). The band at 1570 cm^{-1}, although often referred to as the N—H bending mode, is in reality a heavily mixed mode. The pure in-plane N—H bend naturally falls near 1450 cm^{-1}. Resonance between the carbonyl and the nitrogen lone pair results in a stiffening of the C—N bond. The resulting C—N stretch is raised into the 1400 cm^{-1} region. Mechanical coupling between the N—H bend and the C—N stretch results in first-order coupling and a splitting of the levels. The upper level occurs near 1570 cm^{-1}, whereas the lower level often can be identified near 1300 cm^{-1} as a weak to medium band. The out-of-plane bend of the N—H group is identified as a broad, medium-intensity band centered near 700 cm^{-1}. The symmetric deformation of the methylene group alpha

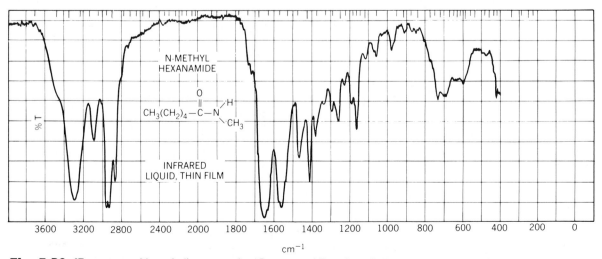

Fig. 7.30 *IR spectrum:* N-*methylhexanamide.* (Courtesy of Bowdoin College.)

Table 7.24 Normal Modes of the Secondary Amide Group

$\tilde{\nu}$ (cm^{-1})	Intensity	Mode Description
3350–3250	Strong	N—H stretch, intensified by H bonding
3125–3075	Medium	Overtone N—H bend in Fermi resonance with N—H stretch
1670–1645	Very strong	C=O stretch, H-bonded
1580–1550	Strong	N—H in-plane bend mixed with C—N stretch
1415–1405	Strong	CH$_2$ symmetric bend alpha to amide carbonyl
1325–1275	Medium	C—N stretch mixed with N—H in-plane bend
725–680	Medium	N—H out-of-plane bend

Table 7.25 Amide Carbonyl: Solution and Solid Phase Data

Amide	Dilute solution (cm^{-1})	Solid (cm^{-1})
R—CO—NH$_2$	~1730	~1690–1650
R—CO—NH**R**	~1700	~1670–1630
R—CO—N**R**$_2$	~1650	~1650

to the carbonyl is located at 1410 cm^{-1}. The unusual intensity of this mode results from interaction with the heavily polarized carbonyl system.

Studies of amide carbonyl frequencies in dilute nonpolar solution indicate that H-bonding effects are largely responsible for the low frequencies observed with primary and secondary amides, but play no role in tertiary amides.

The data (Table 7.25) indicate that when H-bonding effects are removed in primary amides, the inductive effect of the nitrogen dominates over the influence of conjugation, but not as much as in the case of esters. This is consistent with the relative electronegativities involved in esters and amides. In secondary amides with an electron-releasing *N*-alkyl group replacing a hydrogen, conjugation involving the nitrogen lone pair with the carbonyl begins to overcome the inductive effect. In tertiary amides with two *N*-alkyl substituents present, conjugation now dominates the inductive effect. Under these conditions polarized resonance forms make large contributions to the character of the carbonyl and the C=O frequency decreases.

1-HEXYL ISOCYANATE   1-Hexyl isocyanate is obtained by replacement of a terminal hydrogen atom with an —N=C=O group. The out-of-phase stretching mode of the isocyanate group attached to the hexyl chain occurs at 2275 cm^{-1}, as a broad and very strong band (Fig. 7.31). This functional group is representative of a number of

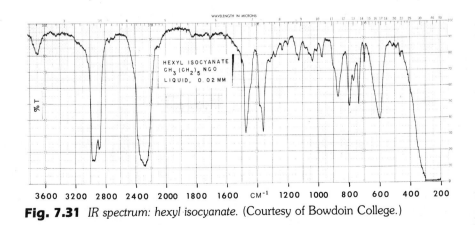

Fig. 7.31 *IR spectrum: hexyl isocyanate.* (Courtesy of Bowdoin College.)

Table 7.26 Normal Mode of the Isocyanate Group

$\tilde{\nu}$ (cm^{-1})	Intensity	Mode Description
2280–2260	Very strong	—N=C=O antisymmetric stretch

cumulated double-bond systems that possess vibrations mechanically identical to that of carbon dioxide ($\tilde{\nu} = 2350$ cm^{-1}). The range of stretching frequencies observed for alkyl-substituted isocyanates is very narrow, $\tilde{\nu} = 2280$–2260 cm^{-1}, which implies little coupling to the rest of the system (Tables 7.26 and B.20). Interestingly, conjugation appears not to have any significant effect on the mode. The symmetric stretching fundamental is not easily observed in the infrared as it is a weak band occurring in the fingerprint region. The remaining group frequencies in the spectrum of 1-hexyl isocyanate are those of the alkyl group.

1-HEXANETHIOL   Insertion of a sulfur atom in a terminal C—H bond of *n*-hexane gives 1-hexanethiol. The spectrum of this material resembles that of hexane itself except for small changes in the fingerprint region and a weak band near 2570 cm^{-1} (Fig. 7.32). The latter absorption is assigned to the S—H stretching fundamental (Table 7.27 or B.21). Although this mode is quite weak, it is not involved in any significant coupling, and it occurs in a region of the spectrum sparsely populated by other absorption bands. The S—H stretch, therefore, can be considered a reliable group frequency. The S—H bending and C—S stretching modes also are weak, and as they fall in the fingerprint region they are not useful as group frequencies. The remaining bands of 1-hexanethiol that can be assigned belong to the alkyl portion of the molecule.

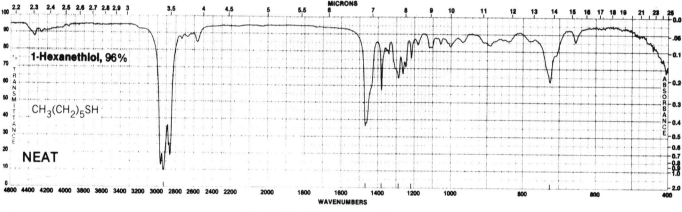

Fig. 7.32 *IR spectrum: 1-hexanethiol.*

Table 7.27 Normal Mode of the Thiol Group

$\tilde{\nu}$ (cm^{-1})	Intensity	Mode Description
2580–2560	Weak	S—H stretch

1-CHLOROHEXANE   Replacement of a terminal hydrogen atom of *n*-hexane by a chlorine atom gives 1-chlorohexane. The massive chlorine atom is connected to the alkyl section by a fairly weak but highly polarized bond that dictates that the C—Cl stretching frequency appears as an intense band at low frequencies (Fig. 7.33 and Table

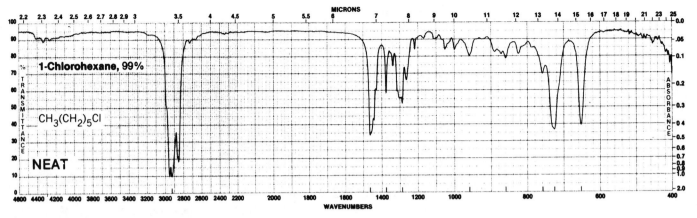

Fig. 7.33 *IR spectrum: 1-chlorohexane.*

Table 7.28 Normal Mode of the Chlorine Group

$\tilde{\nu}$ (cm^{-1})	Intensity	Mode Description
750–650	Strong	C—Cl stretch, rotamers and mixed modes occur

7.28 or B.22). The spectrum of 1-chlorohexane does possess a number of moderately intense absorption bands in the low-frequency region (800–600 cm^{-1}). Some coupling to the main structure adjacent to the C—Cl bond is expected since the carbon atom will be carrying out the majority of the displacement. Reliable assignment of the halogen stretching mode, therefore, is not easy, because the surrounding C—C modes will pick up intensity from the polar C—Cl bond. In the case of 1-chlorohexane, it is possible to assign two C—Cl stretching modes ($\tilde{\nu}_{\text{C—Cl}}$ = 731, 658 cm^{-1}), based on Raman spectral data. The presence of two modes is attributed to the presence of rotamers. The higher frequency is assigned to the anti (trans) conformer, and the lower frequency to the gauche conformer. Note that the stretching frequency of the anti isomer, 731 cm^{-1}, falls at the same frequency as the methylene rocking vibration of the hexyl chain. Thus, without additional data it would have been difficult to assign the C—Cl stretching modes, even in these fairly simple systems. The carbon–halogen stretching vibration must be employed with care as a group frequency!

The bending modes of the halogens usually occur at such low frequencies as to be of little use as conventional group frequencies. The remaining bands are related to the hydrocarbon portion of the molecule.

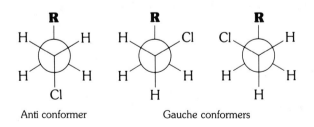

Anti conformer   Gauche conformers

CHLOROBENZENE   The final spectrum to be considered in this section is the spectrum of the C_6 compound chlorobenzene (Fig. 7.34). Here we have introduced a new hydrocarbon platform bearing the functional group. Chlorobenzene is a material in which the oscillators are tightly bound together as a single conformer. The spectrum contains many needle-sharp bands as compared, for example, with

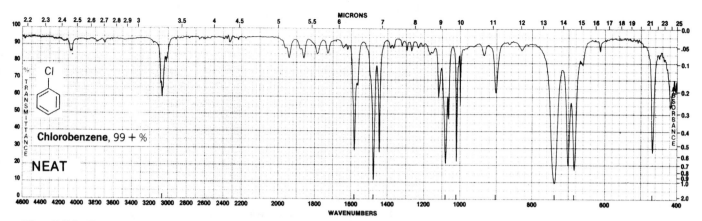

Fig. 7.34 *IR spectrum: chlorobenzene.*

Table 7.29 Group Frequency Assignments for Chlorobenzene

$\tilde{\nu}$ (cm^{-1})	Intensity	Mode Description
3080	Medium	C—H stretching, C—H bonded to sp^2 carbon
1585	Strong	ν_{8a} ring stretching
1575	Weak	ν_{8b} ring stretching
1475	Strong	ν_{19a} ring stretching
1450	Strong	ν_{19b} ring stretching
747	Strong	C—H all in-phase, out-of-plane bend
688	Strong	Ring deformation
1945, 1865, 1788, 1733	All weak	Sum tones, out-of-plane C—H bends, pattern matches monosubstitution of ring

the spectrum of 1-chlorohexane. Thus, it is often possible to detect the presence of either aliphatic or aromatic systems simply on the basis of overall appearance of the spectrum. In this spectrum the group frequencies of the hydrocarbon portion can be assigned as in Tables 7.29 and B.23.

PART D: Strategies for Interpreting Infrared Spectra

1. Divide the spectrum at 1350 cm^{-1}.

2. Above 1350 cm^{-1}, absorption bands have a high probability of being good group frequencies. The interpretation is usually reliable and free from ambiguities. We can be much more confident of our assignments in this region even with rather weak bands.

3. Because of the reliability of the high wavenumber region, we always begin the interpretation of a spectrum at this end.

4. Bands below 1350 cm^{-1} may be either group frequencies or fingerprint frequencies. (The fingerprint region is often considered to begin at slightly higher wavenumber values, close to 1500 cm^{-1}, but for interpretation purposes we will consider the region to be that lying below 1350 cm^{-1}.)

5. Below 1350 cm^{-1} group frequencies are less easily assigned. In addition, even if a reliable group frequency occurs in this region, absorption at that frequency is not necessarily a result of that mode.

6. To make more confident assignments below 1350 cm^{-1}, it is helpful to be able to associate a secondary property such as band shape with the particular mode. For example, the band is very intense, broad, sharp, occurs as a characteristic doublet, gives the correct frequency shift on isotopic substitution, or the like.

7. A good rule to remember is that in the fingerprint region the **absence** of a band is more important than the presence of a band.

8. Before beginning the interpretation, establish the sampling conditions and as much other information about the sample as possible (such as molecular weight, melting point, boiling point, color, odor, elemental analysis, solubility, refractive index).

9. In the interpretation try to assign the most intense bands first. These bands very often will be associated with a polar functional group.

10. Do not try to assign all the bands in the spectrum. Fingerprint bands are unique to a particular system. Occasionally, intense bands will be fingerprint-type absorptions; these bands, generally, will be ignored in the interpretation. They do, however, play an extremely important role when infrared data are employed for identification purposes.

11. The correlation chart (back endpaper) can act as a helpful quick aid for checking potential assignments. It is *not* a substitute for understanding the theory and operation of group frequency logic. *The use of the correlation chart without a good knowledge of group frequencies is the shortest path to disaster!*[12]

12. Try to utilize the so-called *macro group frequency* approach. That is, if the functionality or molecular structural unit requires the presence of more than a single group frequency mode, make sure that all modes are correctly represented. The *macro frequency train* represents a very powerful approach to the interpretation of relatively complex spectral data. This technique is at the core of current work on the automatic computer interpretation of infrared spectra. Contained in the product characterization section of 14 experiments (5A, 5B, 6, 7A, 8B, 8C, 8D, 11B, 14, 16B, 21C, 22, 36, 46) are a series of detailed discussions which demonstrate the operational use of the *macros*. Careful reference to the discussions in these experiments will be helpful in the initial stages of learning this interpretation technique. The student should be able to extend this interpretive approach to other experiments by reference to common infrared library files. In Experiment 52A key peaks are listed without interpretation, which is left to the student. This last suggestion is perhaps the most important strategy to master in learning to interpret infrared spectra. Practice on *macro group frequencies* will pay you big dividends in the future.

QUESTIONS   **7-18.** The form of the C—H out-of-plane bending vibrations of the vinyl group are shown below:

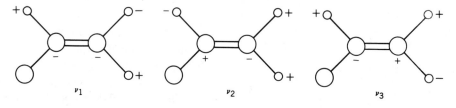

---

[12] Bellamy, L. J. *The Infrared Spectra of Complex Molecules*, 3rd ed.; Chapman and Hall: London, 1975; p 3.

The first two modes give rise to excellent group frequencies, whereas the third fundamental does not lend itself to these correlations.

    a. Explain the factors that lead to the third mode being such a poor group frequency.

    b. Predict the location in the spectrum of the third fundamental.

7-19. In the figure below, the mass of the terminal hydrogen atoms on acetylene is hypothetically varied from zero to infinity. The response of the C—H symmetric stretching (3374 cm^{-1}) and triple-bond stretching (1974 cm^{-1}) modes to the change in mass is shown.

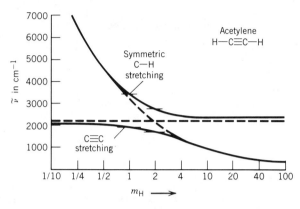

    a. Calculate the expected D-isotopic shift for the C—H symmetric stretching. Is the hypothetical value close to the calculated value? Explain.

    b. Explain why the triple-bond stretching frequency is approximately 100 cm^{-1} higher for high-mass terminal isotopes as compared with low-mass terminal isotopes.

7-20. Acetylene has two C—H groups. It will have two C—H stretching frequencies, the in-phase and out-of-phase stretching modes. The in-phase (symmetric) stretch occurs at 3374 cm^{-1} and the out-of-phase stretch at 3333 cm^{-1}. Explain why the in-phase vibration is located at a higher frequency than the out-of-phase stretch in the case of acetylene.

7-21. Carbonyl stretching frequencies of a series of benzoyl derivatives are listed in the table below:

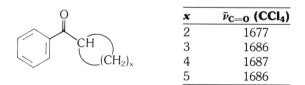

x	$\tilde{\nu}_{C=O}$ (CCl$_4$)
2	1677
3	1686
4	1687
5	1686

If we consider the $\tilde{\nu}_{C=O}$ of acetone at 1715 cm^{-1} as a reference frequency, then identify the factors affecting $\nu_{C=O}$ in the series of compounds listed.

7-22. Explain how mass effects act to lower the carbonyl frequency, as well as how bond angle, inductive and hyperconjugation effects act to raise the carbonyl frequency of aldehydes relative to ketones.

7-23. The carbonyl stretching frequency of aliphatic carboxylic acids in dilute solution is located near 1770 cm^{-1}. This frequency is much higher than the carbonyl frequency of these substances when measured neat (~1720 cm^{-1}). In addition, it is considerably higher than the corresponding simple aliphatic ester (1745 cm^{-1}) value. Explain.

7-24. In a number of cases, dipolar interactions control the frequency shifts found in carbonyl stretching vibrations. Table 7.10 lists wavenumber shifts on going from neat to dilute nonpolar solutions. Explain the observed values.

7-25. The antisymmetric —CH₂—CO—O— stretching vibration in carboxylic acids is heavily mixed with the in-plane bending mode of the O—H group. In alcohols these two vibrations seldom show evidence of mechanical coupling. Explain.

7-26. Conjugation of the functional group in alkyl isocyanates has little impact on the antisymmetric —N=C=O stretching vibration located near 2770 cm⁻¹. Explain.

7-27. In the infrared spectrum of 2-aminoanthraquinone (**I**) two carbonyl stretching frequencies are observed at 1673 and 1625 cm⁻¹. (The spectrum is obtained from a sample mixed with Nujol. Nujol is a complex mixture of saturated hydrocarbons. The Nujol mull is an alternative infrared sampling procedure to KBr disks. This method avoids placing the sample in a highly ionic matrix. The mulling agent, however, does introduce the saturated hydrocarbon group frequencies, which mask a number of regions of the spectrum.)

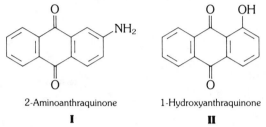

2-Aminoanthraquinone
I

1-Hydroxyanthraquinone
II

a. Assign carbonyl bands in the infrared spectrum to the carbonyl groups in structure **I** and explain your reasoning.

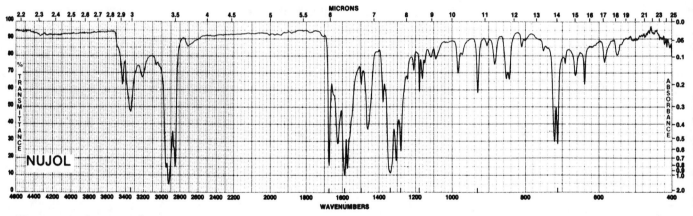

IR spectrum: 2-aminoanthraquinone.

b. The infrared spectrum of 1-hydroxyanthraquinone (**II**) also exhibits two carbonyl frequencies, which are located at 1675 and 1637 cm⁻¹. Assign the carbonyl groups to the related absorption bands. Explain your reasoning.

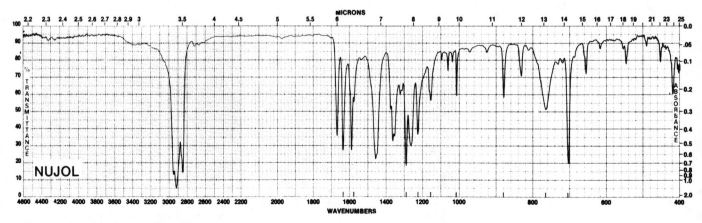

IR spectrum: 1-hydroxyanthraquinone.

7-28. Suggest a possible structure for the hydrocarbon C_6H_{14}, which has the infrared spectrum shown below. Is there more than one correct structure?

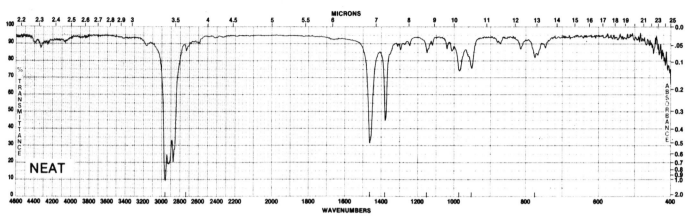

IR unknown spectrum C_6H_{14}.

7-29. The hydroxylamine **I** can be oxidized by MnO_2 to the amide oxohaemanthidine **(II)**. In dilute solution the carbonyl absorption band of **II** occurs at 1702 cm^{-1}. Explain this observation.

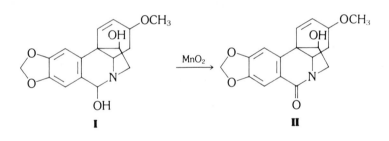

7-30. Identify the following olefinic hydrocarbons (a–e). All samples were obtained from distillation cuts in the C_6 boiling range.

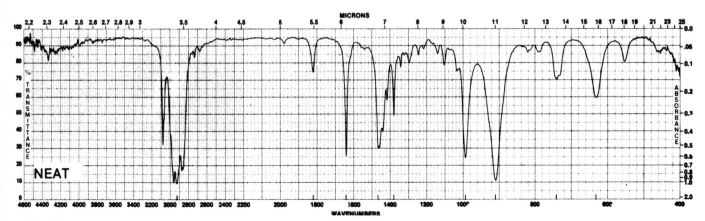

IR unknown spectrum a.

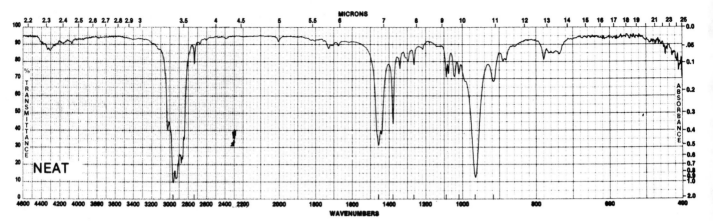

IR unknown spectrum b.

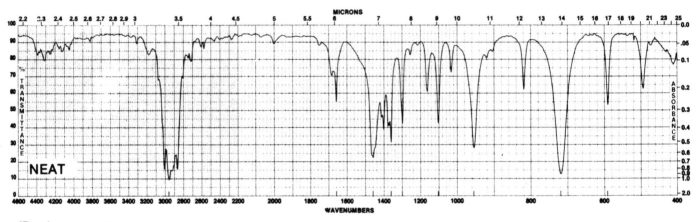

IR unknown spectrum c.

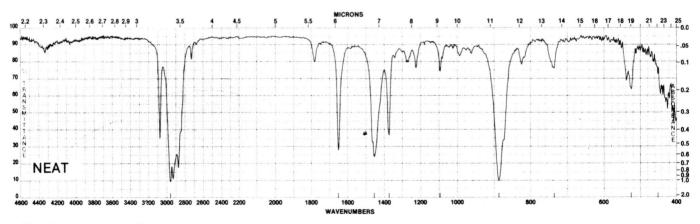

IR unknown spectrum d.

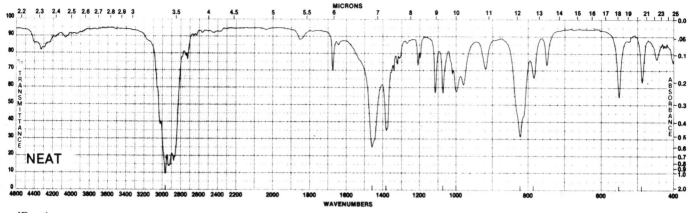

IR unknown spectrum e.

7-31. The infrared spectra of the xylene isomers, and an additional aromatic hydrocarbon, are given below. Assign the spectra and suggest a potential structure for the remaining unknown substance.

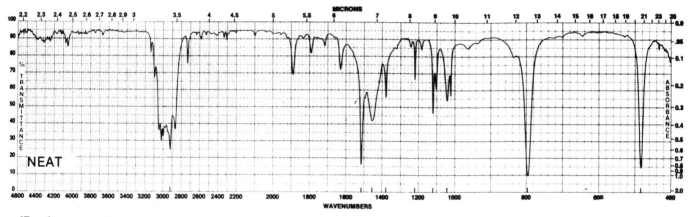

IR unknown spectrum a.

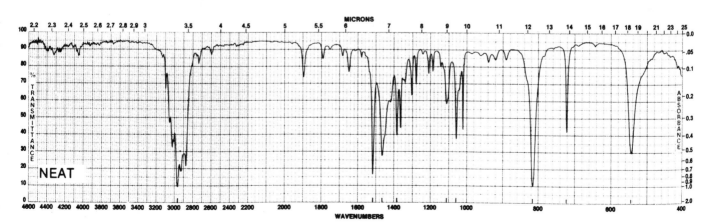

IR unknown spectrum b.

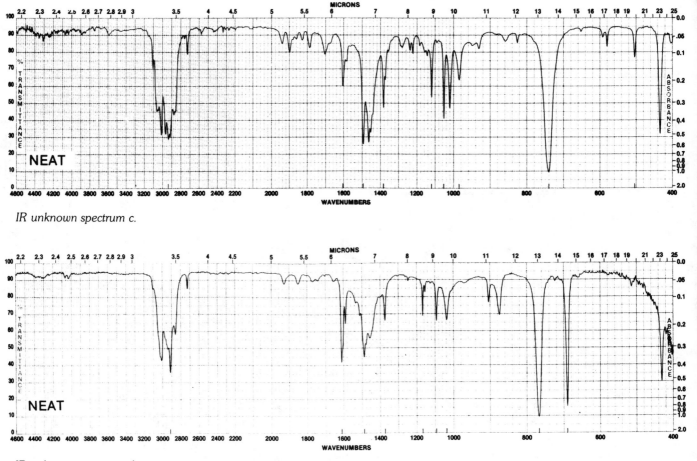

IR unknown spectrum c.

IR unknown spectrum d.

7-32. The C—H stretching mode of chloroform, $CHCl_3$, which occurs at 3022 cm^{-1}, is one of the rare exceptions to the 3000 cm^{-1} rule. What is the rule? Suggest an explanation for this exception.

III. Nuclear Magnetic Resonance

PART A. ¹H NMR Spectroscopy

This section will provide a brief introduction to the use of nuclear magnetic resonance (NMR) spectroscopy for structure determination. A description of the theory underlying this method is given in Appendix C.

INTERPRETATION OF UNKNOWN SPECTRA

Sample Preparation

Spectra may be obtained on quantities of the order of 0.1 mmol in most cases. Fourier transform instruments will yield good spectra on a few micromoles of sample. Spectra are obtained on solutions. The sample tube (generally 5-mm-diameter thin-walled glass) is filled to a depth of about 2 cm, which corresponds to the active portion of the rf coils. Filling the tube to this depth maximizes sample concentration in the active part of the instrument and hence the strength of the signal. Adding more volume just wastes sample. Solvents for ¹H NMR spectroscopy must not contain any hydrogens, since the solvent is present in great excess relative to the solute. Consequently, solvents that contain deuterium in lieu of hydrogen are commonly used. Commercially available deuterated solvents include deuterochloroform, deuteroacetone, deuterobenzene, and deuterium oxide ("heavy water"). Another consideration in the choice of solvent is its ability to dissolve, but not otherwise interfere with, the compound of interest. Since deuterated solvents are somewhat costly, solubilities should first be determined with regular solvents (no deuterium).

Since sampling details are highly dependent on the compound of interest and on the NMR facility available, further details will be provided by your instructor. An example of NMR sample preparation techniques is provided in Experiment 7.

Observed Chemical Shifts

Figure 7.35 summarizes the chemical shifts of protons in a large range of chemical environments. It is, however, a bit dangerous to use such figures without understanding some of the factors that underlie shielding and the chemical shift. To give some flavor of the factors that determine chemical shifts and the range of values observed, we will briefly examine chemical shifts in methyl groups and chemical shifts for protons on sp^2 carbons (relative to TMS).

Methyl groups bonded to an sp^3 carbon generally have chemical shifts in the range 0.8–2.1 ppm as long as there is no more than one electron-withdrawing group attached to the carbon. The shifts generally increase as the strength of the electron-withdrawing group increases or as more electron-withdrawing groups are added. Groups that inductively withdraw electrons reduce the electron density near the methyl group protons. This results in less shielding and a downfield shift of the methyl resonance. This effect is clearly seen in the spectra of 1,1-dibromoethane and 1-nitropropane (Appendix C, pp. 496 and 497, respectively). The chemical shifts for methyl groups bonded to sp^2 carbons fall in the range 1.6–2.7 ppm.

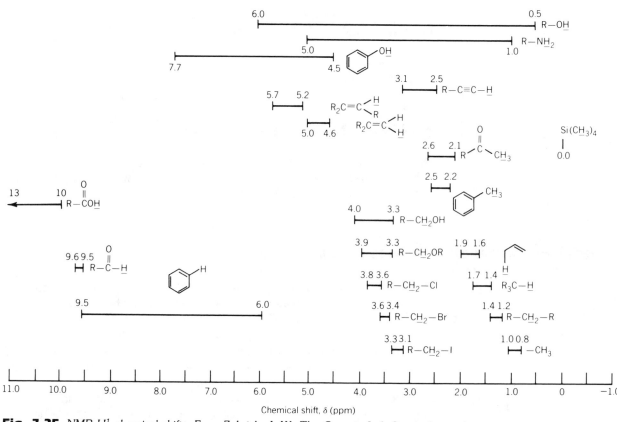

Fig. 7.35 *NMR H¹ chemical shifts.* From Zubrick, J. W. *The Organic Lab Survival Manual*, 2nd ed.; Wiley: New York, 1988. (Reprinted by permission of John Wiley & Sons, New York.)

In the case of a proton bonded to an sp^2 carbon, the location of the proton relative to the π cloud plays an important role in determining the chemical shift. In unconjugated olefins the chemical shifts fall in the range 5–6 ppm. Where more than one proton is bonded to a double-bond system, complex second-order spectra can be expected for low operating frequencies, since the coupling constants are usually fairly large relative to the difference in resonance frequencies (Appendix C). In oximes, **R**CH=NOH, the increased electronegativity of the nitrogen increases the deshielding and the chemical shift falls in the range 6.5–7.5 ppm. This trend continues with aldehydes, which have chemical shifts in the range 9.5–10.5 ppm.

The chemical shift in an aromatic system is generally greater than that for olefins. For example, the chemical shift of benzene is 7.37 ppm, substantially greater than the 4.56 ppm for ethylene. Much of this difference results from the "ring current" effect and the orientation of the proton relative to the aromatic π electrons. If the ring substituents are not strongly electron withdrawing, the chemical shift for ring protons may be shifted to slightly higher fields. Furthermore, these substituents generate only small chemical shift differences among the ring protons. Thus, the 60-MHz spectra for toluene and o- and p-xylene appear to have a single resonance in the aromatic region at about 7.1 ppm. If, on the other hand, the substituents are electron withdrawing, the ring protons will be somewhat deshielded relative to benzene. Electron-withdrawing groups will also cause differentiation of the ring proton chemical shifts, and complex second-order spectra are likely to be seen in the range 6.5–8.5 ppm.

Coupling Constants

The sign of the coupling constant (usually symbolized as J) may be positive or negative. However, first-order spectra are not sensitive to the sign of the coupling constant. In second-order cases, the sign of J may be determined by a careful analysis of the spectrum, though it generally is of little value for organic structure determination.

Proton–proton coupling involves an effect extending over at least two bonds. The first case we shall consider is that of geminal coupling in which two protons are connected to the same atom, as in H—C—H. When the central atom is an sp^3 carbon atom, the coupling constant ranges from 4 to 20 Hz and is sensitive to ring strain and the electronegativity of the other substituents. Geminal coupling constants fall in the range 0 to 2 for protons bonded to the same sp^2 carbon atom.

Coupling extending over three bonds, as in H—C—C—H, is very sensitive to the angle of rotation about the central bond. When the central bond is an sp^3 C—C bond, internal rotation can occur on a time scale that is very short relative to the NMR time scale. Then the effect of internal rotation is completely blurred as far as NMR is concerned and only an average coupling constant is observed. Vicinal coupling constants in ethyl groups are usually observed in the 6.5–8 Hz range, but can be found from 2 to 14 Hz in more complex systems.

When the central carbon–carbon bond is a double bond, internal rotation is restricted and distinct coupling constants for cis and trans protons may be observed. Cis coupling constants fall in the range 5 to 12 Hz, whereas trans coupling constants range from 12 to 20 Hz. As a result of these large coupling constants, second-order effects can be quite noticeable in substituted ethylenes.

Longer-range coupling involving four or more bonds is common in allylic systems and in aromatic rings and other conjugated π-bond systems. These coupling constants are generally smaller than the values considered above (less than 3 Hz).

The NMR user should be alert to the presence of other nuclei that may affect proton spectra. [19]F may be significant because its coupling constant can be quite large. [13]C is present only in about 1% in natural abundance but can have a large effect in enriched compounds because it can be directly bonded to the proton. [31]P is not common among compounds considered in this course, but the coupling constant for a proton bonded to phosphorus can be a few hundred hertz. [14]N has a spin quantum number of 1 and is influenced by quadrupole effects (Appendix C). Its presence when bonded to a proton is often reflected in a broadening of the proton resonance.

QUESTIONS   Several 60-MHz spectra are given on the following pages.[13] You should be able to account for at least one acceptable structure and for all the observed resonances.

---

[13] From Pouchert, C. J. *The Aldrich Library of NMR Spectra;* Aldrich Chemical Company: Milwaukee, WI, 1983.

7-33. C_4H_8O. Spectrum a.

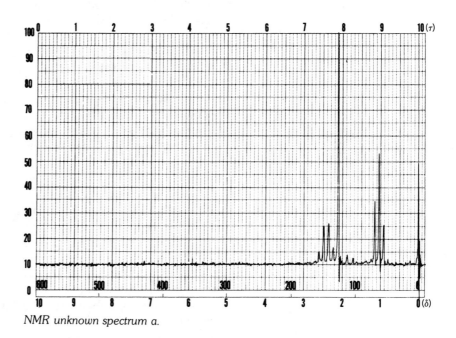

NMR unknown spectrum a.

7-34. $C_3H_6O_2$. Spectra b and c. Two compounds with the same empirical formula.

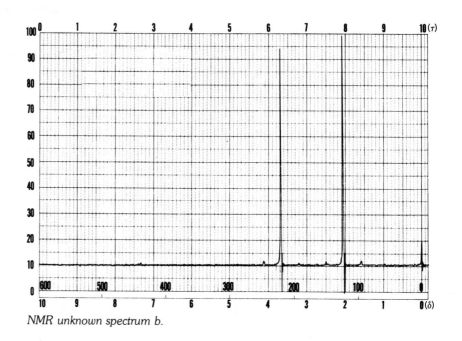

NMR unknown spectrum b.

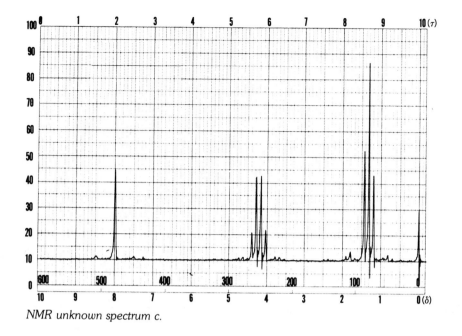

NMR unknown spectrum c.

7-35. C$_4$H$_8$O again, spectrum d. Also give some thought to the weak resonances at 0.5 and 1.8 ppm.

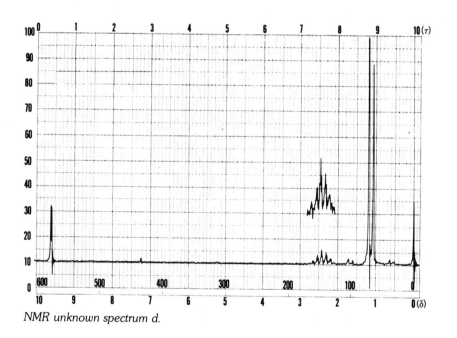

NMR unknown spectrum d.

7-36. C$_7$H$_7$Cl. Spectrum e.

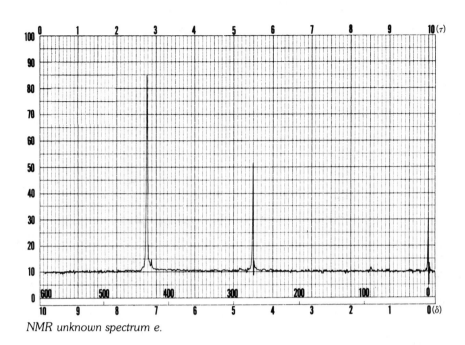

NMR unknown spectrum e.

7-37. C$_8$H$_{10}$O. Spectrum f.

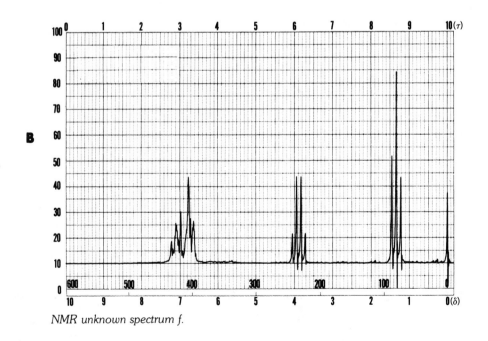

NMR unknown spectrum f.

Appendix A

Ultraviolet–Visible Spectroscopy

There is a common physical basis for the absorption of UV and visible radiation by organic substances. In both wavelength regions, the absorption of energy results in transitions between electronic energy levels of the molecule. The spectral regions observed necessarily restrict the transitions to those involved in the promotion of π and **nonbonding** electrons from ground states to excited states. For example, in the simplest conjugated alkene, 1,3-butadiene, the energy absorbed corresponds to the excitation of a π electron in the highest occupied molecular orbital (HOMO) to the lowest unoccupied molecular orbital (LUMO). In this type of alkene, this corresponds to a $\pi \rightarrow \pi^*$ transition, that is, from a π-bonding molecular orbital to a π^*-antibonding orbital. This excitation is depicted below for ethylene and 1,3-butadiene. Notice that the energy gap between the HOMO and LUMO of ethylene (unconjugated π system) is larger than that in the butadiene molecule. The excitation energy for ethylene, therefore, is greater (shorter wavelength, $\lambda_{max} = 171$ nm) than that for 1,3-butadiene ($\lambda_{max} = 217$ nm).

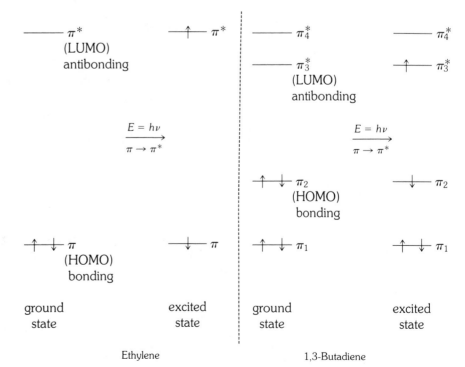

Ethylene   1,3-Butadiene

Table A.1 Absorption Maxima of Conjugated All-trans Alkenes

Name	Structure	λ_{max} (nm)
Ethylene	$CH_2{=}CH_2$	165
1,3-Butadiene	$CH_2{=}CH{-}CH{=}CH_2$	217
1,3,5-Hexatriene	$CH_2{=}CH{-}CH{=}CH{-}CH{=}CH_2$	268
1,3,5,7-Octatetraene	$CH_2{=}CH{-}CH{=}CH{-}CH{=}CH{-}CH{=}CH_2$	290

The wavelength of an absorption band is determined by the energy difference between the π and π^* levels. Experimental evidence has shown that the greater the number of conjugated multiple bonds in a given compound, the longer (greater) will be the wavelength (λ) of the $\pi \rightarrow \pi^*$ absorption in the molecule. This is illustrated in Table A.1.

It is clear that if a compound has sufficient unsaturation, the λ_{max} absorption band will appear in the visible region of the electromagnetic spectrum and the compound will be colored. For example, tetraphenylcyclopentadienone is purple (Experiment 49), the dye methyl red is deep red (Experiment 31), and *trans*-9-(2-phenylethenyl)anthracene is golden yellow (Experiment 21D).

Compounds that contain a carbonyl group, $C{=}O$, also absorb radiation in the near ultraviolet. The π electrons of this functional group undergo a $\pi \rightarrow \pi^*$ transition. The oxygen atom, however, also possesses two pairs of nonbonding electrons. The n electrons reside in orbitals (n) that are higher in energy then the bonding p orbital, but lower in energy than the antibonding π^* orbital. The energy diagram of the carbonyl system is shown below.

excited state resulting from promotion of a nonbonding electron to the antibonding π^* orbital

ground state

excited state resulting from promotion of a π electron to the antibonding π^* orbital

Thus, carbonyl groups absorb ultraviolet radiation in two regions, one corresponding to maxima of $n \rightarrow \pi^*$ transitions and the other to $\pi \rightarrow \pi^*$ transitions. As can be observed, the energy gap of the $\pi \rightarrow \pi^*$ transition is greater (shorter wavelength) than that of the $n \rightarrow \pi^*$ (longer wavelength). For the α,β-unsaturated carbonyl, methyl vinyl ketone, the following data have been recorded:

$$n \rightarrow \pi^* \qquad \lambda_{max} = 324 \text{ nm}, \ \varepsilon_{max} = 24$$

$$\pi \rightarrow \pi^* \qquad \lambda_{max} = 219 \text{ nm}, \ \varepsilon_{max} = 3600$$

The quantitative relationship of absorbance to concentration is expressed by the Beer–Lambert equation (see also Experiment 5),

$$A = \varepsilon c l$$

where A = absorbance
e = molar absorptivity (a constant characteristic of the specific molecule being observed)
c = concentration (mol/liter)
l = length of sample path (cm)

The sections of the molecule that absorb ultraviolet–visible radiation (such as conjugated C=C, carbonyl groups conjugated with C=C, and aromatic rings) are referred to as chromophores. Ultraviolet–visible spectroscopy detects the presence of these chromophoric units and, thus, can impart information concerning the molecular structure of the compound. Often, model compounds are referred to as an aid in the interpretation of the ultraviolet spectrum. A substantial amount of data accumulated from a wide variety of chromophores has lead to a number of very powerful empirical correlations (such as the so-called Woodward–Feiser rules) of substituent effects on the longest wavelength λ_{max} absorption band (see the references at the end of this section for further information and use of these data).

Several important points concerning electronic spectroscopy should be made.

1. Ultraviolet–visible spectra can reveal the presence of conjugated systems in a molecule. These unsaturated structural units are called chromophores.
2. The observed λ_{max} increases (energy gap gets smaller) as the conjugated π-electron system becomes extended. The intensity of the absorption also increases (greater ε_{max}).
3. The λ_{max} is affected by substituents, conformations, and structural characteristics of the conjugated π-electron system.

Table A.2 lists the λ_{max} values of a number of common organic molecules.

Table A.2 Absorption Maxima of Several Unsaturated Molecules

Compound	Structure	λ_{max} (nm)	ϵ_{max}
Ethylene	$CH_2=CH_2$	171	15,530
1,3-Butadiene	$CH_2=CH-CH=CH_2$	217	21,000
Cyclopentadiene		239	3,400
1-Octene	$CH_3(CH_2)_5CH=CH_2$	177	12,600
trans-Stilbene	C_6H_5 H C=C H C_6H_5	295	27,000
cis-Stilbene	H H C=C C_6H_5 C_6H_5	280	13,500
Toluene	CH_3	189 208 262	55,000 7,900 260
4-Nitrophenol	$HO-\!\!\!\!\bigcirc\!\!\!\!-NO_2$	320	9,000
3-Penten-2-one	$CH_3CH=CHCCH_3$ (C=O)	220 311	13,000 35

REFERENCES   **1.** Silverstein, R. M.; Bassler, G. C.; Morrill, T. C. *Spectrometric Identification of Organic Compounds*, 3rd ed.; Wiley: New York, 1981; Chap. 6.

2. Jaffé, H. H.; Orchin, M. *Theory and Application of Ultraviolet Spectroscopy*; Wiley: New York, 1962.

3. Stern, E. S.; Timmons, T. C. J. *Electronic Absorption Spectroscopy in Organic Chemistry*; St. Martin Press: New York, 1971.

4. Lambert, J. B.; Shurvell, H. F.; Verbit, L.; Cooks, R. G.; Stout, G. H. *Organic Structural Analysis*; Macmillan: New York, 1976.

5. American Petroleum Research Institute Project 44: *Selected Ultraviolet Spectral Data*, Vols. I–IV; Thermodynamics Research Center, Texas A&M University: College Station, TX, 1945–1977 (1178 compounds).

6. Grasselli, J. G.; Ritchey, W. M. *Atlas of Spectral Data and Physical Constants*; CRC Press: Cleveland, OH, 1975.

7. *Ultraviolet Reference Spectra*; Sadtler Research Laboratories: Philadelphia.

Introduction to the Theory of Infrared Spectroscopy

I. Interpretation of Infrared Spectra

INTRODUCTION TO THEORY OF EFFECT

The wavelike character of electromagnetic radiation can be expressed in terms of velocity v, frequency ν, and wavelength λ of sinusoidally oscillating electric and magnetic vectors traveling through space (Fig. B.1). Frequency is defined as the number of waves passing a reference point per unit time, usually expressed as cycles per second, $\sec^{-1}$, or hertz, Hz. The velocity of the wave, therefore, equals the product of frequency and wavelength.

$$v = \nu\lambda$$

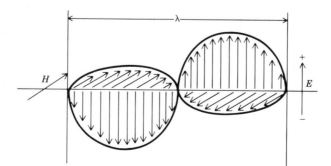

Fig. B.1 *Electromagnetic wave. H magnetic Field, E electric field, λ wavelength.*

479

If the wavelength (the distance between each wave maxima or alternate nodes) is measured in centimeters, v is expressed in centimeters per second (cm/sec). For radiation traveling in a vacuum, v becomes a constant, c ($c \sim 3 \times 10^{10}$ cm/sec), for all wavelengths. When electromagnetic radiation traverses other media, however, the velocity changes. The ratio of the speed in a vacuum, c, to the matrix velocity, v, is termed the *refractive index*, n, of the material:

$$n = c/v$$

Since n is frequency-dependent, the frequency at which the refractive index is measured must be specified. Frequency, however, has been shown to be independent of the medium and, therefore, remains constant. Wavelength thus varies inversely with n.

$$\lambda = c/n\nu$$

Since the velocity of electromagnetic radiation in a vacuum is normally greater than that in any other medium, n will generally be greater than one at all frequencies. Thus, the wavelength must become shorter for a particular frequency when measured in any matrix.

Frequency can be considered to be a more fundamental property of radiation because it is independent of the medium. This property also requires that the energy E associated with the radiation be matrix independent because E is directly proportional to frequency by

$$E = h\nu$$

where E equals the energy of a photon, which is related to frequency ν by Planck's constant (6.6×10^{-27} erg-sec or 6.6×10^{-34} joule-sec).

The vibrational states present in molecules can be excited by absorption of photons. The nuclear masses and bond force constants determine the separation of these states and, therefore, the energies of the photons involved in the absorption process. The corresponding radiation frequencies fall predominantly in the infrared region (10^{14}–10^{12} Hz) of the electromagnetic spectrum.

The infrared spectrum currently is measured in *wavenumbers* $\tilde{\nu}$, which are units proportional to frequency and energy. The wavenumber is defined as

$$\tilde{\nu} = \frac{\nu}{c} = \frac{E}{hc}$$

and as

$$\nu = \frac{c}{n\lambda} \quad \text{then in air} \quad \tilde{\nu} = \frac{\sim 1}{\lambda}$$

The wavenumber, as expressed in units of cm^{-1} (the number of waves per centimeter), offers several advantages.

1. Wavenumbers are directly proportional to frequency and are expressed in much more convenient numbers (in this region of the spectrum), 5000 to 500 cm^{-1}.

2. As shown above, wavenumbers are easily converted to wavelength values. The reciprocal of $\tilde{\nu}$ and conversion of centimeters to wavelength units are all that is required (this is particularly handy because much early infrared data were recorded linearly in wavelength). The wavelength unit employed in

most of these spectra was the micron, μ. The micron has been replaced by a unit expressed in meters, the micrometer, μm ($1\ \mu m = 1 \times 10^{-6}$ m).

3. Because the wavenumber is directly proportional to frequency and energy, the use of wavenumbers allows spectra to be displayed linear in energy. This is a distinct aid in sorting out related vibrational transitions.

Note. *It should also be pointed out that cm^{-1} is not a unit of frequency. Wavenumbers are only proportional to frequency. Thus, it is not correct to refer to a vibrational absorption band as having a frequency of 3000 cm^{-1} or to say that the vibration of the C—H bond possesses a frequency of 3000 cm^{-1}. The C—H oscillator, however, can be said to absorb radiation with an energy of 3000 cm^{-1}, or the C—H bond can be said to vibrate with a frequency of 9 $\times$ 10^{13} Hz.*

MOLECULAR ENERGY

The total molecular energy W may be expressed as the sum of the molecular translational, rotational, vibrational, and electronic energies:

$$W_{mol} = W_{trans} + W_{rot} + W_{vib} + W_{elec}$$

In this approximation, W_{vib} is assumed to be independent of the other types of molecular energy. Translational and rotational motion involve much smaller energies, having little influence on the spectra under observation. W_{elec} is the energy of the electrons (see Appendix A). The energies of these latter transitions are very much larger than vibrational spacings, and their energy changes fall outside the infrared. As a result the infrared region is active mainly to vibrational energy changes of ground electronic state molecules.

MOLECULAR VIBRATIONS

Molecules can be characterized as being in constant vibrational motion. If we are to describe this motion for nuclei of a polyatomic system, we can utilize the Cartesian coordinates x_m, y_m, z_m for nucleus m referred to a fixed coordinate system. Then, for n nuclei, we would generate $3n$ coordinates ($3n$ **degrees of freedom**) to describe the motion of all the atoms. Three of these coordinates, however, may be used to locate the center of mass of the system in space. These three coordinates define the translation of the entire system through space. Because translational energies have a small impact on vibrational spectra, the three coordinates of the center of mass can be dropped from the total required to determine the vibrational degrees of freedom. Therefore, $3n - 3$ coordinates are sufficient to determine the positions of the n nuclei with respect to the center of mass. However, the molecular system is still free to rotate about the center of mass. For nonlinear molecules three additional coordinates are required to fully describe rotational motion about the center of mass. For linear molecules only two coordinates are necessary to define rotation, as all the nuclei lie along one of the principal axes and are considered to be point groups. Thus, for nonlinear molecules $3n - 6$ coordinates fully define the vibrational motion of the nuclei. These coordinates are often referred to as the **vibrational degrees of freedom.** In linear systems one rotational degree of freedom can be considered to have been transformed into a vibrational degree of freedom ($3n - 5$).

The number of vibrational degrees of freedom is directly related to the number of fundamental vibrational frequencies possessed by the molecular system. These fundamental frequencies are often referred to as the **normal modes** of vibration.

To get a feel for the function of the normal modes in the vibrational pattern of a molecular system, let us consider a very simple arrangement of a single nu-

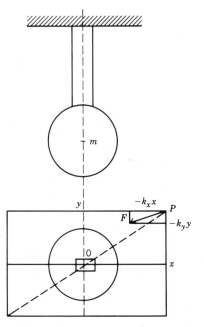

Fig. B.2 *Harmonic vibration in two dimensions.*

cleus vibrating in two dimensions.[1] The nucleus of mass m is held by a rigid but elastic rectangular bar or rod. The mass m can vibrate only in the plane perpendicular to the bar (Fig. B.2). If the nucleus is displaced along the x axis and then released, the system will oscillate with simple harmonic motion at a frequency given by

$$\nu_x = \frac{1}{2\pi}\sqrt{\frac{k_x}{m}}$$

where k_x is the force constant of the bar in the x direction where the restoring force $F = -k_x X$ for displacement X.

If displacement and release are carried out in the y direction, a similar type of oscillation will occur with a frequency given by

$$\nu_y = \frac{1}{2\pi}\sqrt{\frac{k_y}{m}}$$

where k_y is the force constant in the y direction.

The frequency ν_x that results from displacement in the x direction is different from the natural frequency ν_y, which results from displacement in the y direction, because the rectangular bar will possess different force constants k_x and k_y. If the rectangular bar is replaced with a bar of square cross section, then the force constants that result from displacement in the x and y directions are equal. The frequencies of x and y motion also will be equal. In this case the bar of rectangular dimensions can be referred to as having **degenerated** to a bar of square dimensions.

We now consider the displacement and release of the nucleus in a direction not along a principal axis. For example, the corner position P. Now on release, the motion performed by the nucleus is no longer simple harmonic. The restoring force F will have components $-k_x X$ and $-k_y Y$, which are unequal and not directed toward the origin.

The complex motion of the nucleus in the x–y plane, however, will still contain components that are simple harmonic in nature. The motion of the nucleus can be represented as the sum of these "normal modes" of vibration, which are perpendicular to each other. Thus, the position of the nucleus at any point in time after release can be expressed by the two coordinates,

$$x = x_0 \cos 2\pi\nu_x t$$

$$y = y_0 \cos 2\pi\nu_y t$$

where x_0 and y_0 are the coordinates of the initial position P, and t is the time lapse from release.

The complicated pattern of motion performed by the nucleus on release from position P is termed Lissajous motion. This type of motion is the superposition of two simple harmonic motions of differing frequency that are normal to each other. These are termed the *normal modes* or fundamental frequencies of the Lissajous motion of the nucleus of mass m. The x and y coordinates, thus, become the "normal coordinates."

---

[1] Herzberg, G. *Molecular Spectra and Molecular Structure*; Van Nostrand: New York, 1945; Vol. 2, p 62.

In the case of a diatomic molecule the frequency (in wavenumbers) is given by

$$\tilde{\nu} = \frac{1}{2\pi c} \sqrt{\frac{k}{\mu}}$$

where k is the force constant and μ is the reduced mass.

$$\frac{1}{\mu} = \frac{1}{m_1} + \frac{1}{m_2} \quad \text{or} \quad \mu = \frac{m_1 m_2}{m_1 + m_2}$$

QUANTIZED VIBRATIONAL ENERGY

Nature has been kind in distributing vibrational energy in molecules! The vibrational states associated with a particular normal mode are not influenced, to a first approximation, by the energies of adjacent states. The portion of W_{vib} contributed by a particular normal mode ν_1 is given by

$$W_{vib} = (v_i + \tfrac{1}{2})h\nu_i$$

where v_i is the vibrational quantum number for the normal mode and takes the values 0, 1, 2,. . . .

Each normal mode will possess a similar energy–quantum number relationship, and the total vibrational energy scheme can be obtained by summing over all $3n - 6$ fundamental vibrations:

$$W_{vib} = \sum_{n=1}^{3n-6} (v_i + \tfrac{1}{2})h\nu_i$$

The characteristic energy-level pattern for a normal mode as determined by the quantum relationship dictates that the level spacings will be equal with a value of $h\nu_i$. In the upper states, however, the potential energy curve begins to depart from the harmonic values and the vibration becomes anharmonic. In general, anharmonicity results in lower energy transitions or a contraction in the level spacing (Fig. B.3). In rare cases the potential energy well develops steeper sides (quartic terms become important), and the spacing actually becomes greater at higher levels (negative anharmonicity). We will see a few of the more well-known departures of this type.

In addition to the equal spacing of the energy levels associated with each normal mode, the quantization of the vibrational energy requires that the lowest or zero vibrational level (v_0) does not occur at zero energy, but at $\tfrac{1}{2}h\nu$. Thus, the molecule retains, even at absolute zero, some small amount of vibrational energy. This is termed the "zero point" energy, and its origin lies with the Uncertainty Principle.

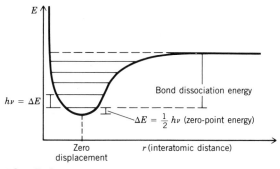

Fig. B.3 *Vibrational energy-level diagram.*

SELECTION RULES

The range of vibrational energy level transitions associated with the fundamental frequencies runs from somewhat under 5000 to approximately 100 cm^{-1}. Thus, the study of the absorption spectra of the normal modes centers on the infrared region of the spectrum. Although most molecules possess a large set of normal modes, the equal spacing of the vibrational levels and the operation of molecular selection rules greatly simplify what would otherwise be a very complex absorption pattern.

For infrared spectra the selection rules define the changes in the vibrational and rotational quantum numbers. The changes for the $3n - 6$ vibrational quantum numbers are given below.

1. Only one quantum number can change during a transition.
2. The change Δv is restricted to $+1$, -1.
3. For certain vibrations, Δv must always be zero.

In a very large percentage of cases, an absorbed photon excites only a single normal mode (this rule also holds for emission spectra). Thus, each frequency observed in the spectrum corresponds to a normal mode present in the molecule. The maximum number of frequencies observed corresponds directly to the $3n - 6$ (or $3n - 5$ for linear molecules) vibrational degrees of freedom.

In normal modes where $\Delta v = 0$, an incomplete set of frequencies will be observed. The symmetry elements present in the molecule largely determine whether $\Delta v = 0$ for a particular normal mode. This results from the further requirement that there be a change in molecular dipole moment during the vibration for the absorption of a photon to occur.

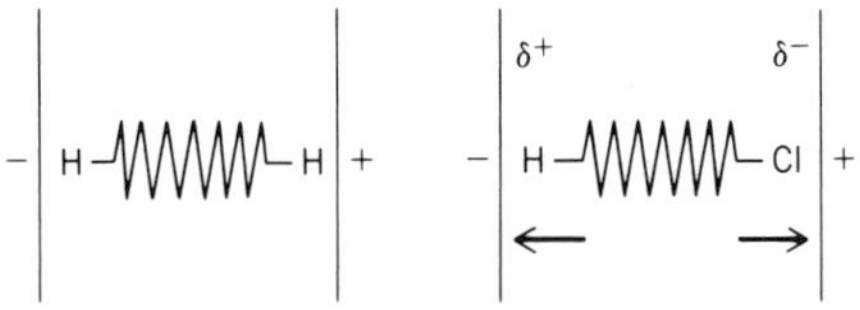

We can see why a variation in the magnitude of the dipole moment is essential to the absorption process by considering hydrogen and hydrogen chloride molecules placed between condenser plates (Fig. B.4).

In the case of HCl, a permanent dipole moment exists along the molecular axis, with the negative pole closer to the chlorine atom and the positive pole closer to the hydrogen atom. When placed between condenser plates, as shown in the figure, the molecule will experience attractive forces at both ends, which will exert a stretching action. If the charge on the condenser plates is quickly reversed, the molecule will then experience repulsive forces at each end and be compressed. If it were possible to alternate the charge on the plates fast enough to match the natural frequency of the HCl molecule, the system would resonate. In the resonance or "tuned" condition, the oscillator will absorb energy from the condenser and expand its vibrational displacements. (If enough energy were absorbed, the molecule would dissociate.) The frequency remains constant, but the amplitude of the vibration increases (much the same way as the input of periodic energy with the correct phase into a child's swing increases the amplitude of the swing, but leaves the frequency constant). Within the band of infrared radiation (5000 to 100 cm^{-1}), oscillating electric fields of the radiation will act on molecules in a fashion similar to alternating condenser fields. The frequencies in this spectral region correspond to the natural vibrational frequencies present in the molecular systems. Thus, at the particular radiation frequency that matches the vibrational frequency of the HCl molecule, resonance will occur and the photon of corresponding energy will be absorbed as the molecule moves to the next-higher vibrational state. In the case of the hydrogen molecule, there is no vibrating electric dipole present because of the symmetry of the system. Thus, no interaction with the oscillating electric vector of the radiation at the natural frequency can occur (this would correspond to no interaction with the condenser plate fields). In this case the normal mode for the hydrogen molecule is not observed in the infrared spectrum, and the selection rule $\Delta v = 0$ applies.

Let us examine two simple examples to illustrate the preceding discussion.

Fig. B.4 H_2 and HCl oscillators.

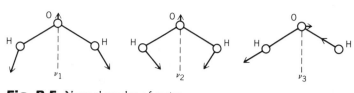

Fig. B.5 *Normal modes of water.*

The Case of Water

Water is a nonlinear three-atom molecule. It has, therefore, three normal modes of vibration or fundamental frequencies ($3n - 6$, where $n = 3$). The displacements of the normal modes can be derived in much the same fashion as the two-dimensional case and are as shown in Figure B.5.

All three vibrations are active in the infrared, and three absorption bands are observed. The arrows represent the relative atomic displacements involved in one phase of the vibration. The atoms all move in phase in simple harmonic motion in each fundamental mode. The high-frequency vibration ν_3 at 3756 cm^{-1} involves predominantly hydrogen motion, with one bond contracting while the other is stretching. This vibration is designated as the "antisymmetric stretching" mode. (Modes of this type are often incorrectly referred to as *asymmetric* vibrations. The vibration is not asymmetric, without symmetry, but a vibration of antisymmetric character, ie opposite symmetry or opposed symmetry.) The other stretching vibration designated ν_1 at 3652 cm^{-1} involves in-phase and identical displacements of the two O—H bonds. It is termed the *symmetric stretching* mode. Finally, the low-frequency mode ν_2 at 1595 cm^{-1} corresponds to a bending of the molecule about the H—O—H bond angle. It occurs at lower frequencies than do the two stretching modes, as it takes less energy to bend a bond than it does to stretch one. Thus, the force constant k_{bend} is considerably smaller than $k_{stretch}$. This bending vibration is often termed the *scissoring vibration* because the symmetric bending motion involved is similar to the action of scissors. The complex molecular vibrational pattern of the water molecule can be resolved into three simple harmonic components or normal modes that correspond directly to three absorption bands in the infrared spectrum. The atomic displacements of these normal modes can be used to characterize the type of fundamental vibration giving rise to the infrared absorption. In the water molecule we have two O—H stretching vibrations and one O—H bending mode. These modes represent the three vibrational degrees of freedom present in the water molecule.

The Case of Carbon Dioxide

Carbon dioxide is a three-atom linear molecule. It will possess four vibrational degrees of freedom (one more than water). The displacements of the normal modes are given in Figure B.6.

In the case of carbon dioxide only two absorption bands are observed in the infrared spectrum even though the molecule has four vibrational degrees of freedom. The high-frequency antisymmetric stretching mode ν_3 occurs at 2350 cm^{-1}. The form of this mode is close to that of the antisymmetric stretching vibration found in water. The symmetric stretching vibration ν_1 of carbon dioxide is similar to its counterpart in water; however, this vibration does not give rise to a change of dipole moment during the vibration. The arrangement of the atoms in the carbon dioxide molecule places a center of symmetry on the carbon atom. This symmetry element remains intact during the symmetric stretching vibration, and as a consequence no change in the dipole moment occurs. Thus, $\Delta v = 0$ and no absorption occurs at this frequency in the infrared. The normal mode still exists in the molecule, but it is infrared inactive. This fundamental has been

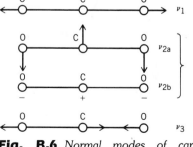

Fig. B.6 *Normal modes of carbon dioxide.*

located near 1334 cm^{-1} in the Raman effect. (The Raman effect is another spectroscopic technique used to observe vibrational energy-level transitions. It does not depend on a change in dipole moment.) There are two bending frequencies present in carbon dioxide, ν_{2a} and ν_{2b}. These two modes are identical except that one is rotated by 90° with respect to the other. The bending motion does produce a change in dipole moment, and as a result it will be infrared active. Since the two modes will have identical frequencies, however, only one absorption band will be observed at 667 cm^{-1}. Carbon dioxide can bend in two mutually perpendicular planes, and therefore two bending vibrations are required to fully characterize the vibrational motion of the molecule. The symmetry of the system dictates that the modes will be identical. Vibrations possessing these characteristics are termed **degenerate vibrations** (remember the *degeneration* of the rectangular bar to a square that gave rise to two mutually perpendicular and identical modes). Thus, although carbon dioxide possesses four vibrational degrees of freedom, the observation of two absorption bands in the infrared spectrum can be satisfactorily explained. Indeed, if we examine the bending vibration present in water, we see that water can bend only in a single plane, and "bending" the molecule out of that plane results in rotation. It is this rotational degree of freedom of water that is translated into a vibrational degree of freedom in carbon dioxide as the nonlinear three-atom system is converted into a linear molecule.

If we place the carbon dioxide molecule between the plates of a condenser (Fig. B.7), we can see why some of the CO_2 vibrations are infrared inactive, and other vibrations are active. As carbon dioxide does not possess a permanent dipole moment, it is important to note that the antisymmetric stretching mode and the degenerate bending vibration will develop dipole moments during these vibrations as a result of the nuclear displacements involved. These dipole moments obviously undergo changes during the different phases of the vibrations, and therefore these modes fit the requirements of the vibrational selection rules for infrared activity. In the symmetric stretching vibration, no dipole moment is developed during the vibration, and we have a case similar to that of the hydrogen molecule. Thus, no absorption band is observed in the infrared spectrum of CO_2 that corresponds to the symmetric stretching mode.

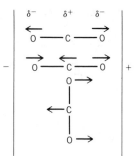

Fig. B.7 *Carbon dioxide oscillator.*

COUPLED OSCILLATORS   Water and carbon dioxide are valuable examples of simple systems in which mechanical coupling between two oscillators is amply demonstrated. If we consider water to be constructed of two O—H oscillators, the individual diatomic systems would be expected to have identical frequencies. When welded together, however, the vibrations of one O—H oscillator interfere with the vibrations of the other O—H oscillator. The coupled oscillators generate two new vibrations, one at higher frequencies and one at lower frequencies (much the same as the resonance that develops between two coupled identical pendulums can be considered to involve two beat frequencies, one higher and one lower than the natural frequency of the pendulum). The coupling interaction is frequency-dependent. The closer the frequencies of the two oscillators, the stronger the interaction. In the case of identical frequencies and a direct mechanical connection, the coupling effect will be maximized. Under the conditions of strong interaction, the form of the new vibrations can be quite different from that of the isolated oscillators. Coupling is also angle-dependent. Oscillators normal to each other couple poorly, whereas colinear oscillators will undergo maximum coupling. [Note the wavenumber separation in water (bent system) of 104 cm^{-1}, as compared with 1016 cm^{-1} for carbon dioxide (back-to-back C=O oscillators)].

SECOND-ORDER COUPLING   The selection rules break down occasionally, particularly in condensed phases, to give overtone bands (Δv greater than 1). These departures may result from anharmonicities. The overtone frequencies are usually somewhat less than double that of the fundamental mode. The drop in expected frequency results from the compression of upper levels on the potential energy curve (Fig. B.3). The absorption bands that result from these transitions are usually very weak, as the mode is formally forbidden.

One of the most spectacular of the second-order events is Fermi resonance. When the first overtone ($\Delta v = 2$) of a fundamental possesses very nearly the same energy as the $\Delta v = 1$ level of another normal mode, an interaction may occur in which the two close-lying levels are split into two new levels, one higher and one lower in frequency than the original modes. As a result of this mixing, the overtone often undergoes a dramatic intensity gain at the expense of the fundamental. The resulting doublet may even possess components of approximately equal intensity. The intensity distribution is dependent to a large extent on the value of the original frequency match. The classic example of the effect is the symmetric stretching frequency (Raman active only, see above) in carbon dioxide, which should occur near 1334 cm^{-1} but which, in fact, exists as a doublet (1388 and 1286 cm^{-1}). The perturbation was explained by Fermi as the interaction of the overtone of the bending fundamental at 667 cm^{-1} with the first exited state of the symmetric stretching mode. For Fermi resonance to occur, (1) the oscillators involved must be so arranged that the anharmonic terms can interact (mechanical interaction can occur) and, in addition, (2) the modes must meet certain symmetry restrictions. The large majority of all complex organic substances are of such low symmetry that the latter condition usually can be assumed to have been met.

In a few cases the overtone of a fundamental, although weak, will occur with higher than usual intensity, and in a region uncluttered by other absorptions. These bands can be utilized as confirmatory evidence in making assignments of fundamentals. In even rarer cases the first overtone will be observed to occur at slightly higher than double the fundamental values. These systems are considered to possess "negative anharmonicity" (see quantized vibrational energy, p. 483).

Another second-order effect is the "sum tone" or combination band. Although forbidden in the harmonic approximation, occasionally there will be absorbed a photon of the appropriate energy to simultaneously excite two normal modes. Combination bands occur as weak absorption bands that possess frequencies near the sum of the two fundamentals. If sum tones occur in regions open to observation, occasionally they can be of importance in group frequency interpretations (see out-of-plane C—H bending modes on aromatic rings; see also Fig. 7.18).

II. Group Frequency Tables

Table B.1 Alkane Normal Modes

C—H Vibrational Modes	$\tilde{\nu} \pm 10$ (cm^{-1})
Methyl groups	
Antisymmetric (degenerate) stretch	2960
Symmetric stretch	2870
Antisymmetric (degenerate) deformation	1460
Symmetric (umbrella) deformation	1375
Methylene Groups	
Antisymmetric stretch	2925
Symmetric stretch	2850
Symmetric deformation (scissor)	1450
Rocking mode (all-in-phase)	720

Table B.2 Substitution Classification of C=C Stretching Frequencies

C=C Normal Modes	$\tilde{\nu}$ (cm^{-1})
trans-, tri-, tetra-Substituted	1680–1665
cis-, vinylidene-(terminal-1,1-), vinyl-Substituted	1660–1620

Table B.3 Alkene Normal Modes

C—H Vibrational Modes	$\tilde{\nu} \pm 10$ (cm^{-1})
Stretching Modes	
Antisymmetric stretch (=CH$_2$)	3080
Symmetric stretch (=CH$_2$)	3020
Uncoupled stretch (=CH)	3030
Out-of-Plane Bending Modes	
Vinyl group	
Trans hydrogens (in-phase)	990
Terminal hydrogens (wag)	910
Vinylidene group	
Terminal (wag)	890
Trans group	
Trans hydrogens (in-phase)	965
Cis group	
Cis hydrogens (in-phase)	~700
Trisubstituted group	
Uncoupled hydrogen	820
Tetrasubstituted group: no modes	

Table B.4 Alkyne Normal Modes

C≡C, C—H Normal Modes	$\tilde{\nu} \pm 10$ (cm^{-1})
Triple bond stretch (monosubstituted)	2120
Triple bond stretch (disubstituted)	2225
C—H bond stretch (monosubstituted)	3300

Table B.5 Arene Out-of-Ring Plane C—H Deformation Modes

Number of Adjacent H's	$\tilde{\nu}$ range (cm^{-1})
5	770–730
4	770–735
3	810–750
2	860–800
1	900–860

Table B.6 Arene Group Frequencies

Arene Fundamentals	$\tilde{\nu}$ (cm^{-1})
C—H stretch	3100–3000
C=C ring stretch (ν_{8a})	1600 ± 10
C=C ring stretch (ν_{8b})	1580 ± 10
C=C ring stretch (ν_{19a})	1500 ± 10
C=C ring stretch (ν_{19b})	1450 ± 10
C—H out-of-plane bend (1H)	900–860
C—H out-of-plane bend (2H)	860–800
C—H out-of-plane bend (3H)	810–750
C—H out-of-plane bend (4H)	770–735
C—H out-of-plane bend (5H)	770–730
C—C ring out-of-plane bend (1; 1,3; 1,3,5)	690 ± 10
C—H out-of-plane bend sum tones	2000–1650

Table B.7 Carbonyl Group Frequencies

Functional Group	$\tilde{\nu}_{C=O}$ (cm^{-1})
Ketones, aliphatic, open chain (**R**$_2$CO)	1725–1700
Ketones, conjugated	1700–1675
Ketones, ring	(see Table 7.8)
Acid halides	>1800
Esters, aliphatic	1755–1735
Esters, conjugated	1735–1720
Esters (conjugated to ether oxygen)	1780–1760
Lactones	(see Table 7.8)
Anhydrides, aliphatic, open chain	1840–1810 and 1770–1740
Acids, aliphatic	1725–1710
Amides	(see Chapter 7, Part C)
Lactams	(see Table 7.8)
Aldehydes	1735–1720

Table B.8 Normal Modes of the Hydroxyl Group

$\tilde{\nu}$ (cm^{-1})	Intensity	Mode Description
3500–3200	Very strong	O—H stretch (only strong when H-bonded)
1500–1300	Medium to strong	O—H in-plane bend (overlaps CH$_2$ and CH$_3$ bend)
1260–1000	Strong	C—C—O antisymmetric stretch
650	Medium	O—H out-of-plane bend

Table B.9 Normal Modes of the Aldehyde Group

$\tilde{\nu}$ (cm^{-1})	Intensity	Mode Description
2750–2720	Weak to medium	C—H stretch in Fermi resonance with C—H bend
1735–1720	Very strong	C=O stretch
1420–1405	Medium	CH$_2$ symmetric bend, —CH$_2$— alpha to —CHO carbonyl
1405–1385	Medium	C—H in-plane bend

Table B.10 Normal Modes of the Ketone Group

$\tilde{\nu}$ (cm^{-1})	Intensity	Mode Description
3430–3410	Very weak	Not fundamental, overtone of carbonyl stretch
1725–1700	Very strong	C=O stretch
1430–1415	Medium	CH$_2$ symmetric bend, —CH$_2$— alpha to ketone carbonyl

Table B.11 Normal Modes of the Ester Group

$\tilde{\nu}$ (cm^{-1})	Intensity	Mode Description
1755–1735	Very strong	C=O stretch
1370–1360	Medium	CH$_3$ symmetric bend alpha to ester carbonyl
1260–1230	Very strong	C—CO—O antisymmetric stretch—acetates
1220–1160	Very strong	C—CO—O antisymmetric stretch—higher esters
1060–1030	Very strong	O—CH$_2$—C antisymmetric stretch—1° acetates
1100–980	Very strong	O—CH$_2$—C antisymmetric stretch—higher esters (may overlap with upper band)

Table B.12 Normal Modes of the Acid Halide Group

$\tilde{\nu}$ (cm^{-1})	Intensity	Mode Description
1810–1800	Very strong	C=O stretch acid chlorides
1415–1405	Strong	CH$_2$ symmetric bend alpha to —COCl carbonyl

Table B.13 Normal Modes of the Carboxylic Acid Group

$\tilde{\nu}$ (cm^{-1})	Intensity	Mode Description
3500–2500	Very, very strong	O—H stretch intensified by H bonding
2800–2200	Very weak	Overtone and sum tones
1725–1710	Very strong	C=O antisymmetric H-bonded dimer stretch
1450–1400	Strong	CH$_2$—CO—O antisymmetric stretch mixed with O—H bend
1300–1200	Strong	CH$_2$—CO—O antisymmetric stretch mixed with O—H bend
950–920	Medium	Out-of-plane O—H bend, acid dimer

Table B.14 Normal Modes of the Anhydride Group (Open Chain)

$\tilde{\nu}$ (cm^{-1})	Intensity	Mode Description
1840–1810	Very strong	C=O in-phase stretch
1770–1740	Very strong	C=O out-of-phase stretch
1420–1410	Strong	CH$_2$ symmetric bend alpha to carbonyls
1100–1000	Very strong	C—O stretch, mixed modes

Table B.15 Normal Mode of the Ether Group

$\tilde{\nu}$ (cm^{-1})	Intensity	Mode Description
1150–1050	Strong	C—O—C antisymmetric stretch, mixed mode

Table B.16 Normal Modes of the Primary Amine Group

$\tilde{\nu}$ (cm^{-1})	Intensity	Mode Description
3400–3200	Weak to medium	NH$_2$ stretch (antisymmetric and symmetric)
1630–1600	Medium	NH$_2$ symmetric bend
820–780	Medium	NH$_2$ wag

Table B.17 Normal Modes of the Nitrile Group

$\tilde{\nu}$ (cm^{-1})	Intensity	Mode Description
2260–2240	Strong	C—N stretch, aliphatic
2240–2210	Strong	C—N stretch, conjugated

Table B.18 Normal Modes of the Primary Amide Group

$\tilde{\nu}$ (cm^{-1})	Intensity	Mode Description
3400–3150	Very strong	NH$_2$ antisymmetric and symmetric stretch, H-bonded
1680–1650	Very strong	C=O stretch, H-bonded
1660–1620	Strong	NH$_2$ symmetric bend (overlap with C=O stretch)
1430–1410	Strong	CH$_2$ symmetric bend alpha to amide carbonyl
750–650	Medium	NH$_2$ wag

Table B.19 Normal Modes of the Secondary Amide Group

$\tilde{\nu}$ (cm^{-1})	Intensity	Mode Description
3350–3250	Strong	N—H stretch, intensified by H bonding
3125–3075	Medium	Overtone N—H bend in Fermi resonance with N—H stretch
1670–1645	Very strong	C=O stretch, H-bonded
1580–1550	Strong	N—H in-plane bend mixed with C—N stretch
1415–1405	Strong	CH$_2$ symmetric bend alpha to amide carbonyl
1325–1275	Medium	C—N stretch mixed with N—H in-plane bend
725–680	Medium	N—H out-of-plane bend

Table B.20 Normal Mode of the Isocyanate Group

$\tilde{\nu}$ (cm^{-1})	Intensity	Mode Description
2280–2260	Very strong	—N=C=O antisymmetric stretch

Table B.21 Normal Mode of the Thiol Group

$\tilde{\nu}$ (cm^{-1})	Intensity	Mode Description
2580–2560	Weak	S—H stretch

Table B.22 Normal Mode of the Chlorine Group

$\tilde{\nu}$ (cm^{-1})	Intensity	Mode Description
750–650	Strong	C—Cl stretch, rotamers and mixed modes occur

Table B.23 Group Frequency Assignments for Chlorobenzene

$\tilde{\nu}$ (cm^{-1})	Intensity	Mode Description
3080	Medium	C—H stretching, C—H bonded to sp^2 carbon
1585	Strong	ν_{8a} ring stretching
1575	Weak	ν_{8b} ring stretching
1475	Strong	ν_{19a} ring stretching
1450	Strong	ν_{19b} ring stretching
747	Strong	C—H all in-phase, out-of-plane bend
688	Strong	Ring deformation
1945, 1865, 1788, 1733	All weak	Sum tones, out-of-plane C—H bends, pattern matches monosubstitution of ring

Appendix C

Nuclear Magnetic Resonance

THEORY AND EXAMPLE CASES

Nuclear Spin

Nuclear spin is a property intrinsic to a nucleus and analogous to the electron spin that plays such an important role in determining electron configurations. Nuclear spins are represented by I and are different for different nuclei. Values range from 0 through $\frac{7}{2}$ in multiples of $\frac{1}{2}$. Of the nuclei of greatest interest to organic chemists, the ^1H, ^{13}C, ^{19}F, and ^{31}P nuclei have spins of $\frac{1}{2}$; the ^{12}C, ^{16}O, and ^{32}S nuclei have spins of 0 (and thus have no NMR spectra); the ^2H (deuterium, D) and ^{14}N nuclei have spins of 1.

The spin may have different orientations with respect to an external magnetic field. These orientations are referred to as the z component of the nuclear spin, m_z. For a nucleus with a spin of $\frac{1}{2}$ the z component of the nuclear spin may be $+\frac{1}{2}$ or $-\frac{1}{2}$. In general, for a nucleus of spin I, the z component takes all values from $-I, -I + 1, \ldots I - 1, I$, or $2I + 1$ different values in all. For this discussion we will limit ourselves to nuclei with spin $\frac{1}{2}$.

In the absence of a magnetic field, the differences in energy of the nuclei with various values of the z component are small enough to be neglected for most chemical purposes. On the other hand, when a sample of protons is placed in a magnetic field the energies of the $m_z = +\frac{1}{2}$ and $-\frac{1}{2}$ states are separated, since in one spin state the nuclear magnetic moment is aligned with the applied magnetic field, and in the other spin state the nuclear magnetic moment is opposed to the applied magnetic field.

The amount of separation of the two states is proportional to the magnetic field and is given by the following expression,

$$E = \mu m_z H_0 / I$$

where H_0 is the strength of the magnetic field at the nucleus, and μ is a characteristic of the nucleus known as the magnetic moment. When protons in the magnetic field are exposed to radiation of the proper frequency, transitions between the two states are stimulated. This occurs when the frequency and the energy difference are related by the Planck relation, $E = h\nu$, and thus the sample will absorb energy of frequency ν. The study of these energy changes is known as *nuclear magnetic resonance* spectrometry, NMR among the cognoscenti.

Chemical Shift

The magnetic field at the nucleus depends not only on H_0, the field generated by the instrument (the external field), but also on the electron density near the nucleus. Electrons are influenced by the external field in such a way that they reduce the actual field at the nucleus. This reduction is very small (relative to the external field) and is of the order of 0.001%, 10 ppm, for most protons. This reduction of the external field is known as *shielding*, and it gives rise to differences in the energy separation for protons in different chemical environments in a molecule. The differences in the energy separation are known as *chemical shifts*.

The magnitude of the chemical shift depends on the nature of the valence and inner electrons of the nucleus and also on electrons that are not directly associated with the nucleus. Chemical shifts are influenced by inductive effects, which reduce the electron density near the nucleus and reduce the shielding. The orientation of the nucleus relative to π electrons also plays an important role in determining the chemical shift. A nucleus located immediately outside a π-electron system (as in the case of the ring protons on substituted benzenes) will be significantly deshielded. In most molecules the chemical shift is determined by a combination of these factors.

A spectrum may in principle be obtained by either changing the magnetic field while holding the frequency constant or by changing the frequency at constant magnetic field. In modern instruments, the magnetic field is held constant. The magnetic field is provided by a large permanent magnet, electromagnet, or superconducting electromagnet. Commercially available NMR spectrometers have magnets that range from 1.4 to 14 tesla (the earth's magnetic field is roughly 5×10^{-5} tesla), and thus operate at frequencies from 60 to 600 MHz for protons. In general, a spectrometer with an operating frequency above 100 MHz will use a superconducting electromagnet.

As the sample is held in a constant magnetic field, the radio frequency (rf) transmitter coil is varied in frequency and the instrument electronics detect the absorption of rf energy. This condition is illustrated in Figure C.1, where the energy change is shown for increasing frequency. In this example we illustrate the case with two different protons, A and X. Since A and X are different, they absorb energy at different frequencies while in the same applied magnetic field.

The actual spectrum in this case would be displayed as shown in Figure C.2. The difference in the resonances is known as a *chemical shift* and is expressed in parts per million (ppm). It is most convenient to measure chemical shifts relative to a reference molecule that has only one type of proton, such as tetramethylsilane, $(CH_3)_4Si$ (TMS). The chemical shift relative to TMS is symbolized by δ,

$$\delta_A = \frac{10^6(\nu_A - \nu_{TMS})}{\nu_o}$$

where ν_o is the operating frequency of the spectrometer. TMS is used as a reference substance for a number of reasons. It is more strongly shielded than most other protons and its resonance is thus well removed from other areas of interest. TMS is also relatively inert and its 12 identical protons per molecule provide a strong signal per mole of TMS.

Fourier transform (FT) NMR spectrometers operate in a different manner. A short powerful pulse of rf energy is applied to the sample, causing the nuclear magnetic moments to be tipped away from their usual alignment with the applied magnetic field. The processes that occur as these moments return to their equilibrium alignment give rise to a sine wave at the frequency of the resonance line in question. Thus, the initial "spectrum," called a free induction decay (FID),

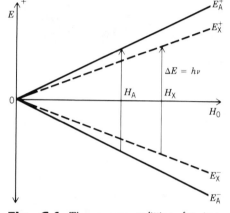

Fig. C.1 *The energy splitting for two chemically different protons. The differences between the A energy levels (solid lines) and the X levels (dashed lines) have been amplified for illustrative purposes. The 60-MHz resonance condition for nucleus A occurs at field H_A and that for X occurs at H_X. Nucleus X is said to be more strongly shielded than A. The resonance for X is said to occur upfield of that for A.*

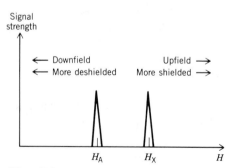

Fig. C.2 *The spectrum for the system in Fig. C.1 as it would be displayed. It is conventional to display the spectrum with magnetic field increasing to the right so that upfield (and more strongly shielded) is toward the right and downfield (and deshielded) is toward the left.*

is a plot of intensity versus time instead of the usual intensity versus frequency. Since time and frequency are mathematically related, a spectrum in either domain can be converted to the other domain. This is accomplished by the Fourier transform technique. The advantage of using an FT spectrometer is that a spectrum can be recorded in about 2 seconds and the process can then be repeated. Noise can be reduced by adding up spectra (since noise is random about a zero level), and thus spectra can be obtained on very small samples (down to ~0.1 mg in the case of ^1H). FT spectrometers are also then capable of obtaining NMR spectra of rare or insensitive nuclei, such as ^{13}C. Since ^{13}C is only 1.1% naturally abundant, there is generally not enough signal to be interpretable in a single spectrum. However, a ^{13}C NMR spectrum of a ~50-mg sample can be obtained in about 15 minutes by adding a few hundred or thousand spectra together.

Spin–Spin Coupling

In a molecule with several protons, the exact frequency at which a proton changes its spin state depends not only on the chemical shift of that proton, but also on the spin states of neighboring protons. The effect of one proton's spin state on another's is known as *coupling*. Coupling is a term used to describe the interaction between two properties we would otherwise expect to act independently.

The spectra resulting from spin–spin coupling depend on the types of nuclei, the distance of separation, the electronic environment, and the total number of spin states possible. This may be illustrated by looking at the spectrum of an imaginary compound that has adjacent protons A and X (Fig. C.3). In the first approximation we would expect one resonance for A and one resonance for X, and the spectrum would resemble that shown in Figure C.2. In the presence of coupling the resonance for A splits into two signals, one of which corresponds to X having $m_z = +\frac{1}{2}$ and the other to $m_z = -\frac{1}{2}$. The coupling effect is symmetric in that the X resonance also splits into two resonances, one for each spin state of A. The magnitude of the separation of the A pair (a doublet) or the X pair is known as the coupling constant. It is usually expressed in frequency units (Hz) since this splitting is independent of the field strength.

The splitting becomes more interesting when there are several nuclei of one type. 1,1-Dibromoethane, CH_3CHBr_2, has three equivalent protons in the methyl group and one proton on the 1-carbon. The chemical shift for the 1-carbon proton is 5.86 ppm and that for the methyl protons is 2.47 ppm. (Here we can see an example of decreased shielding resulting from the presence of electronegative substituents.) The methyl group in this case exhibits rapid internal rotation so that its three protons are equivalent. Equivalent protons do not affect one another (this is an important rule in interpreting spectra), but the methyl protons will effect the proton on the 1-carbon.

To analyze the splitting pattern we need to consider the component of spin parallel to the magnetic field for all three methyl protons. Since each proton may have two spin states, there are $2^3 = 8$ spins states in all for the methyl protons. The component of the total spin directed along the magnetic field axis may have only four different values as shown in Figure C.4. The symbol + is used to

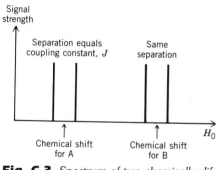

Fig. C.3 *Spectrum of two chemically different protons that are coupled.*

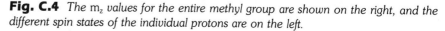

Fig. C.4 *The m_z values for the entire methyl group are shown on the right, and the different spin states of the individual protons are on the left.*

represent $m_z = +\frac{1}{2}$ for a single proton. Thus, $(+)(+)(-)$ means that for protons 1 and 2, $m_z = +\frac{1}{2}$, whereas for proton 3, $m_z = -\frac{1}{2}$.

The number of different m_z states is $2I + 1$, where I is the total spin of the equivalent nuclei ($\frac{3}{2}$ in the case of three protons on the methyl group). In addition to the four different states, we should note that there is only one way to achieve the $+\frac{3}{2}$ and $-\frac{3}{2}$ states and that there are three ways to obtain the $+\frac{1}{2}$ and $-\frac{1}{2}$ states. The number of different ways of having the same m_z value is the *degeneracy* of that state.

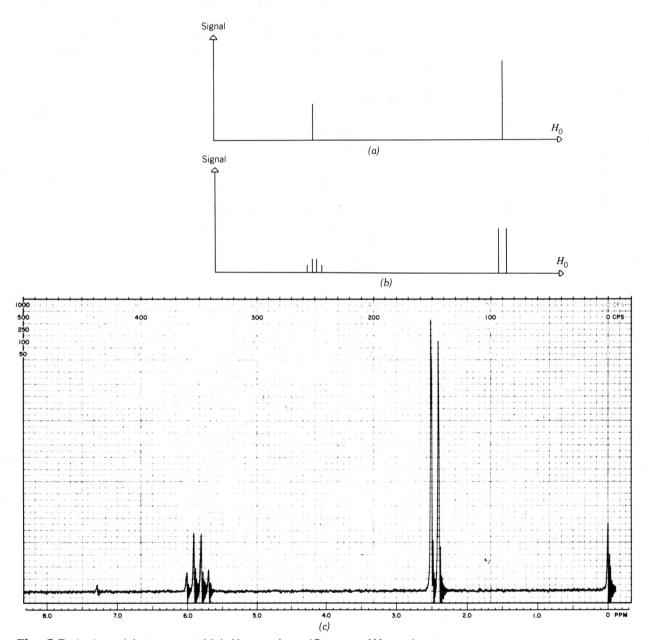

Fig. C.5 *Analysis of the spectrum of 1,1-dibromoethane. (Courtesy of Varian Associates, Palo Alto, CA.) (a) The spectrum without any spin–spin coupling. The resonance for the proton on the 1-carbon is shifted downfield. (b) A "stick figure" spectrum indicating the expected intensities. (c) The actual 60-MHz spectrum. The TMS resonance at 0 ppm is seen as well as a weak resonance at 7.3 ppm, which is not due to this molecule. This spectrum also shows the "ringing," which is a characteristic of a well-tuned spectrometer.*

In the case of 1,1-dibromoethane without coupling, we can arbitrarily represent the 1-carbon proton resonance as having a strength of 8 units. The methyl resonance will then have a strength of 24, three times as great (since there are three times as many methyl protons). This is shown schematically in Figure C.5a. The separation of each of the 1-carbon proton signals (known as a quartet) will equal the separation of the methyl signal. In the presence of coupling, the four m_z values for the methyl group will cause the 1-carbon proton resonance to split into four signals with relative strengths $1:3:3:1$. (Note that they add up to 8.) The methyl resonance will split into two signals (one for each m_z value of the 1-carbon proton. These two resonances will have the same intensity (12 in this case). The proposed spectrum is shown in Figure C.5b. The coupling constant in this case is about 7 Hz. The 60-MHz 1,1-dibromoethane spectrum is shown in Figure C.5c. The "$N + 1$ rule" is used to describe the number of resonances generated by coupling with N equivalent protons. The total spin of a group of N protons is $N/2$ and the number of m_z values is $2I + 1$. Therefore, we can represent the number of m_z values as $N + 1$. This is also the number of resonances generated in neighboring protons as a result of coupling with the N equivalent protons.

A resonance may be affected by more than one group of nuclei. The spectrum of 1-nitropropane, $CH_3CH_2CH_2NO_2$, is shown in Figure C.6. Can you sort out the splitting pattern? Perhaps the first thing to do is to consider the possible m_z states and their degeneracies for the two equivalent methylene protons. The most difficult part of the analysis is sorting out the methylene protons on the 2-carbon. These are first split into a $1:2:1$ triplet as a result of coupling with the 1-

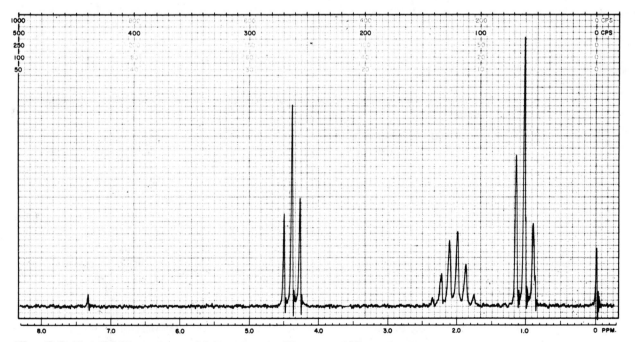

Fig. C.6 The 60-MHz spectrum of 1-nitropropane. (Courtesy of Varian Associates, Palo Alto, CA.) *Starting from the right, the TMS resonance at 0 ppm is seen. Next is a $1:2:1$ triplet at 1.03 ppm. This results from the protons on the 3-carbon and their coupling with the two protons on the 2-carbon. Next is the pattern of at least six resonances centered at 2.07 ppm. These signals result from the protons on the 2-carbon and their coupling with the protons on the 1-carbon and the 3-carbon. Finally we have the signal from the protons closest to the nitro group centered at 4.38 ppm. They appear as a $1:2:1$ triplet, through coupling with the 2-carbon protons.*

carbon protons. Further interaction with the methyl group splits each member of the triplet into a $1:3:3:1$ quartet. Of course, in many cases the separate lines will not be completely resolved or may be too weak to be seen.

Nuclei with a spin of zero have only one spin state and do not produce spin–spin splittings, or any NMR spectrum at all for that matter. Nuclei with spins of 1 or more might be expected to exhibit more complex spin–spin effects.

Intensities

The intensity of a resonance is often represented as the *integral* over a resonance or a group of resonances. As mentioned above, the resonance of a proton that is coupled with a methyl group will exhibit a characteristic $1:3:3:1$ intensity pattern. In more complex spectra the intensities are still useful as a measure of the number of protons of a given type. For instance, in the preceding case the integral over both members of the methyl group doublet will be three times the integral over the quartet of the proton on the 1-carbon. NMR spectrometers have the capability of measuring the integral, though integral data from a Fourier transform spectrometer are often less reliable than those from a continuous-wave spectrometer. This can often provide an additional handle on determining the identity of a compound.

Second-Order Effects

The examples so far considered have all consisted of first-order spectra. These highly symmetric and fairly simple spectra are generally observed when the chemical shift differences (expressed as a frequency) are much greater than the coupling constant. Second-order effects occur when the coupling constants become comparable to or greater than the chemical shift differences.

Second-order effects may be understood in qualitative terms by considering the limiting cases. Let us consider the hypothetical disubstituted ethylene shown in Figure C.7, where R and M are substituents that may be identical or may have very different effects on the olefinic protons. In Figure C.7a, the spectrum is shown for the case in which R and M have very different effects. Perhaps M is a strongly electronegative group and R actually contributes some electron density. In this case we will observe a first-order spectrum consisting of two doublets. The coupling constant is the separation in the doublets and the chemical shift of each nucleus is the midpoint of each doublet.

In Figure C.7b, groups R and M are identical. H_A and H_B are identical in this case and only a single resonance is observed. (There is still coupling between the nuclei, but it is not observed in the spectra of equivalent nuclei.)

In Figure C.7c the difference in the chemical environment of H_A and H_B is very slight. The spectrum shown may be seen as intermediate between the limiting cases in Figures C.7a and C.7b. It may be seen that there is a "leaning in" of the doublets as the central members increase in intensity at the expense of the outer members. For example, see the spectra of substituted cinnamic acids. A full continuum of behavior may be expected with cases observed in which the outer members are lost in the noise and the central members take the appearance of a doublet. This is one example of a class of spectra known as "deceptively simple spectra." It should also be noted that in second-order spectra the coupling constants and the chemical shift differences may not be obtained as simple differences.

The second-order spectra of systems with more than two protons are difficult to describe even in qualitative terms. It is partly for this reason that spectra obtained at high frequencies (and magnetic fields) are useful. As the operating frequency of the instrument is increased, the chemical shift differences (in energy terms) increase while the spin–spin coupling remains constant. Thus, the complicating second-order effects are likely to be less noticeable in high-field spectra.

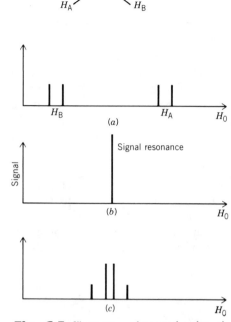

Fig. C.7 *Illustration of second-order effects. (a) The chemical shift difference is much larger than the coupling constant and a first-order spectrum is observed. (b) Protons A and B are equivalent and a single resonance is observed. (c) The chemical shift difference is the same order of magnitude or less than the coupling constant. The "leaning in" of this spectrum relative to that in part a is also seen in second-order spectra systems that have more than two protons.*

The reader is referred to more extensive treatments of NMR for a discussion of second-order cases.

¹³C NMR Spectroscopy

With the advent of Fourier transform (FT) NMR spectrometers, ¹³C NMR spectroscopy is now available as a simple and routine tool for the structure determination of organic molecules. Since ¹³C is of low natural abundance (1.1%), addition of many spectra is required to obtain acceptable signal-to-noise levels. With modern spectrometers, ¹³C spectra can often be acquired simply by issuing software commands; in older FT instruments, a different probe is often inserted into the magnet. ¹³C resonates at roughly 25% of the proton operating frequency of a spectrometer system. Thus, a spectrometer that acquires ¹H spectra at 200 MHz will be reset to 50 MHz for ¹³C work.

Generally, ¹³C NMR spectra are acquired while the entire ¹H frequency range is irradiated by a second rf coil inside the probe assembly. These spectra are referred to as broadband decoupled ¹³C spectra and they do not show the effect of spin–spin coupling to ¹H nuclei. This is done because ¹H–¹³C coupling constants can be quite large (several hundred hertz) relative to chemical shift differences, which leads to multiplets split over a large portion of the spectrum and subsequent confusion. It is often simpler to see a single line for each distinct carbon atom in a molecule. There are several techniques by which one can determine the number of ¹H nuclei attached to a given ¹³C atom and the reader is referred to Reference 6 for a discussion of these methods. ¹³C NMR is often better than ¹H NMR at distinguishing functional groups because typical ¹³C chemical shifts are in the range 0–200 ppm relative to TMS. Since ¹³C spectra are less complex than ¹H spectra, ¹³C spectra are often a better indication of isomeric or other impurities in a sample; it is easy for small peaks to be concealed beneath a complex second-order multiplet in the ¹H NMR spectrum. The pulse technique used to obtain ¹³C spectra provides a spectrum that is not amenable to integration and therefore the number of carbons giving rise to a given signal cannot generally be determined.

REFERENCES

1. American Petroleum Institute Research Project 44, *Selected Nuclear Magnetic Resonance Spectral Data*.
2. Abraham, R. J. *The Analysis of High Resolution NMR Spectra*; Elsevier: New York, 1971.
3. Bhacca, N. S.; Hollis, D. P.; Johnson, L. F.; Pier, E. A.; Shoolery, J. N. *NMR Spectra Catalog*; Varian Associates: Palo Alto, CA, 1963.
4. Cooper, J. W. *Spectroscopic Techniques for Organic Chemists*; Wiley: New York, 1980.
5. McFarlane, W. In *Elucidation of Organic Structures by Physical and Chemical Methods*; Bentley, K. W.; Kirby, G. W., Eds.; Wiley: New York, 1972; Part I, Chapter IV.
6. Derome, A. E. *Modern NMR Techniques for Chemistry Research*; Pergamon Press: Oxford, 1987.

Appendix D

Tables of Derivatives

Table D.1 Derivatives of Carboxylic Acids (Liquids)

Acid	bp (°C)	Melting Point of Derivative (°C)[a]		
		Amide	Anilide	p-Toluidide
Methanoic (formic)	101	—	50	53
Ethanoic (acetic)	118	82	114	153
Propenoic (acrylic)	141	84	104	141
Propanoic	141	81	106	126
2-Methylpropanoic (isobutyric)	155	128	105	109
Butanoic (butyric)	163	115	96	75
2-Methylpentenoic (methacrylic)	163	102	87	—
Pyruvic	165	124	104	109
3-Methylbutanoic	177	135	110	106
Pentanoic (valeric)	186	106	63	74
2-Methylpentanoic	186	79	95	80
2,2-Dichloroethanoic	194	98	118	153
Hexanoic (caproic)	205	100	94	74
Heptanoic (enanthic)	223	96	65; 70	81
Octanoic (caprylic)	239	106; 110	57	70
Nonanoic (pelargonic)	254	99	57	84

[a]Two values are given for those derivatives which may exist in polymorphic forms.

Table D.2   Derivatives of Carboxylic Acids (Solids)

Acid	mp (°C)	Melting Point of Derivative (°C)[a] Amide	Anilide	p-Toluidide
Decanoic	31–32	108	70	78
Lauric	43–45	87	78	100
Myristic	54	103	84	93
Trichloroacetic	54–58	141	97	113
Chloroacetic	61	121	137	162
Palmitic	62	106	90	98
Octadecanoic (stearic)	70	109	95	102
Crotonic	72	158	118	—
3,3-Dimethyl acrylic	69	107	126	—
Phenylethanoic	77	156	65	117
2-Benzoylbenzoic	128	165	195	—
Pentandioic (glutaric)	97	175	223	218
Ethanedioic (oxalic)	101	219	148	169
2-Methylbenzoic (o-toluic)	105	143	125	144
3-Methylbenzoic (m-toluic)	112	94	126	118
Benzoic	122.4	130	160	158
Sebacic	131–134	170 (mono) 210 (di)	122 (mono) 200 (di)	201
trans-Cinnamic	133	147	153	168
2-Acetoxybenzoic (aspirin)	135	138	136	—
cis-Butenedioic (maleic)	137	172	198 (mono) 187 (di)	142 (di)
Malonic	137	—	132 (mono) 230 (di)	86 (mono) 253 (di)
2-Chlorobenzoic	140	—	118	131
3-Nitrobenzoic	140	143	154	162
2-Nitrobenzoic	146	176	155	—
Diphenylacetic	148	168	180	172
2-Bromobenzoic	150	155	141	—
Benzilic	150	153	175	190
Hexanedioic (adipic)	153	125 (mono) 230 (di)	151 (mono) 241 (di)	—
2-Hydroxybenzoic (salicylic)	158	142	136	156
2-Iodobenzoic	162	110	141	—
4-Methylbenzoic (p-toluic)	179	160	144	160; 165
4-Methoxybenzoic (p-anisic)	185	167	170	186
2-Naphthoic	186	192	171	192
Succinic	190	157 (mono) 260 (di)	143 (mono) 230 (di)	180 (mono) 255 (di)
Phthalic	211	149 (mono) 220 (di)	170 (mono) 254 (di)	150 (mono) 201 (di)
3,5-Dinitrobenzoic	205	183	234	—
4-Nitrobenzoic	241	198	204; 211	192; 204

[a]Two values are given for those derivatives which may exist in polymorphic forms.

Table D.3 Derivatives of Alcohols

		Melting Point of Derivative (°C)		
Alcohol	bp (°C)	Phenyl-urethan	α-Naphthyl-urethan	3,5-Dinitro-benzoate
Methyl- (methanol)	65	47	124	108
Ethyl- (ethanol)	78	52	79	93
Isopropyl- (2-propanol)	82	88	106	122
t-Butyl- (t-butanol)	83	136	101	142
Allyl-	97	70	109	49
n-Propyl- (1-propanol)	97	51	80	74
s-Butyl- (2-butanol)	99	65	97	76
t-Pentyl- (2-methyl-2-butanol)	102	42	71	116
Isobutyl- (2-methyl-1-propanol)	108	86	104	87
3-Pentanol	116	48	71	101
n-Butyl- (1-butanol)	118	63	71	64
2,3-Dimethyl-2-butanol	118	—	—	111
2-Pentanol	119	—	76	61
2-Methyl-2-pentanol	121	239	—	72
3-Methyl-3-pentanol	123	50	—	97
2-Methoxyethanol	125	—	113	—
2-Methyl-1-butanol	129	—	—	70
2-Chloroethanol	131	51	101	95
4-Methyl-2-pentanol	132	143	88	65
3-Methyl-1-butanol	132	55	—	61
2-Ethoxyethanol	135	—	67	75
3-Hexanol	136	—	—	77
2,2-Dimethyl-1-butanol	137	—	—	51
1-Pentanol	138	46	68	46
2-Hexanol	139	—	—	39
2,4-Dimethyl-3-pentanol	140	—	—	—
Cyclopentanol	141	132	118	115
2-Ethyl-1-butanol	148	—	—	52
2-Methyl-1-pentanol	148	—	—	51
4-Heptanol	156	—	80	64
1-Hexanol	158	42	59	58
2-Heptanol	159	—	54	49
Cyclohexanol	161	82	128	113
2-Furfuryl-	172	45	129	81
1-Heptanol	177	68	—	47
Tetrahydrofurfuryl-	178	61	—	84
2-Octanol	179	114	—	32
1-Octanol	195	74	—	61
Benzyl-	205	78	—	113
2-Phenylethanol	221	79	—	108
1-Decanol	231	59	—	57
Cinnamyl-	(mp 35)	90	—	121
Benzohydrol	(mp 67)	139	—	141
Cholesterol	(mp 147)	168	—	—

Table D.4   Derivatives of Aldehydes

| Aldehyde | bp (°C) | Melting Point of Derivative (°C)[a] | |
		Semi-carbazone	2,4-Dinitrophenyl-hydrazone
Acetaldehyde	21	162	168
Propionaldehyde	50	89 (154)	154
Isobutyraldehyde	64	125	187 (183)
n-Butyraldehyde	74	104	123
Isovaleraldehyde	92	107	123
n-Valeraldehyde	103	108	107
Crotonaldehyde	104	199	190
n-Hexaldehyde	131	106	104; 107
n-Heptaldehyde	153	109	108
2-Furaldehyde	161	202	212 (230)
Benzaldehyde	179	222	237
Salicylaldehyde	197	231	252 dec
p-Tolualdehyde	204	221	239
2-Chlorobenzaldehyde	215	146 (229)	213
Citral	228	164	116
4-Anisaldehyde	248	210	253
trans-Cinnamaldehyde	252	215	255
4-Chlorobenzaldehyde	(mp 47)	230	254

[a]Two values are given for those derivatives which may exist in polymorphic forms or as syn and anti geometrical isomers.

Table D.5   Derivatives of Ketones

| Ketone | bp (°C) | Melting Point of Derivative (°C)[a] | |
		Semi-carbazone	2,4-Dinitro-phenylhydrazone
Acetone	56	187	126
2-Butanone	80	146	117
3-Methyl-2-butanone	94	113	120
2-Pentanone	102	112	143
3-Pentanone	102	139	156
3,3-Dimethyl-2-butanone	106	157	125
4-Methyl-2-pentanone	119	134	95
2,4-Dimethyl-3-pentanone	124	160	88 (94)
2-Hexanone	129	122	106
Cyclopentanone	131	205	142
4-Heptanone	145	133	75
3-Heptanone	149	101	—
2-Heptanone	151	127	89
Cyclohexanone	155	166	162
2-Octanone	173	122	58
Acetophenone	200	198	240
Benzalacetone	(mp 41)	187	223
Benzophenone	(mp 48)	164	239
Benzalacetophenone	(mp 58)	168; 180	245
Benzil	(mp 95)	175 (182)	189
Benzoin	(mp 133)	206 (dec)	245

[a]Two values are given for those derivatives which may exist in polymorphic forms or as syn and anti geometrical isomers.

Table D.6 Derivatives of Primary and Secondary Amines

| Amine | bp (°C) | Melting Point of Derivative (°C)[a] | | |
		Acetamide	Benzamide	Picrate
Methylamine	−6	—	80	—
Ethylamine	17	—	71	—
Isopropylamine	33	—	71	165
t-Butylamine	45	98	134	198
n-Propylamine	49	47	84	135
Allylamine	53	—	—	140
Diethylamine	55	—	42	155
sec-Butylamine	63	—	76	139
Isobutylamine	69	—	57	150
n-Butylamine	77	—	42	—
Di-isopropylamine	84	—	—	140
Di-n-propylamine	109	—	—	75
Piperidine	106	—	48	152
Ethylenediamine	116	172 (di)	244 (di)	233
Cyclohexylamine	134	104	149	—
Di-isobutylamine	139	86	—	121
Di-n-butylamine	159	—	—	59
Benzylamine	185	60	105	199
Aniline	185	114	163	198
N-Methylaniline	196	102	67	145
2-Methylaniline	199	110	144	213
4-Methylaniline	200 (mp 45)	148	158	—
3-Methylaniline	203	65	125	200
N-Ethylaniline	205	111	147	194
2-Chloroaniline	208	87	99	134
2,5-Dimethylaniline	215	139	140	171
2,6-Dimethylaniline	216	177	168	180
2,4-Dimethylaniline	217	133	192	209
N-Ethyl-3-methylaniline	221	—	72	—
2-Methoxyaniline	225	85	60 (84)	200
4-Chloroaniline	232 (mp 70)	179	192	178
4-Methoxyaniline	243 (mp 57)	130	154	170
2-Ethoxyaniline	229	79	104	—
4-Ethoxyaniline	254	135	173	—
Diphenylamine	(mp 54)	101	180	182
3-Nitroaniline	(mp 114)	152	155	—
4-Nitroaniline	(mp 147)	—	199	—

[a]Two values are given for those derivatives which may exist in polymorphic forms.

Table D.7   Derivatives of Tertiary Amines

Tertiary Amine	bp (°C)	Melting Point of Derivative (°C)[a] Picrate
Trimethylamine	3	216
Triethylamine	89	173
Pyridine	116	167
2-Methylpyridine (2-picoline)	129	169
2,6-Dimethylpyridine (2,6-lutidine)	142	168 (161)
3-Methylpyridine (3-picoline)	143	150
4-Methylpyridine (4-picoline)	143	167
Tripropylamine	157	116
N,N-Dimethylaniline	193	163
Tributylamine	216	105
N,N-Diethylaniline	216	142
Quinoline	237	203
Tri-isopentylamine	245	125

[a]Two values are given for those derivatives which may exist in polymorphic forms.

Table D.8   Derivatives of Acid Chlorides and Anhydrides

Acid Chloride or Anhydride	bp (°C)	mp (°C)	Melting Point of Derivative (°C) Amide
Acetyl chloride	52	—	82
Propionyl chloride	77–79	—	81
Butyryl chloride	102	—	115
Acetic anhydride	138–140	—	82
Propionic anhydride	167	—	81
Butyric anhydride	198–199	—	115
Benzoyl chloride	198	—	130
3-Chlorobenzoyl chloride	225	—	134
2-Chlorobenzoyl chloride	238	—	142
cis-1,2-Cyclohexanedicarboxylic anhydride	—	32	192d (acid)
Benzoic anhydride	—	39–40	130
Maleic anhydride	—	54–56	181 (mono) 266 (di)
4-Nitrobenzoyl chloride	—	72–74	201
Succinic anhydride	—	119–120	157 (mono) 260 (di)
Phthalic anhydride	—	131–133	149 (mono) 220 (di)

Table D.9 Derivatives of Aromatic Hydrocarbons

Aromatic Hydrocarbon	bp (°C)	mp (°C)	Melting Point of Derivative (°C)[a] Picrate
Benzene	80	—	84
Toluene	111	—	88
Ethylbenzene	136	—	96
p-Xylene	138	—	90
m-Xylene	138–139	—	91
o-Xylene	143–145	—	88
Mesitylene	163–166	—	97
1,2,4-Trimethylbenzene	168	—	97
1,2,3,4-Tetramethylbenzene	205	—	92
1-Methylnaphthalene	242	—	142
2-Methylnaphthalene	—	35	116
Pentamethylbenzene	—	51	131
Naphthalene	—	81	149
Acenaphthene	—	94	161
Phenanthrene	—	100	144 (133)
Anthracene	—	216	138

[a]Two values are given for those derivatives which may exist in polymorphic forms.

Table D.10 Derivatives of Phenols

Phenol	mp (°C)	Melting Point of Derivative (°C) Bromo	α-Naphthylurethan
2-Chloro-	7 (bp 175)	48 (mono) 76 (di)	120
Phenol	42	95 (tri)	133
4-Methyl- (p-cresol)	35	49 (di) 108 (tetra)	146
3-Methyl- (m-cresol)	203 (bp)	84 (tri)	128
3,4-Dimethyl-	229 (bp)	171 (tri)	141
2-Methyl- (o-cresol)	33	56 (di)	142
4-Ethyl-	45	—	128
2-Nitro-	45	117 (di)	113
2,6-Dimethyl-	48	79	176
2-Isopropyl-5-methyl- (thymol)	50	55	160
3,4-Dimethyl-	63	117 (tri)	—
3,5-Dimethyl-	64	166 (tri)	—
4-Bromo-	66	95 (tri)	168
2,5-Dimethyl-	73	178 (tri)	173
1-Naphthol	95	105 (di)	152
3-Nitro-	96	91 (di)	—
4-t-Butyl-	98	50 (mono) 67 (di)	110
1,2-Dihydroxy- (catechol)	105	192 (tetra)	175
1,3-Dihydroxy- (resorcinol)	110	112 (tri)	275
4-Nitro-	112	142 (di)	150
2-Naphthol	123	84	157
Pyrogallol	134	158 (di)	173
1,4-Dihydroxy- (hydroquinone)	171	186 (di)	—

Table D.11   Aliphatic Hydrocarbons

Compound	bp (°C)	Compound	bp (°C)
Alkanes			
Pentane	36	2,2,4-trimethyl-	
Cyclopentane	49	pentane	99
2,2-Dimethylbutane	50	trans-1,4-Dimethyl-	
2,3-Dimethylbutane	58	cyclohexane	119
2-Methylpentane	60	Octane	126
3-Methylpentane	63	Nonane	151
Hexane	69	Decane	174
Cyclohexane	81	Eicosane	343 (mp 37)
Heptane	98	Norbornane	(mp 87, subl.)
		Adamantane	(mp 268, sealed)
Alkenes and Alkynes			
1-Pentene	30	3-Hexene	82
2-Methyl-1,3-butadiene	34	Cyclohexene	84
(isoprene)		2-Hexyne	84
trans-2-Pentene	36	1-Heptene	94
cis-2-Pentene	37	1-Heptyne	100
2-Methyl-2-butene	39	2,4,4-Trimethyl-	
Cyclopentadiene	41	1-pentene	102
1,3-Pentadiene	41	2,4,4-Trimethyl-	
(piperylene)		2-pentene	104
3,3-Dimethyl-1-butene	41	1-Octene	123
1-Hexene	63	Cyclooctene	146
cis-3-Hexene	66	1,5-Cyclooctadiene	150
trans-3-Hexene	67	d,l-α-Pinene	155
1-Hexyne	71	(−)-β-Pinene	167
1,3-Cyclohexadiene	80	Limonene	176
		1-Decene	181

Table D.12 Halogenated Hydrocarbons

Compound	bp (°C)	Compound	bp (°C)
Alkyl Halides			
Chlorides		*Bromides*	
n-Propyl-	47	Ethyl-	38
t-Butyl-	51	Isopropyl-	60
sec-Butyl-	68	Propyl-	71
Isobutyl-	69	t-Butyl-	72
n-Butyl-	78	Isobutyl-	91
Neopentyl-	85	sec-Butyl-	91
t-Pentyl-	86	Butyl-	101
Cyclohexyl-	143	t-Pentyl-	108
Hexachloroethane	185 (mp 187, subl.)	Neopentyl-	109
Triphenylmethyl-	(mp 113)	1-Bromoheptane	174; 180
		Iodides	
		Methyl-	43
		Ethyl-	72
		Isopropyl-	90
		Propyl-	102

Compound	bp (°C)	mp (°C)
Aryl Halides		
Chlorobenzene	132	—
Bromobenzene	156	—
2-Chlorotoluene	157–159	—
4-Chlorotoluene	162	—
1,3-Dichlorobenzene	172–173	—
1,2-Dichlorobenzene	178	—
2,4-Dichlorotoluene	196–203	—
3,4-Dichlorotoluene	201	—
1,2,4-Trichlorobenzene	214	—
1-Bromonaphthalene	279–281	—
1,2,3-Trichlorobenzene	—	51–53
1,4-Dichlorobenzene	—	54–56
1,4-Bromochlorobenzene	—	66–68
1,4-Dibromobenzene	—	87–89
1,2,4,5-Tetrachlorobenzene	—	138–140

Table D.13 Nitriles

Compound	bp (°C)	Compound	mp (°C)
Acrylonitrile	77	4-Chlorobenzylcyanide	30.5
Acetonitrile	81	Malononitrile	34
Propionitrile	97	Stearonitrile	40
Isobutyronitrile	108	2-Chlorobenzonitrile	41
n-Butyronitrile	117	Succinonitrile	48
Benzonitrile	191	Diphenylacetonitrile	75
2-Methylbenzonitrile	205	4-Cyanopyridine	80
3-Methylbenzonitrile	212		
4-Methylbenzonitrile	217		
Benzylcyanide	234		
Adiponitrile	295		

Table D.14 Amides[a]

Compound	bp (°C)	mp (°C)
N,N-Dimethylformamide	153	—
N,N-Diethylformamide	176	—
N-Methylformamide	185	—
N-Formylpiperidine	222	—
N,N-Dimethylbenzamide	—	41
N-Benzoylpiperidine	—	48
N-Propylacetanilide	—	50
N-Benzylacetamide	—	54
N-Ethylacetanilide	—	54
N,N-Diphenylformamide	—	73
N-Methyl-4-acetotoluidide	—	83
N,N-Diphenylacetamide	—	101
N-Methylacetanilide	—	102
Acetanilide	—	114
N-Ethyl-4-nitroacetanilide	—	118
N-Phenylsuccinimide	—	156
N-Phenylphthalimide	—	205

[a]Also see Tables D.1, D.2 for amides prepared as derivatives of carboxylic acids.

Table D.15 Nitro Compounds

Compound	bp (°C)	mp (°C)
Nitrobenzene	211	—
2-Nitrotoluene	225	—
2-Nitro-*m*-xylene	225	—
3-Nitrotoluene	231	—
3-Nitro-*o*-xylene	245	—
4-Ethylnitrobenzene	246	—
2-Chloro-6-nitrotoluene	—	36
4-Chloro-2-nitrotoluene	—	38
3,4-Dichloronitrobenzene	—	42
1-Chloro-2,4-dinitrobenzene	—	50
4-Nitrotoluene	—	54
1-Nitronaphthalene	—	56
1-Chloro-4-nitrobenzene	—	84
m-Dinitrobenzene	—	90

Table D.16 Ethers

Compound	bp (°C)	mp (°C)
Furan	32	—
Ethyl vinyl ether	33	—
Tetrahydrofuran	67	—
n-Butyl vinyl ether	94	—
Anisole	154	—
4-Methylanisole	174	—
3-Methylanisole	176	—
4-Chloroanisole	203	—
1,2-Dimethoxybenzene	207	—
4-Bromoanisole	215	—
Anethole	234–237	—
Diphenyl ether	259	—
2-Nitroanisole	273	—
Dibenzyl ether	298	—
4-Nitroanisole	—	50–52
1,4-Dimethoxybenzene	—	56–60
2-Methoxynaphthalene	—	73–75

Table D.17 Esters

Compound	bp (°C)	Compound	bp (°C)
Liquids			
Methyl formate	32	Pentyl formate	132
Ethyl formate	54	Ethyl 3-methylbutanoate	135
Methyl acetate	57	Isobutyl propanoate	137
Isopropyl formate	68	Isopentyl acetate	142
Ethyl acetate	77	Propyl butanoate	143
Methyl propanoate	80	Ethyl pentanoate	146
Methyl propenoate	80	Butyl propanoate	147
Propyl formate	81	Pentyl acetate	149
Isopropyl acetate	91	Isobutyl 2-methylpropanoate	149
Methyl 2-methylpropanoate	93	Methyl hexanoate	151
sec-Butyl formate	97	Isopentyl propanoate	160
t-Butyl acetate	98	Butyl butanoate	165
Ethyl propanoate	99	Propyl pentanoate	167
Propyl acetate	101	Ethyl hexanoate	168
Methyl butanoate	102	Cyclohexyl acetate	175
Allyl acetate	104	Isopentyl butanoate	178
Ethyl 2-methylpropanoate	110	Pentyl butanoate	185
sec-Butyl acetate	112	Propyl hexanoate	186
Methyl 3-methylbutanoate	117	Butyl pentanoate	186
Isobutyl acetate	117	Ethyl heptanoate	189
Ethyl butanoate	122	Isopentyl 3-methylbutanoate	190
Propyl propanoate	122	Ethylene glycol diacetate	190
Butyl acetate	126	Tetrahydrofurfuryl acetate	194
Diethyl carbonate	127	Methyl octanoate	195
Methyl pentanoate	128	Methyl benzoate	200
Isopropyl butanoate	128	Ethyl benzoate	213

Compound	mp (°C)	Compound	mp (°C)
Solids			
d-Bornyl acetate	29 (bp 221)	Ethyl 3,5-dinitrobenzoate	93
Ethyl 2-nitrobenzoate	30	Methyl 4-nitrobenzoate	96
Ethyl octadecanoate	33	2-Naphthyl benzoate	107
Methyl cinnamate	36 (bp 261)	Isopropyl 4-nitrobenzoate	111
Methyl 4-chlorobenzoate	44	Cyclohexyl 3,5-dinitrobenzoate	112
1-Naphthyl acetate	49	Cholesteryl acetate	114
Ethyl 4-nitrobenzoate	56	Ethyl 4-nitrobenzoate	116
2-Naphthyl acetate	71	t-Butyl 4-nitrobenzoate	116
Ethylene glycol dibenzoate	73	Hydroquinone diacetate	124
Propyl 3,5-dinitrobenzoate	74	t-Butyl 3,5-dinitrobenzoate	142
Methyl 4-bromobenzoate	81	Hydroquinone dibenzoate	204

Appendix E

Chapters 5 and 6: Experiments Classified by Mechanism

Index

THE USE OF THE CORRELATION CHART OF INFRARED GROUP FREQUENCIES

This chart is by no means complete. It is scarcely possible to crowd onto one piece of paper of reasonable size all that might possibly be desired; further, as infrared continues to spread in application, new group frequencies are still being discovered. The user of such charts is urged to place his own notations for newfound correlations on the chart.

The chart has been organized into chemical group types, whose designations appear along the left-hand edge. Across the top of the chart have been indicated the various classes of molecular motions that form usable group frequencies.

The short heavy horizontal line under each group symbol indicates the extremes of the frequency region in which such groups are known to have a characteristic absorption. (Often in the past, such regions have had to be extended when a particular group is placed in molecules with less familiar groups in the vicinity, with the result that its characteristic frequencies have strayed.)

The thickness of the line is a very rough index of the intensity of this absorption. A line of tapering thickness indicates the intensity for this group to be quite variable. Often these intensity variations can be correlated with structure, but no clever way to represent this on the limited space of a chart was at hand.

An open, cross-hatched line represents a region in which there is usually more than one absorption characteristic of the particular group. For example, halogenated aliphatic hydrocarbons often have several strong, sharp absorptions in the region 950–1300 cm^{-1} to which it is difficult to ascribe specific vibrational motions, but nonetheless are characteristic of that class of molecules.

The chemical symbols are those of standard organic nomenclature. A few, perhaps, should be amplified: X = halogen *except* fluorine; M = metal; N$^+$—H = hydrogen attached to a positively charged nitrogen atom, as in amine salts of acids; (Σ) = a "summation" band— i.e., combination or overtone—*not* a fundamental; ϕ = phenyl ring; CJ = conjugated.

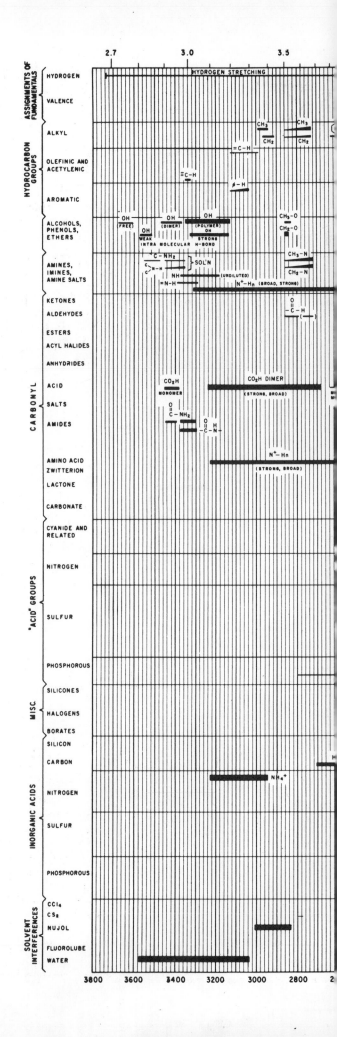